Inhaltsverzeichnis.
Contents. — Table des matières.

Biogenetic-type Syntheses of Natural Products. By E. E. VAN TAMELEN,
Department of Chemistry, The University of Wisconsin, Madison, Wisconsin 242

Medium-ring Terpenes.

By F. Šorm, Prague.

With 1 Figure.

Contents.

I. Introduction.

Work in the field of terpene chemistry has been in many respects of fundamental importance for shaping and developing organic chemistry. The classical unity of the terpene group of compounds is given by their simple composition (limited to the three basic organogenic elements) and by the uniformity of the rules governing their structure. This, together with the great structural diversity and ubiquity of occurrence, has made the terpenes the object of an immense amount of research, often of far-reaching theoretical implications and great practical interest.

Although the study of terpenes has a long history, reaching back to almost the period when organic chemistry was beginning to become an independent branch of science; and although the terpenes are low-molecular weight compounds of comparatively simple structure, this

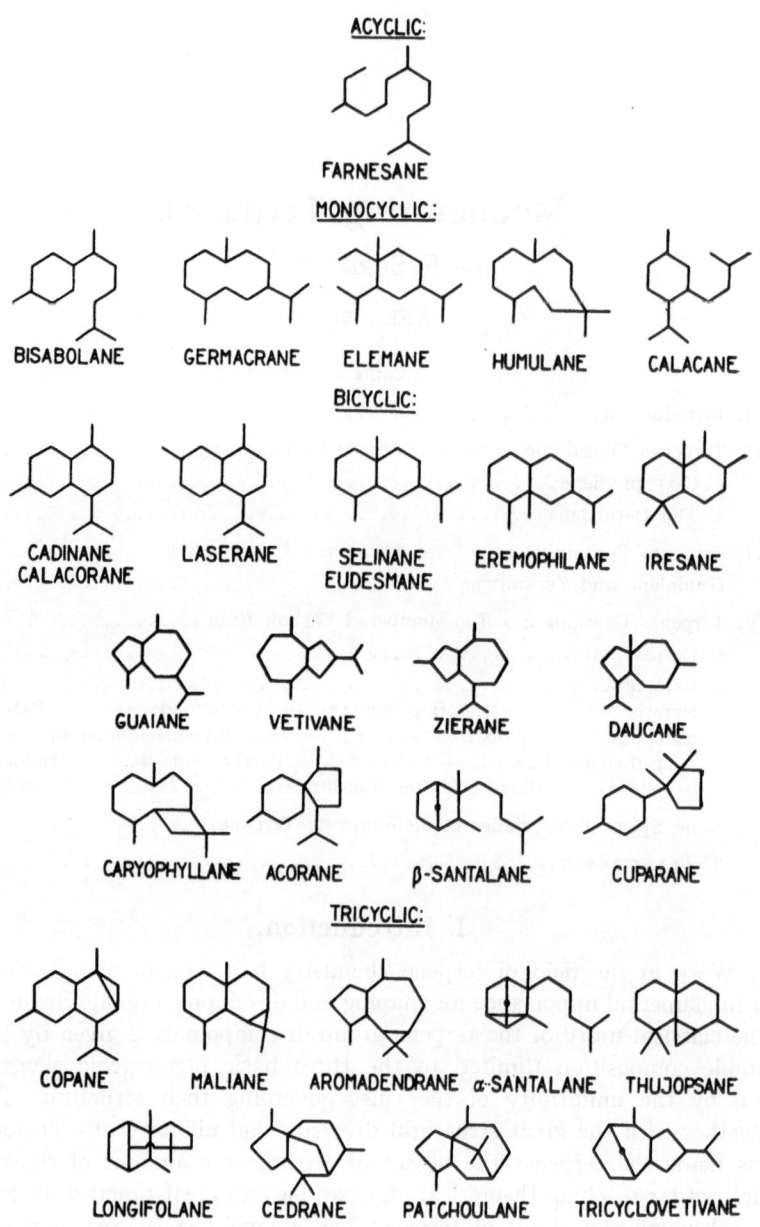

Chart 1. Sesquiterpenic Types.

field is still far from exhausted and remains a rich source of new findings
and fresh inspiration. This is particularly true for sesquiterpenes which

have received great attention only in the last fifteen years (*33, 52, 55, 57*). Today, this group is known to be varied and rich in structural types—at the present time no less than 28 sesquiterpenic types are known which differ in the arrangement of the fundamental fifteen-membered carbon skeleton. It may be noted in passing that only 9 out of the 28 types were kown with certainty previous to 1945; all others were either clarified or newly discovered after that year. Amongst the types were systems previously known only as products of synthesis, such as the simple spirane system of acorone and its stereoisomers (*74–76*), or the medium-ring compounds with which we shall deal in the present article.

The discovery of rings of medium size in common plant consituents, such as caryophyllene, came as a surprise since earlier larger carbon rings had been found in Nature only in the macrocyclic fragrant ketones. Moreover, the difficulty of synthetic access to alicyclic compounds with nine- to eleven-membered carbon rings, which has been overcome only fairly recently by the use of the acyloin condensation (*43, 63, 64*), favoured the view that compounds of this type were purely products of human thought and labour.

The sesquiterpenic types are given in *Chart 1*.

II. Terpenes Containing a Nine-Membered Carbon Ring.

1. Caryophyllene.

Caryophyllene is the first naturally occurring compound in which the presence of a medium sized carbon ring has been unequivocally established. It represents the principal component of the oil of clove and of many other essential oils and has the composition $C_{15}H_{24}$.

It would seem that more effort has been devoted to the elucidation of the structure of this hydrocarbon than of any other sesquiterpenic compound. It has been known for a long time that caryophyllene is a bicyclic hydrocarbon that contains two double bonds and the characteristic four-membered ring carrying geminal methyl groups. The oxidation of caryophyllene gives, as principal products, three dicarboxylic acids containing the intact cyclobutane ring, namely caryophyllenic acid (I) (*2, 10, 47*), nor-caryophyllenic acid (II) (*48–50*) and homocaryophyllenic acid, the structure of which has recently been shown to correspond to formula (III, p. 4) (*16*).

At the time when the work on the caryophyllene structure was started in the writer's laboratory, three alternative formulae, (IV) (*48*), (V) (*80*) and (VI) (*51*), were available for this hydrocarbon; none of them, however, could fully account for its known chemical properties. This work started from the oxido-ketone of TREIBS (VIII) which is readily obtainable from caryophyllene monoxide (VII) by ozonisation and in which the exocyclic vinylidene-type double bond is known to be preserved (*80*). The oxido-ketone was found to exhibit an anomalously low carbonyl frequency (1698 cm.$^{-1}$; melt), a feature characteristic for carbonyl groups located in rings of medium size. To obtain further information on the size of the larger one of the two rings, a degradation

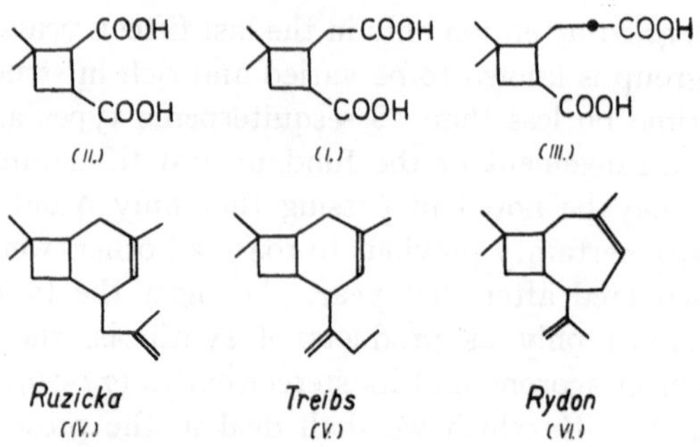

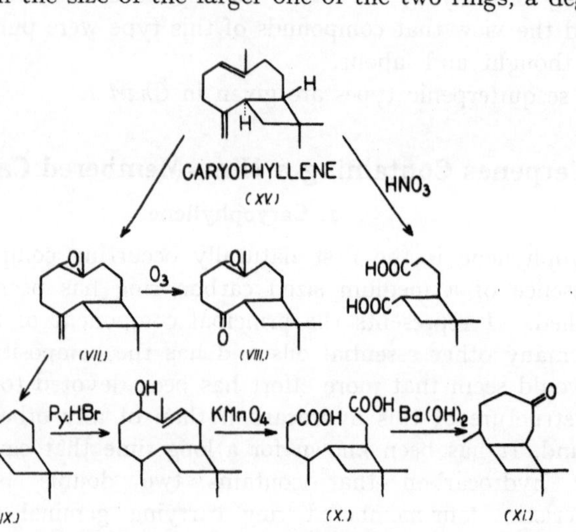

Chart 2. Degradation of Caryophyllene.

of this ring in dihydrocaryophyllene oxide (IX), by unambiguous steps involving the loss of two carbon atoms, was carried out *(Chart 2)* (*56*). Heating of the epoxide with pyridinium hydrobromide gave rise to a mixture of unsaturated alcohols, and when this was oxidised with potassium permanganate a dicarboxylic acid $C_{13}H_{22}O_4$ (X) resulted. Its cyclisation afforded the twelve-carbon ketone (XI) which still had an anomalously low carbonyl absorption frequency (1705 cm.$^{-1}$; liquid), corresponding to a seven-membered or, less probably, to a six-membered cyclic ketone.

These results indicated that the larger ring in caryophyllene must be either nine- or eight-membered. The latter alternative could, however, be excluded since tetrahydrocaryophyllene itself did not exhibit an absorption band at 775 cm.$^{-1}$ characteristic for an ethyl group, which is normally readily distinguished in saturated hydrocarbons. The ketone (XI) therefore contained a seven-membered ring, in addition to the intact four-membered ring, a fact which could subsequently be confirmed by a relay synthesis (20) of the hydrocarbon (XII) starting from monomethyl homocaryophyllenic acid (IIIa) (Chart 3).

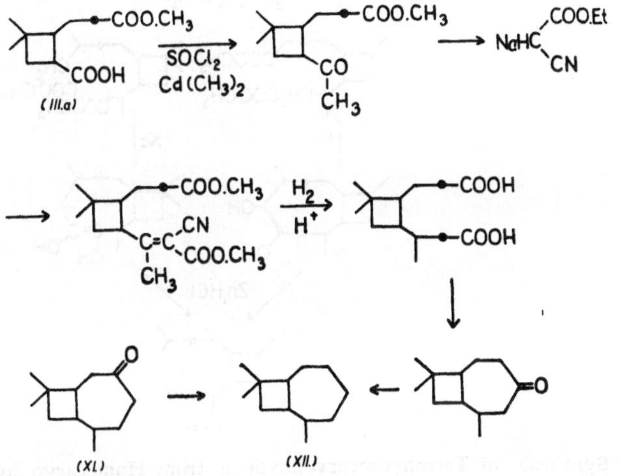

Chart 3. Synthesis of Hydrocarbon (XII) from Monomethyl Homocaryophyllenic Acid.

These findings, together with other results available at that time, made it possible to assign to caryophyllene four alternative formulae, containing a four- and a nine-membered ring (56). When the structure of homocaryophyllenic acid became known, the number of alternative formulae was reduced to two, and these differed only with respect to the location of the double bonds (XVa, XVb).

Additional support for the carbon skeleton of caryophyllene has been provided by the synthesis of tetrahydrocaryophyllene (XXII) starting from homocaryophyllenic acid (III) (29) (Chart 4, next page).

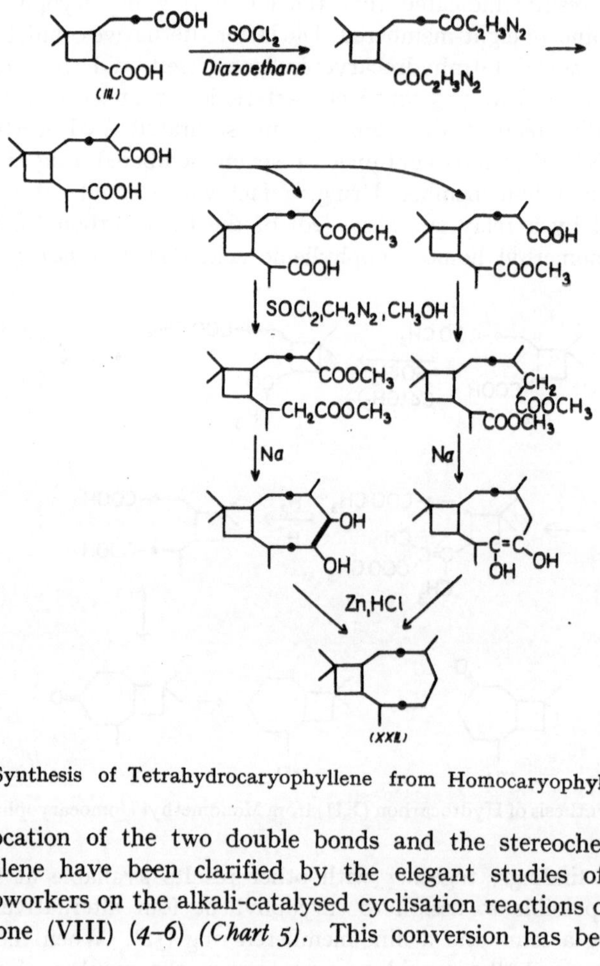

Chart 4. Synthesis of Tetrahydrocaryophyllene from Homocaryophyllenic Acid.

The location of the two double bonds and the stereochemistry of caryophyllene have been clarified by the elegant studies of BARTON and his coworkers on the alkali-catalysed cyclisation reactions of TREIBS' oxido-ketone (VIII) *(4–6) (Chart 5).* This conversion has been shown

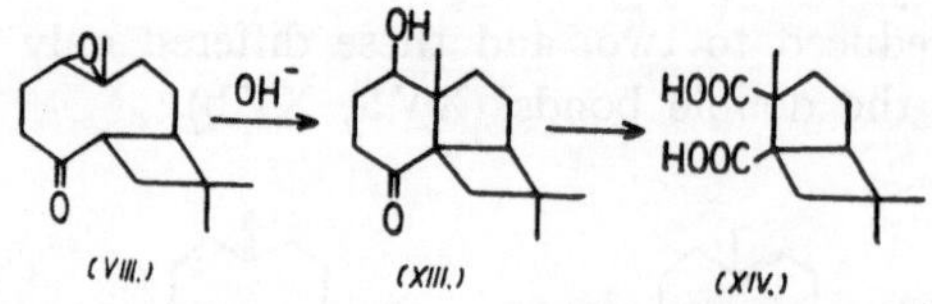

Chart 5. Cyclisation of the Oxido-ketone (VIII).

to lead to a tricyclic hydroxy-ketone (XIII) which on oxidation gave rise to the dicarboxylic acid (XIV). These reactions and the results of X-ray diffraction studies *(44)* have proved that the double bonds in caryophyllene are located as shown in formula (XV) (Chart 2, p. 4). From a comparison of reactivities towards electrophilic reagents AEBI,

BARTON and LINDSEY (*1*) were able to assign the *trans* configuration to the endocyclic double bond in caryophyllene; in the more stable isocaryophyllene, which is formed from caryophyllene by isomerisation, the double bond has *cis* configuration.

2. The Betulenols.

Recently, the presence of the caryophyllene skeleton could be proved in two further sesquiterpenes, α-betulenol (XVIa) and β-betulenol (XVIb) (*28*)*. These are isomeric alcohols, $C_{15}H_{24}O$, which have been isolated from the oil of birch originating from the U. S. S. R. More than twenty years ago it was suggested by TREIBS (*78, 79*) that the alcohols contained in this oil are related to caryophyllene. The compounds were described by him as primary alcohols; however, both the α-betulenol and the isomeric, crystalline β-betulenol as isolated in the writer's laboratory are secondary alcohols.

Both isomers contain two double bonds and on hydrogenation yield mixtures of tetrahydro derivatives (XVII, XVIII) (which on oxidation afford ketones) and the hydrocarbon (XXII) resulting from hydrogenolysis. This hydrocarbon was shown to be identical with tetrahydro-caryophyllene in every respect, and this establishes the structure of the carbon skeleton in both betulenols (cf. *40*). The ketone (XX) obtained from tetrahydro-α-betulenol by chromic acid oxidation proved to be identical with the compound obtained from caryophyllene monoxide (VII) by an unambiguous reaction sequence; and this establishes the position of the hydroxyl group. Caryophyllene monoxide on heating with pyridinium hydrobromide gives rise to a mixture of isomeric alcohols (XXIII) which on hydrogenation are converted into the saturated alcohol (XXIV). Oxidation of this mixture with chromic acid affords the ketone (XX).

The occurrence of homocaryophyllenic acid among the oxidation products of α-betulenol, further the presence of an exocyclic vinylidene-type double bond and the fact that the hydroxyl group undergoes hydrogenolysis establish the structure (XVIa) for α-betulenol. By a similar reaction sequence the tetrahydro derivative of β-betulenol affords a saturated ketone (XXI) which is different from that obtained from α-betulenol. Structure (XVIb) has been assigned to β-betulenol on the basis of its physical properties and reactions (*Chart 6*, next page).

---

* Professor W. TREIBS has recently informed the writer orally that the presence of betulenols containing primary hydroxyl groups has now been proved in his laboratory. Perhaps this may be explained by the fact that the oils investigated in the Leipzig and Prague Laboratories were of different origin (*Betula alba* and *B. lenta*, respectively). Alternatively, it seems also possible that the betulenols of TREIBS are products of allylic rearrangement which might have taken place during hydrolysis of the esters by way of which the betulenols had been purified.

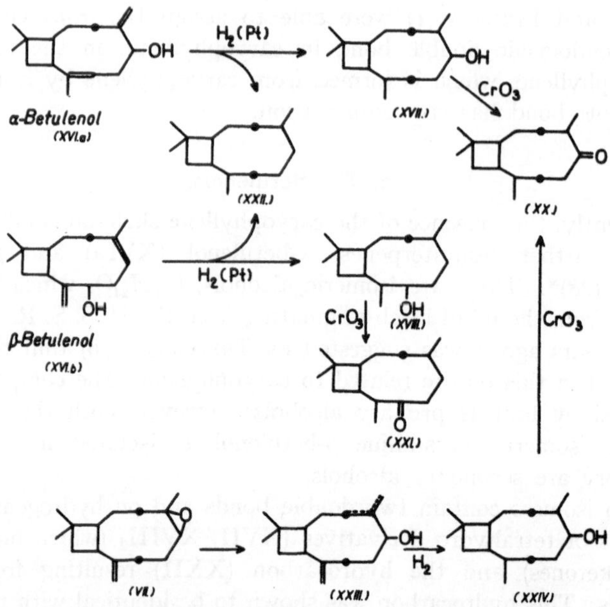

Chart 6. Conversion Products of α- and β-Betulenol.

In Nature, both betulenols very probably originate from caryophyllene by oxidation, presumably via the oxide. This view is supported by the finding that the oil of birch buds contains large amounts of caryophyllene monoxide (*28*).

III. Terpenes Containing an Eleven-Membered Carbon Ring.

Humulene and Zerumbone.

The hydrocarbon humulene, $C_{15}H_{24}$, has been isolated in the pure state from the oil of hops in the writer's laboratory (*58*); subsequently it has been shown to be identical with the so-called α-caryophyllene, which accompanies caryophyllene in the oil of cloves (*61*). Humulene was the second compound in which the presence of a medium ring—in this case an eleven-membered ring—was established. The hydrocarbon contains three double bonds and hence is monocyclic. It can be characterised as the trioxide, the nitrosochloride, or the nitrosite.

Its structure was tentatively derived in the writer's laboratory (*24, 61*), and simultaneously by CLEMO and HARRIS (*12, 13*), from the fact that on oxidation humulene and certain of its derivatives gave, in addition to α,α-dimethylsuccinic acid, only formaldehyde and acetic acid but no higher monofunctional degradation products. A proper confirmation of the suggested carbon skeleton of humulene has been given by the total synthesis of hexahydrohumulene (humulane) (XXVI) (*59, 60*). This synthesis has made use of a modification of the Kolbe anodic synthesis using mono-esters of dicarboxylic acids esterified with different alcohols, as shown in *Chart 7*. The synthetic hexahydrohumulene was identical in every respect with the product of complete hydrogenation of humulene.

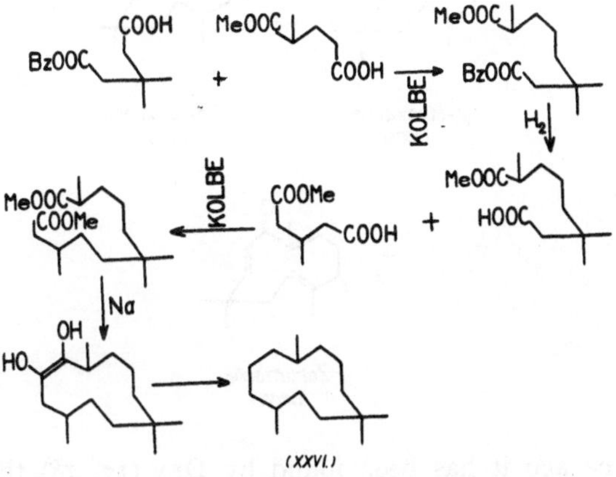

Chart 7. Synthesis of Hexahydrohumulene.

The location of the double bonds in humulene remained unsettled for some time. The structure (XXVa) was suggested by the Prague group, mainly on the basis of the infrared absorption at 890 cm.$^{-1}$, characteristic for vinylidene type double bonds, and from the nature of the oxidation products referred to above. Recently, SUTHERLAND and his coworkers (*26*) have purified humulene by way of its addition product with silver nitrate and found that the infrared spectrum of this adduct did not contain the band characteristic for the vinylidene group. The same conclusion has been reached independently by DEV (*19*). Quite recently, in the course of investigations on humulene from the essential oil of the leaves of *Lindera strychnifolia* (Lauraceae), it has been found (*8*) that natural humulene is a mixture of two isomers, α and β. It would seem that the essential oils of different plants contain these two isomers (XXVa) and (XXVb) in different proportions; however, it is remarkable that α-humulene when chromatographed on alumina is converted into the more stable β-isomer containing an exocyclic, vinylidene type double bond. In this way the β-isomer may be obtained practically pure. On the other hand, in the purification via the silver nitrate adduct, α-humulene is practically freed from the β-isomer, and this accounts for the results of SUTHERLAND et al. (*26*). α-Humulene affords a crystalline trioxide but β-humulene does not. The isomer composition of a mixture of the two humulenes may be determined either by the estimation of the formaldehyde obtained on ozonisation or by quantitative infrared spectroscopy, based on the 888 and 823 cm.$^{-1}$ bands of the exocyclic and trisubstituted endocyclic double bonds, respectively.

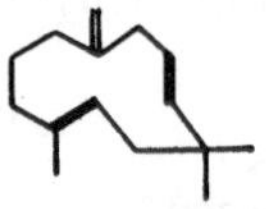

β-Humulene
(XXVa)

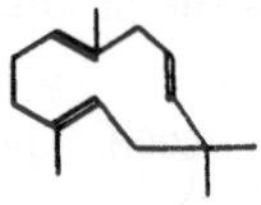

α-Humulene
(XXVb)

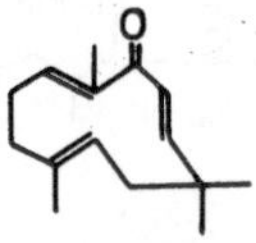

Zerumbone
(XXVII)

Some time ago it has been found by DEV (17, 18) that the ketone zerumbone, isolated from the oil of *Zingiber zerumbet*, also possesses the humulane skeleton, the main evidence being the comparison of its perhydrogenated product with hexahydrohumulene by infrared spectroscopy. Zerumbone has been assigned the structure (XXVII) on the basis of degradation studies and analysis of infrared spectra.

The similarity of the caryophyllene and humulene structures and their frequent common occurrence point to a close biogenetic relationship. Indeed, it seems probable that the bicyclic hydrocarbon arises in Nature from humulene by cyclisation. So far, however, all attempts to bring about this cyclisation in the laboratory have met with failure, possibly because of the relative instability of humulene.

Some characteristic constants of medium-ring sesquiterpenes appear in *Table 1.*

Table 1. Medium-ring Sesquiterpenes.

Compound	m. p.	d_4^{20}	n_D^{20}	$[\alpha]_D^{20}$	Reference
Caryophyllene	—	0.9074	1.5009	— 9.07	(2, 10, 47)
α-Betulenol	—	0.9770	1.5127	— 6.9	(28)
β-Betulenol	~40–41°	—	—	— 55.3	(28)
α-Humulene	—	0.8905	1.5508	± 0	(8)
β-Humulene	—	0.8907	1.5012	+ 0.07	(8)
Zerumbone	66–67°	—	—	0	(17, 18)
Germacrone	56–57°	—	—	0	(36, 37)

IV. Terpenes Containing a Ten-Membered Carbon Ring.

I. Germacrone.

The first natural product for which a ten-membered ring structure has been proposed was the lactone pyrethrosin (7), which will be dealt with below (p. 15). The first compound, however, in which the presence of such a ring was definitely established only a few months later was the ketone germacrone $C_{15}H_{22}O$ — a beautifully crystalline substance forming the principal component of Bulgarian zdravets oil. The compound had been known for some time under the name germacrol and had been studied previously by TREIBS (81) who proposed for it the cyclic ether structure (XXVIII) with a guaiane skeleton. Subsequently it was found that the compound is not an oxide but a monocyclic ketone containing three double bonds (XXIX), and its name was accordingly changed to germacrone (36, 37).

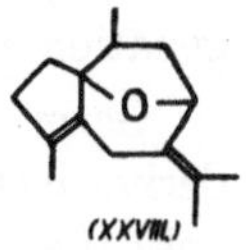

Attempts to determine the carbon skeleton of germacrone gave results which at first appeared contradictory. Thus, hydrogenation of

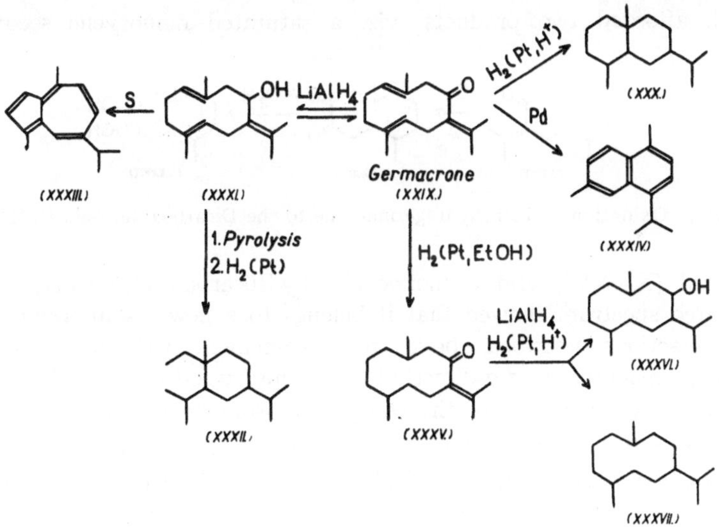

Chart 8. Conversions of Germacrone.

germacrone in acetic acid was accompanied by hydrogenolysis and afforded a bicyclic saturated hydrocarbon which on the basis of its infrared spectrum was shown to be fairly pure selinane (XXX). The reduction of germacrone with aluminium isopropoxide or with lithium aluminium hydride gave rise to the unsaturated alcohol (XXXI) *(Chart 8)*.

This alcohol may be reconverted into germacrone by oxidation, proving that it contains the original carbon skeleton of germacrone. Thermal dehydration of the alcohol and hydrogenation of the resulting hydrocarbon led to a monocyclic hydrocarbon, identified as elemane (XXXII). On the other hand, direct dehydrogenation of the alcohol (XXXI) gave rise to considerable amounts of guaiazulene (XXXIII). It had previously been reported that germacrone yields cadalene (XXXIV) on dehydrogenation under fairly drastic conditions.

Germacrone can thus be converted under various conditions to derivatives belonging to four different sesquiterpene types; of these only the elemane type corresponded to the presumed monocyclic nature of germacrone. Later it was shown that germacrone belongs to none of the four types mentioned but represents a new sesquiterpene structure, termed the germacrane type. The hydrogenation of germacrone in ethanol gave rise to the tetrahydro derivative (XXXV); in this product the isopropylidene double bond which is conjugated with the keto group is preserved, as demonstrated by the formation of 0.7 equivalents of acetone on ozonisation. Reduction of this tetrahydro derivative with lithium aluminium hydride, followed by catalytic reduction in acetic acid, afforded two products, viz. a saturated monocyclic secondary

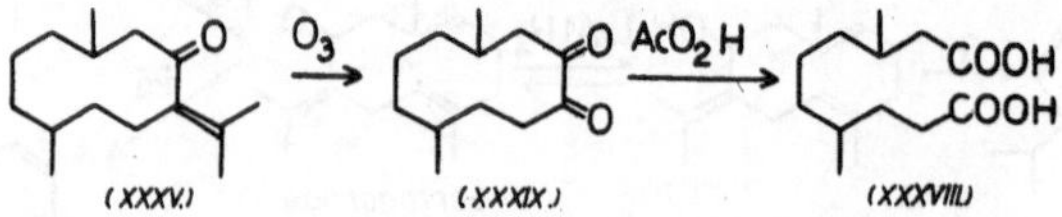

Chart 9. Oxidation of Tetrahydrogermacrone to the Dicarboxylic Acid (XXXVII).

alcohol (XXXVI) and a monocyclic hydrocarbon (XXXVII), whose infrared spectrum showed that it belongs to a new sesquiterpene type. The reactions described above are in agreement with the assumption that germacrane is 1,7-dimethyl-4-isopropylcyclodecane, and contains a regular arrangement of the component isoprene units.

Supporting evidence for this structure was obtained by synthesis of the dicarboxylic acid (XXXVII) (25) which was isolated as the principal product when tetrahydrogermacrone (XXXV) was subjected to oxidation; the diketone (XXXIX) was an intermediate in this reaction sequence *(Chart 9).*

Unequivocal proof of the cyclodecane skeleton in germacrone was obtained by the total synthesis of germacrane (XXXVII) (72) as outlined in *Chart 10.* The synthetic cyclodecane derivative was identified with germacrane, as demonstrated by the infrared spectra and physical constants of the two samples.

References, pp. 27—31.

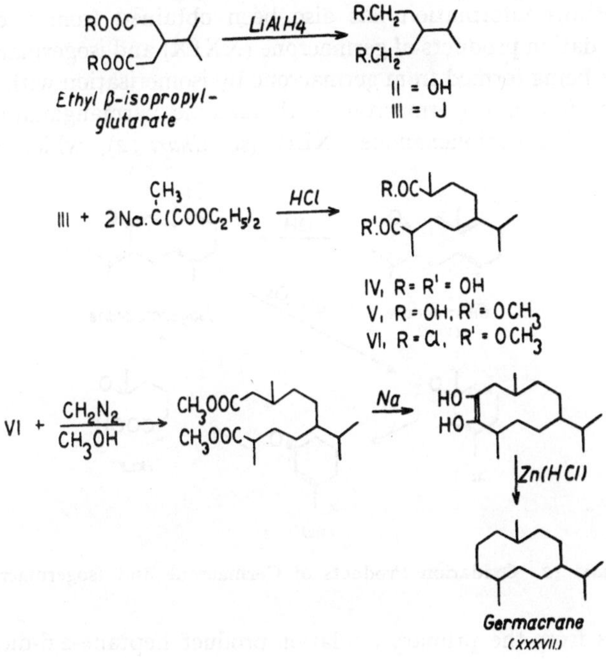

Chart 10. Total Synthesis of Germacrane.

Some difficulties were encountered in the determination of the position of the double bonds in germacrone, particularly in view of the unusual character of both its infrared and ultraviolet spectra. These anomalies have, indeed, led OHLOFF (39) to reject the formula proposed by the Prague group and to suggest for germacrane a structure based on the maaliane skeleton. The spectra of germacrone and other ten-membered ring derivatives will be discussed on p. 25.

Basic information on the location of the double bonds in germacrone originates from a study of its oxidative degradation (Chart 11). Ozonisation

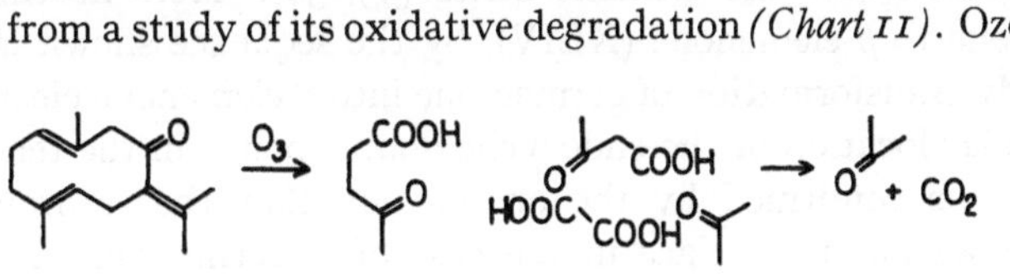

Chart 11. Oxidative Degradation of Germacrone.

has been found to afford 1.4–1.6 equivalents of acetone, together with laevulinic acid; permanganate oxidation gave laevulinic and oxalic acids. The formation of more than one equivalent of acetone may be rationalised on the assumption that it arises from the originally formed acetoacetic acid.

Important information has also been obtained from a comparison of the oxidation products of germacrone (XXIX) and isogermacrone (XL), the latter being formed from germacrone by isomerisation with alkali (69). The oxidation of isogermacrone with alkaline permanganate gives rise to 3-methyl-Δ^2-cyclohexanone (XLI) (see *Chart 12*), which apparently

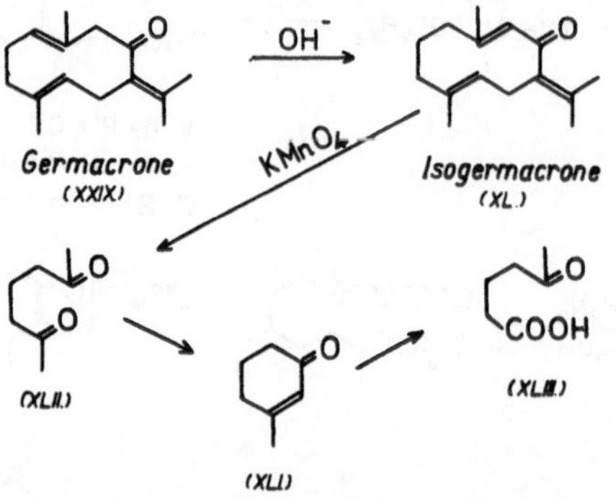

Chart 12. Oxidation Products of Germacrone and Isogermacrone.

originates from the primary oxidation product heptane-2,6-dione (XLII) by aldolisation followed by dehydration. This product is accompanied by homolaevulinic acid (XLIII). Furthermore, ozonisation affords only 0.6 equivalents of acetone. Consequently, isogermacrone differs from germacrone with respect to the location of one double bond, namely the one that is shifted on isomerisation. The presence of cross-conjugation in isogermacrone is brought out by its ultraviolet spectrum.

The germacrone formula (XXIX) is in agreement with the optical inactivity and the readiness with which it undergoes reactions leading to compounds with other carbon skeletons. Particularly noteworthy is the transition into the elemane series (35, 38). Mere heating converts germacrone to β-elemenone (XLIV), by the sequence shown in *Chart 13*. The ready transformation of germacrone into β-elemenone clearly requires a particular location of the endocyclic double bonds in the ten-membered ring; this is confirmed by the observation that the transannular ring closure does not take place in the case of isogermacrone, in which the position of the double bonds is known to be different. The conversions of germacrone to products of other sesquiterpene types, particularly to selinane (XXX), by hydrogenation and dehydrogenation, can be interpreted as transannular cyclisation reactions; such reactions are, as will be shown below, fairly common with compounds of this type. The isomerisation of germacrone to selinane in the course of hydrogenation may be formulated as in Chart 13.

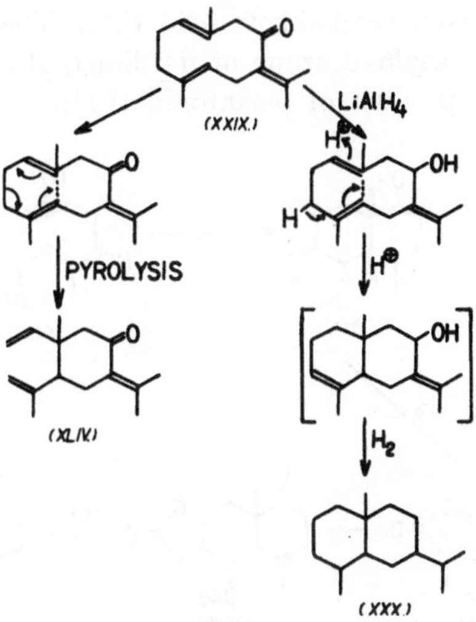

Chart 13. Conversion of Germacrone into β-Elemenone (XLIV) and Selinane (XXX).

2. Lactones.

A detailed examination of the lactones from Compositae was prompted by the earlier discovery of a new group of lactones containing a guaiane skeleton; this has led to the discovery of a considerable number of further compounds containing a ten-membered ring. All the ten-membered lactones hitherto isolated crystallise readily and are well-defined substances, a circumstance which has greatly facilitated their isolation and structural clarification.

Pyrethrosin.

The first study on pyrethrosin (XLV), the sesquiterpenic lactone from *Chrysanthemum cinerariaefolium*, a well known member of the Compositae, has been published by BARTON and DE MAYO in 1957 (7). This compound, $C_{17}H_{22}O_5$, contains a lactone ring, an acetoxy group, two double bonds and one oxygen function which is neither a hydroxyl nor a keto group and was therefore assigned oxide character. These facts establish the monocyclic character of the pyrethrosin molecule. Pyrethrosin was partially hydrogenated, and its more reactive double bond was shown to be of the vinylidene type and conjugated with the lactone carbonyl. The key experiment in the structure determination was an attempt to acetylate in acid solution which led to the opening of the oxide ring and formation of cyclopyrethrosin (XLVI), a bicyclic product of the selinane series. The skeleton of this compound was established by a series of reactions in which dihydrocyclopyrethrosin (XLVII) was converted by the sequence, (XLVIII)–(L), as shown in

Chart 14, to an unsaturated diketo acid (LI). The latter—in the form of the bis-dinitrophenylhydrazone of its dimethyl ester—was correlated with an oxidation product of ψ-santonin (LII).

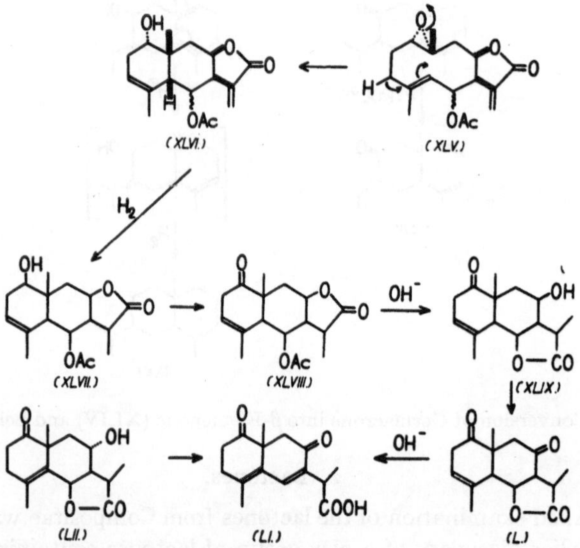

Chart 14. Conversions of Pyrethrosin.

The transformation of pyrethrosin to a derivative of the santonin series can only be accounted for in terms of a ten-membered ring structure. Quite recently, BARTON, BÖCKMAN and DE MAYO (3) were able to prove the presence of the 1,2-epoxide ring in pyrethrosin by isomerisation of its tetrahydro derivative (LIII) to the ketone (LIV), using boron

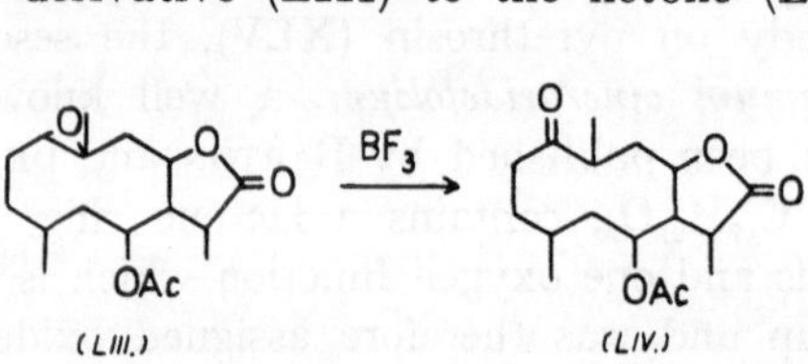

Chart 15. Isomerisation of Tetrahydropyrethrosin.

trifluoride as catalyst *(Chart 15)*. Further conversions of cyclopyrethrosin, in particular the finding that re-lactonisation can take place at the hydroxyl group which was originally acetylated, furthermore, the physical properties of various derivatives have recently enabled BARTON and his coworkers (3) to propose the absolute configurations at most of the asymmetric centres and to suggest the position of the double bonds as shown in (XLV).

Arctiopicrin.

Investigations on the lactone arctiopicrin, the bitter principle of *Arctium minus* (Compositae), were carried out in the writer's laboratory some years ago and revealed that the formula which had been assigned to this compound (77) was incorrect; and that arctiopicrin is the ester of β-hydroxyisobutyric acid and the sesquiterpene hydroxylactone (LV),

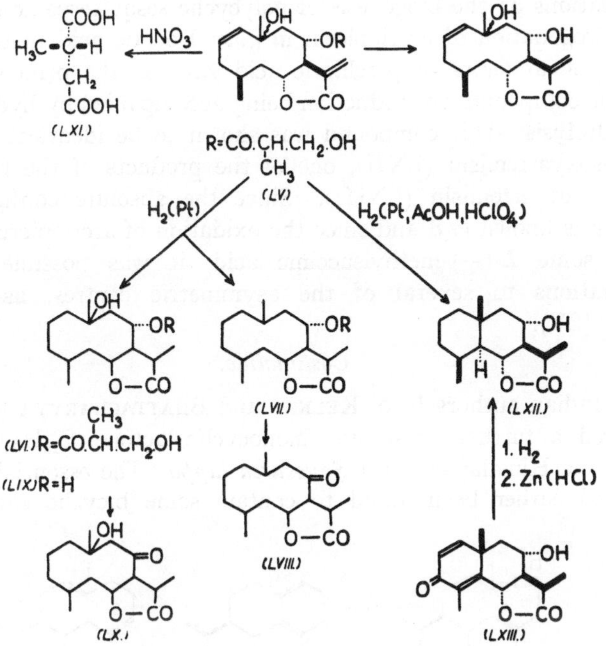

Chart 16. Conversions of Arctiopicrin.

for which the name arctiolide was proposed (66, 68, 71). Hence, the correct formula of arctiopicrin is $C_{19}H_{28}O_6$, and the compound is monocyclic. Hydrogenation of arctiopicrin affords a mixture of stereoisomeric tetrahydro derivatives (LVI) as well as an ester of a monohydroxy-lactone (LVII), formed by hydrogenolysis *(Chart 16)*. The latter product can be converted into two isomeric keto-lactones (LVIII) by hydrolysis and subsequent oxidation. The dihydroxylactone (LIX), obtained by the hydrolysis of (LVI), affords on chromic acid oxidation the hydroxy-ketolactone (LX). These observations prove that the hydroxyl group which is esterified in the original lactone has secondary, and the free hydroxyl group tertiary, character. One of the double bonds is again vinylic and conjugated with the lactonic carbonyl. The location of the other double bond is inferred from the ready hydrogenolysis of the tertiary hydroxyl group and from the oxidation products of arctiopicrin which include, in particular, optically active L-(—)-methylsuccinic acid (LXI).

The most important evidence for the structure of arctiopicrin was its direct steric correlation with compounds of the santonin series—a sequence which represented the first example of stereospecific trans-annular cyclisation; as such it is also of importance in connection with considerations on the biogenesis of polycyclic sesquiterpenic compounds. The hydrogenation of arctiopicrin in glacial acetic acid in the presence of catalytic amounts of perchloric acid gave as the principal product a bicyclic compound, the reduction being accompanied by hydrogenolysis and hydrolysis. This compound was shown to be identical with tetra-hydrodesoxyartemisin (LXII), one of the products of the Clemmensen reduction of artemisin (LXIII). Since the absolute configuration of artemisin is known (73) and since the oxidation of arctiopicrin has given rise to some L-(—)-methylsuccinic acid, it was possible to assign configurations to several of the asymmetric centres, as shown in formula (LV).

Costunolide.

The Indian authors Rao, Kelkar and Bhattacharyya (9, 53) have discovered a further crystalline monocyclic lactone, $C_{15}H_{20}O_2$, in the roots of the Himalayan plant *Saussurea lappa*. The essential oil of this plant had earlier been found to contain some bicyclic sesquiterpenic

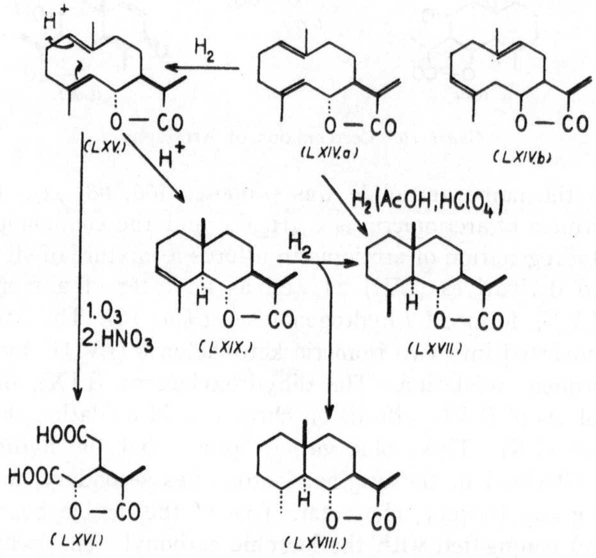

Chart 17. Conversions of Costunolide (LXIVa).

lactones, such as the guaianolide dehydrocostus lactone (*45, 82*). Costunolide contains three double bonds, one of them being vinylic and conjugated with the γ-lactonic carbonyl. On the basis of these data the Indian authors suggested for costunolide the two alternative formulae (LXIVa) and (LXIVb) *(Chart 17)*. At about the same time, the extractives of an Asian species of wormwood, *Artemisia balchanorum*, were studied in the writer's laboratory. The principal lactonic component proved to be identical with costunolide, and its structure was shown to correspond to formula (LXIVa) of the Indian authors. The absolute configuration of costunolide has been established as shown in (LXIVa) (*22*).

Costunolide can be converted into the dihydro derivative (LXV); the double bond which is conjugated with the lactonic carbonyl is reduced preferentially. The ultraviolet spectrum of dihydrocostunolide, like that of germacrone, exhibits a short wave-length maximum at 213 mμ, pointing to an analogous arrangement of the two endocyclic tri-substituted double bonds in the two compounds. The oxidation of dihydrocostunolide gave rise to some laevulinic acid. Further, a lactone dicarboxylic acid, $C_8H_{10}O_6$, (LXVI) which is also formed during the oxidation of santonin (*34*), could be identified amongst the ozonisation products of dihydrocostunolide. Like arctiopicrin, costunolide readily undergoes stereospecific transannular cyclisation which may lead to different products depending on the conditions. Thus, the hydrogenation of dihydrocostunolide in glacial acetic acid in the presence of perchloric acid has been found to give rise to santanolide C (LXVII), one of the hydrogenation products of santonin (*30*). When dihydrocostunolide was refluxed in a mixture of glacial acetic acid and acetic anhydride, it yielded the already known unsaturated bicyclic lactone, 3-santenolide (LXIX) (*14*). Hydrogenation of this product afforded a mixture of santanolide A (LXVIII) and santanolide C (LXVII). Both these saturated lactones had been encountered previously and their stereochemistry was known (*30*). The formula (LXIVa) for costunolide was confirmed later by BHATTACHARYYA and his coworkers (*54*).

Balchanolide, Isobalchanolide, Hydroxybalchanolide and Acetylbalchanolide.

As pointed out above, costunolide is not the only lactonic compound found in *Artemisia balchanorum*. Paper chromatography of the extracts revealed at least four further representatives of this group, and three of these, balchanolide, isobalchanolide and hydroxybalchanolide (*70*) have actually been obtained in crystalline form. The former two have identical composition, $C_{15}H_{22}O_3$; they both contain two double bonds, a γ-lactone ring and one free hydroxyl group. The location of the double bonds in these two lactones was revealed by the characteristic short wavelength maximum at 205 mμ, already found in germacrone and dihydrocostunolide; and further by the fact that oxidation gave laevulinic acid and some other products characteristic for the 1,5-distribution of the double bonds. The assignment made is in agreement with the smooth course of the transannular cyclisation which takes place on treatment of balchanolide (LXX) with acid or by mere heating to 170°. The hydrogenation of the primary product (LXXII) obtained by the pyrolysis of balchanolide afforded tetrahydrodesoxoartemisin (LXII), a compound of known

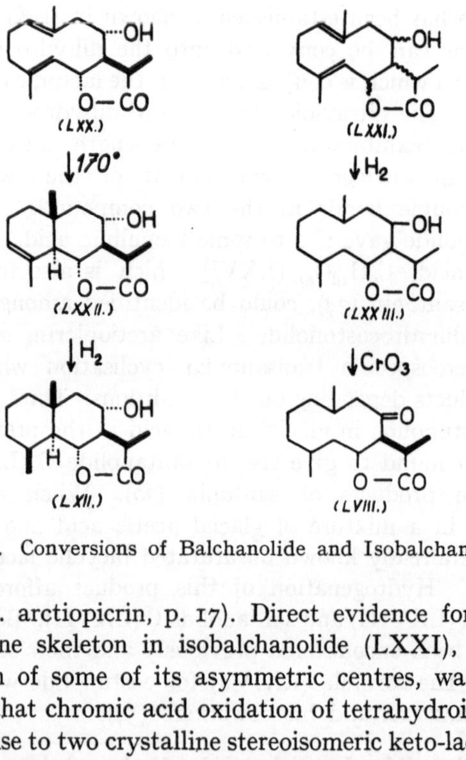

Chart 18. Conversions of Balchanolide and Isobalchanolide.

configuration (cf. arctiopicrin, p. 17). Direct evidence for the presence of the germacrane skeleton in isobalchanolide (LXXI), as well as of the configuration of some of its asymmetric centres, was provided by the observation that chromic acid oxidation of tetrahydroisobalchanolide (LXXIII) gave rise to two crystalline stereoisomeric keto-lactones (LVIII) (Chart 18). The latter could be shown to be identical with products previously obtained from arctiopicrin and cnicin (see below). The mutual relationship between balchanolide and isobalchanolide and some details of their stereochemistry remain to be elucidated.

A lactone closely related to balchanolide was isolated from the common milfoil (Achillea millefolium). The well-known drug is known to contain a chamazulene precursor of the guaianolide type (21) but this unstable compound has not yet been isolated in the pure state. A detailed examination of the extracts has, however, led to the isolation of two crystalline sesquiterpenic lactones (27). One of these has the formula

$C_{17}H_{24}O_4$ and contains two double bonds, a γ-lactone moiety, and an acetoxy group and has been proved to be acetylbalchanolide. Hydrogenation of acetylbalchanolide gave rise to a tetrahydro derivative (LXXV) which on hydrolysis of the acetyl group and chromic acid oxidation of the hydroxylactone (LXXVI) afforded the well-known two isomeric keto-lactones (LVIII), previously obtained from arctiopicrin, cnicine

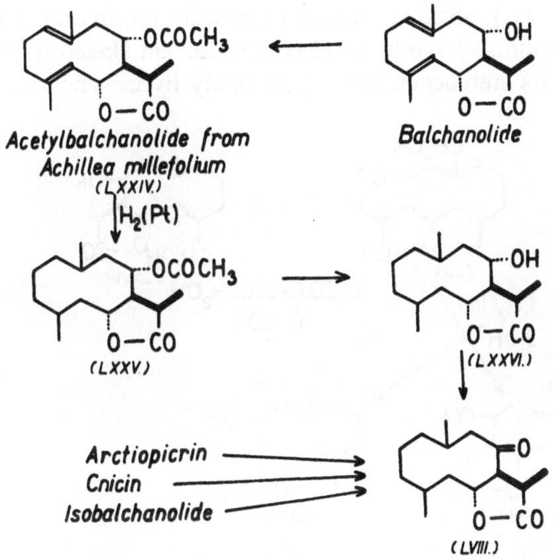

Acetylbalchanolide from
Achillea millefolium
(LXXIV.)

Balchanolide

(LXXV)

(LXXVI.)

Arctiopicrin
Cnicin
Isobalchanolide

(LVIII.)

Chart 19.
Degradation of Acetylbalchanolide to the two Isomeric Keto-lactones (LVIII).

and isobalchanolide (Chart 19). The structure assigned to this lactone was confirmed by direct comparison with the acetylation product of an authentic balchanolide sample.

Cnicin.

Another compound based on the germacrane skeleton is the bitter principle of *Cnicus benedictus*, a substance which has been known for more than a century. It was isolated in the pure state only a few years ago (15) but was still assigned an incorrect formula.

Recently (65, 67) it was shown that this compound, now referred to as cnicin, has the composition $C_{20}H_{28}O_7$ and is the ester of a monocyclic sesquiterpene hydroxylactone, $C_{15}H_{22}O_4$, with a rather uncommon carboxylic acid, i. e. 1,4-dihydroxy-2-butene-2-carboxylic acid. The parent lactone contains two hydroxyl groups and two double bonds, one of which is vinylic and conjugated with the lactone carbonyl. The hydrogenation of cnicin, (LXXVII a) or (LXXVII b), followed by hydrolysis of the saturated product and chromatography, afforded two isomeric crystalline dihydroxy lactones (LXXVIII), as well as a monohydroxy-lactone (LXXX) formed by concurrent hydrogenolysis. Chromic acid oxidation of the dihydroxylactones yielded two different lactone keto-acids (LXXIX) showing that the free hydroxyl group in cnicin is primary and the esterified hydroxyl group, secondary. Oxidation of the mono-hydroxylactone (LXXX) once more afforded the two isomeric keto-lactones (LVIII), obtained earlier from arctiopicrin and other compounds.

These findings have established the carbon skeleton of cnicin, as well as some of its stereochemistry. The ready hydrogenolysis of the primary

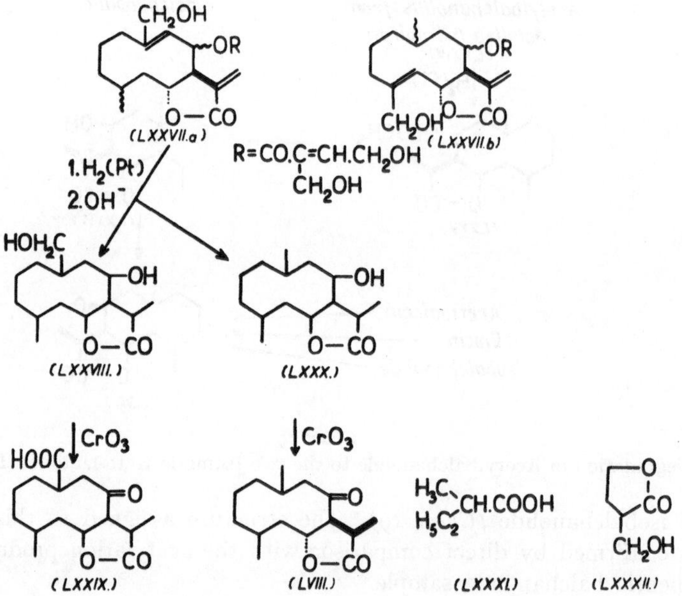

Chart 20. Conversions of Cnicin (LXXVIIa) or (LXXVIIb).

hydroxyl group in cnicin shows it to be located in the allyl position to one of the double bonds. Some methyl ethylacetic acid (LXXXI) is also split off hydrogenolytically, indicating that the esterified hydroxyl group is located next to a double bond. Hence, it is likely that cnicin possesses the structure (LXXVIIa), although the alternative structure (LXXIIb) can not yet be excluded. Further evidence in favour of the proposed structures was obtained from the study of the oxidation products of cnicin, which included succinic, methylsuccinic, α-methylglutaric, and α-methyladipic acids.

Considerable difficulty was encountered in the identification of the acid component of cnicin. The free 1,4-dihydroxy-2-butene-2-carboxylic acid is very unstable and could not be isolated as such. Its structure was therefore determined indirectly, viz. from the nature of the acid products obtained by hydrolysis of the hydrogenation product of cnicin. This process afforded methyl ethylacetic acid (LXXXI), arising from the hydrogenolysis product, and 1,4-dihydroxybutane-2-carboxylic acid, identified as the 3,5-dinitrobenzoyl derivative and as the γ-lactone (LXXXII) into which it is converted on distillation.

Parthenolide.

The study of the extractives of *Chrysanthemum parthenium*, carried out in the course of a systematic examination of Compositae, has resulted in the isolation of another crystalline sesquiterpenic lactone, $C_{15}H_{20}O_3$ (23), which was given the name parthenolide.

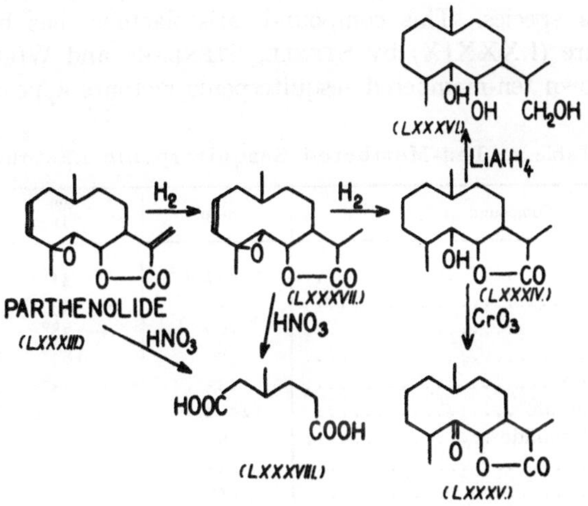

Chart 21. Conversions of Parthenolide.

Parthenolide (LXXXIII) contains two double bonds, one of them conjugated with the lactone carbonyl group. The presence of a γ-lactone ring was revealed by infrared evidence; and the third oxygen atom was shown to be present as a 1,2-epoxide. Hydrogenation of parthenolide gives rise to a hexahydro derivative (LXXXIV) *(Chart 21)* which contains a free hydroxyl group, proving that the reduction process involves a hydrogenolytic cleavage of the epoxide ring. The oxidation of this compound affords the keto-lactone (LXXXV). When (LXXXIV) is reduced with lithium aluminium hydride the triol (LXXXVI) results which can be readily oxidised with periodate, whereby one mole of the reagent is consumed. This establishes the location of the keto group as vicinal to the hydroxyl that originally participipitated in the formation of the lactone ring.

Parthenolide can be partially reduced to dihydroparthenolide (LXXXVII), in which the conjugated methylenic double bond is saturated. Nitric acid oxidation of this compound, and also of the parent lactone, gives rise to a mixture of dicarboxylic acids of which the β-methyladipic acid (LXXXVIII) is the highest homologue. All these observations are in agreement with the parthenolide structure (LXXXIII, p. 23); its stereochemistry has not yet been investigated.

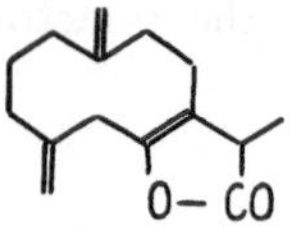

(LXXXVII.)

Aristolactone.

Yet another ten-membered ring lactone of the germacrone type has been obtained some time ago, not from a Compositae but from some Aristolochia species. This compound, aristolactone, has been ascribed the structure (LXXXIX) by STEELE, STENLAKE and WILLIAMS (62).

The known ten-membered sesquiterpenic lactones appear in Table 2.

Table 2. Ten-Membered Sesquiterpenic Lactones.

Compound	m. p.	$[\alpha]_D^{20}$	References
Pyrethrosin	198–200°	− 31°	(7)
Arctiopicrin	115°	+ 133°	(66, 71)
Costunolide	107°	+ 128°	(9, 53)
Balchanolide	150°	+ 183°	(70)
Isobalchanolide	143°	+ 122°	(70)
Acetylbalchanolide	125–126°	− 21.4°	(27)
Hydroxybalchanolide	163°	+ 99°	(70)
Cnicin	143°	+ 158°	(65)
Parthenolide	116°	− 81°	(23)
Aristolactone	110–111°	− 156°	(62)

Biogenetic Relationships.

The readiness with which most of the unsaturated derivatives of the germacrane type undergo transannular cyclisation to give compounds with a selinane skeleton (in the case of lactones, relatives of santonin) suggests a biogenetic significance of such reactions. It may by assumed that the cyclodecane-type lactones are the primary products in Nature, and that lactones of the santonin and guaianolide series are formed from them by a secondary process, analogous to well-known stereospecific transannular cyclisations observed in the laboratory. This hypothesis is supported by the frequent occurrence of lactones of the germacrane type in Compositae which are also the only source of the other two types of sesquiterpenic lactones. However, direct evidence for this biogenetical relationship is still lacking.

V. Some Spectral Anomalies in Medium-ring Terpenes.

In the course of the structural work on compounds of the germacrane type certain anomalies were encountered in the spectral properties which were detrimental to progress in this field or even led some workers to incorrect conclusions. These anomalies evidently result from the specific conformational situation that prevails in medium-ring compounds.

This situation is well exemplified by the spectral properties of germacrone. The latter exhibits three maxima in the ultraviolet region, at 213 mμ (log ε = 4.10), at 240 mμ (log ε = 3.47), and at 315 mμ (log ε = 2.55) *(Fig. 1)*. The second and third maxima may be assigned to the ketone conjugated with the isopropylidene group; the high extinction of the *R* band is in good agreement with the fact that the double bond is located in the β,γ-position to the carbonyl group. On the other hand, the short-

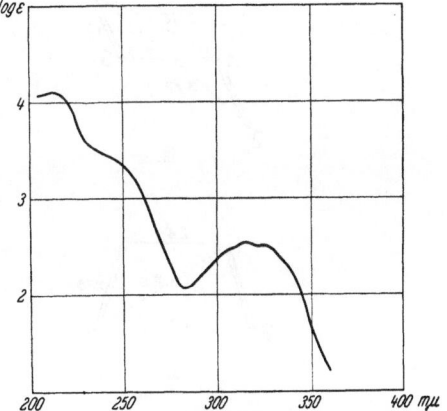

Fig. 1. Molecular extinction curve of germacrone (in cyclohexane).

wavelength maximum at 213 mμ cannot be interpreted on the basis of classical relationships between structure and spectral properties. The same maximum is also found in the alcohol germacrol and in a number of naturally occurring unsaturated lactones such as costunolide, dihydrocostunolide and balchanolide. In contrast, this maximum disappears in both the monoxide and the dioxide of germacrone, in which one or both endocyclic double bonds are saturated. These facts indicate that the 213 mμ-maximum does not result from a transannular orbital overlap between the π-electrons of the double bond and the carbonyl group as in the case of cyclodecenenone studied by LEONARD and OWENS (*31*) but is in some way connected with the two endocyclic double bonds.

The anomalous character of the ultraviolet spectra of germacrone, costunolide and balchanolide indicated the possibility of non-classical interactions of the endocyclic double bonds in the cyclodeca-1,5-diene system. At the suggestion of the writer, this problem has been approached using appropriate quantum mechanical methods by Dr. J. KOUTECKÝ

(Institute of Physical Chemistry, Czechoslovak Academy of Science). From a consideration of scale models, Dr. Koutecký and his coworkers, Dr. R. Zahradník and Dr. J. Paldus, concluded that in all three geometrical isomers of cyclodeca-1,5-diene (*cis,cis*, *cis,trans* and *trans,trans*) conformations could be set up in which the double bonds were roughly parallel. The four spatial arrangements of the relevant four

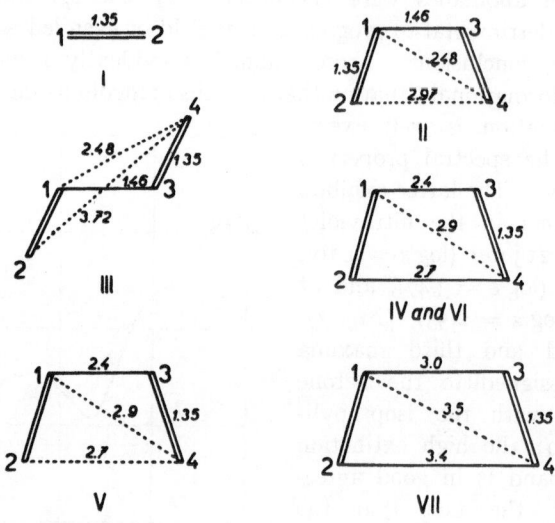

Dotted lines: distances between non-bonded atoms ($\beta_{ij}=0$)

Chart 22. Spatial Arrangements of Ethylene, *cis*- and *trans*-Butadienes, and Four Relevant Carbon Atoms of Cyclodecadiene (1,5).

carbon atoms of the cyclodecadiene molecule for which calculations were made, together with those of ethylene and *cis*- and *trans*-butadiene, are shown in *Chart 22*.

The energies of transition from the highest occupied to the lowest vacant orbit were calculated by a self-consistent molecular orbital method (*32*, *41*, *42*, *46*); the results for ethylene and *cis*- and *trans*-butadiene were used to check the suitability of the chosen parameters. As may be seen from *Table 3*, the delocalisation energies for three of the models are negligible and only for model V, with a decreased β_{24} integral, is the value appreciable. The values obtained for the butadienes are in good agreement with the experimental data.

Table 3 also shows the results of calculations of the absorption frequencies in Clar's classification (*11*); here again the values calculated for ethylene and the butadienes are in good agreement with the experiment.

All the cyclodecadiene models considered show a relatively marked interaction, and models V and VI in particular give values for the singlet p-band very close to the observed absorption maxima of germacrone, costunolide and dihydrocostunolide. The results of these calculations show that the proposed transannular interactions between the endocyclic double bonds of compounds of this type are not only theoretically possible but are in fact very probable.

Table 3. Delocalisation Energy and Frequency of Singlet p-Bands for Models I–VII. [Frequencies in eV; for singlet p-bands also in $m\mu$ (in parentheses).]

Model	Delocalisation energy (kcal/mol)	p
I	—	7.38 (170)
II	13.2	5.18 (239)
III	12.2	5.39 (230)
IV	0.6	4.47 (277)
V	5.5	5.68 (218)
VI	0.2	5.65 (219)
VII	0.2	6.30 (197)

The infrared spectrum of germacrone also possesses some unusual features. This spectrum contains the expected absorption associated with a conjugated carbonyl group (1681 cm.$^{-1}$) and trisubstituted double bonds (1660 cm.$^{-1}$). However, the spectra of ketones containing a tetrasubstituted double bond in conjugation with the carbonyl generally show a very strong absorption at about 1620 cm.$^{-1}$ which is absent from the germacrone spectrum. It would seem that the absence of this band is related to the specific geometry of the germacrane molecule; in particular that the carbon-oxygen double bond is forced out of the plane of the tetrasubstituted ethylenic double bond. This interpretation is supported by the observation that the Raman spectrum of germacrone contains three bands in the carbonyl double bond region; a weak band at 1650 cm.$^{-1}$ is very probably due to the tetrasubstituted double bond, since a band of practically the same frequency (1648 cm.$^{-1}$) is also present in the spectrum of germacrone dioxide.

Up to the present time seventeen medium-ring terpenes have been found in Nature—all of them belonging to the sesquiterpene class. This clearly justifies the great theoretical interest in this group of compounds and promises further interesting discoveries in the future.

References.

1. AEBI, A., D. H. R. BARTON and A. S. LINDSEY: Sesquiterpenoids. III. The Stereochemistry of Caryophyllene. J. Chem. Soc. (London) 1953, 3124.
2. BARTON, D. H. R.: The Constitution of (+)-trans-Caryophyllenic Acid. J. Organ. Chem. (USA) 15, 457 (1950).
3. BARTON, D. H. R., O. C. BÖCKMAN and P. DE MAYO: Sesquiterpenoids. XII. Further Investigations on the Chemistry of Pyrethrosin. J. Chem. Soc. (London) 1960, 2263.

4. Barton, D. H. R., T. Bruun and A. S. Lindsey: Sesquiterpenoids. II. Try-cyclic Derivatives of Caryophyllene. J. Chem. Soc. (London) 1952, 2210.
5. Barton, D. H. R. and A. S. Lindsey: The Constitution of Caryophyllene. Chem. and Ind. 1951, 313.
6. — — Sesquiterpenoids. I. Evidence for a Nine-membered Ring in Caryo-phyllene. J. Chem. Soc. (London) 1951, 2988.
7. Barton, D. H. R. and P. de Mayo: Sesquiterpenoids. VIII. The Constitution of Pyrethrosin. J. Chem. Soc. (London) 1957, 150.
8. Benešová, V., V. Herout and F. Šorm: On Terpenes. CXXVIII. Existence of α- and β-Humulene. Collect. Czech. Chem. Commun. (in press) [in English].
9. Bhattacharyya, S. C., G. R. Kelkar and A. Somasekar Rao: Structure of Costunolide. Chem. and Ind. 1959, 1069.
10. Campbell, A. and H. N. Rydon: The Synthesis of Caryophyllenic Acid. Chem. and Ind. 1951, 312.
11. Clar, E.: Aromatische Kohlenwasserstoffe. Polycyclische Systeme. 2. Aufl. Berlin: Springer-Verlag. 1952.
12. Clemo, G. R. and J. O. Harris: The Structure of Humulene. Chem. and Ind. 1951, 799.
13. — — The Chemistry of Humulene. II. J. Chem. Soc. (London) 1952, 665.
14. Cocker, W. and T. B. H. McMurry: The Chemistry of Santonin. III. The Stereochemistry of Some Reduction Products of Santonin. J. Chem. Soc. (London) 1956, 4549.
15. Corte, F. und G. Bechmann: Über die Bitterstoffe aus Cnicus benedictus. Naturwiss. 45, 390 (1958).
16. Dawson, T. L. and G. R. Ramage: The Caryophyllenes. IX. Homocaryophyllenic Acid. J. Chem. Soc. (London) 1951, 3382.
17. Dev, S.: Zerumbone, a Monocyclic Sesquiterpene Ketone. Chem. and Ind. 1956, 1051.
18. — Studies in Sesquiterpenes. XVI. Zerumbone, a Monocyclic Sesquiterpene Ketone. Tetrahedron 8, 171 (1960).
19. — Studies in Sesquiterpenes. XVIII. The Proton Magnetic Resonance Spectra of Some Sesquiterpenes and the Structure of Humulene. Tetrahedron 9, 1 (1960).
20. Dolejš, L. and F. Šorm: On Terpenes. LIII. Synthesis of 2,8,8-Trimethyl-bicyclo-(5,2,0)-nonane: A Proof of the Constitution of β-Caryophyllene. Collect. Czech. Chem. Commun. 19, 559 (1954) [in English]; Chem. Listy 47, 1849 (1953).
21. Graham, K.: Cytochemical Notes. I. On the Non-preexistence of the Azulene in Milfoil. J. Amer. Pharm. Assoc. 22, 819 (1933).
22. Herout, V. and F. Šorm: Isolation and Structure of Costunolide from Artemisia balchanorum. Chem. and Ind. 1959, 1067.
23. Herout, V., M. Souček and F. Šorm: Parthenolide, another Sesquiterpenic Lactone with a Ten-membered Ring. Chem. and Ind. 1959, 1069.
24. Herout, V., M. Streibl, J. Mleziva and F. Šorm: On Terpenes. XIV. On the Identity of Humulene and "α-Caryophyllene". Collect. Czech. Chem. Commun. 14, 716 (1949) [in English]; Chem. Listy 44, 27 (1950).
25. Herout, V. and M. Suchý: On Terpenes. LXXXIX. Direct Proof of the Carbon Skeleton of Germacrone. Collect. Czech. Chem. Commun. 23, 2169 (1958) [in English]; Chem. Listy 52, 1174 (1958).
26. Hildebrand, R. P., M. D. Sutherland and O. J. Waters: Structure of Humulene. Chem. and Ind. 1959, 489.

27. HOCHMANNOVÁ, J., V. HEROUT and F. ŠORM: On Terpenes. CXXVII. Isolation and Structure of Sesquiterpenic Lactones from Yearrow (*Achillea millefolium* L.). Collect. Czech. Chem. Commun. (in press) [in English].

28. HOLUB, M., V. HEROUT, M. HORÁK and F. ŠORM: On Terpenes. CIV. The Constitution of Betulenols from Oil from the Buds of White Birch (*Betula alba* L.). Collect. Czech. Chem. Commun. 24, 3730 (1959) [in English].

29. JAROLÍM, V., M. STREIBL, L. DOLEJŠ and F. ŠORM: On Terpenes. LXXVI. Synthesis of 4,8,11,11-Tetramethylbicyclo-(0,2,7)-undecane (Caryophyllane). Collect. Czech. Chem. Commun. 22, 1277 (1957) [in English]; Chem. Listy 50, 1299 (1956).

30. KOVÁCS, Ö., V. HEROUT, M. HORÁK and F. ŠORM: On Terpenes. LXVII. Hydrogenation Products of Santonin and Alantolactone. Collect. Czech. Chem. Commun. 21, 225 (1956) [in English]; Chem. Listy 49, 1856 (1955).

31. LEONARD, N. J. and F. H. OWENS: Spectral Properties of Medium- and Large-ring Carbocyclic Compounds: α-Bromoketones, Enol Acetates and Unsaturated Ketones. J. Amer. Chem. Soc. 80, 6039 (1958).

32. MATAGA, N.: Electronic Structure and Spectra of s-Tetrazine. Bull. Chem. Soc. Japan 31, 453 (1958) [in English].

33. MAYO, P. DE: The Chemistry of Perhydroazulene Sesquiterpenoids. Perfumery Essent. Oil Record 48, 18, 68 (1957).

34. NOVOTNÝ, L., V. HEROUT and F. ŠORM: On Terpenes. CX. A Contribution to Stereochemistry of Absinthin and Artabsin. Collect. Czech. Chem. Commun. 25, 1500 (1960) [in English].

35. OGNJANOV, I., V. HEROUT, M. HORÁK und F. ŠORM: Über Terpene. CIII. Über die Konstitution von β-Elemenon aus bulgarischem Geraniumöl. Collect. Czech. Chem. Commun. 24, 2371 (1959) [in German].

36. OGNJANOV, I., D. IVANOV, V. HEROUT, M. HORÁK, J. PLÍVA and F. ŠORM: The Structure of Germacrone (Germacrol). Chem. and Ind. 1957, 820.

37. — — — — — — On Terpenes. LXXXVIII. The Structure of Germacrone, the Crystalline Constituent of Bulgarian "Zdravets" Oil. Collect. Czech. Chem. Commun. 23, 2033 (1958) [in English]; Chem. Listy 52, 1163 (1958).

38. OHLOFF, G.: Zur Pyrolyse des Germacrons. Angew. Chem. 71, 162 (1959).

39. OHLOFF, G., H. FARNOW, W. PHILIPP und G. SCHADE: Über Germacron und seine Umwandlunsprodukte. Liebigs Ann. Chem. 625, 206 (1959).

40. PLÍVA, J. and F. ŠORM: On Terpenes. XI. Infrared Investigations of Terpenes. I. Collect. Czech. Chem. Commun. 14, 274 (1949) [in English].

41. POPLE, J. A.: Electron Interaction in unsaturated Hydrocarbons. Trans. Faraday Soc. 49, 1375 (1953).

42. — The Electronic Spectra of Aromatic Molecules. II. A Theoretical Treatment of Excited States of Alternant Hydrocarbon Molecules Based on Self-consistent Molecular Orbitals. Proc. Phys. Soc. (London) A 68, 81 (1955).

43. PRELOG, V., L. FRENKIEL, M. KOBELT und P. BARMAN: Zur Kenntnis des Kohlenstoffringes. 43. Mitt. Ein Herstellungsverfahren für vielgliedrige Cyclanone. Helv. Chim. Acta 30, 1741 (1947).

44. ROBERTSON, J. M. and G. TODD: The Structure of β-Caryophyllene Alcohol Chloride and Bromide. An X-Ray Determination. Chem. and Ind. 1953, 437.

45. ROMAŇUK, M., V. HEROUT and F. ŠORM: On Terpenes. LXIX. The Constitution of Dehydrocostuslactone. Collect. Czech. Chem. Commun. 21, 894 (1956) [in English]; Chem. Listy 49, 1879 (1955).

46. ROOTHAAN, C. C. J.: New Developments in Molecular Orbital Theory. Rev. Modern Phys. 23, 69 (1951).

47. Ruzicka, L., J. C. Bardhan und A. H. Wind: Höhere Terpenverbindungen. XLVII. Zur Kenntnis der Caryophyllensäure. Helv. Chim. Acta 14, 423 (1931).
48. Ruzicka, L. und W. Zimmermann: Polyterpene und Polyterpenoide. XCII. Zur Kenntnis der Caryophyllensäure und der Nor-caryophyllensäure. Über das Additionsprodukt von Maleinsäure-anhydrid an Caryophyllen. Helv. Chim. Acta 18, 219 (1935).
49. Rydon, H. N.: The Synthesis of cis- and trans-dl-Norcaryophyllenic Acids and of Dehydronorcaryophyllenic Acid. J. Chem. Soc. (London) 1936, 593.
50. — The Resolution of cis- and trans-Norcaryophyllenic Acid. J. Chem. Soc. (London) 1937, 1340.
51. — The Constitution of Caryophyllene. Chem. and Ind. 16, 123 (1938).
52. Simonsen, J. and P. de Mayo: In: J. Simonsen and W. C. J. Ross, The Terpenes. Vol. V. New York: Cambridge Univ. Press. 1957.
53. Somasekar Rao, A., G. R. Kelkar and S. C. Bhattacharyya: Costunolide, a New Sesquiterpene Lactone from Costus Root Oil. Chem. and Ind. 1958, 1359.
54. — — — Terpenoids. XXI. The Structure of Costunolide, a new Sesquiterpene Lactone from Costus Root Oil. Tetrahedron 9, 275 (1960).
55. Šorm, F.: Some Recent Developments in Sesquiterpene Chemistry. Record Chem. Progr. 21, 73 (1960).
56. Šorm, F., L. Dolejš and J. Plíva: On Terpenes. XX. A Note on the Constitution of β-Caryophyllene. Collect. Czech. Chem. Commun. 15, 186 (1950) [in English].
57. Šorm, F., V. Herout and V. Sýkora: Advances in Sesquiterpene Chemistry. Perfumery Essent. Oil Record 50, 679 (1959).
58. Šorm, F., J. Mleziva, Z. Arnold and J. Plíva: On Terpenes. XIII. On the Sesquiterpenes from the Essential Oil of Hops. Collect. Czech. Chem. Commun. 14, 698 (1949) [in English].
59. Šorm, F., M. Streibl, V. Jarolím, L. Novotny, L. Dolejš and V. Herout: Synthesis of 1:1:4:8-Tetramethylcycloundecane (Humulane). Chem. and Ind. 1954, 252.
60. — — — — — — On Terpenes. LVIII. Total Synthesis of 1,1,4,8-Tetramethylcycloundecane: Proof of the Eleven-membered Ring in Humulene. Collect. Czech. Chem. Commun. 19, 570 (1954) [in English]; Chem. Listy 48, 575 (1954).
61. Šorm, F., M. Streibl, J. Plíva and V. Herout: On Terpenes. XXXII. A Contribution to the Constitution of Humulene. Collect. Czech. Chem. Commun. 16, 639 (1952) [in English]; Chem. Listy 46, 30 (1952).
62. Steele, J. W., J. B. Stenlake and W. D. Williams: The Chemistry of the Aristolochia Species. IV. The Structure of Aristolactone. J. Chem. Soc. (London) 1959, 3289.
63. Stoll, M. et J. Hulstkamp: Synthèse de produits macrocycliques à odeur musquée. 2ème commun. Sur une amélioration de la préparation des acyloïnes cycliques. Helv. Chim. Acta 30, 1815 (1947).
64. Stoll, M. et A. Rouvé: Synthèse de produits macrocycliques à odeur musquée. 3ème commun. Sur les acyloïnes cycliques. Helv. Chim. Acta 30, 1822 (1947).
65. Suchý, M., V. Benešova, V. Herout und F. Šorm: Über Terpene. CXIX. Die Struktur des Cnicins, eines Sesquiterpen-Lactones aus Cnicus benedictus L. Chem. Ber. 93, 2449 (1960).
66. Suchý, M., V. Herout and F. Šorm: Terpenes. LXXVI. The Nature of Arctiopicrin, an Unsaturated Lactone from Arctium minus. Chem. Listy 50, 1827 (1956) [in Czech].

67. Suchý, M., V. Herout and F. Šorm: Cnicin: A New Guaianolide-Type Lactone. Chem. and Ind. 1959, 517.

68. — — — On Terpenes: XCVIII. Proof of Structure of Arctiopicrin with a Note on its Stereochemistry. Collect. Czech. Chem. Commun. 24, 1542 (1959) [in English]; Chem. Listy 52, 2110 (1958).

69. — — — On Terpenes. CXXVI. Relation between Germacrone and Isogermacrone. Collect. Czech. Chem. Commun. (in press) [in English].

70. — — — Isolation and Structure of Sesquiterpenic Lactones of Germacrane Type from *Artemisia balchanorum* (Costunolide, Balchanolide, Isobalchanolide and Hydroxybalchanolide). Collect. Czech. Chem. Commun. (in press) [in English].

71. Suchý, M., M. Horák, V. Herout und F. Šorm: Über Terpene. LXXXIV. Über die Struktur des Arctiopikrin, des Sesquiterpenlactons aus *Arctium minus* Bernh. Croat. Chem. Acta 29, 247 (1957) [in German].

72. Suchý, M. and F. Šorm: On Terpenes. XC. Synthesis of 1,7-Dimethyl-4-*iso*-Propylcyclodecane (Germacrane). Collect. Czech. Chem. Commun. 23, 2175 (1958) [in English]; Chem. Listy 52, 1180 (1958).

73. Sumi, M.: The Structure of Artemisin. Proc. Japan Acad. 32, 684 (1956) [in English].

74. Sýkora, V., V. Herout, J. Plíva and F. Šorm: Constitution of Acorone. Chem. and Ind. 1956, 1231.

75. — — — — Über Terpene. LXXXII. Die Konstitution von Acoron. Collect. Czech. Chem. Commun. 23, 1072 (1958) [in German]; Chem. Listy 51, 1704 (1957).

76. Sýkora, V., V. Herout, A. Reiser and F. Šorm: On Terpenes. XCVI. The Stereochemistry of Acorone and its Stereoisomers. Collect. Czech. Chem. Commun. 24, 1306 (1959) [in English]; Chem. Listy 52, 2102 (1958).

77. Tettweiler, K., O. Engel und E. Wedekind: Über die Konstitution des Artemisins. Liebigs Ann. Chem. 492, 105 (1932).

78. Treibs, W.: Konstitutions-Aufklärung einiger Sesquiterpene durch starken oxydativen Abbau, III. Mitt.: Betulol. Ber. dtsch. chem. Ges. 69, 41 (1936).

79. — Beweis der Identität der Betulenolsäure mit der Homo-caryophyllensäure (II. Mitt. über Betulenole). Ber. dtsch. chem. Ges. 71, 612 (1938).

80. — Über das Caryophyllenoxyd, seine Darstellung durch Autoxydation des Caryophyllens und sein Vorkommen in Pflanzenölen. Chem. Ber. 80, 56 (1947).

81. — Über bi- und polycyclische Azulene. XII. Das Germacrol, ein azulenbildendes Sesquiterpen-oxyd aus Geraniumöl. Liebigs Ann. Chem. 576, 116 (1952).

82. Ukita, T.: Costus Oil. J. pharmac. Soc. Japan 59, 80 (1939).

(Received, October 4, 1960.)

Recent Advances in the Chemistry of Azulenes and Natural Hydroazulenes.

By TETSUO NOZOE and SHÔ ITÔ, Sendai, Japan.

Contents.

I. Introduction.

The beautiful blue pigment, named "azulene" from its color a century ago, is a minor constituent of some essential oils and was also obtained by dehydrogenation of various sesquiterpenoids (azulene precursors, proazulenes or azulenogens). Since 1936, when the fundamental structure of azulene was defined by PFAU and PLATTNER (*186*) as the bicyclic ring system (I) with condensed unsaturated five- and seven-membered rings, many chemists have become interested in these compounds because of their unique structure. The structures of most naturally occurring azulenes have since been clarified and many more azulenes have been synthesized.

On the basis of physical and chemical properties it was found more recently that the azulene nucleus (I a ←→ I b) has a fair degree of aromatic

(Ia.)   Azulene.   (Ib.)

character. Azulene occupies an important position in non-benzenoid aromatic chemistry, together with the seven-membered tropylium ion, related troponoids, and five-membered cyclopentadienyl compounds.

The chemistry of azulenes was reviewed by HAAGEN-SMIT (*6*) in this Series and later elsewhere by POMMER (*23*), GORDON (*4*), TREIBS et al. (*28*), REID (*24*), and HAFNER (*8*). Quite recently, extensive reviews have been written by HEILBRONNER (*9*) on the physico-chemical properties and theoretical aspects of azulenes and by KELLER-SCHIERLEIN and HEILBRONNER (*12*) on their synthesis, while NOZOE and ASAO (*20*) have summarized the synthesis and properties of various azulene derivatives.

Although azulene had been known for a long time and synthesized repeatedly during the past twenty years, it had remained a rather rare compound until ZIEGLER and HAFNER (cf. 8) devised a practical synthesis. Thus, azulenic hydrocarbons and some derivatives are now easily available. NOZOE, SETO and MATSUMURA (cf. 16–20) developed a new process by means of which numerous azulenes possessing hitherto not accessible functional groups can be obtained easily from troponoids. This process complements the Ziegler-Hafner method for azulene synthesis. Thus, using a recently improved method of tropone and tropolone synthseis (19), a large number of new azulene derivatives have been synthesized during the last few years.

Furthermore, guaianolides and other sesquiterpenoids found in nature as azulene precursors, have been studied extensively during the last decade, among others by ŠORM and HEROUT, BARTON and DE MAYO, and BÜCHI. Thus, most of the pertinent structures have been elucidated. Since the publication of the reviews (1, 2, 5, 7, 10, 11, 13–15, 25–27) substantial further progress has been achieved.

In the present article mainly the literature of 1958–1960 will be discussed as a continuation of recent reviews on azulenes (8) and natural hydroazulenes (25, 26). Concerning the earlier literature, the reader is referred to (1–28). The survey articles (4, 8, 28) contain tables of the then known individual azulene derivatives. Our Tables (pp. 80–106) include new azulene derivatives reported since 1958 and various unpublished compounds synthesized in the writers' laboratory.

The number of azulenes discussed in earlier reviews (4, 8, 28) totals about 280, while the new azulenic compounds mentioned below number about 550; this demonstrates the rapidity of recent progress.

Although, starting from troponoids, numerous heterocyclic azulenoids that contain hetero atoms in the five-membered ring, have been synthesized in the writers' laboratory (16–18, 21), these are not included in the present article where only azulenes with additional heterocyclic rings appear.

II. Azulenes.

1. Synthetic Methods.

a) Syntheses with Dehydrogenation.

The majority of methods available for azulene synthesis involve dehydrogenation in the final stages (4, 8, 12, 28). These methods have been employed in recent syntheses of zierazulene (83) and other alkyl-azulenes (105, 124, 125, 235, 237a, 247), benzylazulene (225), phenyl-azulenes (57, 190), 1,2,3-triphenylazulene (44), alkoxyazulenes (196), methoxycarbonylazulenes (35, 228), and 1-(triethylsilyl)azulene (II) (231). Rearrangement of phenyl on the seven-membered ring (190) or of alkyl on the five-membered ring (124, 125) was observed during the dehydrogenation of hydroazulene derivatives. The mechanism of such dehydrogenations by means of sulfur or selenium has been studied (*132*). Polybromo derivatives of azulene (*240*) and 5,6-benzazulene (*36*) have been obtained by the interaction of N-bromosuccinimide and hydroazulene derivatives.

b) Syntheses without Dehydrogenation.

An example of azulene synthesis not requiring dehydrogenation (cf. *8*) is the formation of 1,2,3-triphenylazulene (III) by the reaction of diphenylacetylene and 2,4-dinitrophenylsulfenyl chloride in the presence of aluminum trichloride (*44, 46*) and also by dehydration of diphenylcyclopropenyldiphenylcarbinol (IV) (*68*).

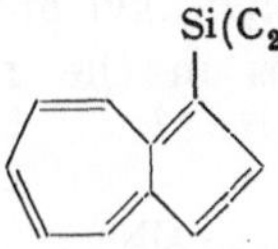

(II.) 1-(Triethylsilyl)azulene. 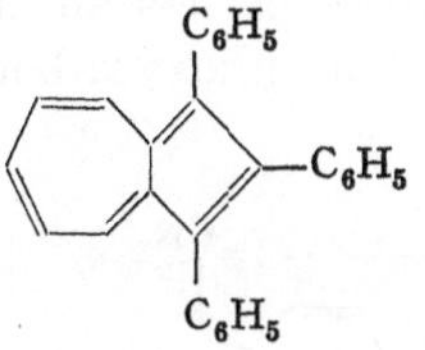   (III.) 1,2,3-Triphenylazulene. 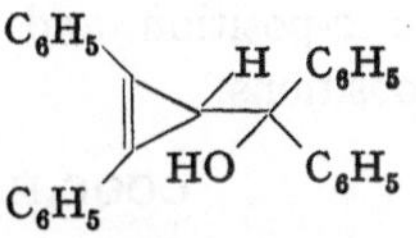   (IV.) Diphenylcyclopropenyl-diphenylcarbinol.

Azulenecarboxylic acids have been obtained by ring enlargement of 1,2-diacetoxyindane (V) followed by distillation (*241*).

By means of the Ziegler-Hafner method (cf. *8*) using cyclopentadiene and pyridinium salt, triphenylazulene (*201*) and benzylazulenes (*225*) were prepared. A detailed description of the Hafner method, using

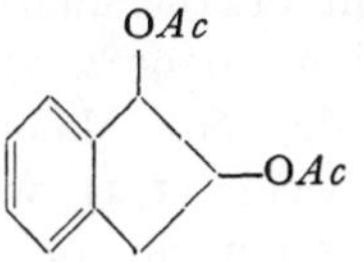

(V.) 1,2-Diacetoxyindane. 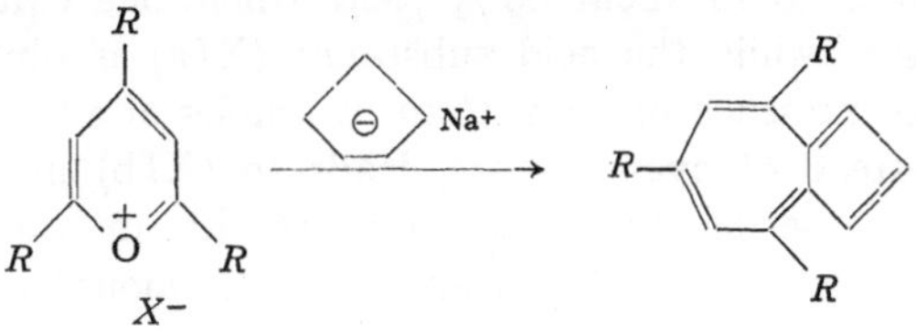   (VI.)

pyrylium salt (VI) instead of pyridinium salt, has appeared (*112*); and this method has made it quite easy to prepare alkyl, aryl,

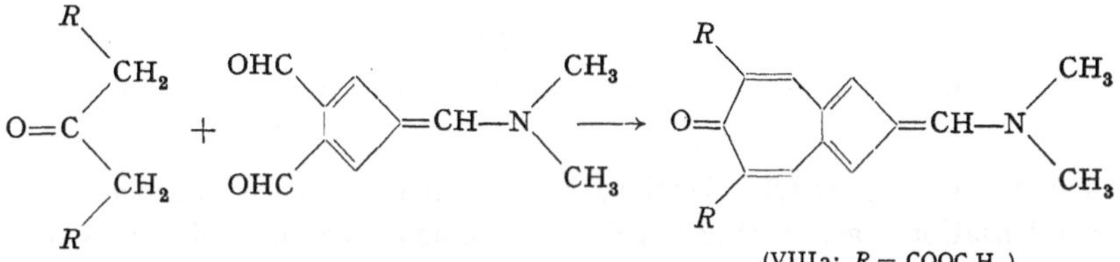

($R = COOC_2H_5$ or C_6H_5.)   (VII.)   (VIIIa; $R = COOC_2H_5$.)
(VIIIb; $R = C_6H_5$.)

and alkoxyl derivatives of azulene. Condensation of the diformyl compound (VII), obtained by the reaction of cyclopentadiene and dimethylformamide with phosphoryl chloride, and acetonedicarboxylate (*115*) or diphenylacetone (*110*) yielded the azulenoquinonederivatives (VIIIa) and (VIIIb).

c) Syntheses Starting from Troponoids and Heptafulvenes.

It was found in the writers' laboratory a few years ago (*164, 171*) that the application of two equivalents of ethyl cyanoacetate to 2-methoxy- or 2-halotropone, in the presence of alkali alkoxide, resulted in facile formation, at room temperature, of the azulene derivatives (IX), (Xa), (Xb), and (XIa); these possess an amino or hydroxyl group in the 2-position and cyano or ethoxycarbonyl groups in the 1- and 3-positions.

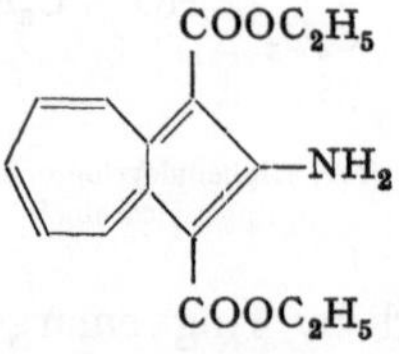

(IX.)

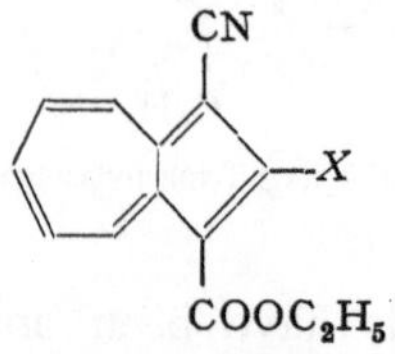

(Xa; $X = NH_2$.)
(Xb; $X = OH$.)

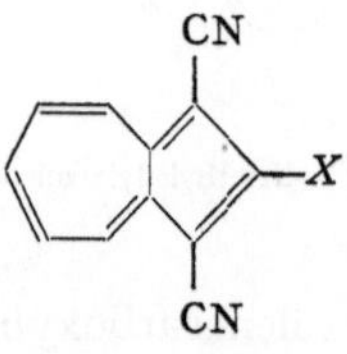

(XIa; $X = OH$.)
(XIb; $X = NH_2$.)

The ratio of these azulenes in the product depends largely on the conditions. For example, from 2-methoxytropone compound (IX) is obtained in about 80% yield when one equivalent of the alkoxide is used, while the acid substance (XIa) is obtained in 80–90% yield in the presence of more than two moles of alkoxide (*167, 169*). Malonitrile, instead of cyanoacetate, leads to (XIb) in good yields (*171*), whereas the use of cyanoacetamide results in the formation of 3-cyano-1-azaazulan-2-one (XII) and a small amount of 2-amino-1,3-dicarbamoyl-

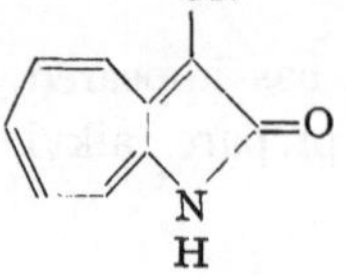

(XII.) 3-Cyano-1-azaazulan-2-one. 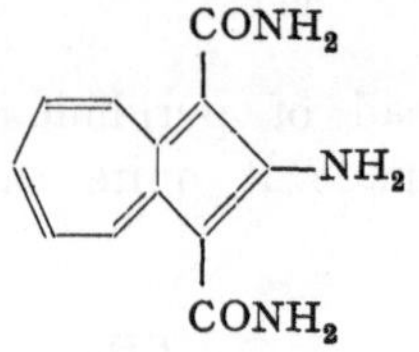   (XIII.) 2-Amino-1,3-dicarbamoylazulene. 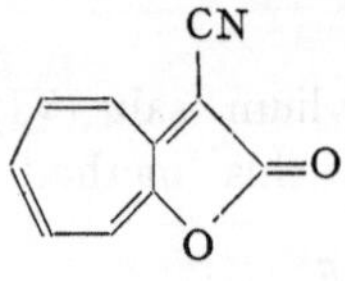   (XIV.) 3-Cyano-1-oxaazulan-2-one.

azulene (XIII) (*174*). Tertiary amine may replace alkoxide as a condensation agent; thus, (IX) is obtained together with 3-cyano-1-oxaazulan-2-one (XIV) from 2-halotropone and cyanoacetate (*169*).

Application of ethyl cyanoacetate to alkyl (*34, 43, 167, 168, 171, 181*), halo (*152, 175*), cyano (*172*), hydroxyalkyl (*170*), acetyl (as a ketal) (*178*), acetamido (*179*), pyrazolyl (*176*), or α-quinolyl (*177*) derivatives of 2-halo-, 2-methoxy-, or 2-tosyloxytropone affords azulene derivatives with a substituent in the seven-membered ring. The use of 2-methoxy-tropone (XVI) in this reaction results in cyclization by "normal" substitution to give (XVIIIa) but the reaction with 2-halo- (XVIIa) or 2-tosyloxytropone (XVIIb) usually results in "abnormal" substitution to give (XVIIIb) (cf. *17, 18*) *(Chart 1)*. Therefore, the isomeric azulene derivatives, (XVIIIa) and (XVIIIb), can be prepared starting from the same tropolone (XV) (*34, 43, 168, 171, 175*). In the case of 6-substituted 2-chlorotropone (XIX), this reaction is very complicated due to the steric interference of the substituents; hence, only a small amount of 5-substituted azulene derivative (XVIIIa) is obtained (*34, 167, 171*) by "normal" substitution.

(XVI.) 2-Methoxytropone.

(XVIIIa.)

(XV.)

(XIX.)

(XVIIa; *X* = Cl.)
(XVIIb; *X* = O*T*s.)

(XVIIIb.)

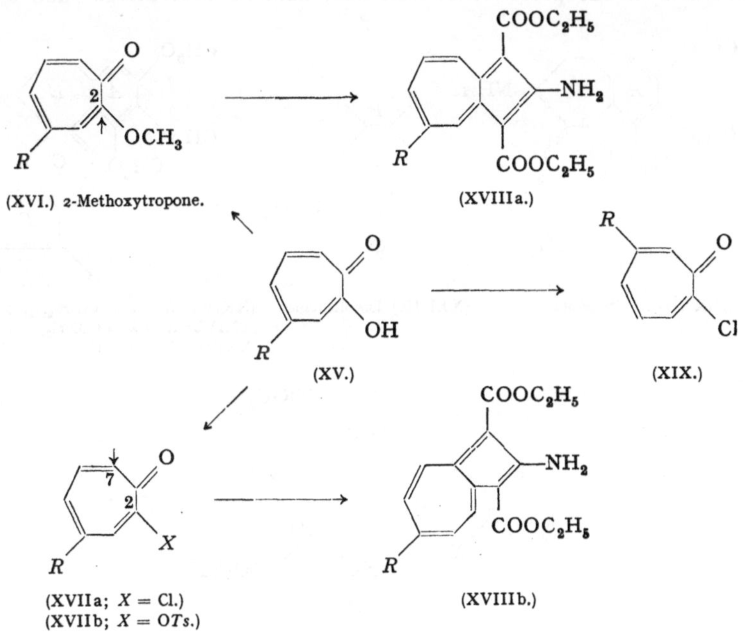

Chart 1. Azulene Synthesis from Troponoid.

In the case of 2-methoxytropone with a methyl or a cyano group in the 3(or 7)-position, the yield of azulenes is generally poor, while the use of 2,7-dihalotropone (XXa) or 2-methoxy-7-halotropone (XXb) fails to yield any azulene; instead rearrangement products such as (XXI), (XXIIa) and (XXIIb) are obtained (*152*).

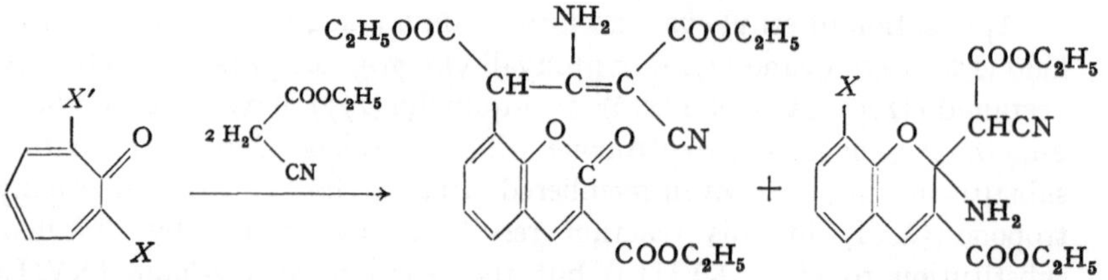

(XXa; $X = X' =$ Br or Cl.)
(XXb; $X =$ Br or Cl, $X' =$ OCH$_3$.)

(XXI.)

(XXIIa; $X =$ Br.)
(XXIIb; $X =$ OCH$_3$.)

The methoxyl group in the methyl ethers of 3,4- and 4,5-benzotropolones and pyridotropolones is inactive and these compounds do not form azulene derivatives at all (cf. *21, 22*), while colchicine (XXIIIa) and iso-colchicine (XXIIIb), having alicyclic *B*-rings, yield the tetracyclic azulenes (XXIVa–c) (*182*).

Even when one equivalent of cyanoacetate is reacted with 2-halo-tropone in the presence of alkoxide, azulene derivatives such as (IX),

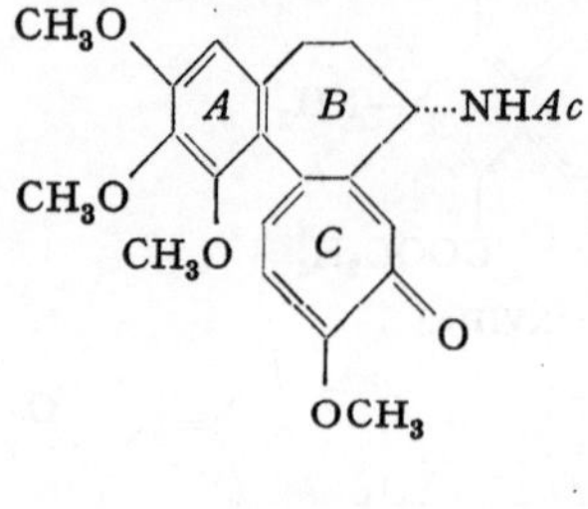

(XXIIIa.) Colchicine.

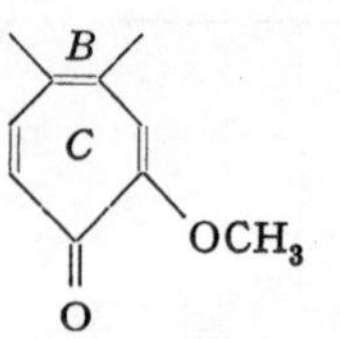

(XXIIIb.) Isocolchicine.

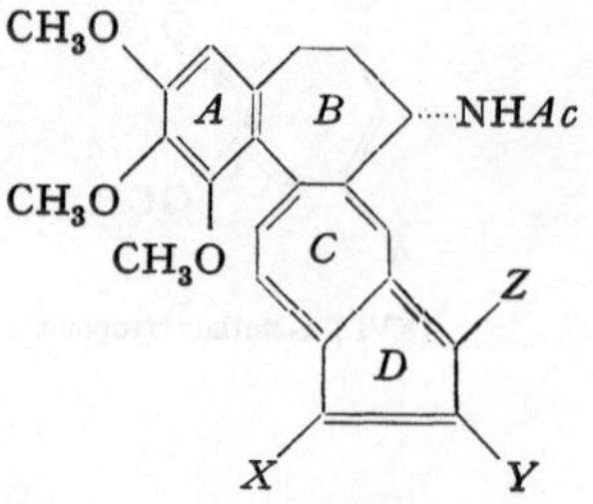

(XXIVa; $X = Z = $ COOC$_2$H$_5$, $Y =$ NH$_2$.)
(XXIVb; $X = Z = $ CONH$_2$, $Y =$ NH$_2$.)
(XXIVc; $X = Z = $ CN, $Y =$ OH.)

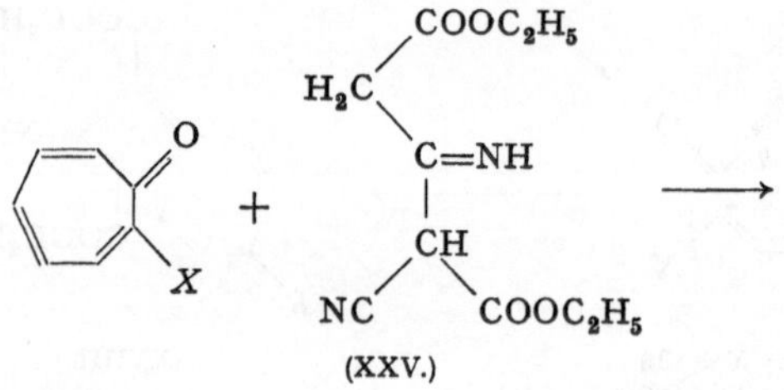

(XXV.)

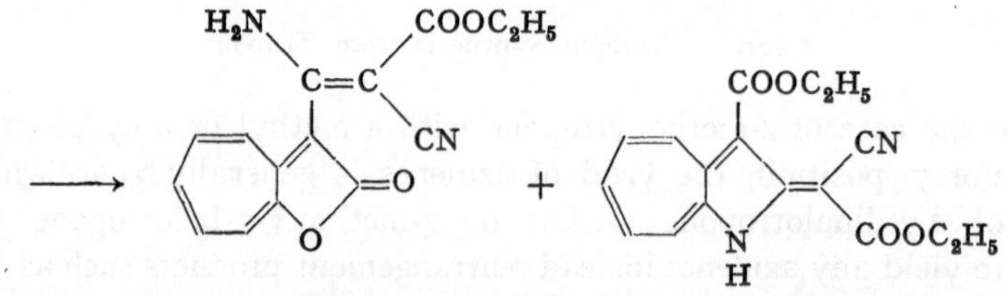

(XXVIa.)

(XXVIb.)

(Xa), (Xb) and (XIa) are obtained. The use of one mole of the dimer (XXV) instead of two moles of cyanoacetate, fails to give azulenes; the products are the heterocyclic azulenoids (XXVIa) and (XXVIb) (*152*).

From the foregoing facts Nozoe and others (*17, 20, 171*) deducted the reaction sequence shown in *Chart 2.*

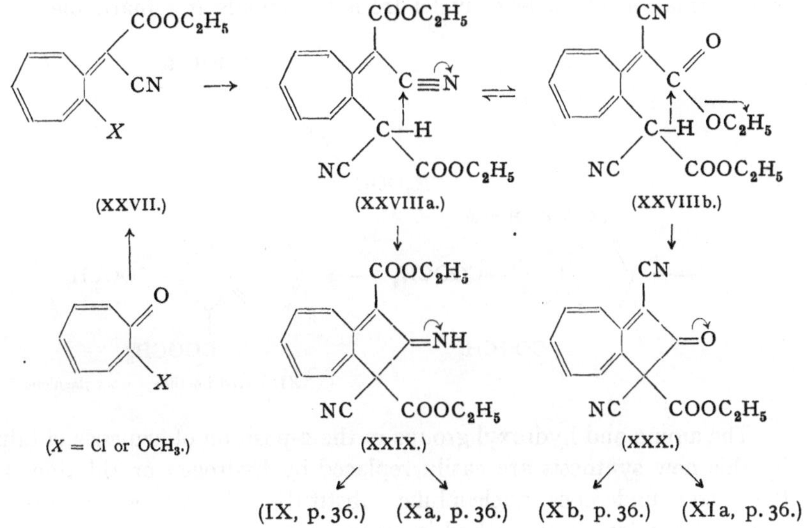

Chart 2. The Mechanism of Azulene Formation.

The first intermediate (XXVII), formed by the interaction of one mole of cyanoacetate and 2-halo- or 2-methoxytropone, is more reactive than the original compound and reacts immediately with a further mole of the reagent to form the second intermediates, (XXVIIIa) and (XXVIIIb), which undergo cyclization by the action of a base to form the third intermediates (XXIX) and (XXX). Their cyano or ethoxy-carbonyl groups are then eliminated to yield a mixture of the final products (IX, Xa, Xb, XIa). The reaction of cyanoacetate with methyl ethers of 4-alkyltropolones results in the formation of all six possible azulenes (*34, 43*) and this may be understood from the foregoing mechanism. More recently, Nozoe and his associates (*165*), and Kitahara et al. (*142 a*) have found that application of cyanoacetate or malononitrile to an 8,8-disubstituted heptafulvene, (XXXIa), XXXIb) or (XXXIc), gave the afore-mentioned azulenes. This confirms (the reaction mechanism given above in Chart 2.

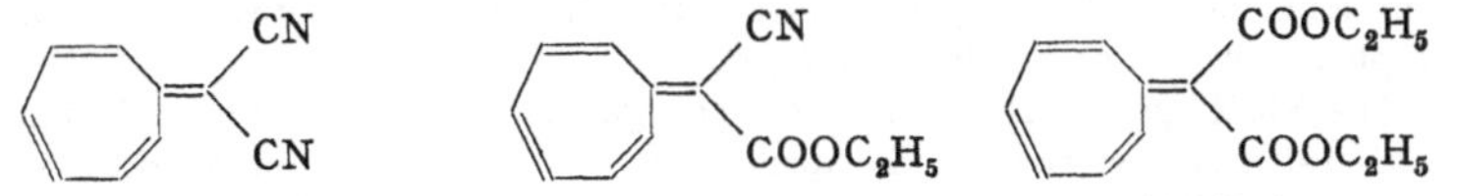

As is well known, heptafulvene (XXXI d) itself, synthesized by
Doering and Wiley (3,254), is extremely labile, but with acetylene
dicarboxylate it affords 1,2-dimethoxycarbonylazulene (XXXII).
Further, Kitahara (142) found that 2-amino-1,3-dicyanoazulene (XI b,
p. 36) resulted from the interaction of tropone and malonitrile. Such
easy formation of azulene nuclei from troponoids is remarkable.

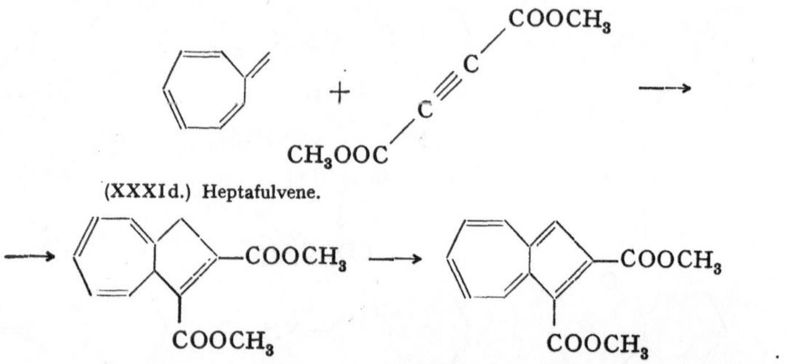

(XXXId.) Heptafulvene.

(XXXII.) 1,2-Dimethoxycarbonylazulene.

The amino and hydroxyl groups in the 2-position of azulenes obtained
by this new synthesis are easily replaced by hydrogen or chlorine; and
the latter undergoes nucleophilic substitution by various groups (see
below). Therefore, this method is convenient for the synthesis of azulene
compounds having various functional groups.

2. Physical Properties.

In Heilbronner's excellent review (9) the theoretical aspects of
ultraviolet, visible and infrared spectra, NMR spectra, dipole moments,
polarographic behavior, and X-ray and electron diffractions of the
azulenes are discussed. Hence, only recent progress in this field will
be presented below.

Dipole moments of azulene (42), 2-substituted (145), 1,3-disubstituted
azulenes (42) and azulene iron carbonyl (72a) have been measured. The observed
value for azulene (1.08 D) (42) is somewhat smaller than the calculated one;
and the value for 2-chloroazulene confirms that the negative pole is directed
toward the five-membered ring (145). On the basis of the dipole moments of
arylazoazulenes (98) their configurations were discussed.

An X-ray diffraction study was carried out with azulene (197), the result of
which showed that the unit cell contains two superimposed azulene molecules with
opposed directions, the space group being P 2_1/a, and the central bond ($C_{(9)}$–$C_{(10)}$)
being longer than the peripheral ones (197). X-ray measurements of chamazulene
complexes were also carried out (135).

The NMR spectra of azulene (60, 199), guaiazulene (59) and benzylazulene (225)
in several solvents, that of the azulenium ion (86) in trifluoroaceric acid and of
azulene-metal carbonyl (72a) were measured. Polarographic studies on acetyl-

azulenes (97) and cyanoazulenes (232) have been reported. The heat of hydrogenation of azulene was estimated as 98.98 kcal./mole, and the resonance energy calculated from this value as 28 kcal./mole (238). Molecular orbital calculations were carried out on azulene (136, 184a) and 1-phenylazoazulene (100).

A discussion of the ultraviolet and visible spectra of 1-aminoazulene (200) and a number of arylazoazulenes (98, 101–103, 157), and the measurement of their relative basicities by spectroscopic methods have been reported. Visible spectra of many azulenes, especially those of 1,3-disubstituted derivatives were investigated in detail (85).

All these observations indicate that the ground state of azulene is represented either as a resonance hybrid of the two Kekulé forms (Ia and Ib, p. 33) or better by (Ic), in which ten π-electrons delocalize around the periphery of the molecule, although a small contribution

(Ic.)   (Id.)

of (Id), in which both rings are π-electron sextets, should not be ignored (24). A major contribution of (Id) in the excited state is, however, to be considered from the results of both electrophilic and nucleophilic substitutions and the formation of azulenium salts (112a, 139a) (see below).

3. Chemical Properties.

a) Electrophilic Substitutions.

The fact that azulenes are liable to electrophilic substitution in the 1- and 3-positions was mentioned in earlier reviews (8, 28); their reactivity which is higher than that of benzenoid hydrocarbons is demonstrated by their facile azo coupling and acylation, even in the absence of catalysts (24). It has been confirmed by treatment with deuterated phosphoric acid that only the positions 1 and 3 are substituted with deuterium (56). Indeed, halogenation with N-chloro-, N-bromo-, or N-iodosuccinimide (41, 42), low temperature nitration with nitric acid or tetranitromethane (40, 194, 232), azo coupling (40, 42, 99, 102), and substitution with thiocyanogen (40) invariably yield 1-mono- or 1,3-di-substituted products.

Azulenes, especially guaiazulene, react easily, without any catalyst, with acetyl bromide (in petroleum ether at room temperature) (194), benzoyl bromide (with heating) (194), and trifluoroacetic anhydride (in carbon tetrachloride at room temperature) (37) to form 1(or 3)-acetyl, -benzoyl, and -trifluoroacetyl derivatives, respectively. Treatment with oxalyl bromide (or chloride) also results in facile reaction at room temperature; hereby a glyoxylic acid (XXXIII) (228, 237) or bis-1-

azulenyl ketone derivative (XXXIV) is formed (*194*); with phosgene azulenoyl halide is obtained (*236*).

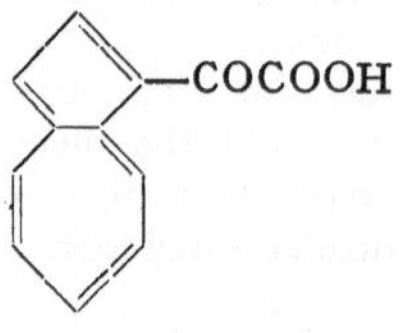

(XXXIII.)

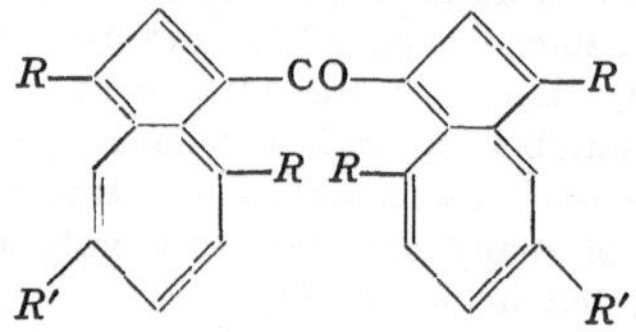

(XXXIVa; $R = R' = $ H.)
(XXXIVb; $R = CH_3$, $R' = $ isoC$_3$H$_7$.)

Azulene undergoes reactions (in the absence of catalysts) with anhydrous formic acid (*227*), acetic acid (*227*), picryl chloride (*227*) or malonyl chloride (*229*), to form, respectively, formyl-, acetyl-, picryl-azulenes, and azulenoylacetic acid derivatives.

Furthermore, acylation can be effected by the Vilsmeier method (*111, 155, 235, 239, 245*) or in a Friedel-Crafts reaction (*38, 41, 156*). Applying the Vilsmeier method, JUTZ (*137*) synthesized azulenyl-polyenals (XXXV) by the interaction of 1-N-methylanilino-1-propen-3-al (or its vinylog) and azulene:

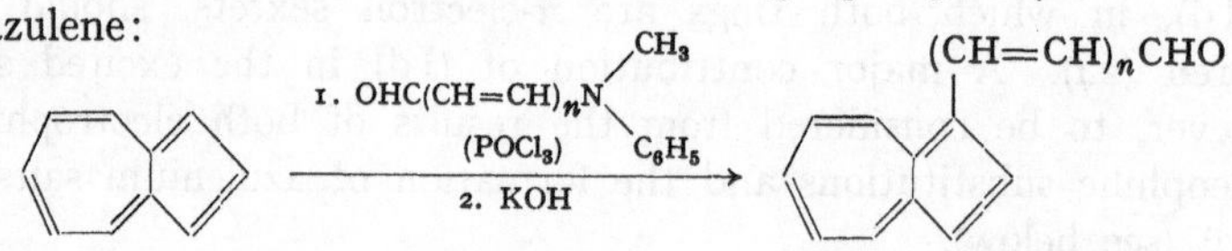

(XXXV.) Azulenyl-polyenals.

Reaction of azulene with cyanogen bromide, in the presence of aluminum chloride in carbon disulfide solution, affords 1-cyanoazulene, whereas in methylene chloride 1-bromomethyl-3-cyano-azulene (XXXVI) is formed (*232*). Azulene and 2,4-dinitrophenylsulfenyl chloride yield, in the presence of aluminum chloride or stannic chloride, 2,4-dinitro-phenylazulenyl sulfide (XXXVII) (*40*).

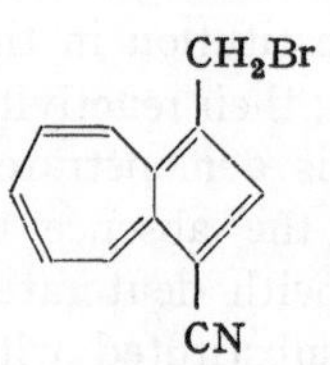

(XXXVI.)
1-Bromomethyl-3-cyanoazulene.

TREIBS and others (*234*) briefly mentioned the formation of 1,3-bis(piperidinomethyl)azulene (XXXVIII; $R = $ piperidino; $X = $ H). By aminomethylation of azulene derivatives NOZOE and SATO (*166*) obtained (XXXVIII and XXXIX; $R = $ dimethylamino, morpholino, or piperidino; $X = $ H, Cl, or acetamido).

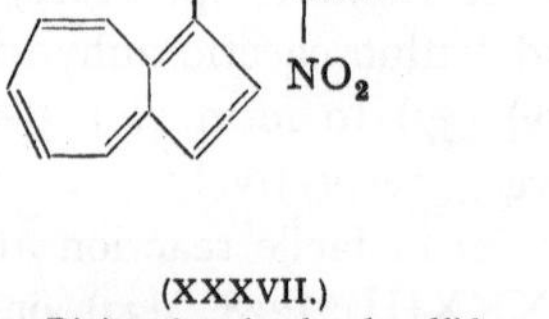

(XXXVII.)
2,4-Dinitrophenylazulenyl sulfide.

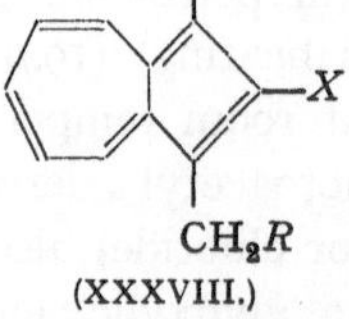

(XXXVIII.)

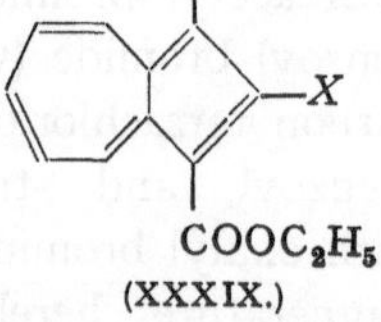

(XXXIX.)

In some instances displacement of 1- or 3-substituents takes place during nitration (*40*) or chlorination (*156*).

It had been assumed earlier that only the 1- and 3-positions can undergo electrophilic substitution; however, azulene derivatives in which these positions are already occupied have been found to suffer such substitution in the seven-membered ring. Recently, ANDERSON et al. (*41*) have obtained the 5-acetyl derivative (XL) by acetylation of 1,3-dichloro-azulene in the presence of stannic chloride; and they assumed that this

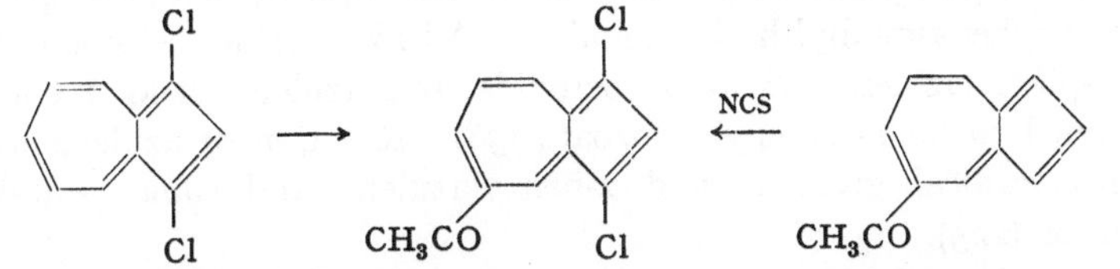

1,3-Dichloroazulene.　　(XL.) 5-Acetyl-1,3-dichloroazulene.　　5-Acetylazulene.

conversion is due to the high electron density (cf. *136*) at the 5(or 7)-position, second only to that at the 1(or 3)-position. On the other hand, experiments in our laboratory have shown that the 6-bromo derivative

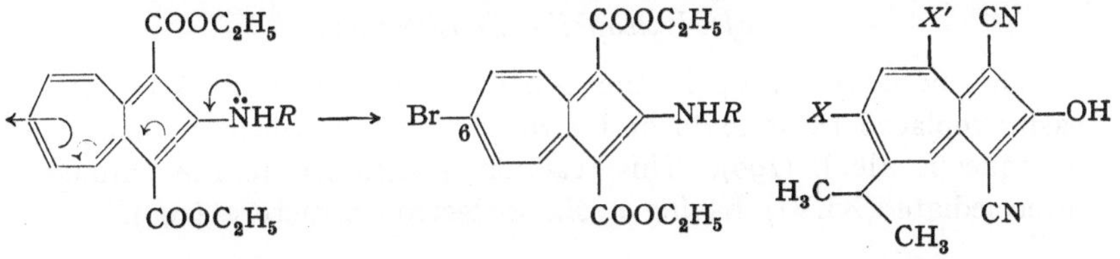

(XLII.) 2-Amino-1,3-diethoxycarbonylazulene　　(XLI.)　　(XLIII.)
　　　　　　　(R = H).

(XLI; R = H) is formed in good yield by treatment of 2-amino-1,3-diethoxycarbonylazulene (XLII; R = H) with bromine in acetic acid solution (*169*). This was considered to be due to the reactivity of the 6-(as well as 4- and 8-)position by the resonance effect of the 2-amino group.

Formation of mono- (XLIII; X = Br, X' = H) and dibromo compounds (XLIII, X = X' = Br) from 1,3-dicyano-2-hydroxy-5-iso-propylazulene (XLIII; X = X' = H) was also reported (*167*). While the 2-methylamino and anilino compounds (XLI; R = CH_3 or C_6H_5) can be brominated in the 6-position, the 2-dimethylamino derivative is resistant to bromination, possibly because of steric interference between the ethoxycarbonyl and dimethylamino groups.

According to UKITA and others (*239*), the Vilsmeier formylation of guaiazulene afforded, besides the 3-formyl compound, a small amount of an isomer, which, they assumed, is substituted in the seven-membered ring. 1,2,3-Triphenylazulene (III, p. 35) is known to form a dibromo compound (44) which might contain bromine at the 5 and 7 positions.

An azulene-metal complex was obtained by heating azulene and iron or molybdenum carbonyl (72a, 73).

b) Free Radical Reactions.

According to ANDERSON and others (38), the interaction of N-nitroso-acetanilide and azulene in benzene solution affords 1-phenylazoazulene besides 1-phenylazulene, while phenylazotriphenylmethane gives rise to 1-triphenylmethyl-hydrazoazulene (XLIV) besides a small amount of 1-phenylazulene. Benzoyl peroxide and azulene yield a compound assumed to be 1-azulenyl benzoate (38). Reaction of azulene with the benzyl radical gives 1- and 2-benzylazulene and some pentabenzyl-azulene (225).

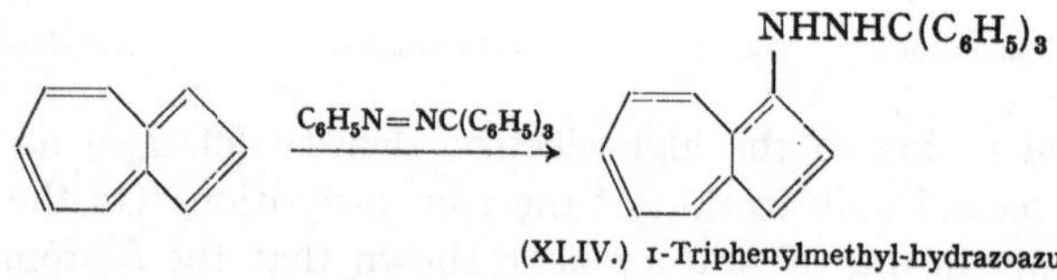

(XLIV.) 1-Triphenylmethyl-hydrazoazulene.

c) Nucleophilic Substitutions.

The methoxyl group in the 4-methoxyazulene derivative (XLV) is easily replaced by ethoxyl or hydroxyl upon treatment with ethanolic or aqueous alkali (196). This reaction is assumed to run through an intermediate (XLVI) having a six π-electron structure (196).

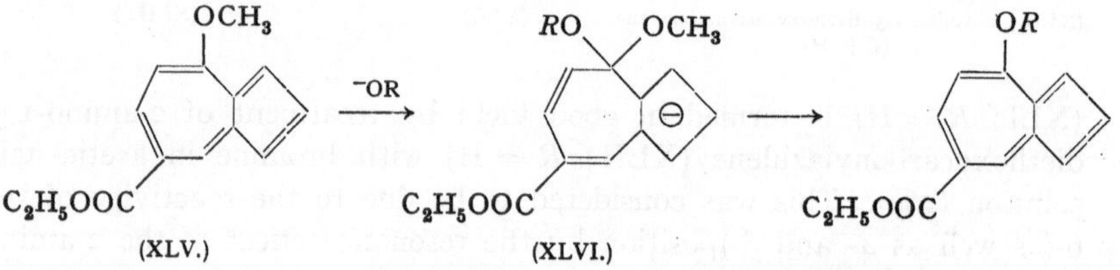

(XLV.)   (XLVI.)

Reaction of 4-methoxyazulene with sodium amide in liquid ammonia results in the formation of a rather labile red substance, probably 4-aminoazulene (XLVII) (196).

2-Haloazulenes, (XLVIIIa) or (XLVIIIb), especially 2-chloro-1,3-diethoxycarbonyl- or 2-chloro-1,3-dicyanoazulenes, (XLIXa) or (XLIXb), react with sodium methoxide, sodium cyanide, sodium mercaptide, alkyl- or arylamines, hydrazine, ethyl cyano- or acetoacetate, and other anionoid reagents to give 2-substituted azulenes (167, 171, 173). 5-Chloro-1,3-diethoxycarbonylazulene also reacts with nucleophilic reagents, such as ethanethiol, and sodium methoxide (175).

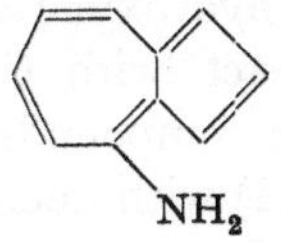

(XLVII.) 4-Aminoazulene.

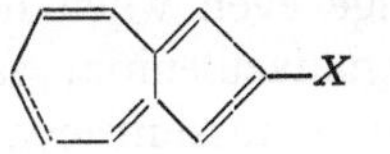

(XLVIIIa; $X =$ Cl.)
(XLVIIIb; $X =$ I.)

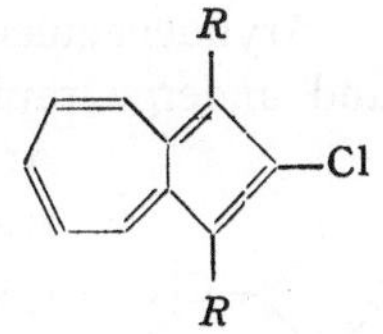

(XLIXa; $R =$ COOC$_2$H$_5$.)
(XLIXb; $R =$ CN.)

d) Arylideneazulenium Salts and Related Compounds.

The methylene group in the azulenium ion (L), a salt of which is sometimes isolated (*112a*, *139a*), is expected to be active and indeed, arylideneazulenium chlorides (LI; $X =$ Cl) are formed when dry hydrogen chloride passes through an ether solution of guaiazulene and arylaldehydes. The chlorides are labile but the picrates are somewhat more stable (*195*). The addition of excess perchloric acid (70%) to the acetic acid solution of guaiazulene and substituted benzaldehydes or heterocyclic aldehydes, such as furyl, thienyl, pyridyl, indolyl and quinolyl aldehydes, resulted in the immediate formation, even at room temperature, of stable crystalline perchlorates (LI; $X =$ ClO$_4$),

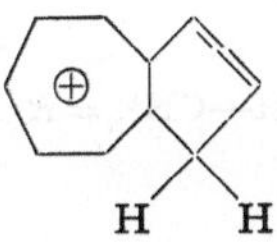

(L.) Azulenium ion. 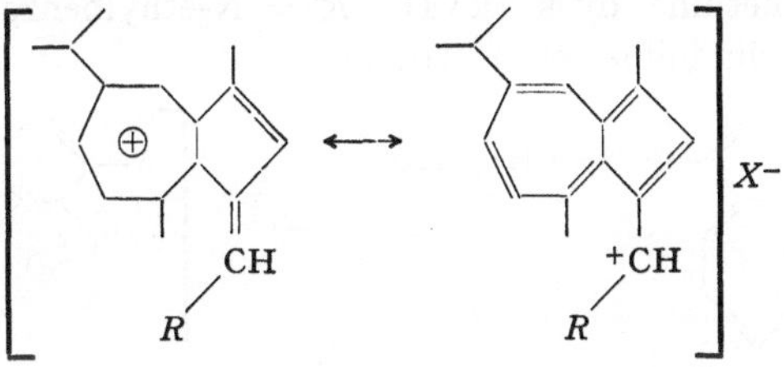   (LI.) Arylideneazulenium salt.

in good yield (*139*). In the case of azulene itself, azulenylmethylene compounds are isolated only when an aldehyde with electron-releasing substituents, such as *p*-hydroxy- or *p*-dimethylaminobenzaldehyde, is used. In the case of aldehydes with electron-withdrawing substituents or aldehydes of the pyridine and quinoline series, the initially formed azulenium salt (LII) undergoes further condensation with azulene to form a green diazulenylmethane (LIII) (*139*).

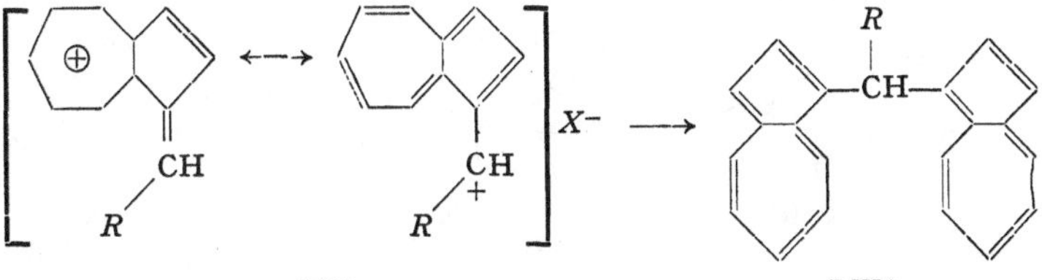

(LII.)   (LIII.)

Arylidenazulenium salts are very reactive with nucleophilic agents and undergo rapid change even when heated in a hydroxylic solvent.

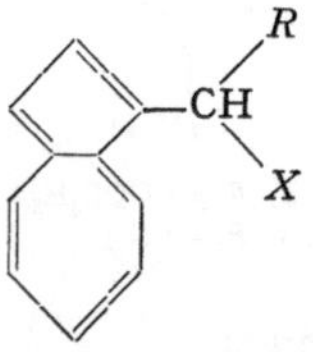

Arylideneguaiazulenium salts (LI) react with Grignard reagents or potassium cyanide to give compounds of the type (LIV). The conversions of salts (LI) with nucleophilic reagents are, however, of complex nature: besides guai-azulene polynuclear compounds having several azulene nuclei are formed (*139*). Azulene undergoes condensation also with heterocyclic formylmethylene compounds and affords deeply colored dyes (LV) (*191*). Further, azulenyl polyenals (LVI; $R =$ H or $NHCOCH_3$) condense with N-alkylated heterocyclic compounds in triethylamine or piperidine solution to form

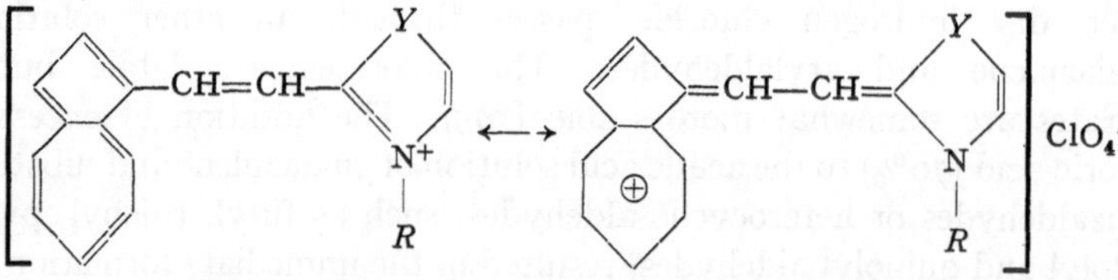

(LIV.)

(LV.) [$R = CH_3$ or C_2H_5; $Y =$ S, Se, CH=CH or $=C(CH_3)_2$.]

polymethine dyes (LVII; $R' =$ N-ethylbenzothiazoline, tri- or tetra-methylindoline, etc.) (*191*).

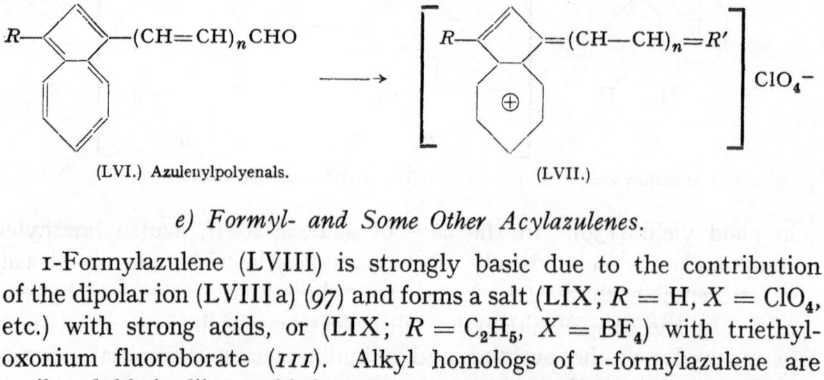

(LVI.) Azulenylpolyenals.   (LVII.)

e) Formyl- and Some Other Acylazulenes.

1-Formylazulene (LVIII) is strongly basic due to the contribution of the dipolar ion (LVIIIa) (*97*) and forms a salt (LIX; $R =$ H, $X =$ ClO_4, etc.) with strong acids, or (LIX; $R = C_2H_5$, $X = BF_4$) with triethyl-oxonium fluoroborate (*111*). Alkyl homologs of 1-formylazulene are easily soluble in dilute acids because of increased ground-state polarization due to the stability of the alkyl group and the perchlorate even in aqueous solution (*111, 139*).

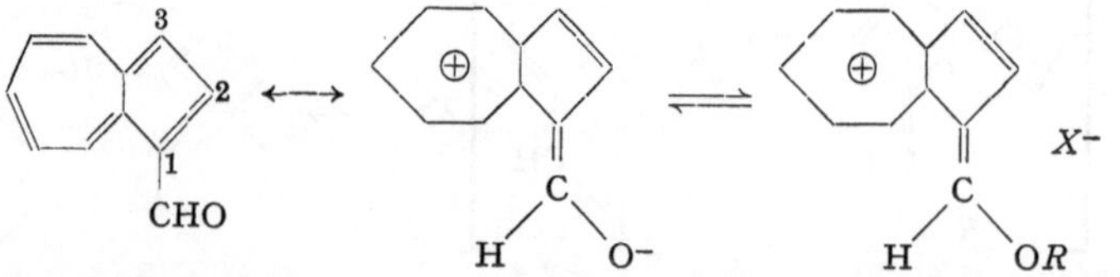

(LVIII.) 1-Formylazulene.   (LVIIIa.)   (LIX.)

1-Formylazulene does not undergo either the Cannizzaro reaction or benzoin condensation, and is not easily oxidized to carboxylic acid; this differentiates it from ordinary aromatic aldehydes (*111*). However, (LVIII) forms a semicarbazone and oxime; it is reduced by lithium aluminum hydride to hydroxymethylazulene; by the Wolff-Kischner reduction to methylazulene (*111*); and affords a secondary carbinol when treated with Grignard reagent (*111, 155*). Nitrile is formed easily by dehydration of the oxime (*111, 194, 232*). 1-Hydroxymethylazulene (LX) is acid-sensitive and tends to form the

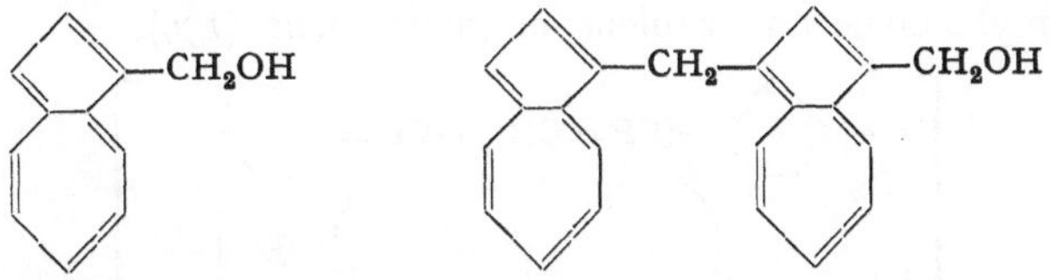

(LX.) 1-Hydroxymethylazulene.   (LXI.)

diazulenylmethane derivative (LXI) or a higher polymer. Hence, ammonium chloride should be used, instead of an acid, for decomposing the complex prepared by the reduction of (LX) with lithium aluminum hydride (*111*).

As stated in an earlier review (*8*), 1-formylazulene, in the presence of a base, undergoes condensation with compounds possessing active methylene groups, such as cyclopentadiene, acetophenone, or nitromethane; hereby α,β-unsaturated azulene derivatives are formed (*111*). The immonium salt, obtained as an intermediate during formylation by the Vilsmeier method (*245*), also reacts with sodium malonate and yields an α,β-unsaturated azulene derivative (*111*).

The interaction of 1-formylazulene and indolizine derivatives affords the azulenylmethylene derivatives (LXIIa) and (LXIIb) (*213*). Azulenyl-

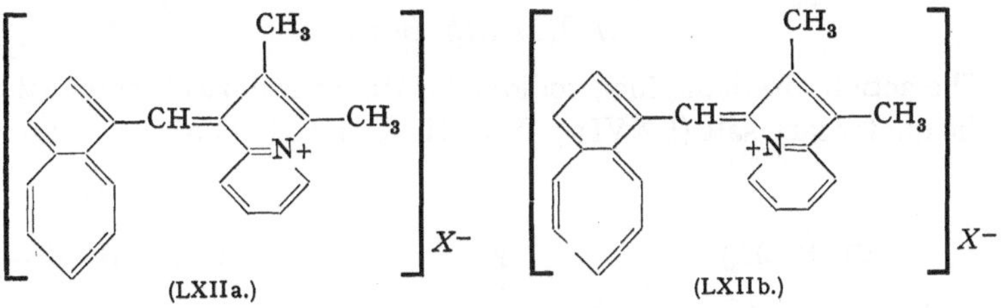

(LXIIa.)   (LXIIb.)

methylene-azulenium salts of the type (LXIII) are obtained by the reacton of azulenes and 1-formylazulene (*108, 139*) or of azulene and orthoformic ester (*45, 245*) or orthothioformic ester (*45*) (in the presence of phosphoryl chloride or a strong acid).

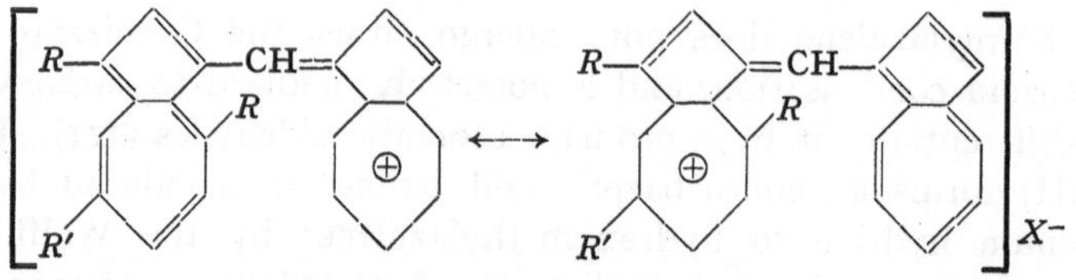

(LXIII.) Azulenylmethylene-azulenium salt.

Condensation of guaiazulenylpolyenals and guaiazulene affords polymethine dyes (LXIV; $n = 0$, 1 or 2) (138). A trinuclear dye of the type (LXV) was obtained, in the form of its diperchlorate, from 1,3-diformylazulene and azulene or guaiazulene (139).

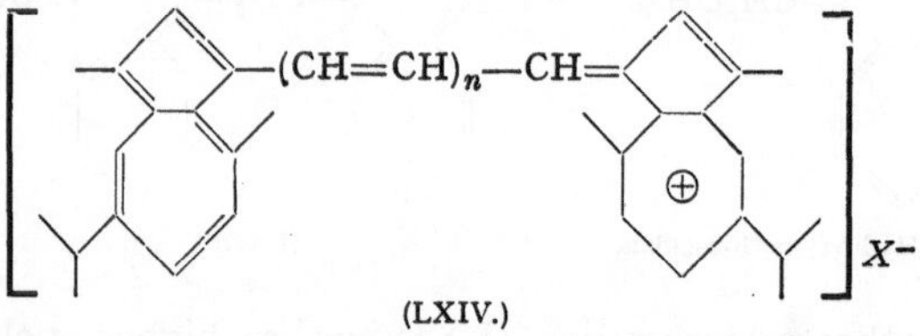

(LXIV.)

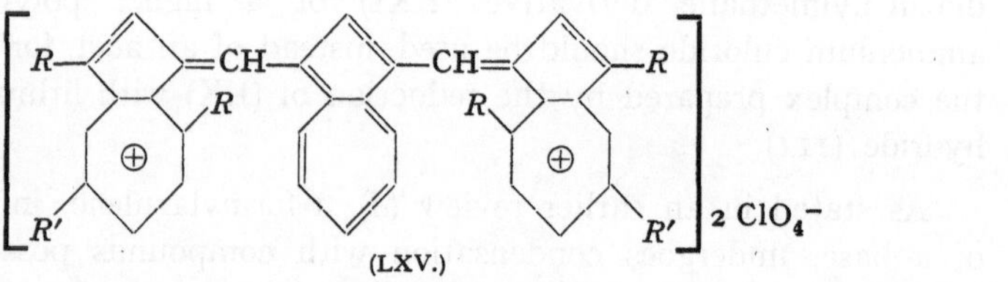

(LXV.)

It is interesting to note that 1-formylazulene, as well as 1-formyl-, 1-acetyl-, 1-benzoyl-, and 1-cyanoguaiazulenes, when heated with strong acid, undergo hydrolysis to azulene or guaiazulene and the corresponding carboxylic acid (111, 194, 227, 235). Acetylazulene oxime gives acetamidoazulene by the Beckmann rearrangement (39). When heated with acetic anhydride 1-formyl-2,4,8-trialkylazulene oximes give 1-acetamidoazulenes and azulenecarbonitrils (232).

f) Mannich Base.

The action of ethanethiol, sodiomalonate, or sodioacetamidomalonate on the quaternary salt (LXVIa; $X = H$, Cl or NHAc) of the 1-dimethyl-

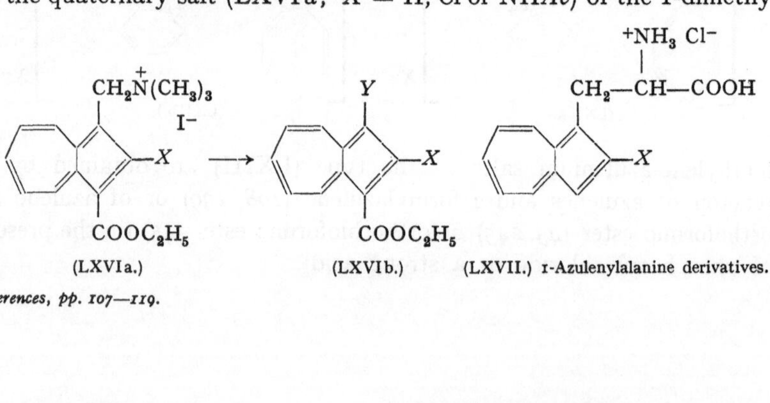

(LXVIa.)   (LXVIb.)   (LXVII.) 1-Azulenylalanine derivatives.

aminomethylazulene derivative (Mannich base) results in the formation of a compound of the type [LXVIb; $Y = CH_2SC_2H_5$, $CH_2CH(COOC_2H_5)_2$, $CH_2C(NHAc)(COOC_2H_5)_2$] (166). 1-Azulenylalanine derivatives (LXVII; $X = H$, Cl or NHAc) can be obtained by this method (166).

g) Amino- and Hydroxyazulenes.

The free forms of 1-, 2-, and 4-aminoazulenes are relatively unstable. The 2-aminoazulene derivative (LXVIII; $R, R' = COOC_2H_5$ or CN), that can be easily prepared from troponoids is deaminated by means of amyl nitrite and one equivalent of sulfuric acid in ethanol to form (LXIX), while 2-chloro derivatives (LXX) are obtained by treatment with hydrochloric acid and nitrite ($164, 171$). These reactions can also be applied to the alkyl ($34, 43, 167, 168, 171$), halo ($152, 175$),

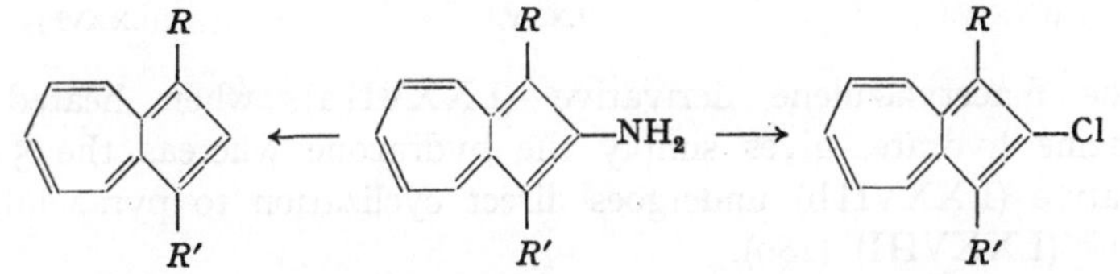

(LXIX.)　　　　　(LXVIII.)　　　　　(LXX.)

acetamido (179) and acetyl (as a ketal) (178) derivatives as well as to compounds containing pyrazole (176) or quinoline rings (177). 2-Amino-azulene derivatives with acetyl or halogen in the seven-membered ring are hardly deaminated by alkyl nitrite in one equivalent of sulfuric acid but are deaminated by alkyl nitrite, in the presence of excess conc. sulfuric acid (178). Deamination of a 2,6-diaminoazulene derivative (LXXI) affords a 6-aminoazulene derivative (LXXII) (179).

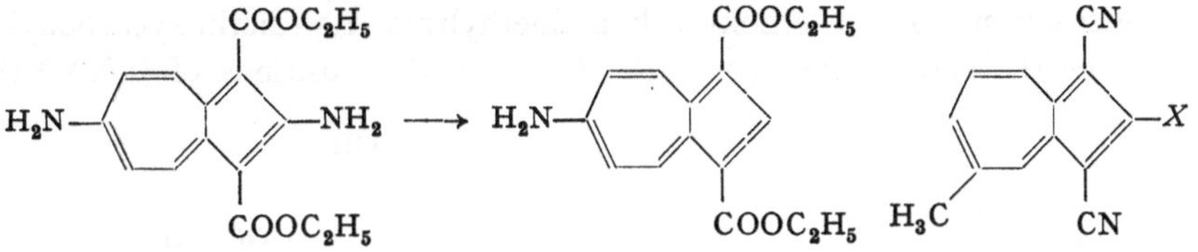

(LXXI.)　　　　　(LXXII.)　　　　　(LXXIII.)

Although 2-aminoazulene itself has not yielded any definite product by the action of nitrous acid, 2-amino-1,3-dicyano-5-methylazulene (LXXIII; $X = NH_2$) forms the 2-hydroxy derivative (LXXIII; $X = OH$) (34). By the action of phosphoryl chloride 1,3-disubstituted 2-hydroxyazulene derivatives are converted into 2-chloro derivatives (LXX) ($34, 43, 167, 168$). 4-Hydroxyazulene-6-carboxylic acid is labile (196), while 4,8-dimethyl-6- hydroxyazulene (112) and 1,3-disubstituted 2-hydroxyazulenes (Xb and XIa, p. 36) are stable and have phenolic character ($34, 43, 167, 169, 171$).

h) Ring Formation on the Azulene Nucleus.

Deamination of β-hydroxy- or α,β-dihydroxy-isopropylazulene derivatives (LXXIV; R = H or OH) with amyl nitrite affords, besides the deaminated product, furo[3,2-f]azulene (LXXV; X = O) or the corresponding dihydro compound (LXXVI). Treatment of the former with liquid ammonia produces a pyrrolo[3,2-f]azulene derivative (LXXV; X = NH) (170).

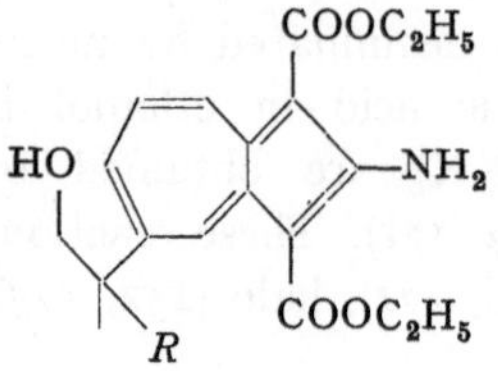

(LXXIV.) 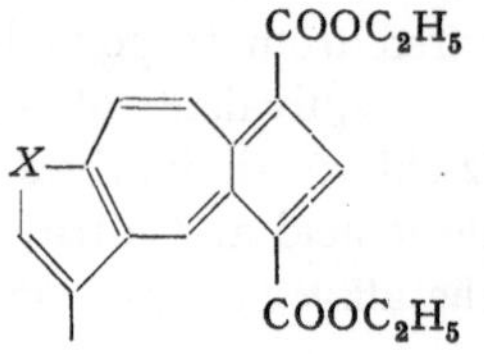   (LXXV.) 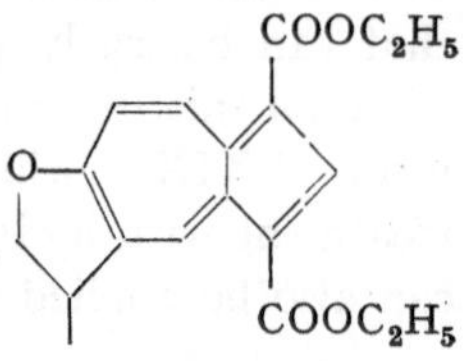   (LXXVI.)

The 6-acetylazulene derivative (LXXVIIa), when heated with hydrazine hydrate, gives simply the hydrazone whereas the 5-acetyl derivative (LXXVIIb) undergoes direct cyclization to pyrazolo[4,5-f]-azulene (LXXVIII) (180).

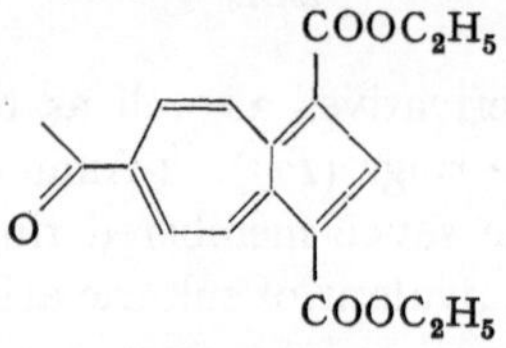

(LXXVIIa.) 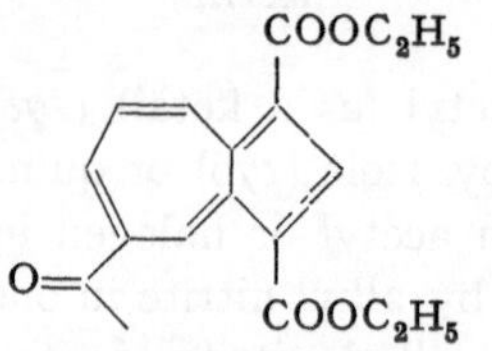   (LXXVIIb.) 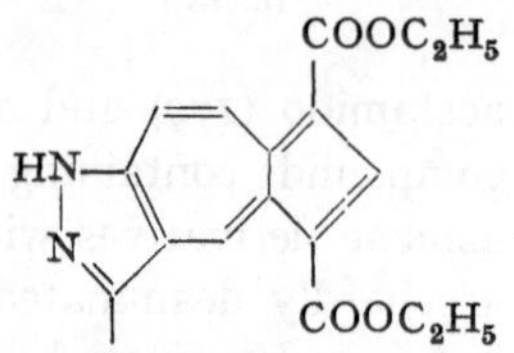   (LXXVIII.)

Thieno[3,2-a]azulene (LXXXI) and its derivatives have been derived from the heterocyclic azulene (LXXX) obtained by Dieckmann condensation of 2-ethoxycarbonylmethylthio-1,3-diethoxycarbonyl-azulene (LXXIX) (146–149, 151). The 2- and 9-positions of (LXXXI)

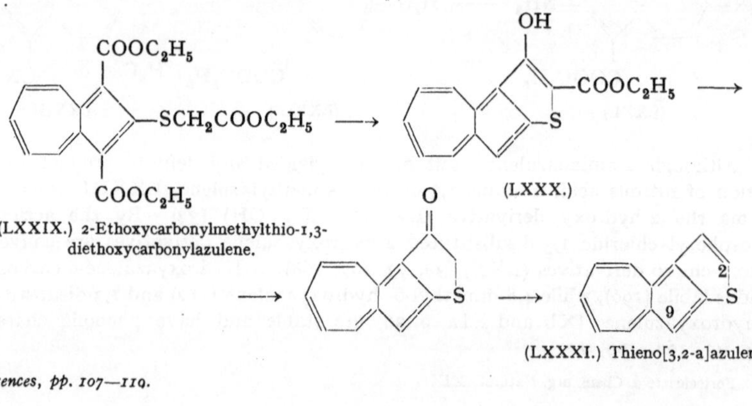

(LXXIX.) 2-Ethoxycarbonylmethylthio-1,3-
diethoxycarbonylazulene.

(LXXX.)

(LXXXI.) Thieno[3,2-a]azulene.

undergo electrophilic substitution, whereby azo or formyl derivatives and Mannich bases are obtained; however, bromination and nitration result in the formation of a black, insoluble substance (*150*). 2-Azulenyl-

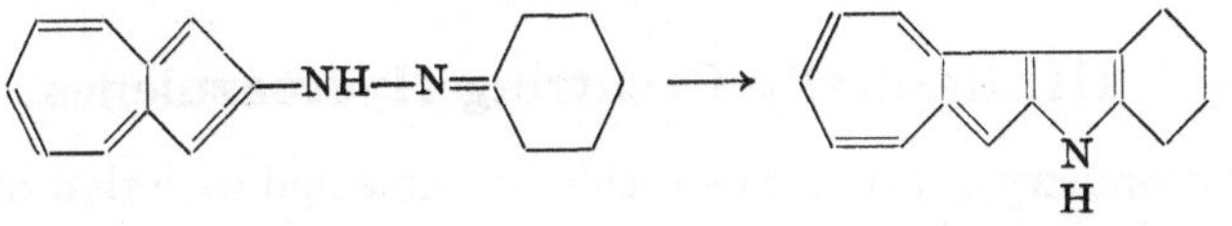

(LXXXII.) Cyclohexanone 2-azulenylhydrazone.   (LXXXIII.)

hydrazone (LXXXII) of cyclohexanone gives the heterocyclic azulene derivative (LXXXIII) by the Fischer indole synthesis (*182*).

The methyl group in the 4-, 6-, and 8-positions of azulene is expected to be active, and HAFNER et al. were able to synthesize pentalene (LXXXIV) and heptalene derivatives (LXXXV) by treating with alkoxide the immonium salts (LXXXVIa) or (LXXXVII), prepared from 4,6,8-

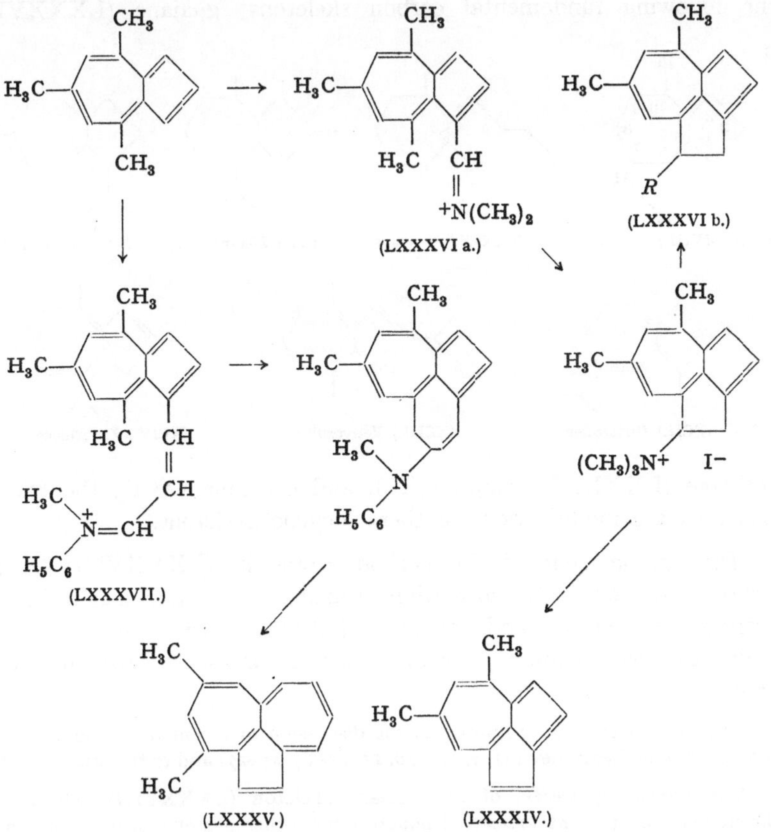

Chart 3. Syntheses of a Pentalene and a Heptalene.

4*

trimethylazulene (*109*, *113*, *114*) *(Chart 3)*. The pentalene derivative is extremely labile, while the heptalene derivative is stable and undergoes electrophilic substitution.

III. Naturally Occurring Hydroazulenes.

The chemistry of azulenoids is closely connected with that of naturally occurring hydroazulenic sesquiterpenoids.

Thanks to the efforts of a number of chemists, structures of the following hydroazulenes could be proposed by the end of 1956: δ guaiene (*204*), aromadendrene, α·chigadamarene (*192*), guaiol (*189*), partheniol (*107*), carotol, ledol, globulol, patchouly alcohol (*71*), kessyl alcohol (*13*), kessyl glycol (*13*), linderene (*223*), lactaroviolin (*188*, *188a*, *202*, *203*), vetivones (*187*), tricyclovetivene, artabsin, matricin, carpesia lactone (*160*), tenulin, helenalin, and arborescine (*153*). Some of these structures had to be revised recently.

All structures so far established for hydroazulenes are derived from the following fundamental carbon skeletons; guaiane (LXXXVIII),

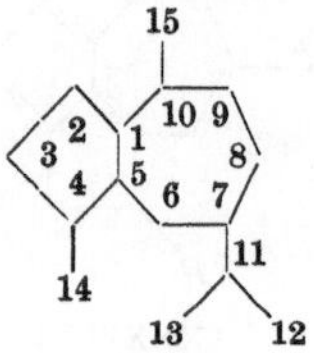

(LXXXVIII.) Guaiane. 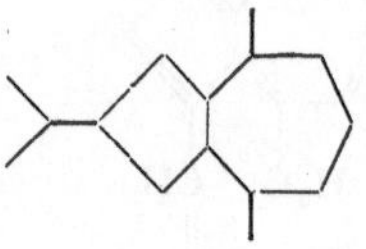   (LXXXIX.) Vetivane. 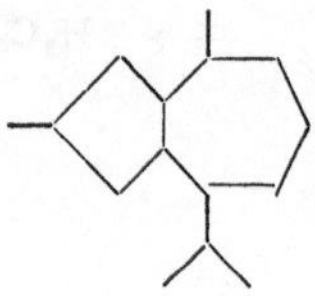   (XC.) Zierane. 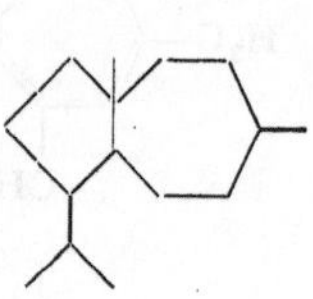   (XCI.) Carotane.

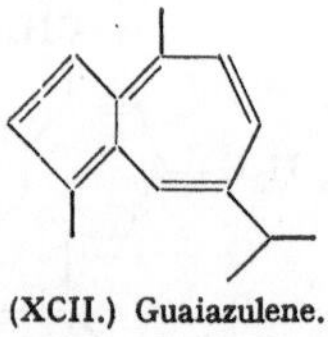

(XCII.) Guaiazulene. 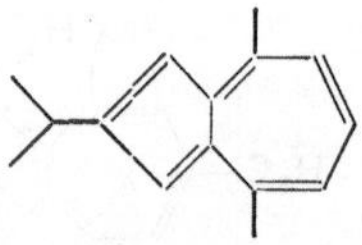   (XCIII.) Vetivazulene. 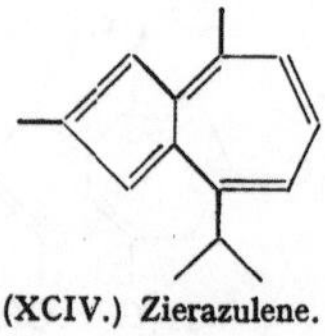   (XCIV.) Zierazulene.

vetivane (LXXXIX), zierane (XC), and carotane (XCI); the tricyclic skeletons are modifications of these bicyclic skeletons.

The arrangement of the carbon atoms in (LXXXVIII)–(XC) is clearly indicated by the formation of azulenes, i. e. guaiazulene (XCII), vetivazulene (XCIII), and zierazulene (XCIV), respectively, from natural hydroazulenes. Carotane, however, is not connected with any of the azulenes.

The present chapter is limited to the discussion of recent data and the reader is referred to earlier reviews (*1*, *2*, *5*, *7*, *10*, *11*, *13–15*, *25–27*) and to the original papers.

The *numbering system* of the guaiane skeleton (LXXXVIII) will be used throughout the present article, although a few other systems also appear in the literature.

1. Non-lactonic Hydroazulenes.

a) Guaiol and Bulnesol.

The stereochemistry of *guaiol* (XCV) (*25*) was partly clarified by EISENBRAUN and others (*96*) in a study of the optical rotatory dispersion, in which they have successfully related liquid, dextrorotatory dihydro-guaiol (XCVI) with a nepetalinic acid (XCVII) of known stereo-chemistry (*55*). The structure (XCVIII) was proposed recently by DOLEJŠ and others (*89*) for *bulnesol*, $C_{15}H_{26}O$, isolated from *Bulnesia sarmienti* together with guaiol. The location of the double bond was determined by ozonolysis that gave the cyclopentanone derivative (IC). Dihydro-bulnesol is identical with crystalline, levorotatory dihydroguaiol (C), which indicates β-configuration for the methyl group at $C_{(10)}$ in guaiol. (See Addendum, p. 79.)

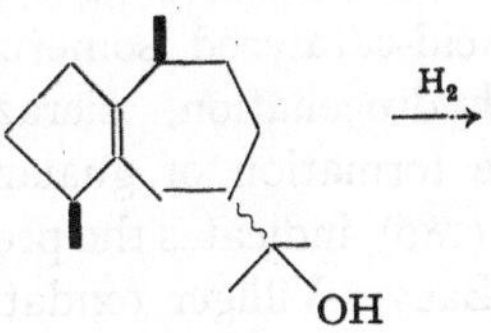

| (XCV.) Guaiol. | (C.) (—)-Dihydroguaiol. | (XCVIII.) Bulnesol. |

| (XCVI.) (+)-Dihydroguaiol. | (XCVII.) Nepetalinic acid. | (IC.) |

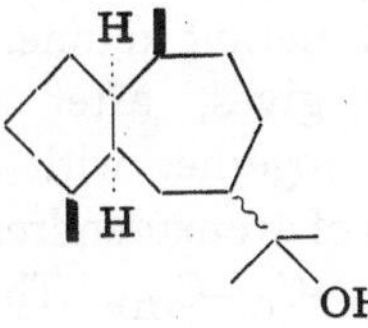

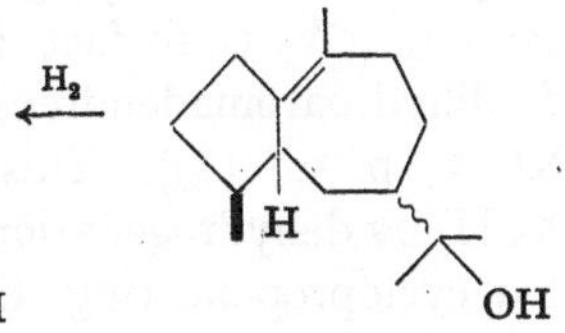

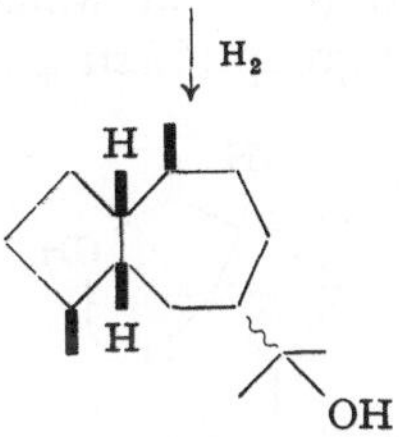

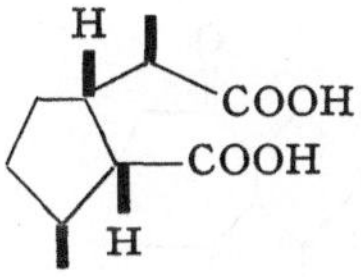

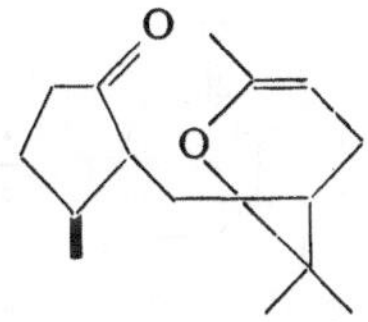

b) Aromadendrene and Alloaromadendrene.

Aromadendrene has been known for a long time as a main constituent of eucalyptus oil (*25, 26*). The structure (CIa), proposed by TREIBS and BARCHET (*230*), was replaced eventually by BIRCH and LAHEY's formulas (CIb, preferred to CIc) (*64*) on the following grounds: (a) The formation of the five-membered ring ketone (CIII) by cleavage of apoaromadendrone (CII), the ozonolysis product of aromadendrene; (b) acid-catalyzed isomerization of apoaromadendrone to its isomer with an exocyclic double bond, termed isoapoaromadendrone; and (c) epimerization (*61*) of "apoaromadendrone", now called allo(or α)-apoaromadendrone, to apoaromadendrone. The latter finding was due to the presence of alloaromadendrene (*63, 87 b, 90, 133*) (see below).

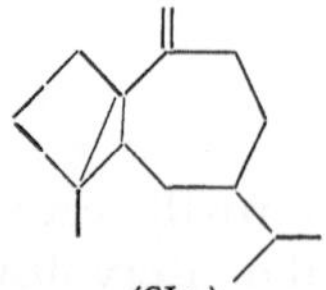

(CIa.) 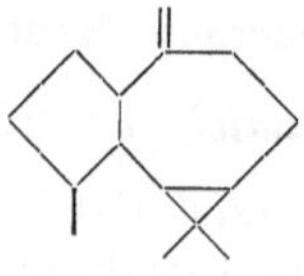   (CIb.) Aromadendrene. 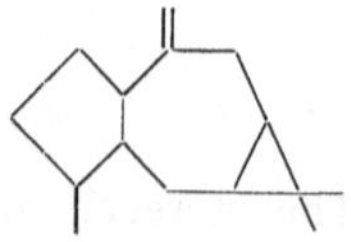   (CIc.)

Recent extensive studies on the structure of this hydrocarbon and related alcohols (ledol, globulol, viridiflorol) have established not only the correctness of (CIb) for aromadendrene but also the stereochemistry of this group of compounds.

The presence and location of a cyclopropane ring in aromadendrene were confirmed by Büchi and his co-workers (70) and independently by Dolejš and Šorm (91) on the basis of the following observations.

Deuteration of aromadendrene and of the keto acid (CIII) indicated the presence of grouping (CIV) and excluded the structure (CIa). The diketone (CV) is, in fact, a methyl ketone. Acid-catalyzed isomerization of dihydroaromadendrene gives, after dehydrogenation, zierazulene (XCIV, p. 52) (83). This, together with the formation of guaiazulene (XCII) by dehydrogenation of aromadendrene (186), indicates the presence of a cyclopropane ring, $C_{(6)}-C_{(7)}-C_{(11)}$. The Baeyer-Villiger oxidation of apoaromadendrone (CII), followed by two successive Barbier-Wieland degradations, yields the hydroxy-carboxylic acid (CVI) in which the cyclopropane ring is in α-position to the carboxyl group (Chart 4).

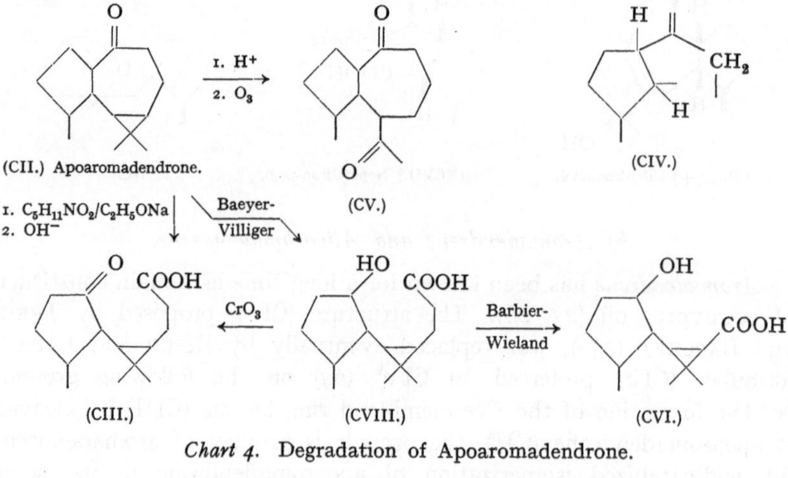

Chart 4. Degradation of Apoaromadendrone.

Alloaromadendrene was first isolated from Metrosideros scandens by Birch (64) who did not know that it is different from aromadendrene (63, 87 b, 90, 133). Alloaromadendrene gives, on ozonolysis or by other oxidation methods, alloapoaromadendrone (CVII) (64, 87 b, 90) which is isomerized by a base to apoaromadendrone (CII). This isomerization can be well understood in terms of epimerization at the asymmetric position α to the carbonyl, apoaromadendrone having the stable configuration and alloapoaromadendrone the labile one. This was clearly demonstrated by the resynthesis of aromadendrene from apoaromadendrone (70, 133).

Contrary to earlier reports (93), the *trans*-fused A/B ring system in aromadendrene follows (70, 92) from the equilibration reaction of the hydroxylic acid (CVIII) and its epimer, prepared from the keto acid (CIII), furthermore, from the catalytic reduction of the α,β-unsaturated ketone (CIX), [prepared by bromination and dehydrobromination of apoaromadendrone (CII)], which gave alloapoaromadendrone (CVII). The configurations at $C_{(4)}$ and $C_{(5)}$ are known to be *trans* from the stability of the keto acid (CIII) (64, 70).

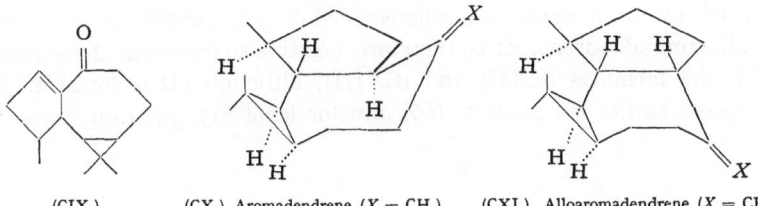

(CIX.) (CX.) Aromadendrene ($X = CH_2$). (CXI.) Alloaromadendrene ($X = CH_2$).
     (CII.) Apoaromadendrone ($X = O$). (CVII.) Alloapoaromadendrone ($X = O$).

Given these relative configurations, the absolute configuration was determined, with the help of the octant rule, from the observation of a single positive Cotton effect for alloapoaromadendrone (CVII) and a negative one for apoaromadendrone (CII), in their rotatory dispersion curves (70).

β-Orientation of the cyclopropane ring was assumed (a) from the analogy with the known configuration of some sesquiterpenoids (50, 91); (b) considering the acid-catalyzed ring-opening (70); and (c) from the study of the dehydration of related alcohols (see below). The formulas (CX) and (CXI) were proposed by BÜCHI (70) and ŠORM (92) for aromadendrene and alloaromadendrene, respectively.

c) Globulol, Ledol and Viridiflorol.

Of the four isomeric alcohols derived from aromadendrene and alloaromadendrene, three are known as *globulol* (25), *ledol* (25) and *viridiflorol* (5, 158). The fourth alcohol, epiglobulol, which has not yet been found in nature, was synthesized (88, 133).

The correlation of the three alcohols mentioned, as well as of aromadendrene and alloaromadendrene, has been established as shown in *Chart 5*, next page (61, 63, 66, 70, 87 b, 90, 133).

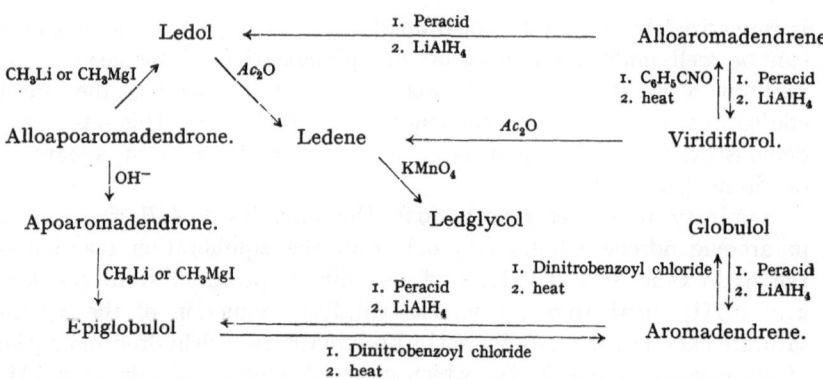

Chart 5. Correlation of Ledol, Globulol and Viridiflorol.

It is clear from this Chart that globulol and epiglobulol are structurally related to aromadendrene, whereas ledol and viridiflorol are related to alloaromadendrene. It is, therefore, possible to represent these alcohols with the formulas (CXII) and (CXIII), although other formulas were proposed earlier for globulol (66) and for ledol (93, 95, 140).

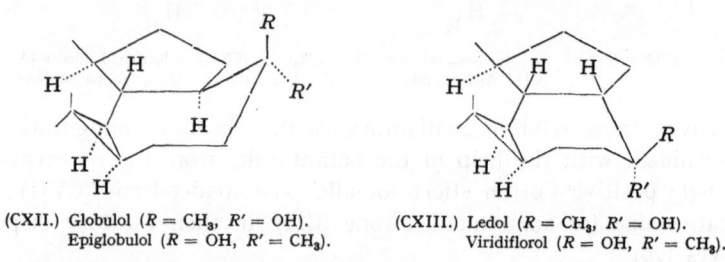

(CXII.) Globulol ($R = CH_3$, $R' = OH$).
Epiglobulol ($R = OH$, $R' = CH_3$).

(CXIII.) Ledol ($R = CH_3$, $R' = OH$).
Viridiflorol ($R = OH$, $R' = CH_3$).

The assignment of these structures to the two alcohol pairs is based on the following observations. Only epiglobulol and ledol were obtained by partial synthesis starting, respectively, from apoaromadendrone and alloapoaromadendrone. The dehydration of globulol with thionyl chloride and pyridine yielded 60% of a hydrocarbon with an exocyclic double bond, while ledol gave only 20% (93). Thus, the following structures are suggested: (CXII, $R = CH_3$, $R' = OH$) for globulol; (CXIII, $R = CH_3$, $R' = OH$) for ledol; and (CXIII, $R = OH$, $R' = CH_3$) for viridiflorol.

   Himbaccol (65), isolated from Himantandra baccata, is identical with viridiflorol (63).

d) Palustrol.

After Kiryalov (141) had isolated palustrol and dehydrated it to ledene it was considered (26) as an epimer of ledol (63, 133, 42) and later as that of globulol (70). However, re-examination of the alcohol (88), isolated by DOLEJŠ et al. from *Baccharis genistelloides*, revealed a structural difference from epiglobulol (CXII, $R = OH$, $R' = CH_3$), although the two compounds are closely related. There are differences in their physical constants, especially in the infrared spectra (stretching vibrational frequencies of hydroxyl).

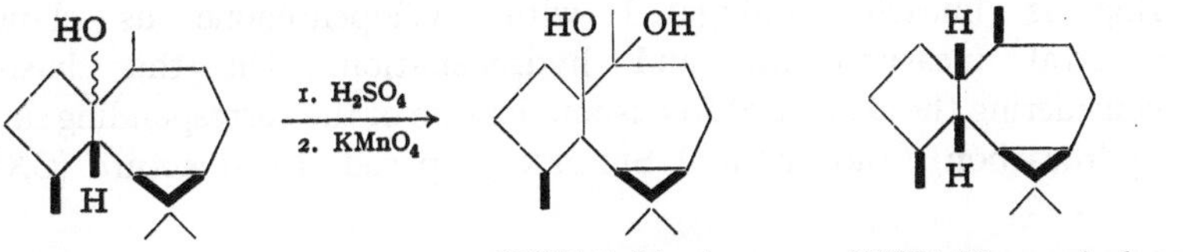

(CXV.) Palustrol.   (CXIV.) Ledglycol.   (CXVI.) Alloaromadendrane.

Palustrol was dehydrated and hydroxylated to give ledglycol (CXIV), also obtained from ledol; this indicates its identity with the alcohols of the aromadendrene type, except for the substituents at $C_{(1)}$ and $C_{(10)}$. Since all four epimeric alcohols of the aromadendrene type are known, the structure (CXV) was proposed for palustrol. Hydrogenolysis of palustrol afforded alloaromadendrane (CXVI).

e) α-Gurjunene.

Reinvestigation of this oily sesquiterpene hydrocarbon (25), an isomer of aromadendrene, was carried out independently by TREIBS (233) and by PALMADE and OURISSON (184) and resulted in the structural proposals (CXVIIa) and (CXVIIb), respectively. From the available evidence, as illustrated in *Chart 6*, the structure (CXVIIb) seems more probable, although (CXVIIa) cannot be excluded.

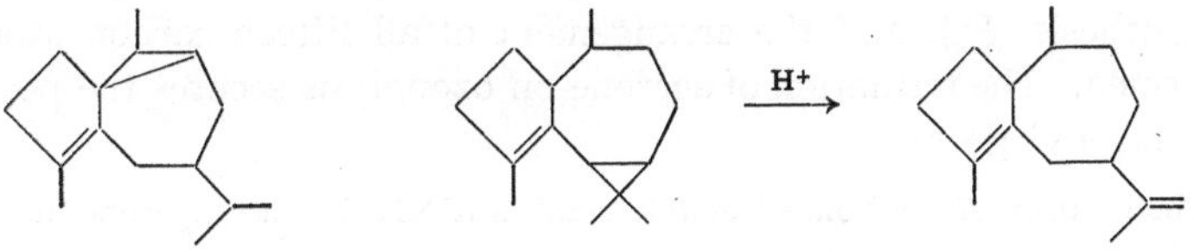

(CXVIIa.)   (CXVIIb.) α-Gurjunene(?).   (CXVIII.)

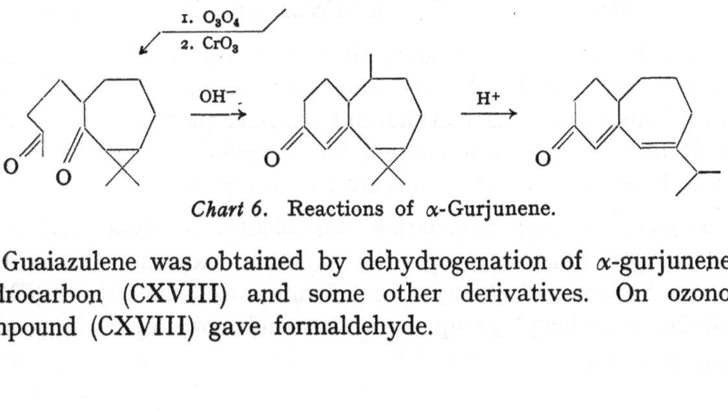

Chart 6. Reactions of α-Gurjunene.

Guaiazulene was obtained by dehydrogenation of α-gurjunene, the hydrocarbon (CXVIII) and some other derivatives. On ozonolysis, compound (CXVIII) gave formaldehyde.

f) Cyclocolorenone.

CORBETT and SPEDEN (*84*) have isolated from the essential oil of *Pseudowintera colorata,* among other terpenoids, a sesquiterpene ketone, *cyclocolorenone,* $C_{15}H_{22}O$. Dehydrogenation of the corresponding alcohol revealed the presence of the guaiane skeleton. A cyclopropane ring is linearly conjugated with cyclopentenone as shown by spectral measurements and hydrogenation. On this basis and considering the acid-catalyzed isomerization of the corresponding saturated hydrocarbon, CORBETT and SPEDEN proposed the structure (CXIX).

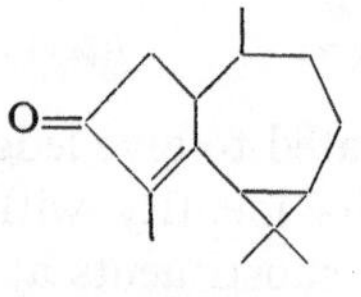

(CXIX.) Cyclocolorenone. 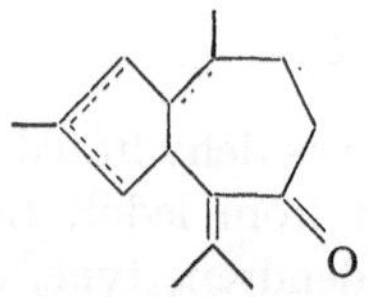   (CXX.) Zierone.

g) Zierone.

The chemical constitution of *zierone* $C_{15}H_{22}O$ (*25*), which is identical with that of *elleryone* (*131*) from *Evodia elleryana* (*134*), has not been fully clarified by the studies of BIRCH (*62*) and HILDEBRAND (*131*). Zierone is an α,β-unsaturated ketone containing another isolated double bond. On dehydrogenation of the conjugated diene, prepared by lithium aluminum hydride reduction and subsequent dehydration, zierone gives zierazulene (XCIV, p. 52). Introduction of one more carbon unit in α-position to the carbonyl group, followed by dehydrogenation, affords 6-methylzierazulene. The structure of these azulenes has been established by synthesis (*83*), and the arrangement of all fifteen carbon atoms was confirmed. The formation of acetone on ozonolysis secures the position of the carbonyl group.

The location of the isolated double bond in (CXX) has no experimental support.

h) Tricyclovetivene, Tricyclovetivenol and Bicyclovetivenol.

It has been known for some time that vetiver oil (from *Vetiveria zizanioides*) contains, besides vetivones (*187*), some bicyclic and tricyclic hydrocarbons as well as bicyclic and tricyclic primary and secondary alcohols (*25*). Recent examination has revealed the structure of one of these hydrocarbons, viz. tricyclovetivene (*77, 80*).

On selenium dehydrogenation *tricyclovetivene* gives vetivazulene (XCIII, p. 52) and its transposition product, eudalene. On ozonolysis it yields, besides formaldehyde and formic acid, the ketone (CXXII) containing a carbonyl group on a six-membered ring. The reactions of (CXXII), as indicated in *Chart 7*, show the correctness of the tricyclovetivene structure (CXXI).

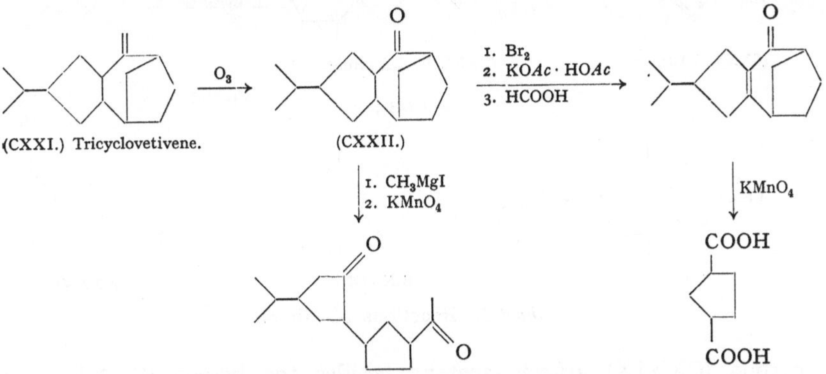

(CXXI.) Tricyclovetivene. (CXXII.)

Chart 7. Degradation of Tricyclovetivene.

CHIURDOGLU and DECOT (*78*) derived the respective structures (CXXIII) and (CXXIV) for *bicyclovetivenol* and *tricyclovetivenol* from the fact that the pyrolysis of dihydrovetivenol acetates gave bicyclovetivene and tricyclovetivene. The former hydrocarbon was claimed to have the structure (CXXV).

The location of the double bond in (CXXIII) and (CXXIV) has no experimental support.

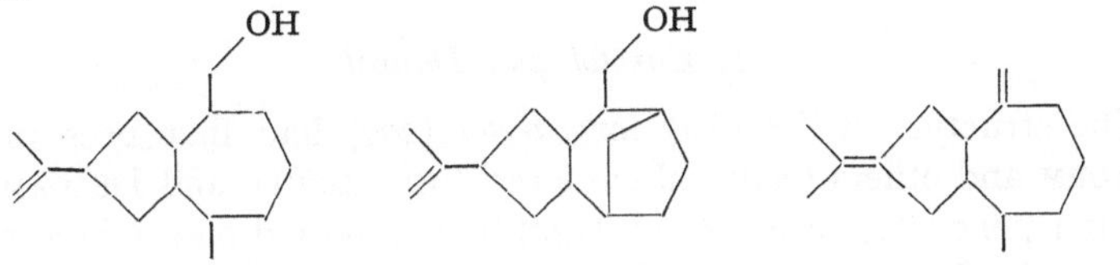

(CXXIII.) Bicyclovetivenol. (CXXIV.) Tricyclovetivenol. (CXXV.) Bicyclovetivene.

i) Hinesol.

Hinesol was recently isolated by YOSHIOKA, HIKINO and SASAKI (*248*) from the essential oil of *Atractilodes lancea*, and the structure (CXXVI) was based on the following experimental results (*249*, *250*).

Hinesol, $C_{15}H_{26}O$, is a tertiary alcohol of the vetivane skeleton containing an ethylenic double bond, as was revealed by hydrogenation, dehydrogenation and dehydration experiments. The reaction sequence that has clarified the structure is given in *Chart 8* (next page).

Resynthesis of dihydrohinesol (CXXVII) from the methyl ketone (CXXVIII) has confirmed the position of the hydroxyl group. The hydroxyketo acid, obtained by ozonolysis of hinesol (CXXVI) has no active methylene next to the keto group. Ozonolysis of the hydro-

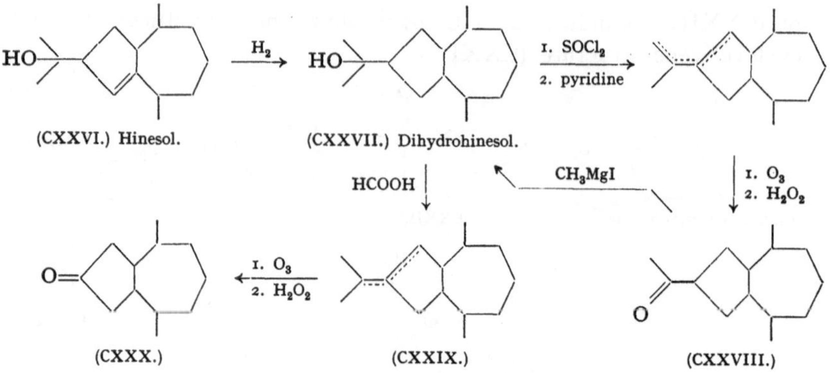

Chart 8. Reactions of Hinesol.

carbon (CXXIX) affords acetone besides the ketone (CXXX). The benzylidene derivative of (CXXX) is identical with that obtained from β-vetivone (*187*); this shows that the configuration of the methyl groups is *cis*. Dihydrohinesol is, however, optically active, which indicates that the juncture of the two rings must be *trans*. Since both the hydrocarbon (CXXIX) and the ketone (CXXX) are optically inactive, Yoshioka and others (*250*) conclude that epimerization has taken place during the dehydration process, (CXXVII) → (CXXIX). This statement is rather surprising and would require a more detailed explanation.

j) Carotol and Daucol.

The structure (CXXXIa) for *carotol* (*207*) had first been proposed by Šorm and others (*208*). Much later, Chiurdoglu and Descamps (*79*) and, independently, Šorm et al. (*222*) have reported on a reinvestigation of carotol. They suggested the structure (CXXXIb) and (CXXXIc), respectively. Some experimental discrepancies in these papers have induced Herout and Šorm (*118*) to present further evidence in favor of their carotol structure (CXXXIc). Both groups of workers now

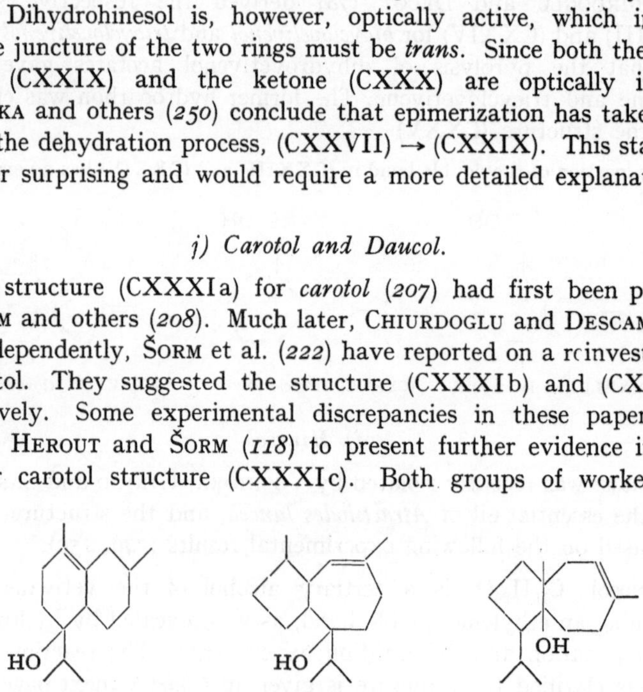

agree that carotol is a hydroazulenic tertiary alcohol with a trisubstituted ethylenic linkage; this excludes the formula (CXXXIa). The azulene(s), formed by dehydrogenation, are not fully characterized (*207*).

The conversions given in *Chart 9* and some other reactions make structure (CXXXIc) more probable.

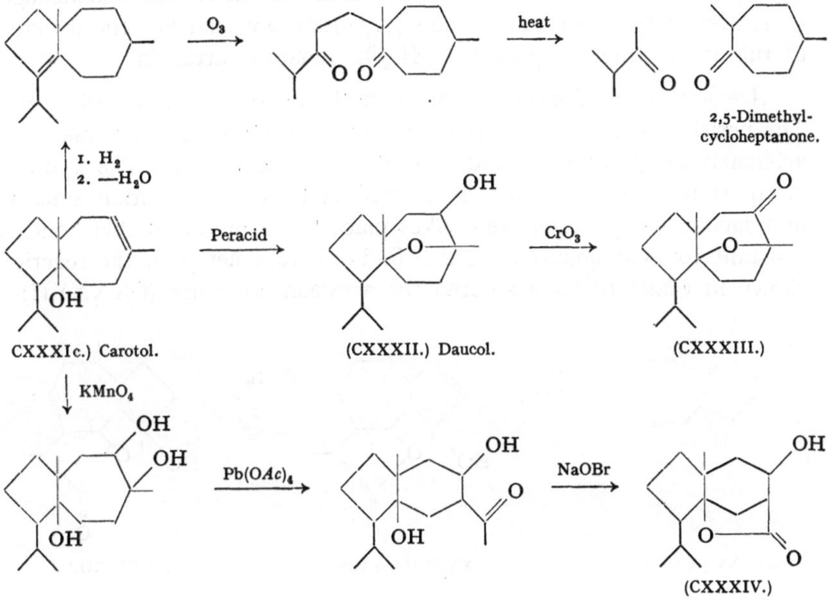

CXXXIc.) Carotol.   (CXXXII.) Daucol.   (CXXXIII.)

(CXXXIV.)

Chart 9. Reactions of Carotol and Daucol.

The formation of 2,5-dimethylcycloheptanone indicates the relative positions of two methyl groups and limits the location of the isopropyl and hydroxyl groups. Peracid oxidation of carotol yielded an oxide (CXXXII), identical with *daucol*. Daucol cannot be a 1,2-epoxide because its oxidation affords an oxide ketone (CXXXIII). The latter has an active methylene adjacent to the keto group. Compound (CXXXIV) is a γ-lactone.

If structure (CXXXIc) is correct, carotol and daucol are unique in the arrangement of their carbon atoms and are the only known members of the carotane series (XCI, p. 52).

2. Lactonic Hydroazulenes.

a) Artabsin, Absinthin and Anabsinthin.

It has been known for a long time that many species of Compositae produce highly colored essential oils. Extensive studies of such an interesting oil which shows pharmacological activity (*10, 14*) have culminated in the isolation of chamazulene (CXXXV) and an unstable, orange-colored chamazulenogen, $C_{15}H_{20}$, now formulated as dihydrochamazulene (*26, 209*). A careful resolution of the oil of wormwood

(Artemisia absinthium) has resulted in the isolation of several hydro-azulenic compounds. One of them gave chamazulenogen on steam distillation in weakly acid medium and was first termed prochamazulenogen (*121, 205*). Later, Šorm and others (*117, 122*) have clarified the structure of this interesting compound, $C_{15}H_{20}O_3$, renamed artabsin.

Artabsin is a hydroxy-γ-lactone containing conjugated double bonds. The position of the lactone ring was established by the conversion into artemazulene (CXXXVI) and guaiazulene (XCII, p. 52). The hydroxyl group is tertiary and must be located in an allylic position since on hydrogenation hydrogenolysis takes place. One of the double bonds is resistant to hydrogenation. These findings, together with the reactions shown in *Chart 10* have secured the artabsin structure (CXXXVII).

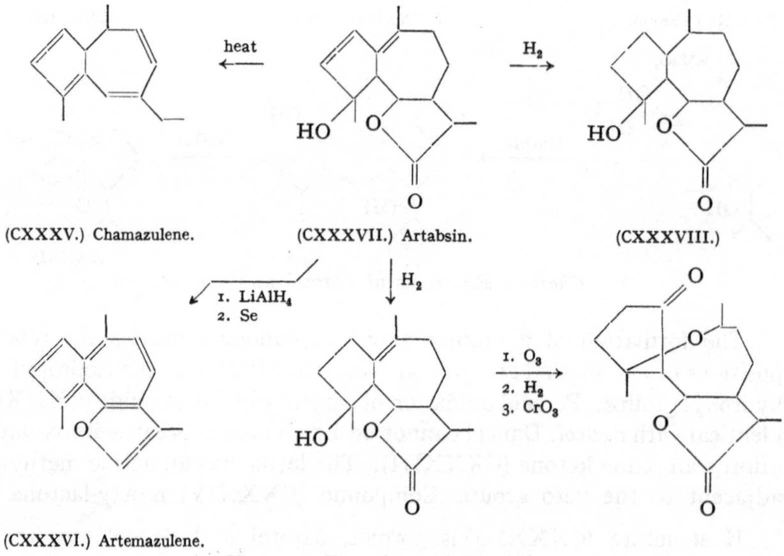

(CXXXV.) Chamazulene.   (CXXXVII.) Artabsin.   (CXXXVIII.)

(CXXXVI.) Artemazulene.

Chart 10. Reactions of Artabsin.

The hydroxyguaianolide (CXXXVIII) is identical with the compound obtained by hydrogenolysis of arborescine, and this confirms one end of the oxide bridge in the latter (*1, 2, 153*).

The separation of the components of *Artemisia absinthium* produced, besides artabsin, a new substance, *anabsinthin*, which was later found to have been formed from *absinthin* during the isolation process (*120, 121, 205*). A recent study by Šorm et al. (*161*) has established the structures (CXXXIX) and (CXL) for absinthin and anabsinthin, respectively. Absinthin was quantitatively converted to anabsinthin by acid or by mere heating.

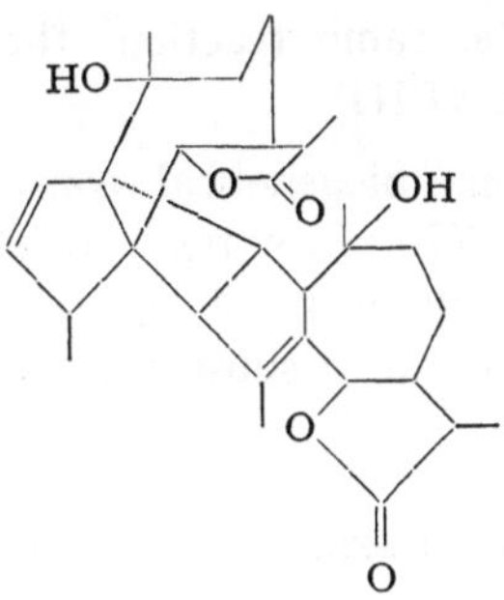

(CXXXIX.) Absinthin.

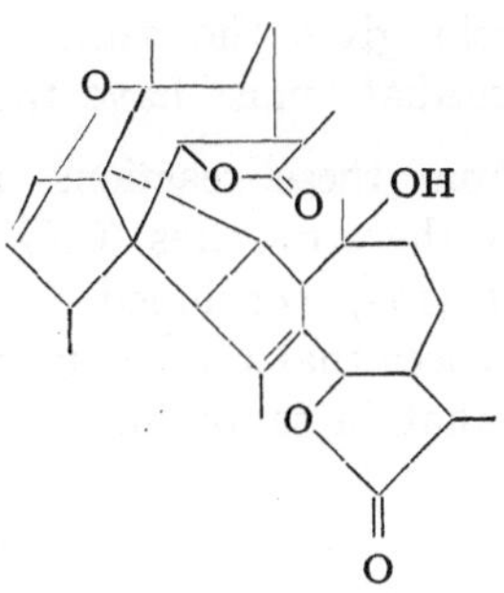

(CXL.) Anabsinthin.

Under mild conditions absinthin gives chamazulene (CXXXV) whereas anabsinthin does not, though the same azulene can be obtained by dehydrogenation of either isomer. The dimeric formula, $C_{30}H_{40}O_6$, has now been adopted for both isomers, after some initial difficulties due to their tendency to undergo hydration.

Both absinthin and anabsinthin display two γ-lactonic functions whose positions were confirmed by the formation of artemazulene (CXXXVI) on reduction and subsequent dehydrogenation (161). The difference between the two isomers is that absinthin has a reducible double bond and two tertiary hydroxyls, while anabsinthin lacks the reducible double bond and has only one hydroxyl group; this implies the formation of an oxide ring from the other hydroxyl and a double bond during isomerization. The presence of another double bond in anabsinthin, hence also in absinthin, was revealed by its dehydration to form a conjugated diene; the possibilities of the location of the diene group were limited when the lactonic acid (CXLI) and the tricarboxylic acid (CXLII) were obtained by oxidation. Both acids have also been obtained from artabsin (117).

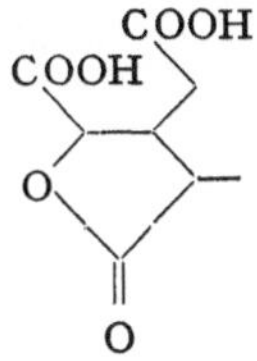

(CXLI.)

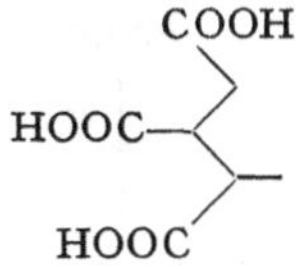

(CXLII.)

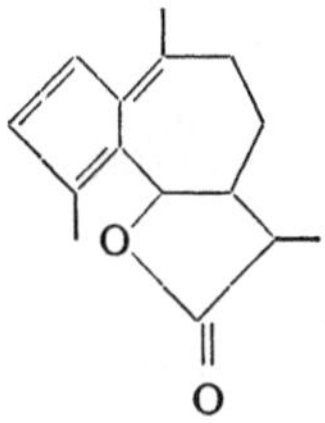

(CXLIII.)

Cleavage of absinthin to a C_{15}-fragment occurs by alkali via a fulvene type intermediate, which is indicated by its facile interactions with aldehydes. After condensation with formaldehyde, for example, 7-ethyl-1,2,4-trimethylazulene was obtained by distillation. Since artabsin gives the same azulene by the same reaction, the fulvene intermediate must have the structure (CXLIII).

From these reactions and from some biogenetical considerations follow the structures (CXXXIX) and (CXL). Novotný, Herout and Šorm (162) concluded from their measurement of optical rotation differences that the configuration at $C_{(6)}$ of these substances is identical with that in santonin.

b) Matricin and Matricarin.

It has been known for some time that *Matricaria chamomilla* contains a chamazulene precursor which undergoes facile transformation to chamazulene under such mild treatment as steam distillation. When searching for the precursor, Šorm and his co-workers (74) obtained a

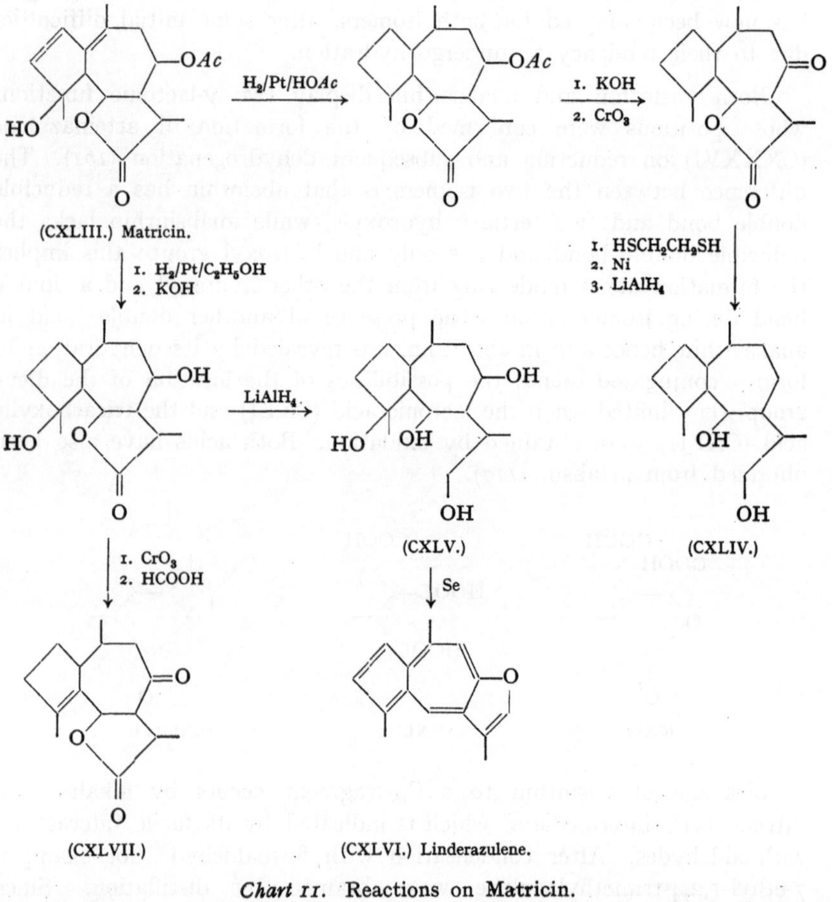

Chart 11. Reactions on Matricin.

crystalline acetoxy-γ-lactone, *matricin*, that also contains conjugated double bonds and a tertiary hydroxyl group. An interesting feature of this chamazulene precursor is that on heating a dilute acidic matricin solution at 50° an azulenic carboxylic acid, guaiazulenic acid is formed which undergoes further decarboxylation to chamazulene at a temperature below 100°.

The matricin structure (CXLIII) was established by the reaction sequence given in *Chart 11 (75)*.

Dehydrogenation of the diol (CXLIV) affords artemazulene (CXXXVI, p. 62), whereas the same reaction of the tetrol (CXLV) affords a mixture of artemazulene and linderazulene (CXLVI). The positions of the hydroxyl group and the conjugated double bonds were suggested by the inactivity of matricin toward maleic anhydride, by the periodate stability of the tetrol (CXLV), and by the circumstance that neither the acetoxyl group nor the lactonic oxygen suffers hydrogenolysis, and that the ketoguaienolide (CXLVII) has no α,β-unsaturated carbonyl.

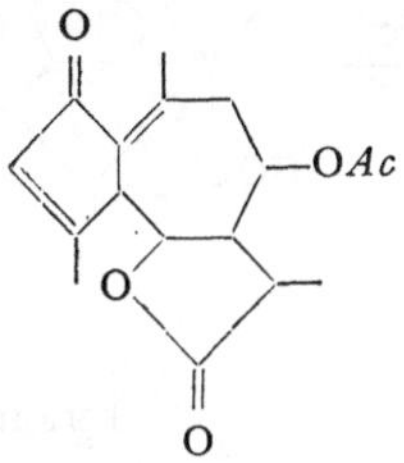

(CXLVIII.) Matricarin.

A second hydroazulenic lactone, isolated from *M. chamomilla*, is *matricarin (76)*. It has two double bonds and a five-membered ring ketonic function. The keto group is part of the chromophore similar to that in lactucin (see below). Artemazulene (CXXXVI) and linderazulene (CXLVI) were also obtained by dehydrogenating the reduction product of matricarin. The structure (CXLVIII) was suggested for matricarin in analogy with matricin (CXLIII). (See Addendum, p. 79.)

c) Lactucin.

In 1958, BARTON and NARAYANAN (54) as well as DOLEJŠ and ŠORM (94) have reported independently on their extensive studies on *lactucin*, $C_{15}H_{16}O_5$, a bitter principle of *Lactuca virosa*; and the structure (CXLIX) was established, after an alternative structure considered by DOLEJŠ and ŠORM had been excluded by BARTON. Lactucin has a methyl group, a five-membered ring carbonyl and two hydroxyls, one of which is primary and the other secondary. The position of the latter is confirmed by the formation of artemazulene and linderazulene from the reduction product of lactucin. The location of these functional groups was supported by the reaction sequence shown in *Chart 12*.

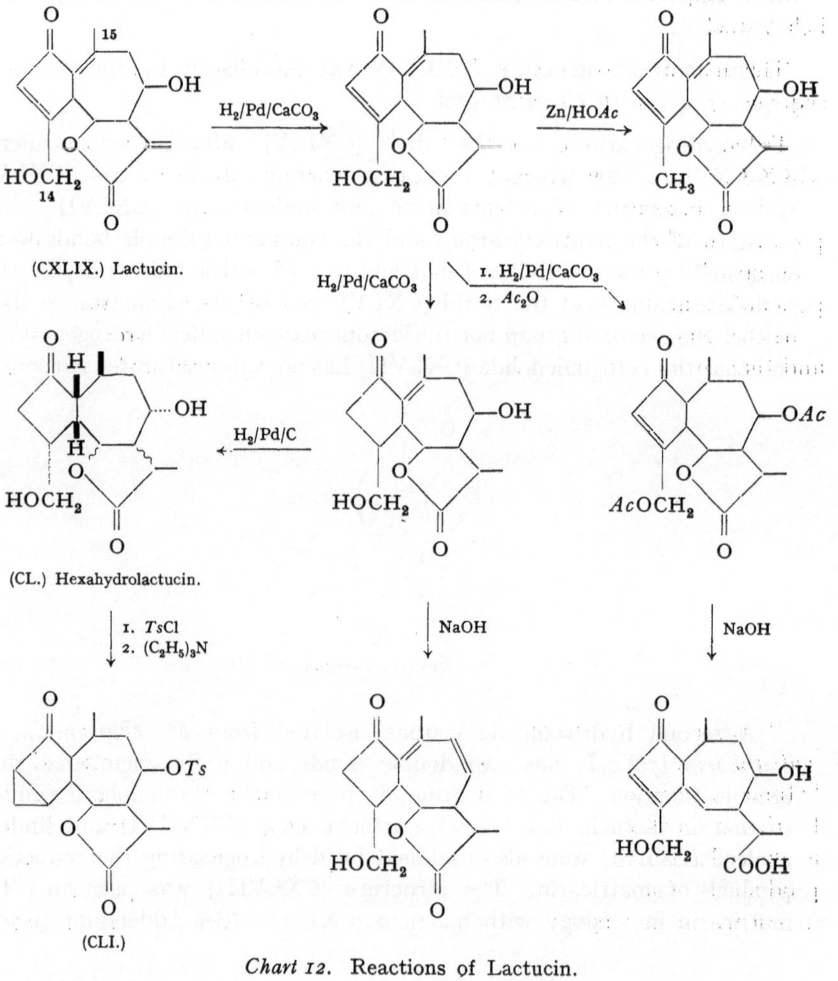

Chart 12. Reactions of Lactucin.

The cyclopropyl ketone (CLI) could not have been formed if lactucin had the alternative structure which contains the primary hydroxyl group at $C_{(15)}$ instead of $C_{(14)}$.

BARTON and NARAYANAN (*54*) have also suggested the partial configuration of hexahydrolactucin (CL) considering the formation of the benzylidene derivative.

d) *Cynaropicrin.*

Quite recently, Šorm and others (*219*) have isolated from *Cynara scolymus* a new sesquiterpene lactone, *cynaropicrin.* This substance, however, is very unstable and polymerizes easily. Hence, all experiments (except the infrared spectral readings) had to be carried out with the fully hydrogenated compound. When heating the lithium aluminum hydride reduction product of cynaropicrin, artemazulene (CXXXVI) was obtained. On hydrogenation of cynaropicrin three compounds were isolated including (CLIa) and (CLIb); all three are isobutyric esters. These reactions are summarized in *Chart 13.*

The cynaropicrin structure [CLII, $R = COC(CH_2OH){=}CH_2$] was proposed recently by Šorm et al. (*219a*).

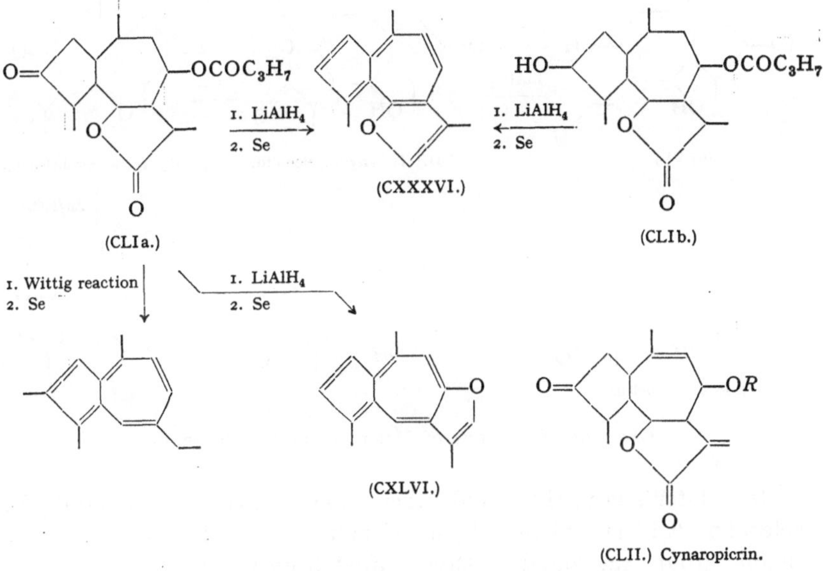
O=
—OCOC₃H₇
1. LiAlH₄
2. Se
(CXXXVI.)
HO—
—OCOC₃H₇
1. LiAlH₄
2. Se
(CLIa.)
(CLIb.)
1. Wittig reaction
2. Se
1. LiAlH₄
2. Se
(CXLVI.)
O=
—OR
(CLII.) Cynaropicrin.

Chart 13. Dehydrogenation of Cynaropicrin Derivatives.

e) *Helenalin, Isohelenalin, Neohelenalin, Tenulin, Balduilin, and Mexicanin C.*

Following investigations by Adams and Herz (*31–33*), Büchi and Rosenthal (*72*) established the complete structure (CLIII) for helenalin that is widely distributed in *Helenium* and related genera. *Helenalin* is a hydroxy-lactone with an exocyclic double bond and a cyclopentenone ring system, position of which was determined by NMR study of dihydrohelenalin (CLIV); that of the hydroxyl group followed from NMR data of tetrahydrohelenalone (CLV). Formation of guaiazulene on dehydrogenation of the tetrol (CLVI) has confirmed the shape of the carbon skeleton. The reactions carried out in order to elucidate the helenalin structure are summarized in *Chart 14*.

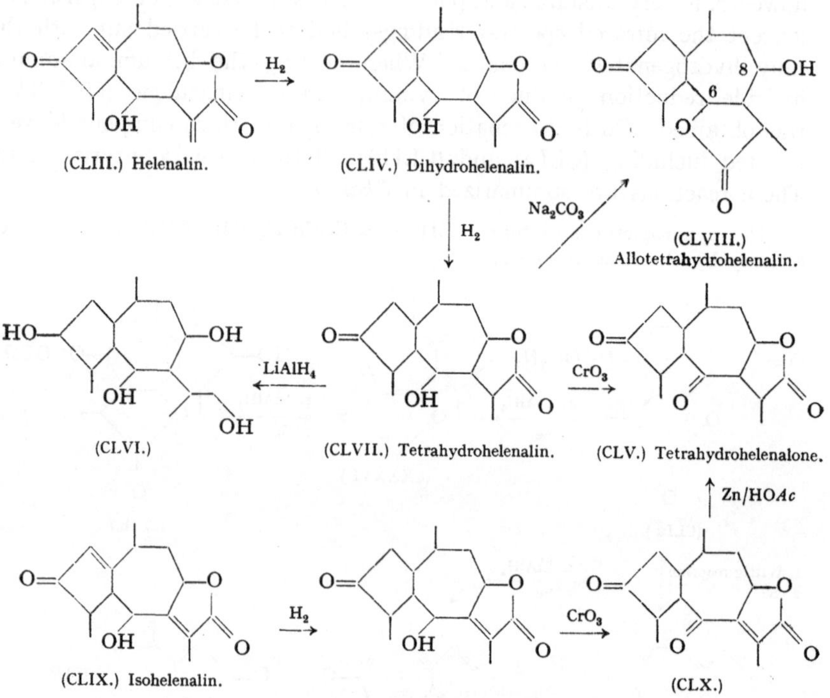

Chart 14. Reactions of Helenalin and Isohelenalin.

In a later study, HERZ and others (*128*) prepared from tetrahydrohelenalin (CLVII) allotetrahydrohelenalin (CLVIII), in which the lactone group was found to have shifted from $C_{(8)}$ to $C_{(6)}$.

The second isomer, *isohelenalin* (CLIX), was isolated by BÜCHI and ROSENTHAL (*72*) from the same source as helenalin. Reduction of its second double bond, which resists direct hydrogenation, was accomplished as shown in Chart 14 via the ketone (CLX); thus, tetrahydrohelenalone (CLV) was formed. This has established the relation of isohelenalin to helenalin and the position of the second double bond.

The third isomer, *neohelenalin*, named flexuosin earlier (*129*), is an artifact formed by isomerization of helenalin on alumina (*126*). Its structure (CLXI) follows from (a) spectral data, (b) the formation of formaldehyde and acetic acid on ozonolysis, and (c) the ozonolysis of the dihydro compound, yielding acetic acid. The configuration of

neohelenalin was determined in connection with that of deacetylneotenulin (CLXV) (87) (see below).

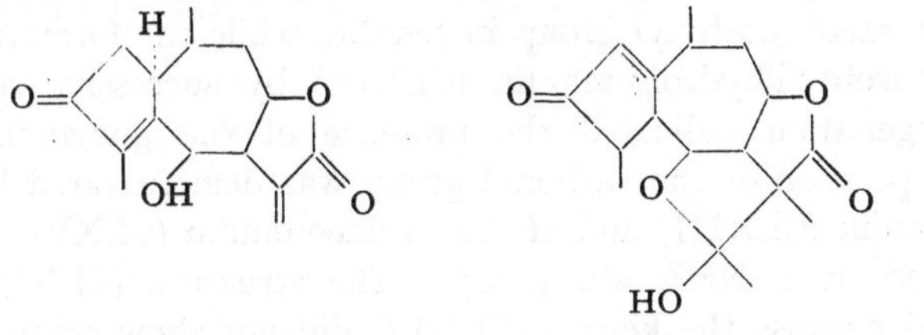

(CLXI.) Neohelenalin.　　　　　　　(CLXIIb.)

The 'structure (CLXIIa) of *tenulin*, $C_{17}H_{22}O_5$ (81, 82, 243), another
bitter principle occurring in various *Helenium* spp., was clarified

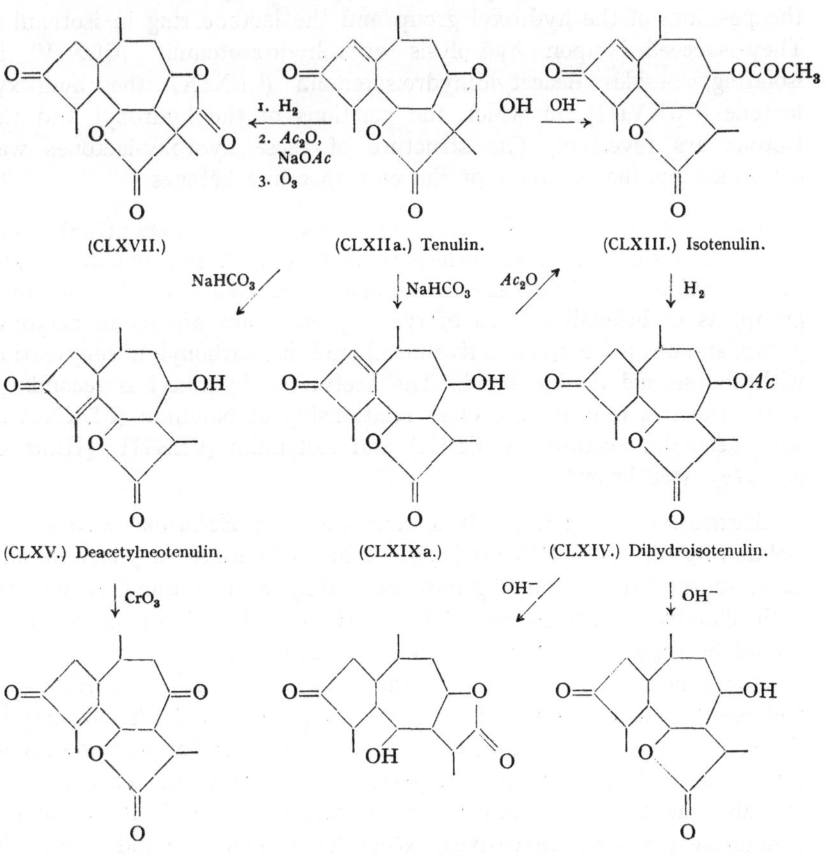

(CLXVII.)　　　　　(CLXIIa.) Tenulin.　　　　　(CLXIII.) Isotenulin.

(CLXV.) Deacetylneotenulin.　　(CLXIXa.)　　　(CLXIV.) Dihydroisotenulin.

(CLXVI.)　　　　　(CLXVIII.)　　　(CLXIX.) Deacetyldihydroisotenulin.

Chart 15. Reactions of Tenulin.

by BARTON and DE MAYO (51) as follows. The IR spectrum showed
the absence of an acetoxyl group. The easy conversion of tenulin
to isotenulin (CLXIII) *(Chart 15)*, an acetate, also indicated the presence
of a "masked" acetoxyl group in tenulin, while the formation of linderazulene from dihydroisotenulin (CLXIV) by successive reduction and
dehydrogenation indicated the presence of the guaianolide skeleton.

The position of the carbonyl group was demonstrated by ozonolysis
of isotenulin (CLXIII) and of deacetylneotenulin (CLXV); this was also
supported by a NMR study (87). The structure (CLXIIb) could be
excluded because the ketone (CLXVI) did not show conjugation of the
newly-formed carbonyl with the α,β-unsaturated ·carbonyl system
originally present. These observations and the formation of the dilactone
(CLXVII) have secured the tenulin structure (CLXIIa) (51). Subsequently, HERZ et al. (67) presented supplementary evidence concerning
the positions of the hydroxyl group and the lactone ring in isotenulin.
They succeeded upon hydrolysis of dihydroisotenulin (CLXIV) in
isolating, besides deacetyldihydroisotenulin (CLXIX) the hydroxy-
lactone (CLXVIII), in which the positions of the hydroxyl and the
lactone are reversed. The structure of these hydroxy-lactones was
confirmed by the behavior of the corresponding ketones.

Balduilin, $C_{17}H_{20}O_5$, isolated by HERZ and others (128) from
Balduina uniflora, is a sesquiterpene lactone with two double bonds,
one of which is an exocyclic methylene conjugated with the γ-lactone
group, as in helenalin. Two of the oxygen atoms are in an acetoxyl
group, and one is located in a five-membered ring carbonyl, in conjugation
with the second double bond. The acetylated hydroxyl is secondary.
These features indicate the close relationship of balduilin (CLXXIVb)
with helenalin acetate (cf. CLIII) and isotenulin (CLXIII) [HERZ et
al. (128)] (see below).

Mexicanin C, $C_{15}H_{20}O_4$, is a constituent of *Helenium mexicanum*,
isolated by VIVAR and ROMO (246). It has a hydroxyl, a γ-lactone and
an α,β-unsaturated carbonyl group. Since dihydromexicanin C is identical
with allotetrahydrohelenalin (CLVIII), the constitution of mexicanin C
should be represented by (CLXIXa) (Chart 15, p. 69).

Stereochemical correlation of the lactones occurring in *Helenium*
and related genera was first achieved by DJERASSI and HERZ (87) in
the course of an optical rotatory dispersion study. The curves showed
clearly that helenalin and tenulin possess not only the same chromophore
but also the same configuration at the ring junctures. This is also the
case in the perhydro derivatives, except for neohelenalin and neotenulin
(87, 126) which have the opposite configuration at $C_{(1)}$. By comparing the
rotatory dispersion curves with those of steroidal reference compounds,

DJERASSI and others could suggest partial stereoformulas for these helenalin and tenulin derivatives as well as for geigerin and photosantonic lactone. Stereochemical correlation of the compounds of this group was also established on chemical grounds by extensive studies of HERZ and his co-workers (128) (Chart 16).

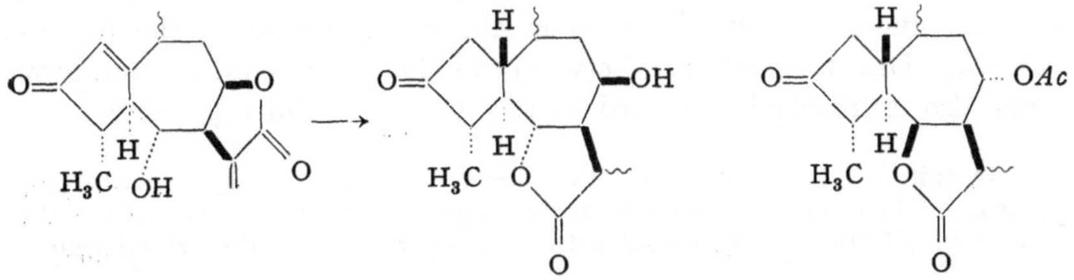

(CLIII.) Helenalin.   (CLVIII.) Allotetrahydrohelenalin.   (CLXIV.) Dihydroisotenulin.

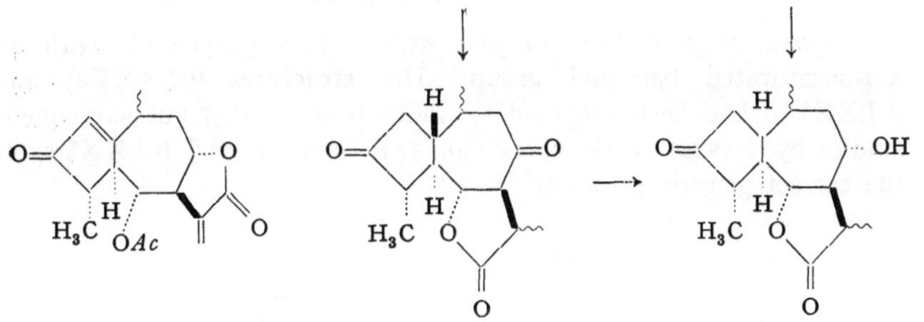

(CLXXIVa.) Balduilin (?).   (CLXX.)   (CLXXIII.) 6-Epidihydroalloisotenulin.

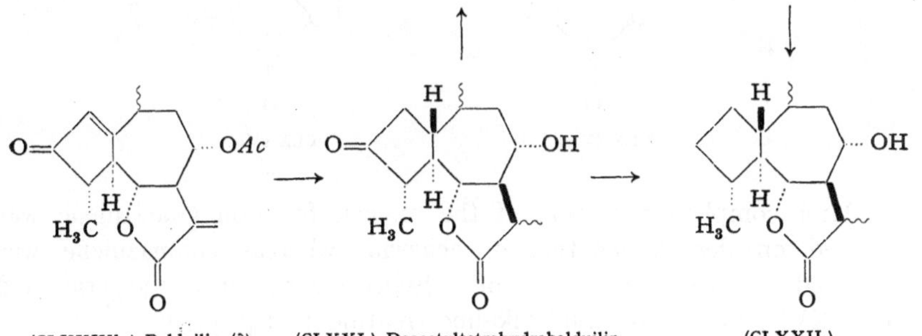

(CLXXIVb.) Balduilin (?).   (CLXXI.) Deacetyltetrahydrobalduilin.   (CLXXII.)

Chart 16. Structural Correlation of Helenalin, Tenulin and Balduilin*.

Accordingly, the ketone (CLXX), obtained from deacetyl-tetrahydro-balduilin (CLXXI), is identical with that from allotetrahydrohelenalin

* The configurations at $C_{(6)}$ and $C_{(8)}$ are given arbitrarily in this Chart in order to visualize the relationship between these groups. Configuration at $C_{(7)}$ was assumed to be β, following some general concepts (50, 54a, 92, 115a).

(CLVIII), thus establishing the stereochemical identity of the two groups, except for the configuration of the hydroxyl at $C_{(8)}$. On the other hand, the hydroxy-lactone (CLXXII), obtained from deacetyl-tetra-hydrobalduilin by elimination of the carbonyl, followed by a treatment with bicarbonate, is identical with that originating from 6-epidihydro-alloisotenulin (CLXXIII); this shows that the difference between the tenulin and balduilin groups is in the configuration of the hydroxyl at $C_{(6)}$. Direct correlation between the helenalin and tenulin groups was also established in regard to $C_{(1)}$, $C_{(4)}$, $C_{(5)}$, and $C_{(10)}$ (*127*).

In spite of these correlations there remains some ambiguity concerning the position of the lactonic oxygen in balduilin. Consequently, the structures (CLXXIVa) and (CLXXIVb) must be considered as equally possible at the present time.

f) Geigerin and Geigerinin.

Geigerin, isolated from *Geigeria aspera*, is a guaianolide with an α,β-unsaturated barbonyl group. The structures (CLXXVa) and (CLXXVb) had been proposed by PEROLD (*1, 2, 185*) but subsequent studies by BARTON and others (*50, 54a*) have secured (CLXXVc) as the correct geigerin structure.

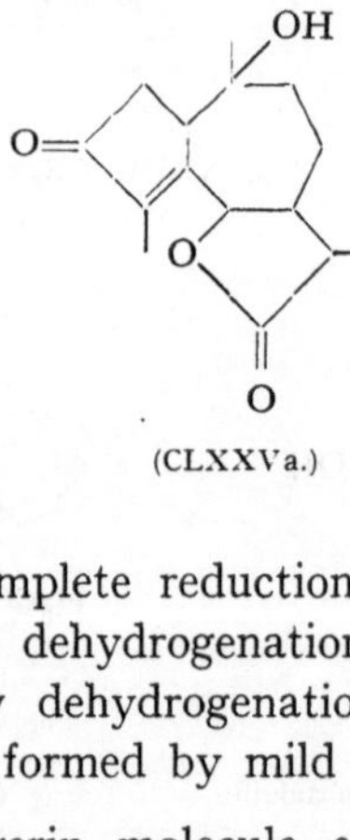

(CLXXVa.) 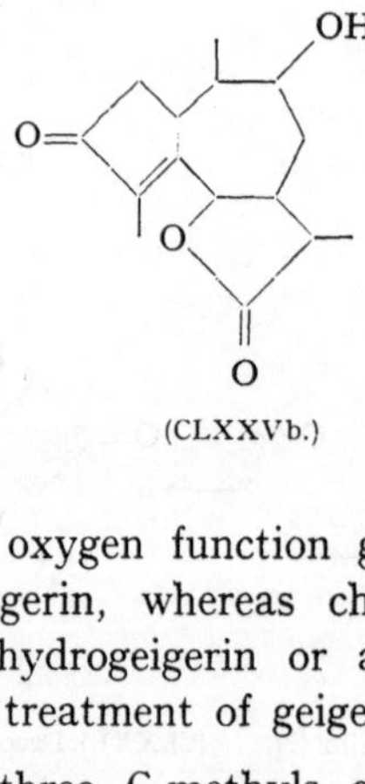   (CLXXVb.)

After complete reduction of the oxygen function guaiazulene was isolated on dehydrogenation of geigerin, whereas chamazulene was obtained by dehydrogenation of dehydrogeigerin or allogeigeric acid (CLXXVI), formed by mild alkaline treatment of geigerin.

The geigerin molecule contains three C-methyls and a hydroxyl whose presence is demonstrated by the formation of an acetate, and by dehydration to anhydrogeigerin, a conjugated dienone (CLXXVII). The keto group is located in the five-membered ring. The position of the double bond follows from the formation of acetic acid on ozonolysis of geigerin. On the basis of these observations and the reaction sequence given in *Chart 17*, the geigerin structure is (CLXXVc) as proposed by BARTON et al. (*50, 54a*).

Resistance of (CLXXVI) to lactonization shows that the hydroxyl and the side-chain at $C_{(7)}$ are *trans*. The configuration at $C_{(1)}$ is assumed to be β considering the rotatory dispersion (87).

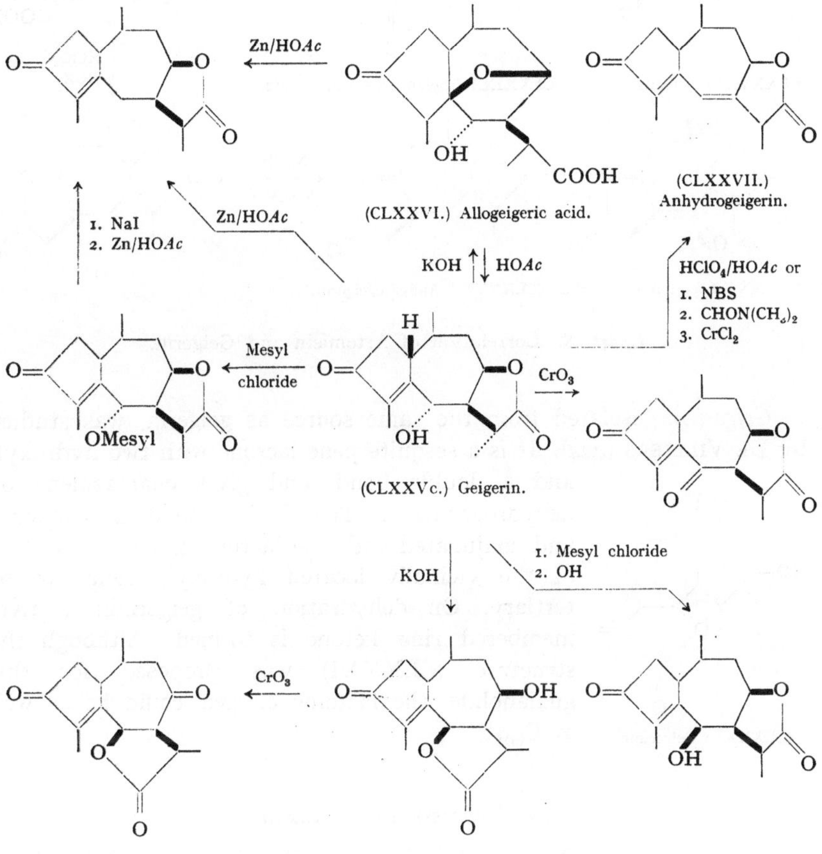

Chart 17. Reactions of Geigerin.

Recently, geigerin and artemisin (CLXXVIII) have been correlated stereochemically [BARTON and PINHEY (54a)]. These authors were successful in degrading isophotoartemisin acetate (CLXXIX) (47, 48) to anhydrogeigerin (CLXXVII). Their results (*Chart 18*, p. 74), together with the X-ray investigation of acetylbromogeigerin (CLXXX) (115a), a bromination product of geigerin acetate, have established the absolute configuration of geigerin (CLXXVc).

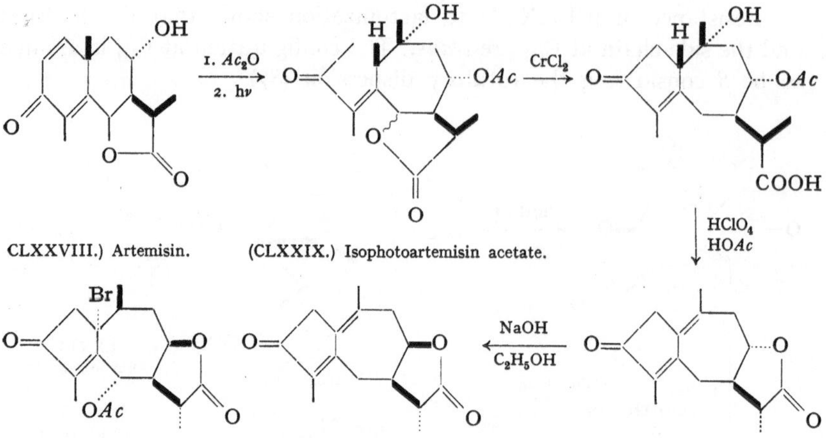

CLXXVIII.) Artemisin.   (CLXXIX.) Isophotoartemisin acetate.

CLXXX.) Acetylbromogeigerin.   (CLXXVII.) Anhydrogeigerin.

Chart 18. Correlation of Artemisin and Geigerin.

Geigerinin, isolated from the same source as geigerin, was studied by DE VILLIERS (*244*). It is a sesquiterpene lactone with two hydroxyls and a double bond, and gives chamazulene on dehydrogenation. The double bond is exocyclic and conjugated with the lactone carbonyl. Neither of the vicinally located hydroxyls seems to be tertiary. On dehydration of geigerinin a five-membered ring ketone is formed. Although the structure (CLXXXI) was proposed for this guaianolide, the lactonic oxygen could be as well at $C_{(8)}$.

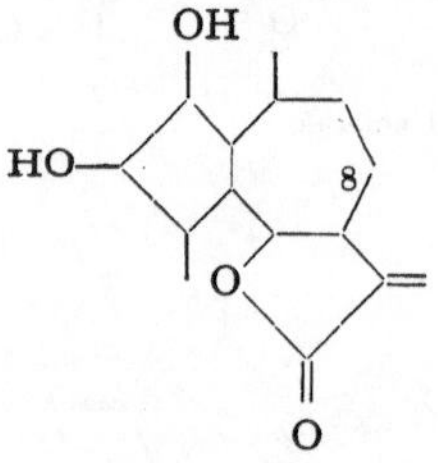

(CLXXXI.) Geigerinin.

g) *Ambrosin and Damsin.*

Ambrosin, $C_{15}H_{18}O_3$, and damsin, $C_{15}H_{20}O_3$, first isolated by ABU-SHADY and SOINE (*29*) from *Ambrosia maritima*, were shown (*30*), after extensive degradative work, to have the same skeleton as helenalin. BERNARDI and BÜCHI (*58*), in the possession of the complete helenalin and tenulin structures, have deduced the ambrosin structure (CLXXXII) that was confirmed by ŠORM et al. (*206*). The shape of the carbon skeleton was revealed by the formation of artemazulene from dihydroambrosin which also located the lactonic oxygen. The presence and position of a five-membered ring carbonyl and of an ethylenic linkage conjugated with it were derived from the infrared maxima of dihydroambrosin and ambrosin, in analogy with those of helenalin and tenulin.

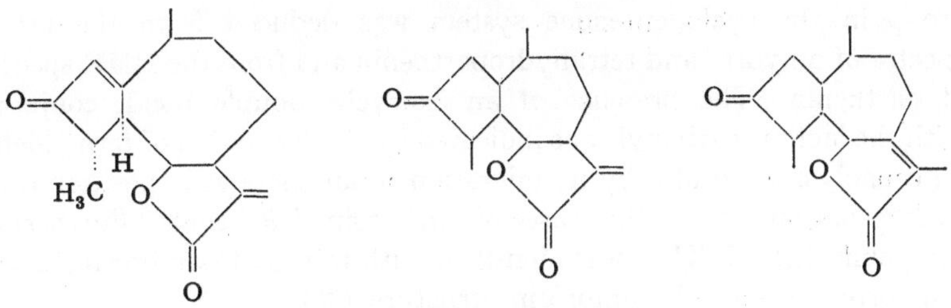

(CLXXXII.) Ambrosin.   (CLXXXIII.) Dihydroambrosin.   (CLXXXIV.) Dihydroisoambrosin.

The formation of formaldehyde on ozonolysis of ambrosin suggested the presence of an exocyclic ethylenic double bond. These combined results established the ambrosin structure (CLXXXII) that was also supported by some experiments carried out with parthenin by HERZ and WATANABE (*130*). On the basis of the rotatory dispersion, the partial configuration (CLXXXII) was suggested (*87*).

"*Damsin*" is likely to be a mixture of dihydroambrosin (CLXXXIII) and dihydroisoambrosin (CLXXXIV).

h) Parthenin.

The structure (CLXXXV) for the guaienolide *parthenin*, $C_{15}H_{18}O_4$, a bitter principle of *Parthenium hysterophorus*, was recently proposed by HERZ and WATANABE (*130*) (*Chart 19*). The location of the ketone

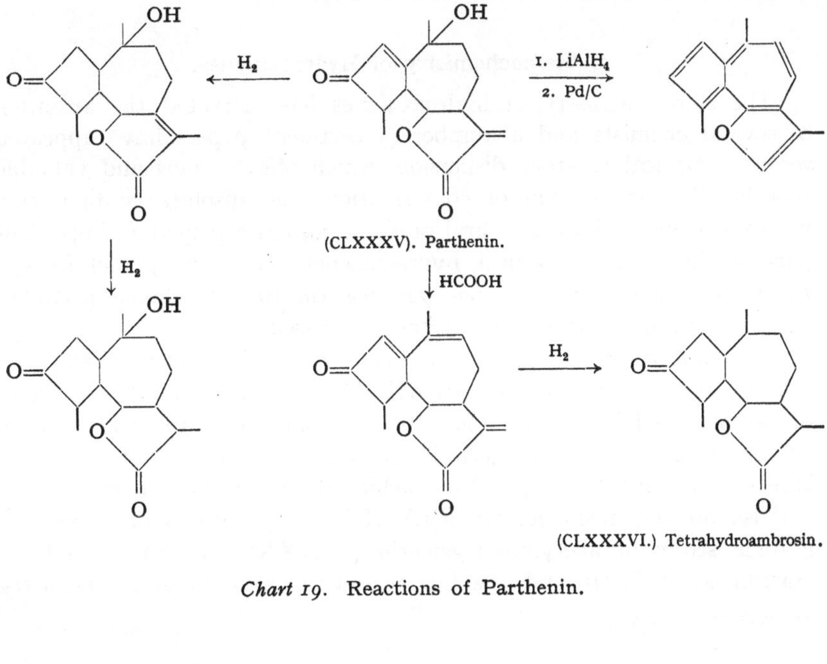

(CLXXXV). Parthenin.

(CLXXXVI.) Tetrahydroambrosin.

Chart 19. Reactions of Parthenin.

group in the cyclopentenone system was deduced from the infrared spectra of dihydro- and tetrahydroparthenin and from the NMR spectrum of parthenin. The presence of an exocyclic double bond, conjugated with the lactone carbonyl, was indicated by the formation of formaldehyde on ozonolysis, and also by its migration to an endocyclic position during hydrogenation as in the case of ambrosin (*58, 206*). Furthermore, compound (CLXXXVI) was identified with tetrahydroambrosin, showing the correctness of the ambrosin structure (*58*).

i) Dehydrocostus Lactone.

This lactone was first isolated by UKITA (cf. *25*) from the oil of costus (*Saussurea lappa*) and also by CRABALONA and by NAVES (cf. *25*). Later, ROMAŇUK and his colleagues (*198*) secured the structure (CLXXXVII). Dehydrocostus lactone consumes three moles of hydrogen to give a hexahydro compound, and quantitative ozonolysis indicates the presence of three exocyclic double bonds and no other unsaturation. Dehydrogenation of dehydrocostus lactone gives, in a low yield, chamazulene, whereas the same treatment of the hexahydro compound leads to a mixture of several azulenes, including guaiazulene. The position of the lactonic oxygen was deduced from the resemblance of the infrared spectra of the hexahydro compound and deoxodihydro-carpesia lactone [NAITO (*159*)].

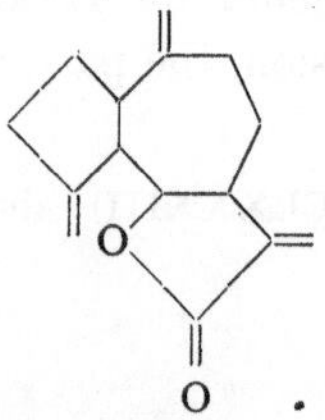

(CLXXXVII.)
Dehydrocostus lactone.

3. Stereochemistry of Hydroazulenes.

The stereochemistry of hydroazulenes has attracted the attention of several chemists and a number of pertinent papers have appeared recently. Optical rotatory dispersion, which offers a new and valuable tool for the clarification of conformation and absolute configuration problems connected with natural products, has also played an important part in the field of natural hydroazulenes [cf., e. g., guaiol (p. 53), *Helenium* guaianolides (pp. 67—72), geigerin (p. 72)]. Some pertinent chemical reactions have already been discussed.

It is known from rotatory dispersion data [DJERASSI et al. (*87*)] that isophotosantonic lactone (CLXXXIX) (*53*), an irradiation product of santonin (CLXXXVIII) (*69*), has the same absolute configuration at the ring juncture as geigerin (CLXXVc, p. 73). Quite recently, BARTON and PINHEY (*54a*) have achieved the correlation of geigerin and isophotoartemisin acetate (CLXXIX, p. 74). An X-ray crystallographic study of acetylbromogeigerin (CLXXX) was carried out by HAMILTON, McPHAIL and SIM (*115a*) and revealed the stereochemistry of all the substituents at $C_{(6)}$, $C_{(7)}$, $C_{(8)}$ and $C_{(10)}$. Thus, geigerin is the first hydroazulenic sesquiterpene whose absolute configuration has been established rigorously.

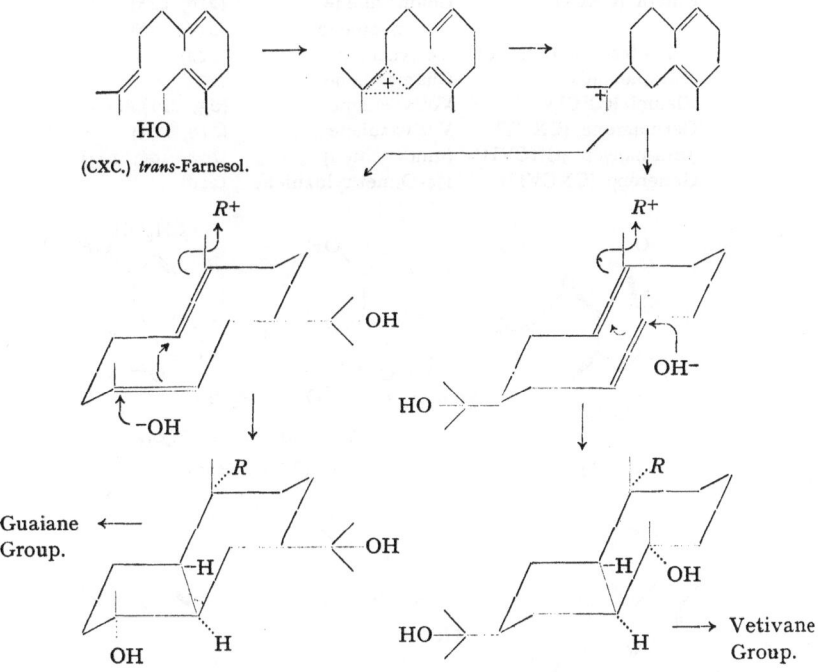

(CLXXXVIII.) Santonin.   (CLXXXIX.) Isophotosantonic lactone.

BARTON has conducted photochemical reactions using all isomeric santonins (48), whereby the substituents of the seven-membered ring seem to retain their configurations (47). Accordingly, more correlations of these photolysis products with naturally occurring sesquiterpenoid lactones can be predicted for the near future, viz. the determination of their configurations (e. g. in the parthenin, helenalin and tenulin series), as well as their synthesis starting from suitable santonin derivatives.

(CXC.) trans-Farnesol.

R^+

OH

−OH

R^+

OH

OH−

HO

Guaiane ←
Group.

R

−H

OH   H

OH

R

−H   OH

HO

H

→ Vetivane
Group.

HO

Chart 20. Biogenesis of Hydroazulenes.

β-Configuration of the substituents at $C_{(7)}$ was suggested by Hendrickson (*116*) from biogenetic considerations. He has demonstrated ingeniously the biogenetic path leading from *trans*-farnesol (CXC) to cyclic sesquiterpenoids; he made use of some stereochemical concepts developed by Ruzicka and his school (*96a, 198a, 198b*) and by Barton (*2*) (*Chart 20*, p. 77).

4. Azulene Precursors of Some Other Types.

Azulene formation from natural hydroazulenes is very helpful in pertinent structural studies, because it clarifies the arrangement of almost all carbon atoms. In contrast, dehydrogenation, a drastic process, may cause skeletal rearrangements and may mislead the chemist as has been the case in the study of jatamanson (*104, 106, 144*) and of germacrone (*119, 183, 226*), both of which were considered erroneously as hydroazulenes for some time. Some other examples have already been mentioned.

A few compounds are listed below which do not possess hydroazulenoid structures and nevertheless yield azulenes on dehydrogenation.

Natural Product	Azulene	References
Pyrethrosin (CXCI)	(unidentified)	(*49, 52*)
Cnicin (CXCII)	Chamazulene	(*216, 218*)
	Artemazulene	(*216, 218*)
Aristolactone (CXCIII)	Vetivazulene	(*212*)
Ketoplenoids	Chamazulene	(*123*)
Elemol (CXCIV)	Vetivazulene	(*25, 221*)
Germacrone (CXCV)	Vetivazulene	(*119, 183*)
Jatamanson (CXCVI)	(unidentified)	(*104, 106, 144*)
Geijerene (CXCVII)	1,4-Dimethylazulene	(*220*)

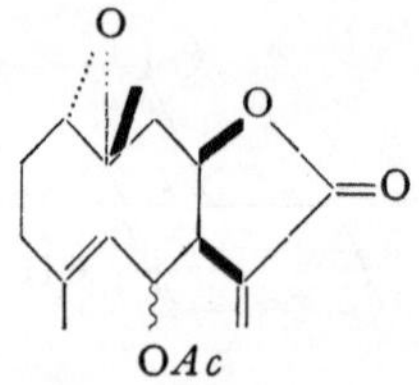

(CXCI.) Pyrethrosin.

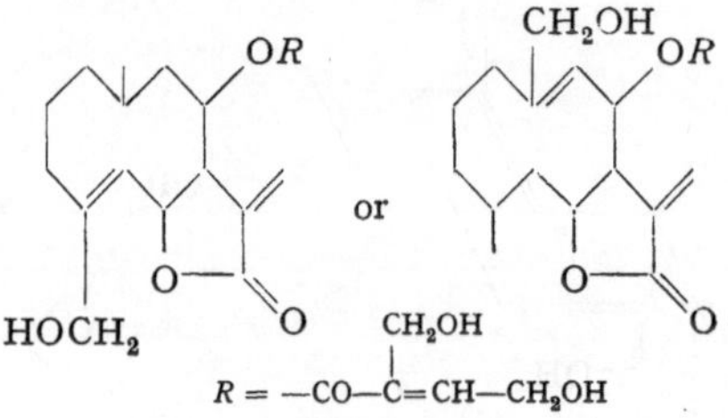

$R = -CO-\underset{\underset{CH_2OH}{|}}{C}=CH-CH_2OH$

(CXCII.) Cnicin.

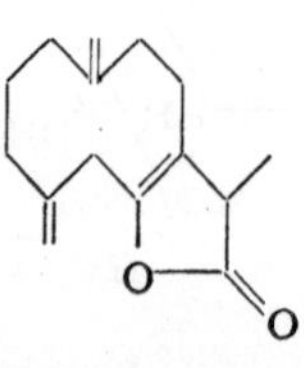

(CXCIII.) Aristolacton. 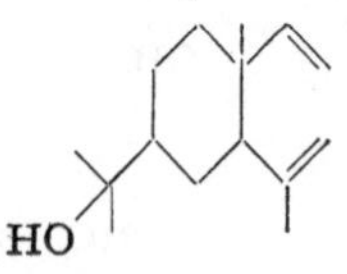 (CXCIV.) Elemol. 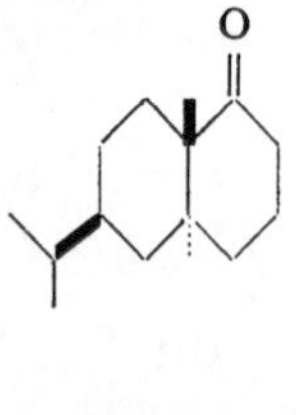 (CXCVI.) Jatamanson.

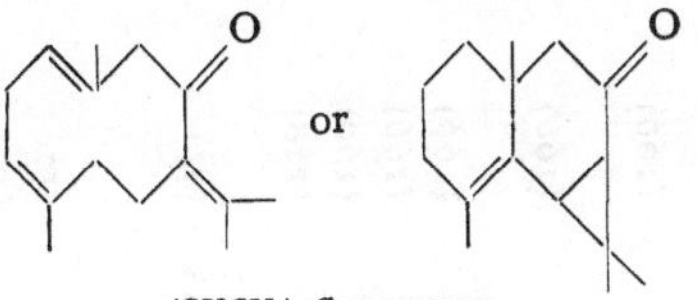

(CXCV.) Germacrone. 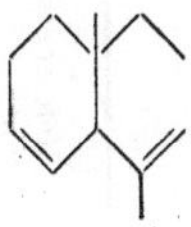   (CXCVII.) Geijerene.

Addendum.

Since the completion of the manuscript, among others, the following data have become available.

Alkylazulene was prepared from cyclopentadiene and tropylium iron carbonyl (*184b*). A detailed paper appeared on the infrared spectra of azulenes (*132a*). Cyanin dyes (*69a*) were obtained from 2-methyl-4(azulenyl-1)-thiazole, and also azulene-pyrrole-methin dyes (*237b*) from azulene and pyrrole aldehyde or pyrrole and formylazulene.

TAKEDA and MINATO (*223a*) have confirmed the configuration of guaiol by preparing optically active (+)-α-methylglutaric acid and (—)-γ-methylbutylolactone by a series of oxidation processes.

ROMAŇUK and HEROUT (*197a*) have isolated several bicyclic hydrocarbons containing the vetivane skeleton (LXXXIX, p. 52). The pertinent structures were assigned on the basis of ozonolysis and catalytic reduction experiments as well as spectral measurements.

A new prochamazulene, globicin, was isolated by ŠORM et al. (*76a*) from *Matricaria globifera* together with arborescine (*153*). Globicin possesses, besides γ-lactone grouping, a double bond, an acetoxyl group and an oxidic ring, which probably occupies a position allylic to the double bond. A partial structure was proposed (*76 a*).

IV. Tables.

Table 1. Azulene Derivatives (since 1958).

(Abbreviations, see next page.)

Compound	m. p. (°)	Color	Derivative m. p. (°)	Spectral maxima (mμ)	References
(a) *Azulene Derivatives:*					
Azulene					
1-Acetamido-3-S-acetylthio-....	192–193	green		608	(40)
2-Acetamido-1,3-bis-dimethyl-aminomethyl-....	125–126	blue violet		680, 620, 575; IR	(166)
2-Acetamido-1-dimethylamino-methyl-....	96–97	violet	MeI > 210*	660, 590, 555; IR	(166)
2-Acetamido-1-morpholino-methyl-....	169–170	blue violet		650, 590, 555; IR	(166)
2-Acetamido-1-piperidinomethyl-....	119–120	blue violet		660, 590, 556; IR	(166)
2-Acetamido-1-p-tolylazo-....	182	black violet			(169)
5-Acetyl-2-amino-....	166.5	violet red	Ac 177, Pic 167.5	490; IR	(178)
5-Acetyl-1,3-dichloro-....	95, 103–104 (double m. p)	green		769, 738, 687, 664, 655, 627, 606, 578; IR	(41)
5-Acetyl-1,3-dicyano-2-hydroxy-....	223–225	light brown	Ethylene ketal 239	503; IR	(178)
1-p-Acetylphenylazo-....	118	blue		655, 625, 596	(101)
1-S-Acetylthio-....	oil	blue		682, 658, 618, 591, 572; IR	(40)
1-S-Acetylthio-3-bromo-....	oil	blue		713, 643, 595; IR	(40)
1-S-Acetylthio-3-nitro-....	185–186.5*	red		519, 403	(40)
1-Amino-....	94–95	dark green	Ac 153–154	719; IR	(200)
4-Amino-....	cryst.	bright red	TNB 218*	(curve)	(196)

Compound	M.p.	Color	Derivative	Absorption maxima (mμ)	Ref.
2-Amino-6-bromo-	135	red violet	Ac 186		(169)
2-Amino-5-chloro-	86–87	crimson	Pic 182		(175)
2-Amino-1,3-dicyano-	260	orange yellow	Ac 170–171*	420	(171)
2-Anilino-	144	red			(173)
1-α-Anthranylazo-	185–188		TNB 166–167	(curve)	(99)
1-β-Anthranylazo-	247–248			(curve)	(99)
1-Benzoyl-	117–118	violet red	Ox 156–157	648, 586, 543	(38, 111)
1-Benzoyloxy- (?)	oil	blue		735, 679, 662, 631, 605, 580, 558, 537; IR	(38)
1,3-Bis-dimethylaminomethyl-	oil	blue	MeI > 200*		(166)
5-Bromo-	57–58	blue		602	(169)
1-Bromomethyl-3-cyano-	169–170	violet		702, 672, 633, 609, 582	(232)
1-o-Bromophenylazo-	105			658, 630, 600, 576, 553	(102)
1-p-Bromophenylazo-	116			660, 631, 602, 577, 554	(102)
1-Bromo-3-phenylazo-	114–115	green black		629	(42)
1-Bromo-3-thiocyano-	100–101*	dark green		683, 618, 595, 572	(40)
1-p-Carboxyphenylazo-	220*			660, 629, 600	(101)
5-Chloro-	68	blue	TNB 105–106		(175)
2-Chloro-1-dimethylamino-methyl-	oil	blue violet	Pic 151–152; MeI > 180*	670, 600, 565	(166)
2-Chloro-1,3-bis-dimethylamino-methyl-	oil	blue	MeI > 215*	680, 620, 575	(166)
2-Chloro-1,3-dimorpholino-methyl-	116–117	blue		680, 615, 575	(166)
2-Chloro-1-formyl-	132–133	red	Ox 147–148	602, 550, 517	(173)
1-Chloro-3-nitro-	185–186	red brown		666, 603, 560	(40)
2-Chloro-1-morpholinomethyl-	97–98	blue violet		665, 605, 565; IR	(166)

Abbreviations: * = Point of decomposition, TNB = 1,3,5-trinitrobenzene complex, TNT = 2,4,6-trinitrotoluene complex, Pic = picrate (for Table 1); picryl anion (for Table 2), Ac = acetate, Ox = oxime, Semi = semicarbazone, DNP = 2,4-dinitrophenylhydrazone, MeI = methyliodide, MeCl = methylchloride.

(Table 1, continued.)

Compound	m.p. (°)	Color	Derivative m.p. (°)	Spectral maxima (mμ)	References
Azulene					
1-o-Chlorophenylazo-	106			658, 630, 600, 576, 553	(102)
1-m-Chlorophenylazo-	105			659, 630, 601, 576, 522	(102)
1-p-Chlorophenylazo-	129			660, 631, 602, 577, 554	(102)
1-Chloro-3-phenylazo-	115–116	green		634, 426	(42)
2-Chloro-1-piperidinomethyl-	98–99	blue violet		665, 605, 566; IR	(166)
1-Chloro-3-trifluoroacetyl-	110–111	red		536	(37)
2-Cyano-	78	blue		697, 679, 660, 635, 621, 586	(173)
2-Cyano-1-formyl-	182	blue violet		645, 629, 583, 543	(173)
1-p-Cyanophenylazo-	192			649, 620, 590	(101)
2,5-Dichloro-	61–62	violet		679, 664, 614, 571; IR	(175)
1,3-Dicyano-	245–260 (sublimes)	red		560, 523	(232)
1,3-Dideuterio-	194–195	red		(curve); IR	(56)
1,3-Diformyl-	>100	deep blue green	diOx 206*	548, 508	(111, 235)
1,3-Diiodo-	99	orange yellow		752, 674, 616	(42)
2-Dimethylamino-	125			418, 398, 380	(173)
1-p-Dimethylaminophenylazo-	88–89	blue	MeCl 189	708, 642, 614, 588	(102, 103)
1,3-Dimorpholinomethyl-	158–159	deep maroon			(166)
2,4-Dinitrophenylthio-	149–150	rose red		658, 590, 548	(40)
1,3-Dithiocyano-		red		535	(40)
2-Ethoxy-	oil	violet		(curve)	(173)
4-Ethoxy-	117		TNB 143	654, 625, 595, 572	(196)
1-p-Ethoxycarbonylphenylazo-	oil	blue			(101)
2-Ethylthiomethyl-	oil				(166)
1-Formyl-	oil	red violet	Ox { α 121–122, β 135.5–136.5, Semi 217* }	649, 589, 542	(111, 235)
2-Hydrazino-	128	red			(173)

	m.p.	color	derivative		references
2-Hydroxymethyl-	115–116	violet	TNB 132–134	677, 617, 570	(228)
5-Hydroxymethyl-	76–76.5	blue	TNB 117–119	710, 642, 588	(228)
6-Hydroxymethyl-	127–128	blue	TNB 129.5–130.5	690, 625, 580	(196, 228)
1-o-Hydroxyphenylazo-	135			678, 648, 617, 591, 566	(102)
1-m-Hydroxyphenylazo-	156			666, 636, 608, 581, 556	(102)
1-p-Hydroxyphenylazo-	221			680, 650, 619, 593, 567	(102)
2-Iodo-	126	violet		660, 601, 560	(173)
1-o-Iodophenylazo-	108			658, 630, 600, 576, 553	(102)
1-p-Iodophenylazo-	145			660, 631, 602, 577, 554	(102)
1-Iodo-3-phenylazo-	148.5–149.5	turquoise		626	(42)
1-Lauroyl-	27–28	violet	Ox 39–40	653, 592, 546	(111)
2-Methoxy-	82	red		613, 592, 561, 550, 544	(196)
4-Methoxy-	oil	violet	TNB 163	(curve)	(196)
5-Methoxy-	oil	blue	TNB 127–128	(curve)	(175, 196)
1-o-Methoxyphenylazo-	122			680, 650, 618, 592, 567	(102)
1-p-Methoxyphenylazo-	96			683, 652, 620, 594, 569	(102)
1-Morpholinomethyl-	40–41			724, 654, 622, 596	(166)
1-α-Naphthylazo-	189	blue		672, 642, 613, 586, 560	(99)
1-β-Naphthylazo-	170			680, 649, 618, 592	(99)
1-o-Nitrophenylazo-	126			641, 612, 584	(102)
1-m-Nitrophenylazo-	183			647, 618, 591, 566	(102)
1-p-Nitrophenylazo-	204			642, 613, 586	(102)
1-Nitro-3-phenylazo-	153.5–154.5	black		562	(42)
1-Nitro-3-thiocyano-	176–177	orange red		500	(40)
1-p-Phenylazo-phenylazo-	179			672, 644, 612	(101)
1-Phenylazo-3-thiocyano-	139–140*	green		580, 403	(40)
1,2,3,5(1,3,5,6)-Tetrabromo-	158–160	green		628	(240)
1,3,5,7-Tetrabromo-	247	green		680	(240)
1-Thiocyano-	76.5–77.5	purple		653, 594, 567, 552	(40)
1-o-Tolylazo-	80			672, 643, 613, 587, 562	(102)
1-p-Tolylazo-	94			674, 644, 614, 588, 562	(102)
1,3,5-Tribromo-	113–114	green		656	(240)

(Table 1, continued.)

Compound	m.p. (°)	Color	Derivative m.p. (°)	Spectral maxima (mµ)	References
Azulene					
1-Triethylsilyl-	oil	pale brown	TNB 102	717, 644, 597	(231)
1-Trifluoroacetyl-	62.5–63	red		525	(37)
1-β-Triphenylmethylhydrazo-	115–116	blue		700, 662, 632, 606, 580, 558, 535	(38)
N-(2-Azulenyl)-N'-benzoylthiourea	197	red violet		618, 581, 543	(182)
N-(2-Azulenyl)-N'-phenylthiourea	182–183	violet			(182)
1-(2-Azulenyl)thiosemicarbazide	215	brown		507	(182)
N-(2-Azulenyl)thiourea	201–202	red violet		617, 579, 539	(182)
Bis-(3-nitroazulenyl-1)-disulfide	200*	deep maroon		515, 465	(40)
Bis-(3-phenylazoazulenyl)-disulfide	191–193*	blue green		658, 598, 422	(40)
Di-1-azulenyldisulfide	118–119	dark green		690, 623, 575	(40)
1-Acetamido-3-N-methyl-carbamoylazulene	213–214	dark green		590	(39)
1-Acetyl-3-N-methylcarbamoyl-azulene	172	red	Ox 181*	510	(39)
2-Amino-1,3-dicarbamoylazulene	262–263	orange			(174)
2-Amino-1,3-diethoxycarbonyl-5-methoxycarbonylazulene	124	orange red		428	(172)
2-Amino-1,3-diethoxycarbonyl-6-methoxycarbonylazulene	136–137	orange red		427	(172)
2-Amino-1,3,6-triethoxycarbonyl-azulene	124	orange		428	(172)
1-Carboxyazulene	188–190*	red		620, 570, 533	(37, 236)
2-Carboxyazulene	195*, 218–220	green		732, 661, 610	(173, 228)
6-Carboxy-4-ethoxyazulene	234–236*	green			(196)
1-Carboxy-3-ethoxycarbonyl-azulene	240*	red	Ethylene ketal 164*		(169)
5(7)-Acetyl-2-amino-					(178)

Compound	M.p.	Color	Derivative	Absorption (nm); IR	Ref.
2-Amino-	190*	yellow			(171)
2-Anilino-	140*	red			(173)
2-Chloro-	193*	green			(169)
6-Carboxy-4-hydroxyazulene	236*	orange		(curve)	(196)
1-Carboxy-2-methoxy-3-methoxy-carbonylazulene	145*				(173)
1,3-Dicarboxyazulene	204.5*	black	Ethylene ketal 175*	482; IR	(178)
5-Acetyl-2-amino-	> 300				(175)
2-Amino-5-chloro-	> 270				(175)
2,5-Dichloro-					
1,3-Diethoxycarbonylazulene	182–183				
5-Acetamido-2-amino-	117	yellow orange		453; IR	(179)
5-Acetyl-	129–130	red	Ox 162.5–164	540, 508; IR	(178)
6-Acetyl-	124	black		(curve); IR	(178)
5-Acetyl-2-amino-	139–140	orange	Ethylene ketal 113	489; IR	(178)
6-Acetyl-2-amino-	235–235.5	orange	Ethylene ketal 154	414; IR	(178)
6-Amino-	117–118	orange		413; IR	(179)
2-Amino-5-bromo-	162–163	orange red			(169)
2-Amino-6-bromo-	114–115	orange		460, 390; IR	(169)
2-Amino-5-chloro-	149–150.5	orange red		473, 410; IR	(175)
2-Amino-6-chloro-	129	orange		464	(175)
2-Amino-4-cyano-	163	orange			(163)
2-Amino-5-cyano-	202–203	orange		428	(172)
2-Amino-6-cyano-	142	orange red		432	(172)
2-Anilino-	125	orange		420	(173)
5-Bromo-	168	red violet		(curve); IR	(169)
2-p-Bromoanilino-	116	orange			(173)
6-Bromo-2-methylamino-	100	orange			(169)
2-Carboxymethylthio-	78	red		521; IR	(146, 151)
2-Chloro-	130–131	red		502; IR	(171)
5-Chloro-	135–137	red violet		640, 579, 490	(175)
5-Chloro-2-hydrazino-		red orange		(curve)	(175)

(Table 1, continued.)

Compound	m.p. (°)	Color	Derivative m. p. (°)	Spectral maxima (mμ)	References
1,3-Diethoxycarbonylazulene 5-Cyano-	146	red violet	Pic 181	523	(172)
2,6-Diamino-	208	dark yellow	2-Ac 216–217 6-Ac 211 diAc 188–189 2-Benzal 141	467; IR	(179)
2,5-Dichloro-	84–85 106–107 (double m.p.)	red		556, 525	(175)
2-Dimethylamino-	82	red		430	(173)
2-Ethoxy-	46–47	red		516, 478	(173)
2-Ethoxycarbonylmethylthio-	oil	red			(146, 151)
2-Ethylthio-	oil	red			(173)
5-Ethylthio-	96–97	red violet		642, 584, 497	(175)
2-Hydrazino-	oil	orange	Ac 131		(173)
Acetone condensate	134				(173)
2-Methylamino-	62	orange red		405	(173)
1,2-Dimethoxycarbonylazulene	oil	violet		622, 583; IR	(87a, 228)
1,3-Dimethoxycarbonylazulene 5-Chloro-	176–177	red violet		640, 579, 490	(175)
2-Methoxy-	62	orange		517, 479	(173)
5-Methoxy-	185–186	violet		633, 577, 493	(175)
2,5-Dimethoxycarbonylazulene	95–96	blue		705, 638, 589	(228)
2,6-Dimethoxycarbonylazulene	144–144.5	green		802, 718, 556	(228)
1,3-Di(N-methylcarbamoyl)-azulene	227–228	violet		525	(39)
1-Ethoxycarbonylazulene	oil	violet		645, 588, 545	(236)

Compound	m.p.	Color	Derivatives	Spectral	Ref.
5(7)-Acetyl-2-amino-	131–132	orange brown	Ethylene ketal 145	492; IR	*(178)*
5(7)-Acetyl-2-amino-3-cyano-	199–200	orange	Ethylene ketal 182	490, 424; IR	*(178)*
5(7)-Acetyl-3-cyano-2-hydroxy-	204.5–206	yellow orange	Ethylene ketal 220–221	493; IR	*(178)*
6-Acetyl-3-cyano-2-hydroxy-	198	pale red	Ethylene ketal 205–206		*(178)*
2-Acetamido-3-dimethylamino-methyl-	108–109	blue violet		525; IR	*(166)*
2-Acetamido-3-ethylthiomethyl-	140–141	violet		532; IR	*(166)*
2-Acetamido-3-morpholino-methyl-	126–127	purple		528; IR	*(166)*
2-Acetamido-3-piperidinomethyl-	108–109	purple	TNB 114–115	528; IR	*(166)*
2-Anilino-	oil	red	Pic 91–92	405	*(173)*
2-Chloro-	oil	red		600, 550, 525	*(173)*
2-Chloro-3-dimethylamino-methyl	oil	violet	Pic 182–183* MeI 205*		*(166)*
2-Chloro-3-morpholinomethyl-	77–78	purple		630, 568, 534: IR	*(166)*
3-Cyano-2-hydroxy-	189	yellow	Ac 142		*(169, 171)*
2-Dimethylamino-	oil	red	TNB 151		*(173)*
3-Dimethylaminomethyl-	oil	purple	MeI > 197*		*(166)*
2-Ethoxy-	85	orange red		574, 527, 488, 464	*(173)*
3-Ethylthiomethyl-	oil	violet		680, 615, 566	*(166)*
3-Formyl-	79–81	red		603, 550, 510	*(169)*
3-Morpholinomethyl-	oil	violet		670, 602, 560	*(166)*
3-Piperidinomethyl-	oil	violet		673, 609, 563	*(166)*
4-Ethoxycarbonyl-8-methoxy-azulene	oil	blue	TNB 123	(curve)	*(196)*
5-Ethoxycarbonyl-7-methoxy-azulene	oil	blue	TNB 70	(curve)	*(196)*
6-Ethoxycarbonyl-4-methoxy-azulene	63.5–64.5	dark blue	TNB 90–92	(curve)	*(196)*
7-Ethoxycarbonyl-4-methoxy-azulene	76.5–77.5	violet	TNB 102–103	(curve)	*(196)*

(Table 1, continued.)

Compound	m.p. (°)	Color	Derivative m.p. (°)	Spectral maxima (mμ)	References
5-Hydroxymethyl-1-methoxy-carbonylazulene	74.5–75.5	violet	TNB 82–83	645, 592, 540	(228)
7-Hydroxymethyl-1-methoxy-carbonylazulene	111.5–113	violet	TNB 105–106	625, 575, 538	(228)
2-Methoxycarbonylazulene	110	blue		728, 713, 656, 604	(173, 228)
5-Methoxycarbonyl-7-methoxy-azulene	oil	red violet	TNB 118	(curve)	(196)
1-N-Phenylcarbamoylazulene	149–150			548	(236)
N,N'-Bis-1,3-diethoxycarbonyl-2-azulenylhydrazine	164	orange			(173)
2,2'-Bis-(1,3-diethoxycarbonyl-azulenyl)sulfide	189–190	red		550, 430	(173)
1-Oxalylazulene	165	yellow red		532	(228)
1-Methoxalylazulene	72	red		540	(228)
1-Methoxymalonoylazulene	70				(229)
1-o-Ethoxycarbonylheptanoyl-azulene	oil				(111)
Azulenoquinone (VIIIa)	169–170	violet orange	Ox 63–64	653, 592, 546	(115)
(b) Alkyl Derivatives:					
2-Acetamido-1-ethoxycarbonyl-3-methylazulene	117–118	blue violet		540; IR	(166)
2-Acetylethoxycarbonylmethyl-1,3-diethoxycarbonylazulene	54	red		(curve); IR	(173)
1-Acetyl-2-methylazulene			TNB 102–104		(97)
5-Chloro-2-cyanoethoxycarbonyl-methyl-1,3-diethoxycarbonyl-azulene	117	violet			(175)

Compound	M.p. (°C)	Color	TNB	Spectrum; IR	Ref.
2-Cyanoethoxycarbonylmethyl-1,3-diethoxycarbonylazulene	117	red	TNB 163.5–164.5	(curve); IR	(173)
1-Cyano-3-methylazulene	101–102	blue violet		640, 613, 588, 565, 545	(232)
1-Methyl-3-phenylazoazulene	38–40	dark brown		633, 612, 582	(42)
5-Methylazulene					(174)
2-Amino-1,3-dicarbamoyl-	287–290	orange			(34)
2-Amino-1,3-dicyano-	262–263	orange yellow		452; IR	(34)
2-Amino-1,3-diethoxycarbonyl-	110–111	orange yellow			(34)
2-Amino-1(3)-ethoxycarbonyl-	oil	red orange			(34)
2-Chloro-1,3-dicyano-	263–264	red		508; IR	(34)
2-Chloro-1,3-diethoxycarbonyl-	76–77	red		504	(34)
1,3-Dicarboxy-	263	red		IR	(34)
1,3-Dicarboxy-2-hydroxy-	>240	red brown			(34)
1,3-Dicarboxy-2-methoxy-	>240	orange yellow			(34)
1,3-Dicyano-	240	red brown		524; IR	(34)
1,3-Dicyano-2-ethoxy-	193–194.5	red orange			(34)
1,3-Dicyano-2-hydroxy-	259–260	brown		(curve); IR	(34)
1,3-Dicyano-2-mercapto-	>270	orange		(curve)	(34)
1,3-Dicyano-2-methoxy-	211–212	yellow		(curve)	(34)
1,3-Dicyano-2-methylamino-	202–203	red orange		426	(34)
1,3-Dicyano-2-phenoxy-	192–193.5	red		477	(34)
1,3-Diethoxycarbonyl-	70.5–71.5	red		(curve); IR	(34)
1,3-Diethoxycarbonyl-2-methoxy-	67–68	red		(curve)	(34)
1,3-Diethoxycarbonyl-2-phenoxy-	106–108	red		487	(34)
1,3-Dimethoxycarbonyl-2-methoxy-	oil	red			(34)
2-Hydroxy-	300	light purple			(34)
5(7)-Methylazulene					
2-Amino-1-carbamoyl-3-carboxy-	200.5–202	yellow			(34)
	215–217.5	yellow			(34)

(*Table 1, continued.*)

Compound	m. p. (°)	Color	Derivative m. p. (°)	Spectral maxima (mμ)	References
5(7)-Methylazulene					
2-Amino-1-carbamoyl-3-cyano-	248–248.5	brown yellow		(curve); IR	(34)
2-Amino-1-carboxy-3-cyano-	170.5–171	brown yellow		(curve)	(34)
	172–174	dark red		(curve)	(34)
2-Amino-1-carboxy-3-ethoxycarbonyl-	130–132	yellow			(34)
2-Amino-1-cyano-3-ethoxycarbonyl-	215–216	yellow			(34)
	174–175	yellow			(34)
2-Amino-1-cyano-3-methoxycarbonyl-	222–224	yellow			(34)
2-Chloro-1-cyano-3-ethoxycarbonyl-	192–194	red		508	(34)
	125–126.5	red		504	(34)
6-Methylazulene					
2-Amino-1,3-dicyano-	> 300	orange yellow		455	(34)
2-Amino-1,3-diethoxycarbonyl-	164	orange yellow		(curve); IR	(34)
2-Chloro-1,3-dicyano-	247–248	red		505	(34)
1,3-Dicarboxy-	250	red			(34)
1,3-Dicyano-	198–200	yellow		450	(34)
1,3-Diethoxycarbonyl-	69	red			(34)
2-Hydroxy-1,3-dicyano-	195				(34)
2-Methoxy-1,3-dicyano-					(34)
1-Ethylazulene	oil	deep blue	TNB 112–113.5 Pic 92–94	740, 668, 610	(193)
1-(β-Acetyl-β-ethoxycarbonylethyl)-2-acetamido-3-ethoxycarbonyl-azulene	oil	red			(166)

Compound	m.p. (°C) / oil	Color	Derivative	Absorption maxima	Ref.
1-(β-Acetamido-β,β-diethoxy-carbonylethyl)-azulene	90–91	blue		718, 653, 624, 595	(166)
2-Acetamido-	151–152	blue violet		660, 600, 557; IR	(166)
2-Acetamido-3-ethoxycarbonyl-	244–245	violet		IR	(166)
2-Chloro-	125–126	blue violet		665, 605, 566; IR	(166)
2-Chloro-3-ethoxycarbonyl-	150–151	red violet		630, 575, 540	(166)
3-Ethoxycarbonyl-	151–152	violet		680, 613, 585, 564	(166)
1-(β,β-Diethoxycarbonylethyl)-3-ethoxycarbonylazulene	oil	violet		674, 610, 560	(166)
2-Acetamido-	94–95	red		526; IR	(166)
5-Ethylazulene	oil	blue	TNB 106–106.5 Pic 89.5–91	720, 679, 650, 618, 594, 570	(181)
2-Amino-1,3-diethoxycarbonyl-	87.5–88	orange		449	(181)
1,3-Dicarboxy-	> 300	dark red		513	(181)
1,3-Dicyano-2-hydroxy-	222–223	orange		461, 412	(181)
1,3-Diethoxycarbonyl-	79–81.5	red		608, 556, 520	(181)
2-Amino-1-cyano-3-ethoxycarbonyl-5(7)-ethylazulene	130–132	orange		448	(181)
2-Acetylethoxycarbonylmethyl-1,3-dicyano-5-methylazulene	202–204	red		(curve)	(34)
2-Cyanoethoxycarbonylmethyl-1,3-dicyano-5-methylazulene	222–224	red		507	(34)
1,3-Dicyano-2-diethoxycarbonyl-methyl-5-methylazulene	230–238	red		505	(34)
1,3-Diethoxycarbonyl-2-diethoxy-carbonylmethyl-5-methylazulene	75–78	brown violet		(curve)	(34)
1-Cyano-4,7-dimethylazulene	132.5–133.5	dark red brown		660, 598, 554	(232)
1-Formyl-4,7-dimethylazulene	47–48	red		633, 580, 541	(235)
1-Cyano-4,8-dimethylazulene	105–106			632, 608, 581, 555, 537	(232)
1-Ethoxycarbonyl-4,8-dimethyl-azulene	oil	red violet		537	(236)

(Table 1, continued.)

Compound	m.p. (°)	Color	Derivative m.p. (°)	Spectral maxima (mμ)	References
1-Formyl-4,8-dimethylazulene	79–80	red	Ox 169–172	533	(235)
5-Isopropylazulene					
2-Amino-5-bromo-1,3-diethoxy-carbonyl-	94	orange	Pic 136–137	462; IR	(43, 167)
2-Amino-1,3-dicyano-	198–199	orange		455	(43, 167)
2-Amino-1,3-diethoxycarbonyl-	oil	green		450	(43, 167)
1(3)-Bromo-	247	red		634	(43, 167)
6(8)-Bromo-1,3-dicyano-2-hydroxy-	oil	red	Ac 195–196	(curve); IR	(43, 167)
2-Carboxymethylthio-1,3-diethoxycarbonyl-					
2-Chloro-	oil	violet		(curve)	(148)
2-Chloro-1(3)-cyano-	oil	red		492	(43, 167)
2-Chloro-1,3-dicarboxy-	250	pale red		502	(43, 167)
2-Chloro-1,3-dicyano-	163	red leaflets		528, 510	(43, 167)
2-Chloro-1,3-diethoxycarbonyl-	< 30	red		680, 615, 585, 565	(43, 167)
1-Cyano-	174–175.5	blue violet		670, 600, 546	(43, 167)
3-Cyano-	116.5–117	blue violet		690, 638	(43, 167)
1,3-Dibromo-					
6,8-Dibromo-2-hydroxy-1,3-dicyano-	262–263	dark green	Ac 216–218	483, 428; IR	(43, 167)
1,3-Dicyano-2-dimethylamino-	162–163	orange		490	(43, 167)
1,3-Dicyano-2-ethoxy-	171–172	orange			(43, 167)
1,3-Dicyano-2-hydrazino-	166–168	orange			(43, 167)
Acetone condensate	181–182	orange			(43, 167)
1,3-Dicyano-2-hydroxy-	208–209	orange	Ac 163–165	456, 410; IR	(43, 167)
1,3-Dicyano-2-methoxy-	167–168	red			(43, 167)
1,3-Dicyano-2-p-toluidino-	200–201	orange			(43, 167)
1,3-Diethoxycarbonyl-	61	deep red		553, 520	(43, 167)

Compound	M.p.	Colour	Deriv.	Absorption; IR	Ref.
1,3-Diethoxycarbonyl-2-dimethylamino-	140–141	red orange			(43, 167)
1,3-Diethoxycarbonyl-2-ethoxy-carbonylmethylthio-	oil	red			(148)
1,3-Diethoxycarbonyl-2-hydrazino-	oil	red			(43, 167)
Acetone condensate	115–116				(43, 167)
1(3)-Formyl-	oil	red	TNB 106–107.5	(curve); IR	(43, 167)
5(7)-Isopropylazulene					
2-Amino-1-ethoxycarbonyl-3-cyano-	115	orange		440; IR	(43, 167)
	173	orange		440; IR	(43, 167)
1-Bromo-3-ethoxycarbonyl-	77	red violet		680, 618, 571	(43, 167)
	83	blue violet		690, 619, 575	(43, 167)
1-Carboxy-2-chloro-3-cyano-	193–194	yellow brown		470	(43, 167)
1-Carboxy-2-chloro-3-ethoxy-carbonyl-	187.5	red		498	(43, 167)
	166–167	orange red		502	(43, 167)
	223	red brown		525	(43, 167)
1-Carboxy-3-cyano-	240–244	red		533	(43, 167)
1-Carboxy-3-ethoxycarbonyl-	184–185	red orange		514	(43, 167)
	222–222.5	red orange		510	(43, 167)
2-Chloro-1-ethoxycarbonyl-	oil	red	Pic 118–119	610, 545, 526	(43, 167)
	64	red			(43, 167)
2-Chloro-1-ethoxycarbonyl-3-cyano-	171–173	red violet		538, 512	(43, 167)
	131–132	deep red		590, 543, 510	(43, 167)
1-Ethoxycarbonyl-	oil	red violet		(curve)	(43, 167)
	63	red violet		533	(43, 167)
1-Ethoxycarbonyl-3-cyano-	122–123	red violet		533	(43, 167)

(Table 1, continued.)

Compound	m. p. (°)	Color	Derivative m. p. (°)	Spectral maxima (mμ)	References
5(7)-Isopropylazulene					
1-Ethoxycarbonyl-3-cyano-2-hydroxy-	152	orange		460; IR	(43, 167)
	200	orange		455; IR	(43, 167)
1-Ethoxycarbonyl-3-formyl-	oil				(43, 167)
1,3-Dicarboxy-5-β-hydroxyisopropylazulene	oil				(43, 167)
1,3-Diethoxycarbonyl-5-α,β-dihydroxyisopropylazulene	ca. 240*				(170)
2-Amino-	oil	red			(170)
1,3-Diethoxycarbonyl-5-β-hydroxyisopropylazulene	120	orange			(170)
2-Amino-	118		Ac oil		(170)
2-Chloro-	oil		Ac 91		(170)
5-β-Hydroxyisopropylazulene	oil	blue	Ac 70		(170)
6-Isopropylazulene					(170)
2-Carboxymethylthio-1,3-diethoxycarbonyl-	oil	red			(149)
1-Cyano-3-ethoxycarbonyl-2-hydroxy-	230	yellow		(curve)	(167, 168)
2-Chloro-1,3-dicarboxy-	200	pale red		(curve)	(167, 168)
1,3-Dicyano-2-hydroxy-	133–134	yellow			(167, 168)
1,3-Diethoxycarbonyl-2-ethoxycarbonylmethylthio-	oil	red			(167, 168)
3-Cyano-1,4,7-trimethylazulene	146–146.5	dark blue		698, 632, 610, 582	(232)
3-Formyl-1,4,7-trimethylazulene	148–149	dark blue violet		683, 615, 573	(235)

	m.p.	Color	TNB	Absorption bands	Ref.
2,4,8-Trimethylazulene					
1-Acetamido-	229–230	blue violet		550	(232)
1-Cyano-	115–116	red		615, 598, 550, 530, 520	(232)
1-Ethoxycarbonyl-	oil	red violet		565, 535	(236)
1-Formyl-	76–77.5	red		550, 520	(235)
1-Nitro-	84–85	red		523	(232)
1-Benzoyl-4,6,8-trimethylazulene	113–114	red		518	(111)
1-Hydroxymethyl-4,6,8-trimethyl-azulene	107–108*		TNB 144–145	647, 589, 548	(111)
1-Nitro-4,6,8-trimethylazulene	95	red		516, 397	(111)
1-tert.-Butylazulene	oil	blue violet	TNB 123.5–124.5	742, 710, 668, 659, 637, 627, 608, 583, 565, 535	(124)
2-Isopropyl-4-methylazulene	oil	violet	TNB 129.5–131	660, 630, 625, 598, 580, 566, 556	(125)
3-Isopropyl-4-methylazulene	oil	deep blue	TNB 109–112	721, 680, 653, 635, 623, 596, 576, 555, 535	(125)
1,3-Diethoxycarbonyl-2-ethoxy-carbonylmethyl-5-isopropyl-azulene	72	red		488	(43, 167)
2-Amino-1-cyano-3-ethoxy-carbonyl-7-isopropyl-4-methyl-azulene	139–140	orange		444; IR	(43, 167)
(2-Amino-3-cyano-1-ethoxy-carbonyl-7-isopropyl-4-methyl-azulene)					
1,3-Dicyano-2-hydroxy-7-iso-propyl-4-methylazulene	258–260*	orange		456, 406; IR	(43, 167)
3-Acetyl-7-ethyl-1,4-dimethyl-azulene	36	dark violet	TNB 123		(217)
7-α-Carboxyethyl-1,4-dimethyl-azulene (Chamazulene carboxylic acid)	86–88	blue	TNB 150–152		(210, 211)

(Table 1, continued.)

Compound	m. p. (°)	Color	Derivative m. p. (°)	Spectral maxima (mμ)	References
7-α-Methoxycarbonylethyl-1,4-dimethylazulene	52	dark blue	TNB 122		(210, 211)
2,4,6,8-Tetramethylazulene	100–101	violet	TNB 211–212	626, 608, 558, 533	(112)
2-Isopropyl-4,8-dimethylazulene (Vetivazulene)					
1-Acetamido-	207–208	violet		548	(232)
1-Carbamoyl-	134–135	violet		575, 542	(236)
1-Cyano-	82.5–83.5	bright red		550, 523	(232)
1-Ethoxycarbonyl-	oil	red violet		570, 538	(236)
1-Methoxycarbonyl-	oil	red			(228)
1-Nitro-	95–96	brown red		528	(232)
1-N-Phenylcarbamoyl-	220–221	red violet		539	(236)
7-Isopropyl-1,4-dimethylazulene (Guaiazulene)					
3-Acetyl-	85–86.5	violet	TNB 124 Ox 140–141.5 DNP 109–110.5	580	(156, 194, 242)
3-α-Anthranylazo-	168–170			(curve)	(99)
3-β-Anthranylazo-	195–196			(curve)	(99)
3-Benzylaminomethyl-	oil	blue	HCl salt 130*	640, 586	(154)
3-Carbamoyl-	242–243	blue		590	(236)
3-Chloro-	oil	blue	TNB 123–124	629	(156)
3-Cyano-	72–73	blue violet		690, 625, 580; IR	(194, 232)
3-Cyclohexylaminomethyl-	solid	blue	HCl salt 180*	640, 590	(154)
3-Ethoxycarbonyl-	oil (52–53)	blue violet		578	(236)

References, pp. 107–119.

Compound	m.p.	Colour	Derivatives / notes	Spectra	Ref.
3-Formyl-	85–86	violet black	TNB 92–93, TNT 60–61*, Ox { α 128–129, β 162, Semi 191–192, DNP 285*	680, 604, 570, 548; IR	(111, 155, 194, 235, 239)
Condensate with benzylamine	68–69	blue		645, 604, 485; IR	(154)
with cyclohexylamine	121–122	blue		655, 604, 438; IR	(154)
with β-hydroxy-tert.-butylamine	126–127*	blue		660, 604, 485; IR	(154)
5-Formyl- (?)	124.5–125.5	red	TNB 86–87, Ox 155–156	610, 556, 519; IR	(155, 239)
3-Hydroxymethyl-	oil	blue		622	(228)
3-β-Hydroxy-α,α-dimethyl-ethyl-aminomethyl-	oil	blue	HCl salt 157–158.5*	640, 590	(154)
3-Methoxalyl-	97.5–98	red		(curve)	(194)
3-Methoxycarbonyl-	oil	violet		572	(228)
3-α-Naphthylazo-	139			(curve)	(99)
3-β-Naphthylazo-	123			(curve)	(99)
3-Oxalyl-	115–116	red brown		(curve)	(194)
3-N-Phenylcarbamoyl-	179–180	blue violet		590	(236)
7-β-Hydroxyisopropyl-1,4-dimethylazulene	oil	dark blue	TNB 149–150, Phenylurethane 96–98, Tosylate oil		(161, 210, 211)
7-Isopropyl-2,4-dimethylazulene (Isoguaiazulene, Se-Guaiazulene)	196–197	violet		539	(236)
1-Carbamoyl-	170–172	violet		523	(228)
1-Carboxy-	198–199*	red		518	(236)
1-Cyano-	148–149	red		562, 535	(232)
1,3-Dicyano-	214–215	red		609, 594, 557, 545, 515	(232)

(Table 1, continued.)

Compound	m. p. (°)	Color	Derivative m. p. (°)	Spectral maxima (mμ)	References
7-Isopropyl-2,4-dimethylazulene (Isoguaiazulene, Se-Guaiazulene)					
1-Ethoxycarbonyl-	oil	red violet	diOx 206–207	555, 527	(236)
1,3-Diformyl-	137–138	dark red	Ox 152–153	496	(235)
1-Formyl-	124–125	red	Semi 166–167	522	(235)
1-ω-Hydroxyacetyl-	109–110	dark red	TNB 158–162	545, 518; IR	(228)
1-Methoxalyl-	72.4–73.5	red brown	TNB 121.5	545, 513; IR	(228)
1-Methoxycarbonyl-	oil	violet			(228)
1-Oxalyl-	128–130	red brown		513	(194, 228)
1-N-Phenylcarbamoyl-	148–149	red violet		570, 539, 535	(236)
2-Isopropyl-1,7-dimethylazulene		blue	TNB 174	622, 592, 532	(105)
2-Isopropyl-4,6-dimethylazulene		red violet	TNB 176	582, 565, 545	(105)
2-Isopropyl-4,7-dimethylazulene		blue violet	TNB 184	665, 602, 565	(105)
4-Isopropyl-2,7-dimethylazulene (?) (Jatazulene)		deep blue violet	TNB 126–128	582, 562	(105)
1-Cyano-6-isopropyl-2,4-dimethyl-azulene	91–92	red		612, 596, 556, 542, 520	(232)
1-Formyl-6-isopropyl-2,4-dimethyl-azulene	125–125.5	bright red		508	(235)
8-Isopropyl-2,4-dimethylazulene (Zierazulene)	oil	deep violet	TNB 122.5; Pic 118	632, 582, 545; IR	(62, 70, 83)
1-α-Hydroxyethyl-4,6,8-trimethyl-azulene	87–88	violet	TNB 153	653, 556 (curve); IR	(111)
7-Ethyl-1,2,4-trimethylazulene			TNB 126	(curve); IR	(161, 215)
7-sec.-Butyl-1,4-dimethylazulene					(217)
6-tert.-Butyl-4,8-dimethylazulene	33–34	violet black	TNB 160–161	644, 585, 562, 546	(112)
7-Isopropyl-1,2,4-trimethylazulene	44–45	blue	TNB 169–170	588	(235)
3-Cyano-	146–147	violet		600, 568	(232)

	M.p.	Colour	Derivative	Absorption max.	Ref.
3-Formyl-.............................	111—112	dark brown	Ox 154—155	563	(235)
7-Isopropyl-1-α-hydroxy-α-methoxycarbonylmethyl-2,4-dimethylazulene...	oil	violet blue	TNB 126—127	578, 562; IR	(228)
2-Isopropyl-1,4,8-trimethylazulene	79—79.5	dark blue	TNB 155.5—156	577	(235)
3-Acetamido-.........................	110—111	brown		590, 555	(232)
3-Cyano-.............................	oil	red brown		561	(232)
3-Formyl-.............................	oil	blue	TNB 148—149	622	(235)
7-Isopropyl-1,3,4-trimethylazulene	155—157	red brown	TNB 155	622, 570, 534	(235, 242)
8-Isopropyl-2,4,6-trimethylazulene			Pic 138—140		(62, 83)
4-Isopropyl-2,6,7-trimethyl-azulene (?).........................			TNB 142—143	572, 552	(106)
2,7-Diethyl-1,4-dimethylazulene...		blue	TNB 133	(curve); IR	(161, 215, 217)
3,7-Diethyl-1,4-dimethylazulene...		blue	TNB 148	(curve)	(217)
7-Isopropyl-3-α-hydroxyethyl-1,4-dimethylazulene.................	70—71	blue		740, 660, 644, 611	(155)
7-Isopropyl-3-α,β-dihydroxyethyl-1,4-dimethylazulene...............	116—117				(194)
7-Isopropyl-1-α,β-dihydroxyethyl-2,4-dimethylazulene...............	112—113				(228)
3-Ethyl-7-isopropyl-1,4-dimethyl-azulene.........................	oil	dark blue	TNB 114—116	600, 560; IR	(155)
1,3-Di-tert.-butylazulene (?).......		blue	TNB 141—142	775, 692, 632	(124)
2,7-Diisopropyl-1,4-dimethyl-azulene............................			TNB 101—112	774, 690, 662, 643, 631, 604, 579	(242)
(c) *Alkenyl Derivatives:*					
1-(β-Formylethenyl)azulene	84—85	blue	TNB 155—157	588	(137)
1-(δ-Formylbutadienyl)azulene....	94—95	green			(137)

7*

(Table 1, continued.)

Compound	m. p. (°)	Color	Derivative m. p. (°)	Spectral maxima (mμ)	References
4,6,8-Trimethyl-1-(β-nitroethenyl)-azulene	168–169	black brown		539, 439, 421	(111)
4,6,8-Trimethyl-1-(β-benzoyl-ethenyl)azulene	142–144	dark red		562, 422	(111)
4,6,8-Trimethyl-1-(β,β-dicarboxy-ethenyl)azulene	161–162*	dark red			(111)
3-β-Formylethenyl-7-isopropyl-1,4-dimethylazulene	125	blue			(137)
3-δ-Formylbutadienyl-7-isopropyl-1,4-dimethylazulene	139–140	gray			(137)
7-Isopropyl-1,4-dimethyl-3-vinyl-azulene	oil	blue	TNB 99–100.5		(155)
2-Isopropenyl-7-isopropyl-1,4-dimethylazulene		blue	TNB 74–75	582	(242)
3-Isopropenyl-7-isopropyl-1,4-dimethylazulene	oil	blue	TNB 85–86	620	(242)
(d) Aralkyl Derivatives:					
2-Benzylazulene	46–47	blue	TNB 125–126	677, 650, 614, 605, 590, 570, 564, 523; IR	(225)
5-Benzylazulene	54.5–55.5	blue	TNB 103–105	712, 674, 644, 612, 589, 567, 547, 526; IR	(225)
6-Benzylazulene	99–100	blue	TNB 109–110	687, 649, 622.5, 595, 571.5, 551, 532, 512; IR	(225)
1-α-Methylbenzylazulene	83	blue	TNB 115–116	735, 665, 606, 580	(195)
Pentabenzylazulene	83.5–84		TNB 79–80	702, 675, 633.5, 611.5, 586, 545; IR	(225)
7-Isopropyl-1,4-dimethylazulene 3-Benzyl-				(curve)	(195)

	M.p.	Colour	Absorption maxima	Ref.
3-m-Bromobenzyl-	77–78	blue	(curve)	(195)
3-m-Bromo-α-methylbenzyl-	oil	blue		(195)
3-m-Bromo-α-phenylbenzyl-		blue		(195)
3-α-Cyanobenzyl-	96–97.5	purple		(195)
3-α-Cyano-o-methoxybenzyl-	160	blue		(195)
3-α-Methylbenzyl-	oil	blue		(195)
3-α-Methyl-3',4'-methylene-dioxybenzyl-	oil	blue green	(curve)	(195)
(e) Aryl- and Heterocyclic Derivatives:				
Dibromo-1,2,3-triphenylazulene	258–259	black		(44)
4,8-Dimethyl-6-phenylazulene	100–101	bright brown	663, 604, 564	(112)
8-Methyl-4,6-diphenylazulene	glassy	blue	680, 618, 579	(112)
2-Amino-1,3-diethoxycarbonyl-5-(3'-ethoxycarbonyl-1'-methyl-pyrazol-5'-yl)azulene	126	yellow brown	465	(176)
2-Amino-1,3-diethoxycarbonyl-5-(3'-ethoxycarbonyl-1'-phenyl-pyrazol-5'-yl)azulene	131	brown yellow	468	(176)
1,3-Diethoxycarbonyl-5-(3'-ethoxy-carbonyl-1'-methylpyrazol-5'-yl)-azulene	144	red violet	517, 440	(176)
5-α-Quinolylazulene	oil	blue violet	708, 640, 589	(177)
2-Amino-1,3-diethoxycarbonyl-	125–126	orange	475, 435	(177)
1,3-Diethoxycarbonyl-	166	red violet	485	(177)
(f) Polycyclic Azulenes:				
1,2-Benzazulene				
3-Carboxy-	218–229	green	620, 570, 534, 490	(236)
3-Cyano-	154–155	dark green	742, 720, 663, 642, 599, 552, 510, 495	(232)

(Table 1, continued.)

Compound	m. p. (°)	Color	Derivative m. p. (°)	Spectral maxima (mμ)	References
1,2-Benzazulene					
3-Formyl-	101–102	black green	Ox 172–173	740, 711, 657, 637, 595, 545, 510	(235)
6-Hydroxymethyl-	181–183.5	dark green	TNB 133–134	735, 655, 598	(228)
3-Triethylsilyl-	oil	pale brown	TNB 104–105.5	638, 598	(231)
Dibromo-5,6-benzazulene	360	deep purple	TNB 118–120	585	(36)
Tetrabromo-5,6-benzazulene	224–226	green violet		585, 575, 560, 550, 510–505	(36)
4,6-Dimethyl-(cyclopenteno-1',5',4':1,9,8-azulene) (LXXXVIb)	64	blue	TNB 181–182	678, 647, 612, 587, 561	(113, 114)
2'-Methoxy-	39–40	red violet	TNB 115–116	642, 582, 560, 540	(113, 114)
2'-Dimethylamino-	oil	blue violet		653, 592, 568, 546	(109, 113, 114)
2'-Phenyl-	113–114	blue violet		665, 602, 577, 554	(113, 114)
4,6-Dimethyl-(cyclopentadieno-1',2',3':1,9,8-azulene)	77*	yellow brown		476	(109, 114)
1,1,2,2-Tetracyanoethylene adduct	190			644, 586, 544	(114)
4,6-Dimethyl-(cycloheptatrieno-1',2',3':1,9,8-azulene)	85	red brown	TNB 207–208	1073, 1027, 897, 793, 453, 421	(109, 113, 114)
Maleic anhydride adduct	239	blue		722, 657, 600	(114)
2'(3')-Formyl-	162	yellow brown			(114)
Tetracyclic azulene from colchicine	147–150			953, 849, 760, 462, 438	(114)
Compound (XXIVa)	217–218				(182)
Compound (XXIVb)	197–200				(182)
Compound (XXIVc)					(182)

(g) Azulenes with Condensed Heterocycles:					
3,5,8-Trimethylfuro[3,2-f]azulene (Linderazulene)	106.5	violet	TNB 151–152		(143, 224)
3,6,9-Trimethylfuro[2,3-f]azulene (Artemazulene)	oil	blue	TNB 191–192		(122)
3-Methyl-2,3-dihydrofuro[3,2-f]azulene	oil	violet		648, 615, 585, 540 (curve); IR	(170)
5,7-Diethoxycarbonyl-	86	orange		662, 627, 600, 572, 550, 527	(170)
3-Methylfuro[3,2-f]azulene	56	violet			(170)
5,7-Diethoxycarbonyl-	118	red		571, 522, 488; IR	(170)
5(7)-Ethoxycarbonyl-	85	red violet		(curve)	(170)
3-Methylpyrrolo[3,2-f]azulene	oil	red		(curve)	(170)
5,7-Diethoxycarbonyl-	245	red orange		(curve): IR	(170)
3-Methylpyrazolo[4,5-f]azulene	186–187	dark purple red	Ac 154.5–156	480	(180)
5-Carboxy-7-ethoxycarbonyl-(7-carboxy-5-ethoxycarbonyl-)	247–250*	orange light brown		462	(180)
5,7-Diethoxycarbonyl-	252–253	orange	Ac 121–122.5	463	(180)
5(7)-Ethoxycarbonyl-	202–203	red		485	(180)
1,2,3,4-Tetrahydroindolo[3,2-a]-azulene	143–144	dark green		415	(182)
Thieno[3,2-a]azulene	174	blue green	TNB 165	740, 660, 612, 575; IR	(147, 151)
9-Carbamoyl-	247–248	dark red		568; IR	(150)
9-Cyano-	182–183	dark red		698, 643, 567; IR	(150)
2,9-Diethoxycarbonyl-3-hydroxy-	139	brown		515; IR	(146, 151)
2,9-Diethoxycarbonyl-3-hydroxy-5(7)-isopropyl-	106–115	brown		515; IR	(148)
2,9-Diethoxycarbonyl-3-hydroxy-6-isopropyl	oil	brown	Benzoate 111	515	(149)
2,9-Diethoxycarbonyl-6-isopropyl-3-methoxy-	oil	brown		515	(149)

(Table 1, continued.)

Compound	m.p. (°)	Color	Derivative m.p. (°)	Spectral maxima (mμ)	References
Thieno[3,2-a]azulene					
2,9-Diethoxycarbonyl-5(7)-isopropyl-3-methoxy-	87–89 / oil	brown		515; IR	(148), (148)
9-Ethoxycarbonyl-	oil	dark brown	TNB 148–149	560	(150)
9-Ethoxycarbonyl-3-methoxy-	141	brown		567; IR	(146)
3-Ethyl-	oil	blue green	TNB 130		(146, 151)
2,9-Diformyl-3-methoxy-	oil	brown	diOx > 280; DNP > 280; Ox 194–195		(150)
9-Formyl-	oil	brown			(150)
9-β-Benzoylethenyl-	243–244	brown			(150)
9-β,β-Dicyanoethenyl-	267–268	brown			(150)
2,3-Dihydro-3-hydroxy-	108–112	violet			(151)
9-Hydroxymethyl-	145–146	dark green			(150)
5-Isopropyl-	119–120	blue green	TNB 137–138	762, 681, 617, 568	(148)
6-Isopropyl-	156–157	blue green	TNB 109–110	760, 665, 621, 573; IR	(149)
7-Isopropyl-	oil	blue green	TNB 147–148	608	(148)
5(7)-Isopropyl-3-methoxy-	oil	blue green	TNB 179–184	755, 663, 617, 563; IR	(148)
6-Isopropyl-3-methoxy-	oil	green brown	TNB 150–152	604	(149)
9-Dimethylaminomethyl-	53–54	blue		760, 675, 618, 580	(150)
2,9-Diethoxycarbonyl-3-methoxy-	126	brown		515; IR	(146)
3-Methoxy-	103	blue green		718, 661, 608, 568	(146)
3-Methyl-	77–78	blue green	TNB 163	750, 672, 612, 578	(147, 151)
9-Morpholinomethyl-	141–142	blue	TNB 149–150	750, 680, 618, 582	(150)
9-Piperidinomethyl-	105–106	blue		758, 678, 618, 578	(150)
9-p-Tolylazo-	135–136	dark red			(150)
9-Methylthiomethyl-	64–65	blue			(150)
9-Phenylthiomethyl-	119	blue			(150)
9-p-Tolylthiomethyl-	112–113	blue	TNB 138–139		(150)

(continuation of preceding table)

Compound	m.p.	Color	Derivative	spectral	References
2-Benzal-	247	red			(147)
2-Benzal-9-ethoxycarbonyl-	199	red orange			(147)
2-Carboxy-9-ethoxycarbonyl-	100*	brown			(146)
5(7)-Isopropyl-	88–90 / oil	orange	DNP 230–260	480; IR	(148) (148)
9-Ethoxycarbonyl-2-p-dimethyl-aminophenylanil-	245	brown			(146)
9-Ethoxycarbonyl-	184	yellow	DNP 255	482; IR	(146, 147)
6-Isopropyl-	oil	red	DNP 232–234	480	(149)
6-Isopropyl-2-p-nitrobenzal-	247				(149)
2-p-Dimethylaminophenylanil-	254	brown			(147)

Table 2. Azulenium Salts and Other Related Polymethine Dyes.

(For abbreviations see p. 81; and for the structures see pp. 45–48.)

Compound	X⁻	m.p. (°)	Color	Spectral maxima (mμ)	References
(LI): R = 3-Bromophenyl	Pic	74–75*	orange	452	(195)
R = 4-Chlorophenyl	ClO_4	212–220*	orange	454	(139)
R = 2,4-Dihydroxyphenyl	ClO_4	207–210*	dark red	563	(139)
R = 4-Dimethylaminophenyl	Pic	159–160.5	copper	640	(195)
R = 4-Dimethylaminophenyl	ClO_4	236.5–240*	copper	647	(139)
R = 2'-Furyl	ClO_4	220–225*	red	505	(139)
R = 2-Hydroxyphenyl	ClO_4	203–211*	red brown	499	(139)
R = 3-Hydroxyphenyl	ClO_4	> 230*	orange red	475	(139)
R = 4-Hydroxyphenyl	ClO_4	242*	brown	523	(139)
R = 3'-Indolyl			violet black	582	(139)
R = 2-Methoxyphenyl	Pic	193–106*	orange	496	(195)
R = 4-Methoxyphenyl	Pic	150*	red brown	518	(195)
R = 1'-Naphthyl	ClO_4	201–203	red brown	515	(139)
R = 3-Nitrophenyl	ClO_4	195–220*	red	503	(139)
R = 4-Nitrophenyl	ClO_4	195–200*	orange	438	(139)
R = 4-Nitrophenyl	ClO_4	195–201*	orange	435	(139)

(Table 2, continued.)

Compound	X^-	m.p. (°)	Color	Spectral maxima (mμ)	References
(LI): R = Phenyl	Pic	101–102.5*	orange	452	(195)
	ClO_4	195–200*	orange	456	(139)
R = Piperonyl	Pic	104–105*	red	530	(195)
	ClO_4	236–240*	crimson	531	(139)
R = Pyrenyl	ClO_4	>230*	green		(139)
R = 4'-Pyridyl	$2 \cdot ClO_4$	195–198*	yellow brown	420	(139)
R = 2'-Pyridyl	$2 \cdot ClO_4$	188–190.5*	orange	410	(139)
R = 4'-Quinolyl	$2 \cdot ClO_4$	>203*	yellow	435	(139)
R = 2'-Quinolyl	$2 \cdot ClO_4$	>210*	orange red	425	(139)
R = 2'-Thienyl	ClO_4	222–226	dark red	505	(139)
(LII): R = 4-Dimethylaminophenyl	ClO_4	ca. 210	blue	635	(139)
R = 2'-Furyl	ClO_4	179–181	orange red	496	(139)
R = 4-Hydroxyphenyl	ClO_4	210–215*	red brown	500	(139)
R = 3'-Indolyl	ClO_4	ca. 270	violet black	560	(139)
(LVII): R = H, R' = N-Ethylbenzthiazolin, $n = 1$	ClO_4	258–259*	red		(191)
R = H, R' = 5.6-Dimethyl-N-ethylbenzthiazolin, $n = 1$	ClO_4	295–297*	red		(191)
R = H, R' = N-Ethylbenzselenazolin, $n = 1$	ClO_4	276*	red		(191)
R = H, R' = 1,3,3-Trimethylindolin, $n = 1$	ClO_4	243–245*	red violet		(191)
R = H, R' = 1,3,3,5-Tetramethylindolin, $n = 1$	ClO_4	240*	red violet		(191)
R = NHCOCH₃, R' = N-Ethylbenzthiazolin, $n = 2$	ClO_4	284*	blue violet		(191)
R = H, R' = N-Ethylbenzthiazolin, $n = 2$	ClO_4	193–195*	blue		(191)
(LXIIa)				568	(213)
(LXIIb)				580	(213)
(LXIII): R = R' = H	ClO_4, Cl, BF_4	>300	black	618	(139)
(LXIV): R = CH₃, R' = Isopropyl	ClO_4	199–200.5	green	645	(139)
n = 0	ClO_4, Cl			680	(137)
n = 1	ClO_4, Cl			765	(137)
n = 2	ClO_4, Cl			870	(137)
(LXV): R = R' = H	$2 \cdot ClO_4$	>325	black	638	(139)
R = CH₃, R' = Isopropyl	$2 \cdot ClO_4$	325*	black	664	(139)

References.

General References.

1. BARTON, D. H. R.: Some Recent Advances in the Chemistry of Sesquiterpenoid Lactones. Record Chem. Progr. **18**, 125 (1957).

2. BARTON, D. H. R. and P. DE MAYO: Recent Advances in Sesquiterpenoid Chemistry. Quart. Rev. Chem. Soc. (London) **11**, 189 (1957).

3. DOERING, W. v. E.: Tropylium and Related Molecules. In: Theoretical Organic Chemistry, p. 35. London: Butterworths Sci. Publ. 1958.

4. GORDON, M.: The Azulenes. Chem. Rev. **50**, 127 (1952).

5. GUENTHER, E.: The Essential Oils. Vol. IV. New York: Van Nostrand Co. 1950.

6. HAAGEN-SMIT, A. J.: Azulenes. Fortschr. Chem. organ. Naturstoffe **5**, 40 (1948).

7. — Sesquiterpenes and Diterpenes. Fortschr. Chem. organ. Naturstoffe **12**, 1 (1955).

8. HAFNER, K.: Neuere Ergebnisse der Azulen-Chemie. Angew. Chem. **70**, 419 (1958).

9. HEILBRONNER, E.: Azulenes. In: D. GINSBURG, Non-Benzenoid Aromatic Compounds, p. 171. New York: Interscience Publ. 1959.

10. HEROUT, V.: Über die Prochamazulene. Parfumerie und Kosmetik **40**, 12 (1959).

11. — Les progrès récents de la chimie des composés sesquiterpéniques. France et ses parfums **3**, (14) 15 (1959).

12. KELLER-SCHIERLEIN, W. and E. HEILBRONNER: Pathways to Azulenes. In: D. GINSBURG, Non-Benzenoid Aromatic Compounds, p. 277. New York: Interscience Publ. 1959.

13. MAYO, P. DE: The Chemistry of the Perhydroazulene Sesquiterpenoids. Part I. Perfumery Essent. Oil Record **48**, 18 (1957).

14. — The Chemistry of the Perhydroazulene Sesquiterpenoids. Part II. Perfumery Essent. Oil Record **48**, 68 (1957).

15. — Mono- and Sesquiterpenoids. New York: Interscience Publ. 1959.

16. NOZOE, T.: Natural Tropolones and Related Troponoids. Fortschr. Chem. organ. Naturstoffe **13**, 234 (1956).

17. — Synthesis of Azulenoids from Troponoids. Croat. Chem. Acta **29**, 207 (1957).

18. — Tropones and Tropolones. In: D. GINSBURG, Non-Benzenoid Aromatic Compounds, p. 339. New York: Interscience Publ. 1959.

19. — Tropylium and Related Compounds. Progr. Organ. Chem. **5**, 132 (1961).

20. NOZOE, T. and T. ASAO: Azulenes and Heptafulvenes. Dai Yuki Kagaku (Comprehensive Organic Chemistry), Asakura Shoten (Tokyo) **13**, 439 (1960).

21. NOZOE, T. and K. KIKUCHI: Troponoids with Heterocyclic Ring. Dai Yuki Kagaku (Comprehensive Organic Chemistry), Asakura Shoten (Tokyo) **13**, 535 (1960).

22. NOZOE, T., K. TAKASE and H. MATSUMURA: Tropylium Ion and Troponoids. Dai Yuki Kagaku (Comprehensive Organic Chemistry), Asakura Shoten (Tokyo) **13**, 1 (1960).

23. POMMER, H.: Über den Stand der Forschung auf dem Gebiet der Azulene. Angew. Chem. **62**, 281 (1950).

24. REID, D. H.: Azulene and Related Substances. Chem. Soc. Symposia, Bristol, 1958. Spec. Publ. No. 12. London: Chem. Soc. 1958.

25. SIMONSEN, J. L. and D. H. R. BARTON: The Terpenes. Vol. III. Cambridge: Univ. Press. 1952.

26. Simonsen, J. L. and W. C. J. Ross: The Terpenes. Vol. V. New York: Cambridge Univ. Press. 1957.
27. Šorm, F., V. Herout and V. Sýkora: Advances in Sesquiterpene Chemistry. Perfumery Essent. Oil Record 50, 679 (1959).
28. Treibs, W., W. Kirchhof und W. Ziegenbein: Fortschritte der Azulen-chemie seit 1950. Fortschr. chem. Forsch. 3, 334 (1955).

Special References.

29. Abu-Shady, H. A. and T. O. Soine: The Chemistry of *Ambrosia maritima*. I. The Isolation and Preliminary Characterization of Ambrosin and Damsin. J. Amer. Pharm. Assoc. 42, 387 (1953).
30. — — The Chemistry of *Ambrosia maritima*. II. Hydrogenation, Oxidation, and Dehydrogenation of Ambrosin and Damsin. J. Amer. Pharm. Assoc. 43, 365 (1954).
31. Adams, R. and W. Herz: Helenalin. I. Isolation and Properties. J. Amer. Chem. Soc. 71, 2546 (1949).
32. — — Helenalin. II. Helenalin Oxide. J. Amer. Chem. Soc. 71, 2551 (1949).
33. — — Helenalin. III. Reduction and Dehydrogenation. J. Amer. Chem. Soc. 71, 2554 (1949).
34. Akino, H.: Synthesis of Various Azulenes from 4-Methyltropolone. Bull. Chem. Soc. Japan (to be published).
35. Alder, K., R. Muders, W. Krane und P. Wirtz: Über die Konstitution photochemisch dargestellter Norcaradien-carbonsäureester. Liebigs Ann. Chem. 627, 59 (1959).
36. Amiel, Y. and D. Ginsburg: Alicyclic Studies. XI. Attempted Syntheses of 5:6-Benzazulene and Benzheptalenes. Tetrahedron 1, 9 (1957).
37. Anderson, A. G., Jr., R. G. Anderson and L. L. Replogle: The Reaction of Azulenes with Trifluoroacetic Anhydride. Proc. Chem. Soc. (London) 1960, 72.
38. Anderson, A. G., Jr. and G. M-C. Chang: Reaction of Azulene with *N*-Nitrosoacetanilide, Phenylazotriphenylmethane, and Benzoyl Peroxide. J. Organ. Chem. (USA) 23, 151 (1958).
39. Anderson, A. G., Jr., C. G. Fritz and R. Scotoni, Jr.: Azulene. VII. A Study of the Beckmann Rearrangement of 1,3-Diacetylazulene Dioxime and 1,3-Diacetylazulene Dioxime Diacetate. J. Amer. Chem. Soc. 79, 6511 (1957).
40. Anderson, A. G., Jr. and R. N. McDonald: Azulene. IX. Synthesis of Some Derivatives of 1-Azulenethiol and 1,3-Azulenedithiol. J. Amer. Chem. Soc. 81, 5669 (1959).
41. Anderson, A. G., Jr. and L. L. Replogle: Electrophilic Substitution of 1,3-Dichloroazulene. J. Organ. Chem. (USA) 25, 1275 (1960).
42. Anderson, A. G., Jr. and B. M. Steckler: Azulene. VIII. A Study of the Visible Absorption Spectra and Dipole Moments of Some 1- and 1,3-Substituted Azulenes. J. Amer. Chem. Soc. 81, 4941 (1959).
43. Asano, T.: Syntheses of Azulenic Compounds from Hinokitiol. Thesis, Tohoku Univ., Sendai (Japan), 1959.
44. Assony, S. J. and N. Kharasch: Derivatives of Sulfenic Acids. XXXII. The Synthesis of Azulenes via the Interactions of Arylacetylenes with Sulfenyl Halides. Part I. 1,2,3-Triphenylazulene. J. Amer. Chem. Soc. 80, 5978 (1958).
45. Bach, G., E.-J. Poppe und W. Treibs: Darstellung von stickstoffreien Azulenmonomethin-Farbstoffen. Naturwiss. 45, 517 (1958).
46. Balaban, A. T.: Possible Formation of Tetraphenylcyclobutadiene. Tetrahedron Letters 1959, No. 5, 14.

47. BARTON, D. H. R.: Some Aspects of Sesquiterpenoid Chemistry. Proc. Chem. Soc. (London) 1958, 61.

48. — Some Photochemical Rearrangements. Helv. Chim. Acta 42, 2604 (1959).

49. BARTON, D. H. R., O. C. BÖCKMAN and P. DE MAYO: Sesquiterpenoids. Part XII. Further Investigations on the Chemistry of Pyrethrosin. J. Chem. Soc. (London) 1960, 2263.

50. BARTON, D. H. R. and J. E. D. LEVISALLES: Sesquiterpenoids. Part XI. The Constitution of Geigerin. J. Chem. Soc. (London) 1958, 4518.

51. BARTON, D. H. R. and P. DE MAYO: Sesquiterpenoids. Part VII. The Constitution of Tenulin, a Novel Sesquiterpenoid Lactone. J. Chem. Soc. (London) 1956, 142.

52. — — Sesquiterpenoids. Part VIII. The Constitution of Pyrethrosin. J. Chem. Soc. (London) 1957, 150.

53. BARTON, D. H. R., P. DE MAYO and M. SHAFIQ: Photochemical Transformations. Part I. Some Preliminary Investigations. J. Chem. Soc. (London) 1957, 929.

54. BARTON, D. H. R. and C. R. NARAYANAN: Sesquiterpenoids. Part X. The Constitution of Lactucin. J. Chem. Soc. (London) 1958, 963.

54a. BARTON, D. H. R. and J. T. PINHEY: The Stereochemical Correlation of Artemisin and Geigerin. Proc. Chem. Soc. (London) 1960, 279.

55. BATES, R. B., E. J. EISENBRAUN and S. M. McELVAIN: The Configurations of the Nepetalactones and Related Compounds. J. Amer. Chem. Soc. 80, 3420 (1958).

56. BAUDER, A. und Hs. H. GÜNTHARD: Deuterierte Azulene. I. Herstellung und spektroskopische Eigenschaften von Azulen-d$_2$-(1,3). Helv. Chim. Acta 41, 889 (1958).

57. BERGMANN, E. D. and R. IKAN: 4- and 5-Phenylazulenes. J. Amer. Chem. Soc. 80, 3135 (1958).

58. BERNARDI, L. and G. BÜCHI: The Structures of Ambrosin and Damsin. Experientia 13, 466 (1957).

59. BERNSTEIN, H. J., J. A. POPLE and W. G. SCHNEIDER: The Analysis of Nuclear Magnetic Resonance Spectra. 1. Systems of Two and Three Nuclei. Canad. J. Chem. 35, 65 (1957).

60. BERNSTEIN, H. J., W. G. SCHNEIDER and J. A. POPLE: The Proton Magnetic Resonance Spectra of Conjugated Aromatic Hydrocarbons. Proc. Roy. Soc. London A 236, 515 (1956).

61. BIRCH, A. J.: The Volatile Oil of *Metrosideros scandens*. J. Chem. Soc. (London) 1953, 715.

62. BIRCH, A. J., D. J. COLLINS and A. R. PENFOLD: Zierone: Derivative of a New Natural Azulene. Chem. and Ind. 1955, 1773.

63. BIRCH, A. J., J. GRIMSHAW, R. N. SPEAKE, R. M. GASCOIGNE and R. O. HELLYER: Aromadendrene and Viridiflorol. Tetrahedron Letters 1959, No. 3, 15.

64. BIRCH, A. J. and F. N. LAHEY: The Structure of Aromadendrene. I. Austral. J. Chem. 6, 379 (1953).

65. BIRCH, A. J. and K. M. C. MOSTYN: A New Sesquiterpene Alcohol from *Himantandra baccata* BAIL. Austral. J. Chem. 8, 550 (1955).

66. BLUMANN, A., A. R. H. COLE, K. J. L. THIEBERG and D. E. WHITE: The Constitution of Globulol. Chem. and Ind. 1954, 1426.

67. BRAUN, B. H., W. HERZ and K. RABINDRAN: The Structure of Tenulin. J. Amer. Chem. Soc. 78, 4423 (1956).

68. BRESLOW, R. and M. BATTISTE: An Unusual Rearrangement in the Cyclopropene Series. J. Amer. Chem. Soc. 82, 3626 (1960).

69. BRUDERER, H., D. ARIGONI und O. JEGER: Zur Kenntnis der Sesquiterpene und Azulene, 116. Mitt. Über die absolute Konfiguration des α-Santonins. Helv. Chim. Acta **39**, 858 (1956).

69a. BRUNKEN, J. und E.-J. POPPE: Cyaninfarbstoffe aus 2-Methyl-4(azulenyl-1)-thiazol. Chem. Ber. **93**, 2572 (1960).

70. BÜCHI, G., S. W. CHOW, T. MATSUURA, T. L. POPPER, H. H. RENNHARD and M. S. v. WITTENAU: Terpenes. XII. The Constitutions of Aromadendrene, Globulol, Ledol and Viridoflorol. Tetrahedron Letters **1959**, No. 6, 14.

71. BÜCHI, G. and R. E. ERICKSON: Terpenes. V. The Structure of Patchouly Alcohol. J. Amer. Chem. Soc. **78**, 1262 (1956).

72. BÜCHI, G. and D. ROSENTHAL: Terpenes. VI. The Structures of Helenalin and Isohelenalin. J. Amer. Chem. Soc. **78**, 3860 (1956).

72a. BURTON, R., L. PRATT and G. WILKINSON: Transition-metal Complexes of Seven-membered Ring Systems. Part II. Azulenemetal Carbonyls. J. Chem. Soc. (London) **1960**, 4290.

73. BURTON, R. and G. WILKINSON: An Azulene-Metal Complex. Chem. and Ind. **1958**, 1205.

74. ČEKAN, Z., V. HEROUT and F. ŠORM: A Chamazulene Precursor from Chamomile (*Matricaria chamomilla* L.). Chem. and Ind. **1954**, 604. — On Terpenes. LXII. Isolation and Properties of the Pro-chamazulene from *Matricaria chamomilla* L., a Further Compound of the Guaianolide Group. Collect. Czech. Chem. Commun. **19**, 798 (1954).

75. — — — Structure of Matricin. Chem. and Ind. **1956**, 1234. — Über Terpene. LXXX. Die Struktur von Matricin, ein Guajanolid aus der Kamille (*Matricaria chamomilla* L.). Collect. Czech. Chem. Commun. **22**, 1921 (1957).

76. ČEKAN, Z., V. PROCHÁZKA, V. HEROUT and F. ŠORM: On Terpenes. CI. Isolation and Constitution of Matricarin, another Guaianolide from Camomile (*Matricaria chamomilla* L.). Collect. Czech. Chem. Commun. **24**, 1554 (1959).

76a. — — — — On Terpenes. CXV. Isolation of Globicin, a Guaianolide from *Matricaria globifera* (THUNB.) DRUCE. Coll. Czech. Chem. Commun. **25**, 2553 (1960).

77. CHIURDOGLU, G., A. COPET and P. TULLEN: Study of the Structure of Tricyclovetivene of *Vetiveria zizanioides* Oil. Bull agr. Congo Belge **48**, 1503 (1957) [Chem. Abstr. **52**, 20906 (1958)].

78. CHIURDOGLU, G. et J. DECOT: Contribution à l'étude des composés sesquiterpéniques. II. Étude de la structure du bicyclovétivénol et du tricyclovétivénol, alcools primaires de l'essence de vetiver. Tetrahedron **4**, 1 (1958).

79. CHIURDOGLU, G. and M. DESCAMPS: Structure of Carotol. Chem. and Ind. **1959**, 1377. — Contribution à l'étude des composés sesquiterpéniques. IV. Étude de la structure du carotol, alcool $C_{15}H_{26}O$ de l'essence de *Daucus carota*. Tetrahedron **8**, 271 (1960).

80. CHIURDOGLU, G. and P. TULLEN: Structure of Tricyclovetivene. Chem. and Ind. **1956**, 1094. — Sesquiterpenes. I. Structural Study of Tricyclovetivene from Belgian Congo Vetiver Oil. Bull. soc. chim. Belges **66**, 169 (1957) [Chem. Abstr. **52**, 10010 (1958)].

81. CLARK, E. P.: The Constituents of Certain Species of Helenium. II. Tenulin. J. Amer. Chem. Soc. **61**, 1836 (1939).

82. — The Constituents of Certain Species of Helenium. III. The Ester Nature of Tenulin. J. Amer. Chem. Soc. **62**, 597 (1940).

83. COLLINS, D. J.: The Structure of Zierone. Part I. The Synthesis of Zierazulene and 6-Methylzierazulene. J. Chem. Soc. (London) **1959**, 531.

84. CORBETT, R. E. and R. N. SPEDEN: The Volatile Oil of *Pseudowintera colorata*. Part II. The Structure of *cyclo*Colorenene. J. Chem. Soc. (London) 1958, 3710.

85. COWLES, E. J.: The Effects of Substituents at the 1- and 3-Positions on the Visible Absorption Spectrum of Azulene. J. Amer. Chem. Soc. 79, 1093 (1957).

86. DANYLUK, S. S. and W. G. SCHNEIDER: Proton Resonance Spectrum and Structure of the Azulinium Ion. J. Amer. Chem. Soc. 82, 997 (1960).

87. DJERASSI, C., J. OSIECKI and W. HERZ: Optical Rotatory Dispersion Studies. XIII. Assignment of Absolute Configuration to Certain Members of the Guaianolide Series of Sesquiterpenes. J. Organ. Chem. (USA) 22, 1361 (1957).

87a. DOERING, W. v. E. and D. W. WILEY: Heptafulvene (Methylenecyclohepta-triene). Tetrahedron 11, 183 (1960).

87b. DOLEJŠ, L., V. HEROUT, O. MOTL, F. ŠORM and M. SOUČEK: Epimeric Aroma-dendrenes: Stereoisomerism of Ledol, Viridiflorol and Globulol. Chem. and Ind. 1959, 566.

88. DOLEJŠ, L., V. HEROUT and F. ŠORM: Structure of Palustrol. Chem. and Ind. 1960, 267.

89. DOLEJŠ, L., A. MIRONOV and F. ŠORM: Structure of Bulnesol. Stereochemistry of Guaiol, Nepetalinic Acids and Iridomyrmecins. Tetrahedron Letters 1960, No. 11, 18.

90. DOLEJŠ, L., O. MOTL, M. SOUČEK, V. HEROUT and F. ŠORM: On Terpenes. CVIII. Epimeric Aromadendrenes. Stereoisomerism of Ledol, Viridiflorol and Globulol. Collect. Czech. Chem. Commun. 25, 1483 (1960).

91. DOLEJŠ, L. and F. ŠORM: Position of the Cyclopropane Ring in Aromadendrene. Tetrahedron Letters 1959, No. 10, 1.

92. — — Stereochemistry of Aromadendrene, Alloaromadendrene, Globulol, Ledol and Viridiflorol. Tetrahedron Letters 1959, No. 17, 1. — On Terpenes. CXIII. Structure of Aromadendrene, *allo*Aromadendrene, Globulol, Ledol and Viridiflorol. Collect. Czech. Chem. Commun. 25, 1837 (1960).

93. DOLEJŠ, L., F. ŠORM and M. SOUČEK: Structure of Ledol and its Stereoisomerism with Globulol. Chem. and Ind. 1959, 160.

94. DOLEJŠ, L., M. SOUČEK, M. HORÁK, V. HEROUT and F. ŠORM: Structure of Lactucin. Chem. and Ind. 1958, 530.

95. DOLEJŠ, L., M. SOUČEK, M. HORÁK and F. ŠORM: The Structure of Ledol. Chem. and Ind. 1958, 494. — On Terpenes. XCVII. The Constitution of Ledol. Collect. Czech. Chem. Commun. 24, 1353 (1959).

96. EISENBRAUN, E. J., T. GEORGE, B. RINIKER and C. DJERASSI: Terpenoids. XLIII. On the Absolute Configuration on Guaiol. Correlation with Nepetalinic Acid. J. Amer. Chem. Soc. 82, 3648 (1960).

96a. ESCHENMOSER, A., L. RUZICKA, O. JEGER und D. ARIGONI: Zur Kenntnis der Triterpenene, 190. Mitt. Eine stereochemische Interpretation der bio-genetischen Isoprenregel bei den Triterpenen. Helv. Chim. Acta 38, 1890 (1955).

97. GERDIL, R. und E. HEILBRONNER: Zur Kenntnis der Sesquiterpene und Azulene, 122. Mitt. Azulenaldehyde und Azulenketone: Die polarographische Reduktion von sterisch gehinderten Azulenketonen. Helv. Chim. Acta 40, 141 (1957).

98. GERSON, F., T. GÄUMANN und E. HEILBRONNER: Elektronenstruktur und physikalisch-chemische Eigenschaften von Azo-Verbindungen. Teil III: Dipol-momente substituierter Phenyl-azo-azulene. Helv. Chim. Acta 41, 1481 (1958).

99. GERSON, F. und E. HEILBRONNER: Elektronenstruktur und physikalisch-chemische Eigenschaften von Azo-Verbindungen. Teil I: Aryl-azo-azulene und Aryl-azo-guaj-azulen. Helv. Chim. Acta 41, 1444 (1958).

100. Gerson, F. und E. Heilbronner: Elektronenstruktur und physikalisch-chemische Eigenschaften von Azo-Verbindungen. Teil IV: LCAO-MO-Modelle des Phenyl-azo-azulens. Helv. Chim. Acta **41**, 2332 (1958).

101. — — Elektronenstruktur und physikalisch-chemische Eigenschaften von Azo-Verbindungen. Teil V: Über den Einfluß von Substituenten auf das Absorptions-Spektrum des Phenyl-azo-azulens, ein Beitrag zur Kenntnis der Plattner'schen Verschiebungsregel. Helv. Chim. Acta **42**, 1877 (1959).

102. Gerson, F., J. Schulze und E. Heilbronner: Elektronenstruktur und physikalisch-chemische Eigenschaften von Azo-Verbindungen. Teil II: Substituierte Phenyl-azo-azulene. Helv. Chim. Acta **41**, 1463 (1958).

103. — — — Elektronenstruktur und physikalisch-chemische Eigenschaften von Azo-Verbindungen. Teil VII: Notiz über das Absorptions-Spektrum und die Basizität des *p*-Trimethylammonium-phenyl-azo-azulen-Kations. Helv. Chim. Acta **43**, 517 (1960).

104. Govindachari, T. R., B. R. Pai, K. K. Purushothaman and S. Rajadurai: Structure of Jatamansone. Chem. and Ind. **1960**, 1059.

105. Govindachari, T. R., K. K. Purushothaman und S. Rajadurai: Struktur von Jatamanson. II. Synthese einiger Trialkylazulene. Chem. Ber. **92**, 1662 (1959).

106. Govindachari, T. R., S. Rajadurai und B. R. Pai: Struktur von Jatamanson. I. Chem. Ber. **91**, 908 (1958).

107. Haagen-Smit, A. J. and C. T. O. Fong: Chemical Investigation in Guayule. II. The Structure of Partheniol, a Sesquiterpene Alcohol from Guayule. J. Amer. Chem. Soc. **70**, 2075 (1948).

108. Hafner, K.: Neuere Untersuchungen über Azulene. Angew. Chem. **70**, 413 (1958).

109. — Neue quasi-aromatische Verbindungen. Angew. Chem. **71**, 378 (1959).

110. — Neue Reaktionen des Cyclopentadiens. Angew. Chem. **72**, 574 (1960).

111. Hafner, K. und C. Bernhard: Zur Kenntnis der Azulene, IV. Azulen-Aldehyde und -Ketone. Liebigs Ann. Chem. **625**, 108 (1959).

112. Hafner, K. und H. Kaiser: Zur Kenntnis der Azulene, III. Eine einfache Synthese substituierter Azulene. Liebigs Ann. Chem. **618**, 140 (1958).

112a. Hafner, K. und H. Pelster: Synthese und Reaktionen von Azuleniumsalzen. Angew. Chem. **72**, 781 (1960).

113. Hafner, K. und J. Schneider: Synthese eines Heptalens. Angew. Chem. **70**, 702 (1958).

114. — — Darstellung und Eigenschaften von Derivaten des Pentalens und Heptalens. Liebigs Ann. Chem. **624**, 37 (1959).

115. Hafner, K. und K.-H. Vöpel: Fulven-Aldehyde. Angew. Chem. **71**, 672 (1959).

115a. Hamilton, J. A., A. T. McPhail and G. A. Sim: The Structure of Acetyl-bromogeigerin. Proc. Chem. Soc. (London) **1960**, 278.

116. Hendrickson, J. B.: Stereochemical Implications in Sesquiterpene Biogenesis. Tetrahedron **7**, 82 (1959).

117. Herout, V., L. Doleǰš and F. Šorm: The Structure of Artabsin, the Prochamazulenogen from *Artemisia absinthium* L. Chem. and Ind. **1956**, 1236. — On Terpenenes. LXXIX. On the Structure of Artabsin, a Prochamazulenogen from Wormwood (*Artemisia absinthium* L.). Collect. Czech. Chem. Commun. **22**, 1914 (1957).

118. Herout, V., M. Holub, L. Novotný, F. Šorm and V. Sýkora: Constitution of Carotol and Daucol. Chem. and Ind. **1960**, 662.

119. HEROUT, V., M. HORÁK, B. SCHNEIDER and F. ŠORM: Location of the Double Bonds in Germacrone. Some Properties of its Spectra. Chem. and Ind. 1959, 1089.

120. HEROUT, V., L. NOVOTNÝ und F. ŠORM: Über Pflanzenstoffe. V. Die Isolierung von weiteren kristallinen Substanzen aus Wermut (*Artemisia absinthium* L.). Collect. Czech. Chem. Commun. 21, 1485 (1956).

121. HEROUT, V. and F. ŠORM: On the Composition of Wormwood (*Artemisia absinthium* L.) and the Isolation of a Crystalline Pro-chamazulenogen. Collect. Czech. Chem. Commun. 18, 854 (1953).

122. — — On Terpenes. LXI. Contribution to the Constitution of Pro-chamazulenogen, the Natural Precursor of Chamazulene in *Artemisia absinthium* L. Collect. Czech. Chem. Commun. 19, 792 (1954); Chem. Listy 48, 706 (1954).

123. — — Über Terpene. LXX. Monocyclische Lactone aus Wermut (*Artemisia absinthium* L.). Collect. Czech. Chem. Commun. 21, 1494 (1956).

124. HERZ, W.: Azulenes. VIII. 1- and 2-*t*-Butylazulene. Migration of the *t*-Butyl Group. J. Amer. Chem. Soc. 80, 1243 (1958).

125. — Azulenes. IX. Migration of the Isopropyl Group during the Synthesis of 1-Isopropyl-8-methylazulene. J. Amer. Chem. Soc. 80, 3139 (1958).

126. HERZ, W., P. JAYARAMAN and H. WATANABE: Constituents of Helenium Species. IX. The Sesquiterpene Lactones of *H. flexuosum* RAF. and *H. campestre* SMALL. J. Amer. Chem. Soc. 82, 2276 (1960).

127. HERZ, W. and R. B. MITRA: Constituents of Helenium Species. VI. Correlation of Helenalin and Alloisotenulin. J. Amer. Chem. Soc. 80, 4876 (1958).

128. HERZ, W., R. B. MITRA and P. JAYARAMAN: Constituents of Helenium Species. VIII. Isolation and Structure of Balduilin. J. Amer. Chem. Soc. 81, 6601 (1959).

129. HERZ, W., R. B. MITRA, K. RABINDRAN and W. A. ROHDE: Constituents of Helenium Species. VII. Bitter Principles of *H. pinnatifidum* (NUTT.) RYDB., *H. vernale* WALT., *H. brevifolium* (NUTT.) A. WOOD and *H. flexuosum* RAF. J. Amer. Chem. Soc. 81, 1481 (1959).

130. HERZ, W. and H. WATANABE: Parthenin, a New Guaianolide. J. Amer. Chem. Soc. 81, 6088 (1959).

131. HILDEBRAND, R. P. and M. D. SUTHERLAND: Terpenoid Chemistry. I. Zierone and Elleryone. Austral. J. Chem. 12, 436 (1959) [Chem. Abstr. 53, 22061 (1959)]

132. HOUSE, W. T. and M. ORCHIN: A Study of the Selenium Dehydrogenation of Guaiol and Related Compounds. Selenium as a Hydrogen Transfer Agent. J. Amer. Chem. Soc. 82, 639 (1960).

132a. HUNT, G. R. and I. G. ROSS: Spectrum of Azulene. Part I. Infrared Spectrum. J. Mol. Spectr. 3, 604 (1959).

133. JEFFERIES, P. R., G. J. H. MELROSE and D. E. WHITE: Structure of Aromadendrene, Globulol and Ledol. Chem. and Ind. 1959, 878.

134. JONES, T. G. H. and S. E. WRIGHT: Essential Oils from the Queensland Flora. XXI. The Essential Oil of *Evodia elleryana*. Univ. Queensland Papers, Dept. Chem. 1, (27) 7 (1946) [Chem. Abstr. 41, 1391 (1947)].

135. JOST, K. H.: Röntgenstrukturanalyse des Chamazulens. Naturwiss. 43, 224 (1956).

136. JULG, A.: Étude de l'azulène par la méthode du champ moléculaire seltonsistant. J. chim. phys. 52, 377 (1955).

137. JUTZ, C.: Synthese von Azulen-polyenalen. Angew. Chem. 70, 270 (1958).

138. — Zwei neue Synthesen von Polyenalen und ungesättigten Ketonen durch Reaktion vinyloger Acylamide. (Darstellung des *cis*-Nonadien-(2,6)als, Veilchenblätteraldehyds.) Angew. Chem. 71, 380 (1959).

139. Kirby, E. C. and D. H. Reid: Conjugated Cyclic Hydrocarbons and Their Heterocyclic Analogues. Part II. The Condensation of Azulenes with Homocyclic and Heterocyclic Aromatic Aldehydes in the Presence of Perchloric Acid. J. Chem. Soc. (London) 1960, 494.

139a. — — 4,6,8-Trimethylazulenium Perchlorate. Chem. and Ind. 1960, 1217.

140. Kiryalov, N. P.: Structure of Ledol. Sbornik Statei Obshchei Khim. 2, 1617 (1953) [Chem. Abstr. 49, 5389 (1955)].

141. — Structure of Palustrol. Zhurn. Obshchei Khimii 24, 1271 (1954) [Chem. Abstr. 49, 13944 (1955)].

142. Kitahara, Y. and K. Doi: Studies on Heptafulvenes. II. Condensation Reaction of Tropone and Some Active Methylene Compounds. Bull. Chem. Soc. Japan (to be published).

142a. Kitahara, Y., K. Doi and Y. Sato: Studies on Heptafulvenes. III. (to be published).

143. Kondo, H. und K. Takeda: Über die Bestandteile der Wurzel von Lindera strychnifolia Vill. (III. Mitt.). J. pharmac. Soc. Japan 59, 504 (1939).

144. Křepinský, J., M. Romaňuk, V. Herout and F. Šorm: Constitution of Valeranone. Tetrahedron Letters 1960, No. 7, 9.

145. Kurita, Y. and M. Kubo: The Dipole Moments and Electronic Structures of Some Azulene Derivatives. J. Amer. Chem. Soc. 79, 5460 (1957).

146. Matsui, K.: Synthesis of 3-Methoxythieno[3,2-a]azulenes. J. Chem. Soc. Japan 82 (1961) (in press).

147. — Synthesis of Thieno[3,2-a]azulenes. J. Chem. Soc. Japan 82 (in press) (1961).

148. — Synthesis of 5- and 7-Isopropylthieno[3,2-a]azulenes. J. Chem. Soc. Japan 82 (1961) (in press).

149. — Synthesis of 6-Isopropylthieno[3,2-a]azulenes. J. Chem. Soc. Japan 82 (1961) (in press).

150. — Cationoid Substitution of Thieno[3,2-a]azulenes. J. Chem. Soc. Japan 82 (1961) (in press).

151. Matsui, K, and T. Nozoe: Synthesis of Thieno[3,2-a]azulene. Chem. and Ind. 1960, 1302.

152. Matsumura, S.: Reaction of Troponoid with Ethyl Cyanoacetate and Diethyl α-Cyano-β-iminoglutarate. Thesis, Tohoku Univ., Sendai (Japan). 1960.

153. Mazur, Y. and A. Meisels: The Structure of Arborescine, a New Sesquiterpene from Artemisia arborescens L. Chem. and Ind. 1956, 492.

154. Miyazaki, M.: Studies on Azulenes. X. 3-Aminomethyl-S-guaiazulenes. Pharm. Bull. Japan 8, 146 (1960).

155. Miyazaki, M., M. Hashi and T. Ukita: Studies on Azulenes. IX. 3-(1-Hydroxy-ethyl)-S-guaiazulene. Chem. Pharm. Bull. Japan 8, 146 (1960).

156. Miyazaki, M., H. Watanabe, M. Hashi and T. Ukita: Studies on Azulenes. V. S-Guaiazulene-3-sulfonic Acid. Pharm. Bull. Japan 5, 417 (157).

157. Mörikofer, A. und E. Heilbronner: Elektronenstruktur und physikalisch-chemische Eigenschaften von Azo-Verbindungen. Teil VI: Die relative Basizität von Aryl-azo-azulenen in den Systemen Äthanol/Salzsäure und Methyliso-butylketon/Schwefelsäure. Helv. Chim. Acta 42, 1909 (1959).

158. Motl, O., V. Herout and F. Šorm: On Terpenes. CXII. The Composition of the Oil from Juniperus oxycedrus L. Berries. Collect. Czech. Chem. Commun. 25, 1656 (1960).

159. Naito, S.: Studies on the Components of Carpesium abrotanoides. IV. Chemical Structure of Carpesia Lactone (3). J. pharmac. Soc. Japan 75, 39 (1955).

160. — Studies on the Components of Carpesium abrotanoides. V. Chemical Constitution of Carpesia Lactone (4). J. pharmac. Soc. Japan 75, 325 (1955).

161. NOVOTNÝ, L., V. HEROUT and F. ŠORM: A Contribution to the Structure of Absinthin and Anabsinthin. Chem. and Ind. 1958, 465. — On Terpenes. CIX. A Contribution to the Structure of Absinthin and Anabsinthin. Collect. Czech. Chem. Commun. 25, 1492 (1960).

162. — — — On Terpenes. CX. A Contribution to Stereochemistry of Absinthin and Artabsin. Collect. Czech. Chem. Commun. 25, 1500 (1960).

163. NOZOE, T., Y. KITAHARA, K. TAKASE, I. MURATA and R. HAYASHI: The Reaction of 3-Carboxytropolone and 3-Cyanotropolone with Hydrazine. Bull. Chem. Soc. Japan 34 (1961) (in press).

164. NOZOE, T., S. MATSUMURA, Y. MURASE and S. SETO: Synthesis of 2-Amino-azulene Derivatives from 2-Halogenotropones. Chem. and Ind. 1955, 1257.

165. NOZOE, T., T. MUKAI and K. OSAKA: Synthesis and Reactions of 8,8-Di-substituted Heptafulvenes. Bull. Chem. Soc. Japan 34 (1961) (in press).

166. NOZOE, T. and A. SATO: Synthesis of Amino Acids with Azulene Nucleus. Bull. Chem. Soc. Japan (to be published).

167. NOZOE, T., S. SETO and T. ASANO: Synthesis of Azulene Derivatives from Hinokitiol Methyl Ether. Bull. Chem. Soc. Japan (to be published).

168. — — — 6-Isopropylazulene and its Derivatives. Bull. Chem. Soc. Japan (to be published).

169. NOZOE, T., S. SETO and S. MATSUMURA: Synthesis of Azulene Derivatives from Troponoids and Ethyl Cyanoacetate. Bull. Chem. Soc. Japan 34 (1961) (in press).

170. — — — Synthesis of Furo[3,2-f]azulene Derivatives. Bull. Chem. Soc. Japan 34 (1961) (in press).

171. NOZOE, T., S. SETO, S. MATSUMURA and T. ASANO: Synthesis of Azulene Derivative from Troponoids and Cyanoacetic Ester. Proc. Japan Acad. 32, 339 (1956).

172. NOZOE, T., S. SETO, S. MATSUMURA and T. KUSUNOSE: Synthesis of Azulenes with Cyano- and Carboxyl Group at 5- and 6-Positions. Bull. Chem. Soc. Japan (to be published).

173. NOZOE, T., S. SETO, S. MATSUMURA and A. SATO: Synthesis of 2-Substituted Azulene Derivatives. Bull. Chem. Soc. Japan (to be published).

174. NOZOE, T., S. SETO and S. NOZOE: Synthesis of Azulene and 1-Azaazulanone Derivatives by the Application of Cyanoacetamide to Tropolone and 4-Methyl-tropolone Methyl Ethers. Proc. Japan Acad. 32, 472 (1956).

175. NOZOE, T., S. SETO and T. SATO: Synthesis of Azulene Derivatives from 5-Chlorotropolone. Bull. Chem. Soc. Japan 34 (1961) (in press).

176. NOZOE, T., K. TAKASE and T. KUSUNOSE: On 4-(5-Pyrazolyl)tropolones. Bull. Chem. Soc. Japan (to be published).

177. — — — On 4-(2-Quinolyl)tropolones. Bull. Chem. Soc. Japan (to be published).

178. NOZOE, T., K. TAKASE and M. TADA: Synthesis of 5- and 6-Acetylazulene Derivatives. Bull. Chem. Soc. Japan (to be published).

179. — — — Synthesis of 5- and 6-Aminoazulene Derivatives. Bull. Chem. Soc. Japan (to be published).

180. — — — Pyrazolo[4,5-f]azulene Derivatives. Bull. Chem. Soc. Japan (to be published).

181. NOZOE, T., K. TAKASE and K. UMINO: Synthesis of 5-Ethylazulene Derivatives. Bull. Chem. Soc. Japan (to be published).

182. NOZOE, T. and co-workers: unpublished data.

183. OHLOFF, G., H. FARNOW, W. PHILIPP und G. SCHADE: Über Germacron und seine pyrolytische Umwandlung. Liebigs Ann. Chem. 625, 206 (1959).

184. PALMADE, M. et G. OURISSON: La structure de l'α-gurjunène (Note préliminaire). Bull. soc. chim. France **1958**, 886.
184a. PARISER, R.: Electronic Spectrum and Structure of Azulene. J. Chem.
Physics **25**, 1112 (1956).
184b. PAUSON, P. L.: Hydrocarbon Metal Carbonyls. (Tilden Lecture.) Proc.
Chem. Soc. (London) **1960**, 297.
185. PEROLD, G. W.: The Structure of Geigerin. J. Chem. Soc. (London) **1957,** 47.
186. PFAU, A. ST. und PL. A. PLATTNER: Zur Kenntnis der flüchtigen Pflanzenstoffe.
IV. Über die Konstitution der Azulene. Helv. Chim. Acta **19**, 858 (1936).
187. — — Études sur les matières végétales volatiles. XI. Sur la constitution
de la β-vétivone. Helv. Chim. Acta **23**, 768 (1940).
188. PLATTNER, PL. A. und E. HEILBRONNER: Über die Konstitution des Lactaroviolins. Experientia **1**, 233 (1945).
188a. PLATTNER, PL. A., E. HEILBRONNER, R. W. SCHMID, R. SANDRIN and
A. FÜRST: The Structure of Lactaroviolin. Chem. and Ind. **1954**, 1202.
189. PLATTNER, PL. A. und L. LEMAY: Zur Kenntnis der Sesquiterpene. Über
das Kohlenstoff-Gerüst des Guajols und des Guaj-azulens. Helv. Chim. Acta **23,**
897 (1940).
190. POMMER, H. und K.-D. MÖHLE: Über das 4-, 5- und 6-Phenyl-azulen. Arch.
Pharmaz. **291**, 23 (1958).
191. POPPE, E.-J. und W. TREIBS: Darstellung von Azulenstyryl-Farbstoffen.
Naturwiss. **45**, 517 (1958).
192. RAO, A. S., K. B. BUTT, S. DEV and P. C. GUHA: Sesquiterpenes. XI. Sesquiterpenes of the Essential Oil of *Lansium annamalayanum.* Structure of
α-Chigadmarene. J. Indian Chem. Soc. **29**, 620 (1952) [Chem. Abstr. **47,**
8698 (1953)].
193. RAO, A. S. and M. S. MUTHANA: Synthesis of 1-Ethyl- and 5-Ethylazulenes.
J. Indian Inst. Sci. **37** A, 79 (1955) [Chem. Abstr. **50**, 3377 (1956)].
194. REID, D. H., W. H. STAFFORD and W. L. STAFFORD: The Azulene Series.
Part IV. The Synthesis and Properties of 3-Acetylguaiazulenes. J. Chem.
Soc. (London) **1958**, 1118.
195. REID, D. H., W. H. STAFFORD, W. L. STAFFORD, G. McLENNAN and A. VOIGT:
The Azulene Series. Part III. The Synthesis and Properties of 3-Benzylidene-
guaiazulenium Chloride. J. Chem. Soc. (London) **1958**, 1110.
196. REID, D. H., W. H. STAFFORD and J. P. WARD: The Azulene Series. Part II.
The Synthesis and Properties of Alkoxyazulenes. J. Chem. Soc. (London)
1958, 1100.
197. ROBERTSON, J. M., H. M. M. SHEARER, G. A. SIM and D. G. WATSON: A
Revision of the Azulene Structure. Nature (London) **182**, 177 (1958).
197a. ROMAŇUK, M. and V. HEROUT: On Terpenes. CXIV. On Stereoisomeric Vetivanes
and Sesquiterpenic Hydrocarbons of Vetiver Oil. Coll. Czech. Chem.
Commun. **25**, 2540 (1960).
198. ROMAŇUK, M., V. HEROUT and F. ŠORM: On Terpenes. LXIX. The Constitution
of Dehydrocostuslactone. Collect. Czech. Chem. Commun. **21**, 894 (1956).
198a. RUZICKA, L.: The Isoprene Rule and the Biogenesis of Terpenic Compounds.
Experientia **9**, 357 (1953).
198b. — Bedeutung der theoretischen organischen Chemie für die Chemie der
Terpenverbindungen. In: A. TODD, Perspectives in Organic Chemistry, p. 265.
New York: Interscience Publ. 1959.
199. SCHNEIDER, W. G., H. J. BERNSTEIN and J. A. POPLE: The Proton Magnetic
Resonance Spectra of Azulene and Acepleiadylene. J. Amer. Chem. Soc. **80**,
3497 (1958).

200. SCHULZE, J. und E. HEILBRONNER: Zur Kenntnis der Sesquiterpene und Azulene, 124. Mitt. Basizität und Absorptions-Spektren des 1-Aminoazulens. Helv. Chim. Acta 41, 1492 (1958).

201. SLOBODKIN, N. R.: A Fifth Route to 1,2,3-Triphenylazulene. J. Organ. Chem. (USA) 25, 273 (1960).

202. ŠORM, F., V. BENEŠOVÁ und V. HEROUT: Über Terpene. LIV. Über die Struktur des Lactarazulens und Lactaroviolins. Collect. Czech. Chem. Commun. 19, 357 (1954).

203. ŠORM, F., V. BENEŠOVÁ, J. KRUPIČKA, V. ŠNEBERK, L. DOLEJŠ, V. HEROUT and J. SICHER: The Structure of Lactaroviolin. Chem. and Ind. 1954, 1511.

204. ŠORM, F., L. DOLEJŠ, O. KNESSL and J. PLÍVA: On Terpenes. XVI. On a Bicyclic Sesquiterpene and a new Azulene from the Oil of Pogostemon patchouli P. Collect. Czech. Chem. Commun. 15, 82 (1950).

205. ŠORM, F., L. NOVOTNÝ and V. HEROUT: A Further Chamazulene Precursor: The Bitter Principle of Artemisia absinthium L. Chem. and Ind. 1955, 569.

206. ŠORM, F., M. SOUCHÝ and V. HEROUT: On Terpenes. C. The Structure of Ambrosin. Collect. Czech. Chem. Commun. 24, 1548 (1959).

207. ŠORM, F. and L. URBÁNEK: On Terpenes. IV. On the Constitution of Carotol. Collect. Czech. Chem. Commun. 13, 49 (1948).

208. — — On Terpenes. VI. On the Constitution of Carotol, II. Collect. Czech. Chem. Commun. 13, 420 (1948).

209. ŠORM, F., F. VONÁŠEK and V. HEROUT: On Terpenes. VII. On a New Coloured Hydrocarbon from the Oil of Wormwood (Artemisia absinthium L.). Collect. Czech. Chem. Commun. 14, 91 (1949).

210. STAHL, E.: Über das Cham-Azulen und dessen Vorstufen, II. Mitt.: Cham-Azulencarbonsäure aus Kamille. Chem. Ber. 87, 505 (1954).

211. — Über das Cham-Azulen und dessen Vorstufen, III. Mitt.: Zur Konstitution der Cham-Azulencarbonsäure. Chem. Ber. 87, 1626 (1954).

212. STEELE, J. W., J. B. STENLAKE and W. D. WILLIAMS: The Chemistry of the Aristolochia Species. Part IV. The Structure of Aristolactone. J. Chem. Soc. (London) 1959, 3289.

213. STEPANOW, F. N. und N. A. ALDANOWA: Polymethin-Farbstoffe der Azulen-Reihe. Angew. Chem. 71, 125 (1959).

214. — — Novie proizvodnie azulena. (A New Preparation of Azulene.) Zhur. Obshchei Khimii 29, 339 (1959).

215. SUCHÝ, M : On Terpenes. XCV. The Constitution of Two Alkylchamazulenes. Collect. Czech. Chem. Commun. 24, 1303 (1959); Chem. Listy 52, 2099 (1958).

216. SUCHÝ, M., V. BENEŠOVÁ, V. HEROUT and F. ŠORM: Contribution on the Structure of Cnicin, the Bitter Principle from Cnicus benedictus L. Tetrahedron Letters 1959, No. 10, 5.

217. SUCHÝ, M., V. HEROUT and F. ŠORM: On Terpenes. LXVIII. Formation of Two Tetraalkylazulenes in the Working up of Wormwood. Collect. Czech. Chem. Commun. 21, 477 (1956); Chem. Listy 49, 1870 (1955).

218. — — — Cnicin: A New Guaianolide-Type Lactone. Chem. and Ind. 1959, 517.

219. — — — On Terpenes. CVI. On Hydrogenation Products of Cynaropicrin, the Bitter Principle of Artichoke (Cynara scolymus L.). Collect. Czech. Chem. Commun. 25, 507 (1960).

219a. — — — On Terpenes. CXVI. Structure of Cynaropicrin. Coll. Czech. Chem. Commun. 25, 2777 (1960).

220. SUTHERLAND, M. D.: Structure of the Terpenoid Geijerene. Chem. and Ind. 1959, 1220.

221. Sýkora, V., V. Herout and F. Šorm: On Terpenes. LXIV. The Constitution of Elemol. Collect. Czech. Chem. Commun. **20**, 220 (1955).

222. Sýkora, V., L. Novotný and F. Šorm: Constitution of Carotol and Daucol. Tetrahedron Letters **1959**, No. 14, 24.

223. Takeda, K.: Components of the Root of *Lindera strychnifolia* Vill. VI. Structure of Linderene. Pharm. Bull. Japan **1**, 244 (1953).

223a. Takeda, K. and H. Minato: Absolute Configuration of Guaiol. Tetrahedron Letters **1960**, No. 22, 33.

224. Takada, K. and W. Nagata: Components of the Root of *Lindera strychnifolia* Vill. V. Azulenes Isolated from Linderene by Zinc-Dust Distillation. Pharm. Bull. Japan **1**, 164 (1953).

225. Tilney-Bassett, J. F. and W. A. Waters: The Substitution of Azulene by Benzyl Radicals. J. Chem. Soc. (London) **1959**, 3123.

226. Treibs, W.: Über bi- und polycyclische Azulene. XII. Das Germacrol, ein azulenbildendes Sesquiterpen-oxyd aus Geraniumöl. Liebigs Ann. Chem. **576**, 116 (1952).

227. — Über elektrophile Substitutionen am Azulen. Naturwiss. **45**, 336 (1958).

228. — Über bi- und polycyclische Azulene. XL. Azulenglyoxylsäuren und Azulencarbonsäuren, ihre Darstellung und Reduktion mit Lithiumalanat. Chem. Ber. **92**, 2152 (1959).

229. — Über substituierende und zyklisierende Malonylierung ohne Katalysatoren. Naturwiss. **47**, 179 (1960).

230. Treibs, W. und H.-M. Barchet: Über bi- und polycyclische Azulene. IV. Mitt.: Das Aromadendren, sein chemischer Bau und seine Überführung in 5 Azulene. Liebigs Ann. Chem. **566**, 89 (1950).

231. Treibs, W. und K. Gründel: Über bi- und polycyclische Azulene, XXX. Synthese von siliciumhaltigen Azulenen. Chem. Ber. **91**, 143 (1958).

232. Treibs, W., J. Hiebsch und H. J. Neupert: Über bi- und polycyclische Azulene, XXXVIII. Synthesen von Azulencarbonsäure-nitrilen. Chem. Ber. **92**, 606 (1959).

233. Treibs, W. und D. Merkel: Über die Gurjunene, II. α-Gurjunen. Liebigs Ann. Chem. **617**, 129 (1958).

234. Treibs, W., M. Mühlstadt und K. D. Köher: Die Aminomethylierung der Azulene. Naturwiss. **45**, 336 (1958).

235. Treibs, W., H. J. Neupert und J. Hiebsch: Über bi- und polycyclische Azulene, XXXVII. Synthesen und Eigenschaften von Azulen-aldehyden. Chem. Ber. **92**, 141 (1959).

236. — — — Über bi- und polycyclische Azulene, XXXIX. Carboxychlorierung von Azulenen. Chem. Ber. **92**, 1216 (1959).

237. Treibs, W. und H. Orttmann: Über Carboxylierungen mit Oxalylbromid. Naturwiss. **45**, 85 (1958).

237a. Treibs, W., C. Vollrad und M. Reimann: Über bi- und polycyclische Azulene. XLI. Die Gültigkeit der Plattnerschen Regel für 6- und 7-substituierte Azulene. Liebigs Ann. Chem. **634**, 111 (1960).

237b. Treibs, A., R. Zimmer-Galler und C. Jutz: Über Azulenpyrrol-methin-farbstoffe. Chem. Ber. **93**, 2542 (1960).

238. Turner, R. B., W. R. Meador, W. v. E. Doering, L. H. Knox, J. R. Mayer and D. W. Wiley: Heats of Hydrogenation. III. Hydrogenation of Cyclooctatetraene and of Some Seven-membered Non-benzenoid Aromatic Compounds. J. Amer. Chem. Soc. **79**, 4127 (1957).

239. Ukita, T., M. Miyazaki and M. Hashi: Studies on Azulenes: S-Guaiazulene-aldehydes. Chem. Pharm. Bull. Japan **6**, 223 (1958).

240. UKITA, T., M. MIYAZAKI and H. WATANABE: Studies on Azulenes. II. Synthesis of Polybromoazulenes. Pharm. Bull. Japan 3, 199 (1955).

241. UKITA, T., H. WATANABE and M. ICHIGE: Studies on Azulenes. VI. Synthesis of Azulenes from Hydrindenglycol Diacetate without Dehydrogenation. Pharm. Bull. Japan 5, 422 (1957).

242. UKITA, T., H. WATANABE and M. MIYAZAKI: Studies on Azulenes. I. 2- and 3-Substituted S-Guaiazulene. J. Amer. Chem. Soc. 76, 4584 (1954).

243. UNGNADE, H. E., E. C. HENDLEY and W. DUNKEL: Tenulin. II. Anhydrotenulin and Pyrotenulin. J. Amer. Chem. Soc. 72, 3818 (1950).

244. VILLIERS, J. P. DE: The Isolation and Structure of Geigerinin, a Guaianolide. J. Chem. Soc. (London) 1959, 2412.

245. VILSMEIER, A. und A. HAACK: Über die Einwirkung von Halogenphosphor auf Alkyl-formanilide. Eine neue Methode zur Darstellung sekundärer und tertiärer *p*-Alkylamino-benzaldehyde. Ber. dtsch. chem. Ges. 60, 119 (1927).

246. VIVAR, A. R. DE and J. ROMO: Constituents of *Helenium mexicanum* H. B. K. Chem. and Ind. 1959, 882.

247. WATANABE, H.: Studies on Azulenes. VII. Oxidation of Hydroazulenes with Selenium Dioxide. Pharm. Bull. Japan 5, 426 (1957).

248. YOSHIOKA, I., H. HIKINO and Y. SASAKI: Studies on the Constituents of Atractylodes. III. Separation of Atractylol into Eudesmol and Hinesol. Chem. Pharm. Bull. Japan 7, 319 (1959).

249. — — — Studies on the Constituents of Atractylodes. IV. The Structure of Hinesol (1). The Skeleton. Chem. Pharm. Bull. Japan 7, 817 (1959).

250. — — — Structure of Hinesol. Abstract, Chem. Soc. Japan Symposium Nat. Prod., Kyoto, 1960; Chem. Pharm. Bull. Japan 9, 84 (1961).

(Received, November 14, 1960.)

Chemistry of the Natural Pyrethrins.

By **L. CROMBIE**, London, and **M. ELLIOTT**, Harpenden, Herts.

With 4 Figures.

Contents.

Acknowledgement. We wish to acknowledge a scientific debt to Professor S. H. HARPER who introduced us both to pyrethrum chemistry.

I. Introduction.

The flowers of the herbaceous perennial *Chrysanthemum cinerariae-folium* VIS. (synonym, *Pyrethrum cinerariaefolium* TREV.), a member of the Compositae family, are valued for their insecticidal properties: the history of their employment has been discussed by GNADINGER (77, 78). Commercial supplies were originally obtained from Dalmatia and Japan, but to-day the principal world source is Kenya where a rationally administered industry provides standardised products. Other Compositae flowers have insecticidal activity but of these only *Chrysanthemum coccineum* Willd. (synonyms, *C. carneum* Steud and *C. roseum* Adam) has commercial value, though it is cultivated on a restricted scale. The active insecticidal principles, the pyrethrins, are contained mostly in the achenes of *C. cinerariaefolium* (*19–21*). Harvesting the crop requires much hand labour as maximum yields are obtained by gathering the

Fig. 1. *Chrysanthemum cinerariaefolium* Vis.

flower *(Fig. 1)* when four or five rows of disc florets are open. The flowers are artificially dried and baled and are known commercially as "pyrethrum". As insecticides, the pyrethrins have the advantage of low mammalian toxicity and so far insects have not easily developed resistance towards them. They are also well known for their rapid "knock-down" or paralytic properties which are valuable in dealing with flying insects.

Early investigation of the pyrethrins by Fujitani (*65*) and Yamamoto (*194–196*) led to their recognition as esters, but the basic form of their structures was not appreciated until the distinguished investigations of STAUDINGER and RUZICKA, carried out in the years 1910 to 1916, but not published until 1924 (*174*). They recognised two esters, pyrethrin I and pyrethrin II, and formulated them. These structures have since required modification and two more esters, cinerin I and cinerin II, have been added as the outcome of a careful investigation by LaFORGE and BARTHEL (*116, 117*). Below are represented the structures, with stereochemical details, now accepted for the four insecticidal esters.

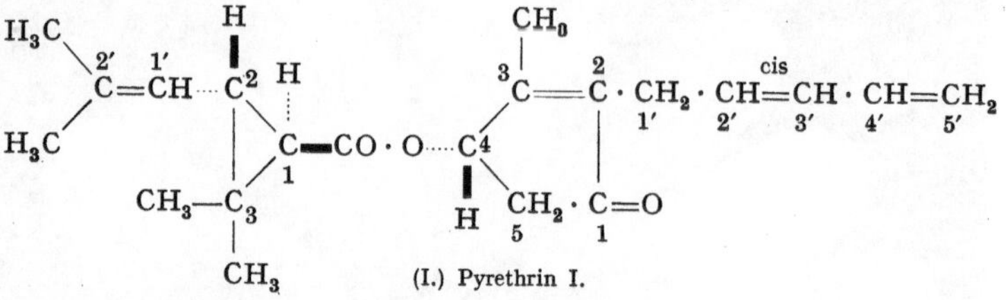

(I.) Pyrethrin I.

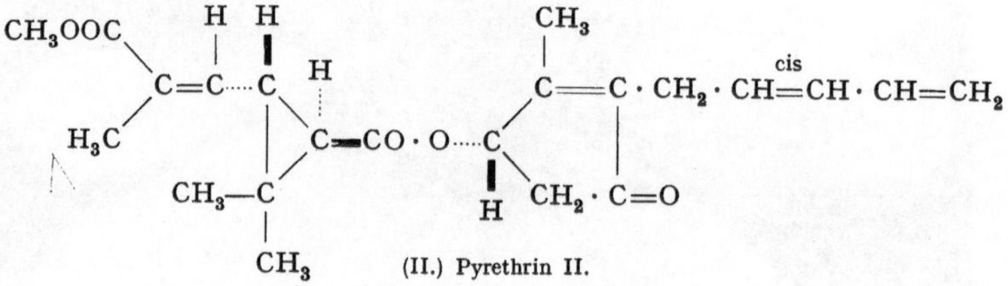

(II.) Pyrethrin II.

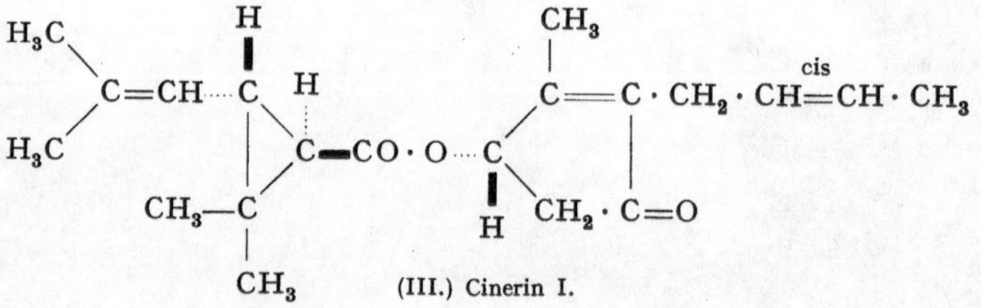

(III.) Cinerin I.

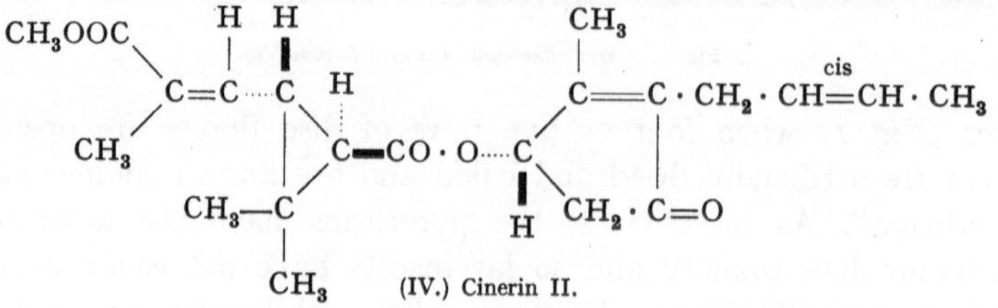

(IV.) Cinerin II.

In pyrethrin I and cinerin I, besides two optically active features, there are a double bond and a cyclopropane ring both able to show geometrical isomerism. Each natural compound is thus one of sixteen possible stereoisomers. In pyrethrin II and cinerin II there is still another double bond capable of geometrical isomerism so each natural ester is

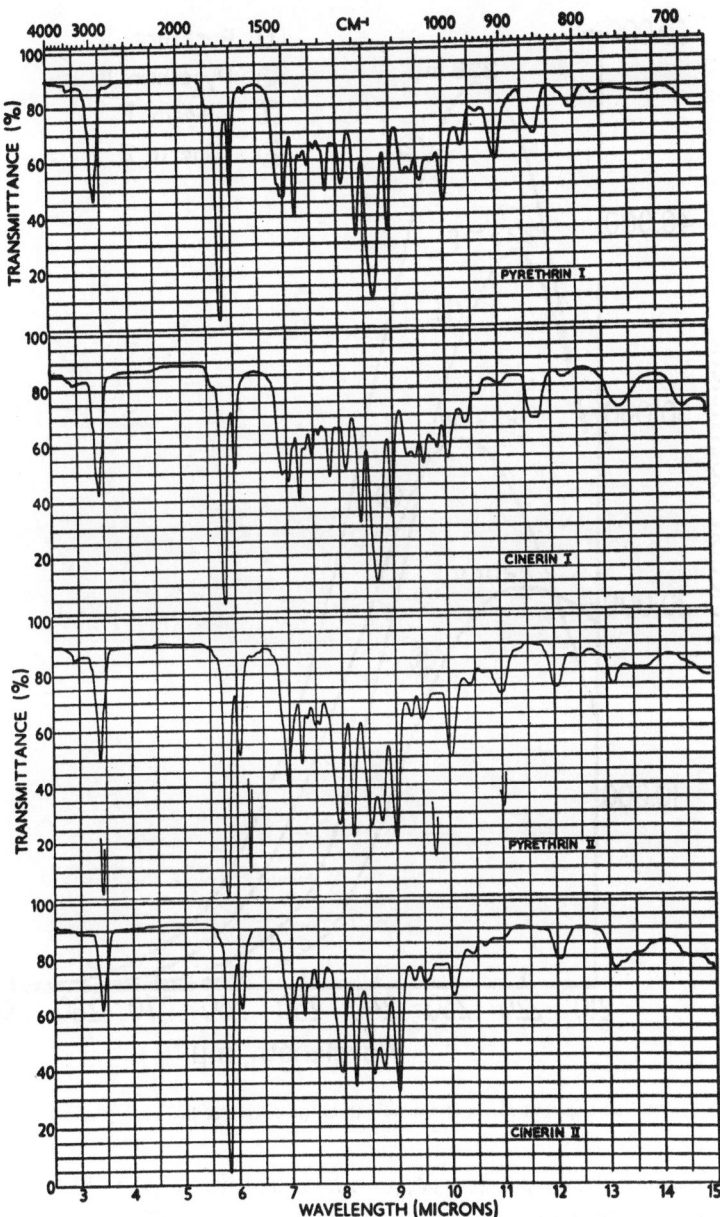

Fig. 2. Infrared spectra of the four natural rethrins [ELLIOTT (52)].

one of 32 possible stereoisomers. All four esters are high-boiling oils with the properties listed below (45, 49, 50).

8 a*

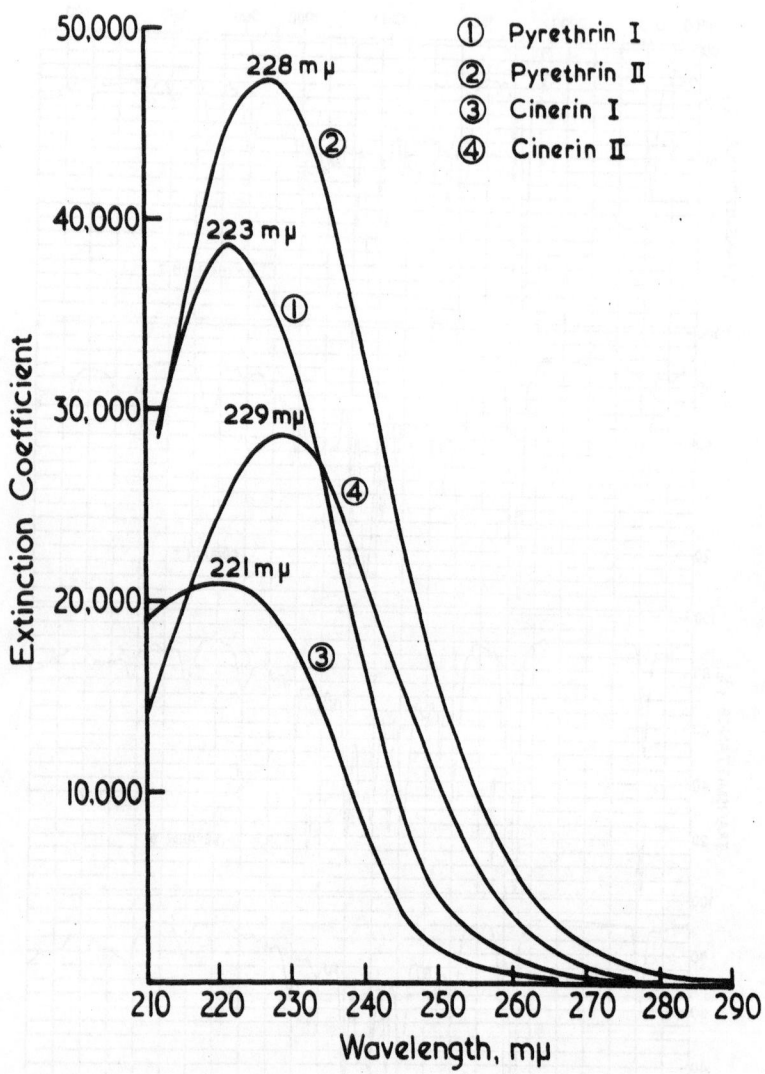

Fig. 3. Ultraviolet spectral curves of natural rethrins (in hexane or isooctane) [ELLIOTT (*45*)].

Pyrethrin I (I): b. p. 146–150°/5 × 10⁻⁴ mm., n_D^{20} 1.5242, $[\alpha]_D^{20}$ — 14° (in isooctane). 2,4-Dinitrophenylhydrazone m. p. 129–131°, $[\alpha]_D^{20}$ — 204° (c. 0.16 in benzene), λ_{max} (ethanol) 222, 380 mμ (ε 40,900 and 29,300).

Cinerin I (III): b. p. 136–138°/8 × 10⁻³ mm., n_D^{20} 1.5064, $[\alpha]_D^{20}$ — 22° (in hexane). 2,4-Dinitrophenylhydrazone m. p. 112–113° (from ethanol), m. p. 92–95° (from hexane), λ_{max} (ethanol) 224, 380 mμ (ε 22,000 and 28,000).

References, pp. 155—164.

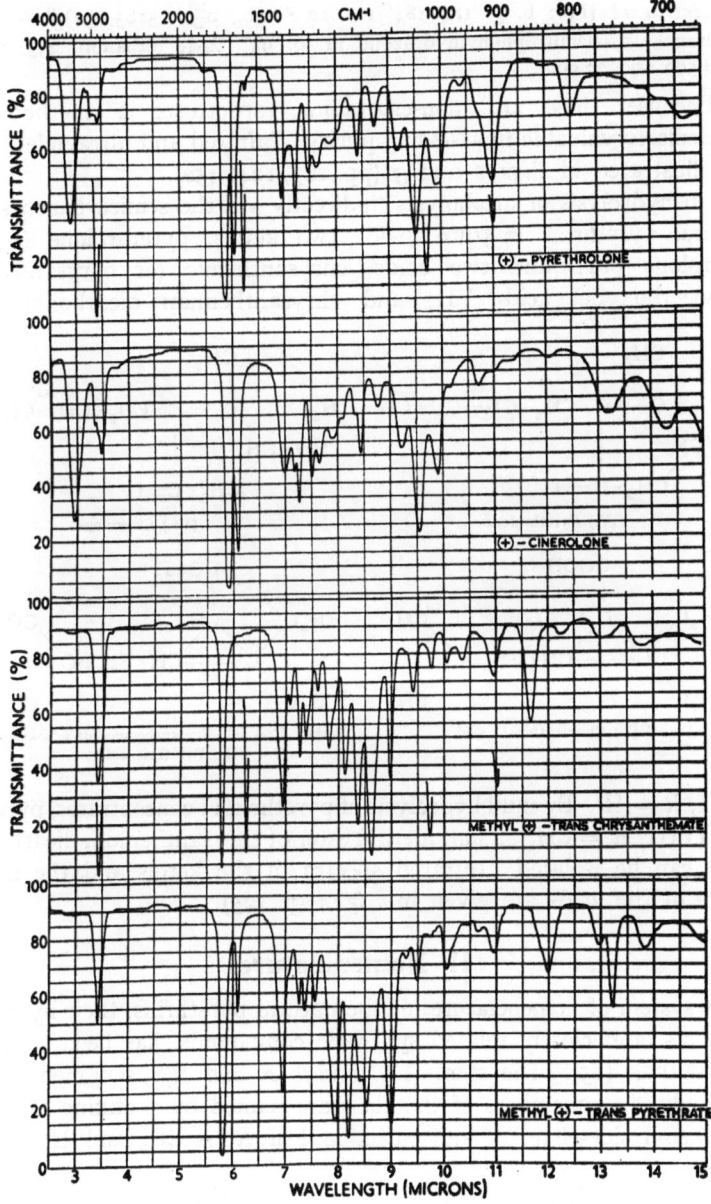

Fig. 4. Infrared spectra of natural rethrolones and chrysanthemum esters [ELLIOTT (52)].

Pyrethrin II (II): b. p. 192–193°/7 × 10⁻³ mm., n_D^{20} 1.5355, $[\alpha]_D^{19} + 14.7°$ (in isooctane-ether). 2,4-Dinitrophenylhydrazone m. p. 65–66°, λ_{max} (ethanol) 229, 380 mμ (ε 48,000 and 28,600).

Cinerin II (IV): b. p. 182–184°/1 × 10⁻³ mm., n_D^{20} 1.5183, $[\alpha]_D^{16}$ + 16° (in isooctane). 2,4-Dinitrophenylhydrazone m. p. 93°, λ_{max} (ethanol) 235, 380 mμ (ε 28,500 and 27,000).

Because they are non-crystalline, spectroscopic methods are particularly valuable in assessing purity. Infrared and ultraviolet curves are illustrated in *Figs. 2* and *3* (pp. 123, 124) (*52*).

On hydrolysis or alcoholysis (indirectly via the semicarbazones), the natural pyrethrins mixture yields two ketols, pyrethrolone (V) and cinerolone (VI), and two acids, chrysanthemic (or chrysanthemum monocarboxylic) acid (VII) and chrysanthemum dicarboxylic acid

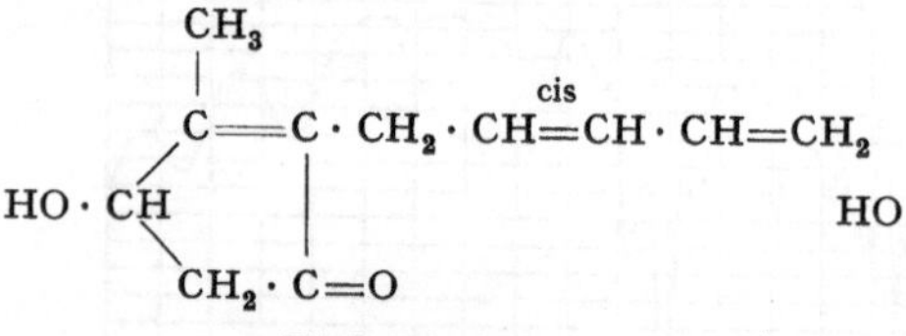

(V.) Pyrethrolone.

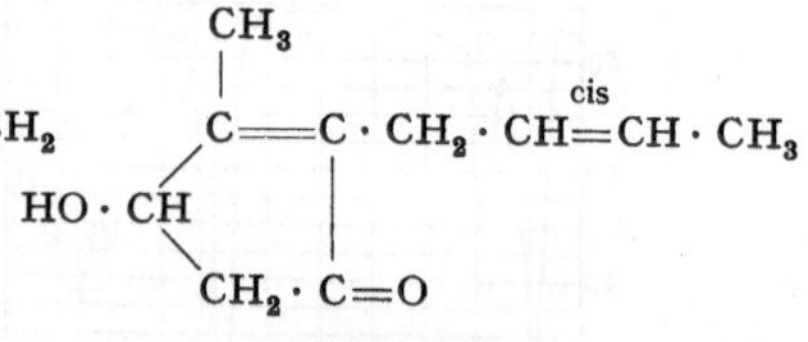

(VI.) Cinerolone.

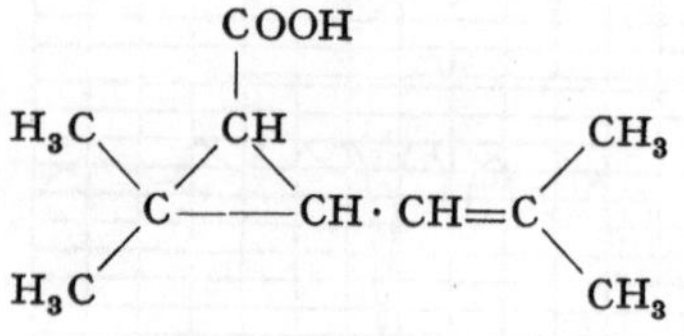

(VII.) Chrysanthemic acid.

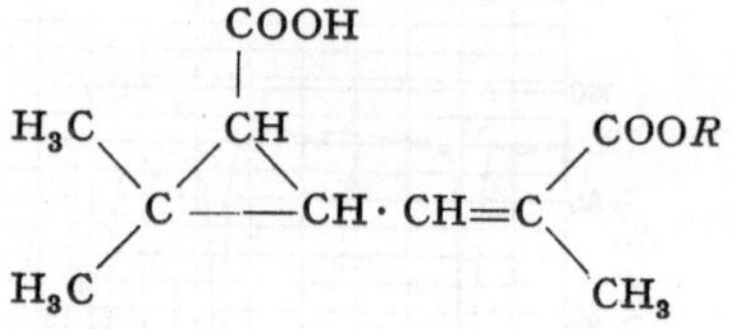

(VIII.) R = H. Chrysanthemum dicarboxylic acid.
R = CH₃. Pyrethric acid.

(VIII, R = H). By mild hydrolysis of pyrethrin II concentrates, pyrethric acid (VIII, R = CH₃), a half methyl ester of chrysanthemum dicarboxylic acid, can be isolated. Infrared spectra of the ketols and the methyl esters of the acids are given in *Fig. 4* (p. 125).

II. Nomenclature.

A system of nomenclature for the natural pyrethrins, their hydrolysis products, and closely related synthetic compounds, has been suggested by Harper (*85*). The stem (IX, *minus R and R′*) is named *-rethrin* and individual members of the class are named by prefixing the alkyl or alkenyl side-chain R and suffixing I if $R′$ is CH₃ and II if $R′$ is

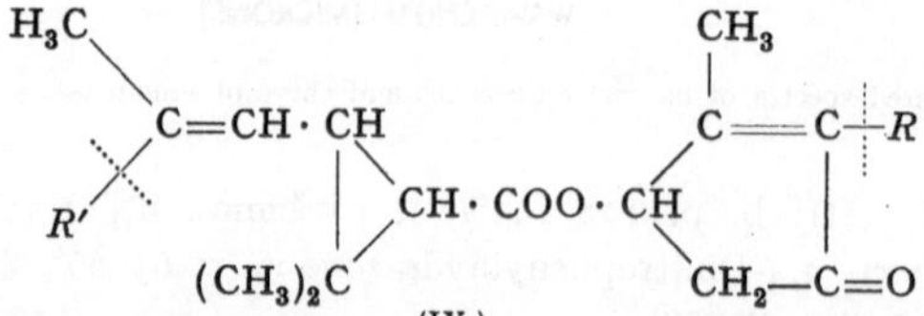

(IX.)

COOCH₃. Thus cinerin I (III, p. 122) is but-2'-enylrethrin I and
pyrethrin II (II, p. 122) is penta-2',4'-dienylrethrin II. The stem (X
minus R) is termed *-rethrolone* and individual ketols are named by
prefixing the radical *R*. The ketones (XI *minus R*) themselves are named
as *-rethrones*.

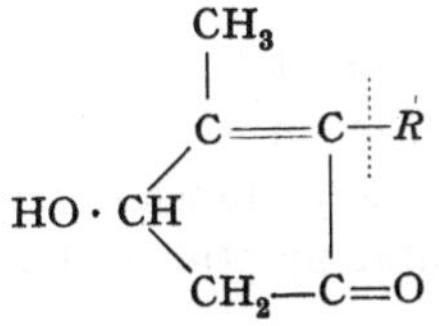

$$CH_3$$
$$|$$
$$C{=}C{\mid}R$$
$$HO\cdot CH$$
$$CH_2{-}C{=}O$$

(X.)

$$CH_3$$
$$|$$
$$C{=}C{\mid}R$$
$$H_2C$$
$$CH_2{-}C{=}O$$

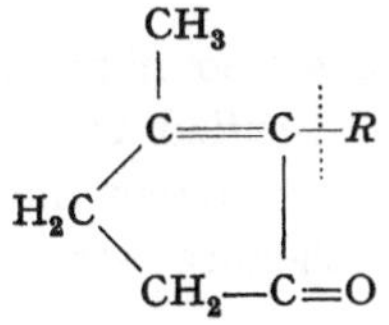

(XI.)

By expanding the system to involve the names of the cyclopropane
acids, stereochemical details can be expressed. Thus the unnatural
ester (XII), made by esterifying (±)-pent-*cis*-2'-enylrethrolone with

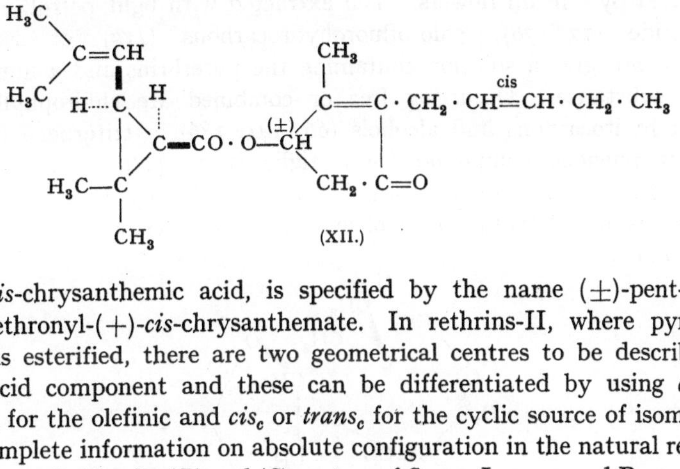

$$H_3C$$
$$C{=}CH$$
$$H_3C \quad H{\cdots}C \quad H$$
$$\qquad C{-}CO\cdot O{-}CH$$
$$H_3C{-}C \qquad (\pm)$$
$$|$$
$$CH_3$$

$$CH_3$$
$$|$$
$$C{=}C\cdot CH_2\cdot CH{=}CH\cdot CH_2\cdot CH_3 \quad cis$$
$$CH_2\cdot C{=}O$$

(XII.)

(+)-*cis*-chrysanthemic acid, is specified by the name (±)-pent-*cis*-2'-
enylrethronyl-(+)-*cis*-chrysanthemate. In rethrins-II, where pyrethric
acid is esterified, there are two geometrical centres to be described in
the acid component and these can be differentiated by using cis_o or
$trans_o$ for the olefinic and cis_c or $trans_c$ for the cyclic source of isomerism.
As complete information on absolute configuration in the natural rethrins
is now available, the (*R*) and (*S*) system of CAHN, INGOLD and PRELOG (*16*)
can be used and then description of the geometrical arrangement about the
cyclopropane ring is not needed. Thus natural pyrethrin I is fully specified
as penta-*cis*-2',4'-dienyl-4(*S*)-rethronyl-1(*R*),2(*R*)-chrysanthemate and
pyrethrin II as penta-*cis*-2',4'-dienyl-4(*S*)-rethronyl-1(*R*),2(*R*)-*trans*-
pyrethrate, the last *trans*-designation referring to the cyclopropane
trans-olefinic side-chain.

For rapid specification of stereochemical details, FELDMAN'S
notation (*59*) is useful: (*R*) or *cis*- is set equal to 0 and (*S*) or *trans*- equal
to 1. The following conventions are suggested for rethrins: (a) optical
isomerism is dealt with first, then geometrical isomerism (this follows
the precedence accorded by the numbering), but (b) the *entire* ketol

portion is dealt with before the acid. Natural pyrethrin I (I, p. 122) is thus:

$$\text{Centres} \begin{array}{|cc|}\hline \text{Ketol} & \\ 4(S), & 2'(cis) \\ \hline \end{array} \begin{array}{|cc|}\hline \text{Acid} & \\ 1(R), & 2(R) \\ \hline \end{array}$$
$$1\times2^3+0\times2^2 \quad + \quad 0\times2^1+0\times2^0$$

making the stereo number /8/ and the complete configuration of (I), is uniquely specified as /8/-pyrethrin I. Any other stereoisomer can be given a unique stereonumber which is readily translatable back to the exact stereochemical structure (59). Natural pyrethrin II (II, p. 122) is:

$$\text{Centres} \begin{array}{|cc|}\hline \text{Ketol} & \\ 4\ (S), & 2'\ (cis) \\ \hline \end{array} \begin{array}{|ccc|}\hline \text{Acid} & & \\ 1\ (R), & 2\ (R), & 1'\ (trans) \\ \hline \end{array} \text{, i. e. } /17/\text{-pyrethrin II.}$$
$$1\times2^4+0\times2^3 \quad + \quad 0\times2^2 + 0\times2^1 + 1\times2^0$$

III. Isolation of Natural Rethrins.

Dried pyrethrum flowers, when extracted with light petrol, ethylene dichloride (77, 78), chlorofluorohydrocarbons (112) or isopropyl ether (168), give a solution containing the pyrethrins and a number of other substances. Occurring free or combined are chlorophylls (14), higher hydrocarbons and alcohols (66, 141, 186), a triterpene (66), an irritant phenolic compound (181), tiglic (114), palmitic and linoleic acids (1), and pyrethrosin formulated as (XIII) by BARTON and his colleagues (6). Palmitic and linoleic acids occur esterified with pyrethrolone (128).

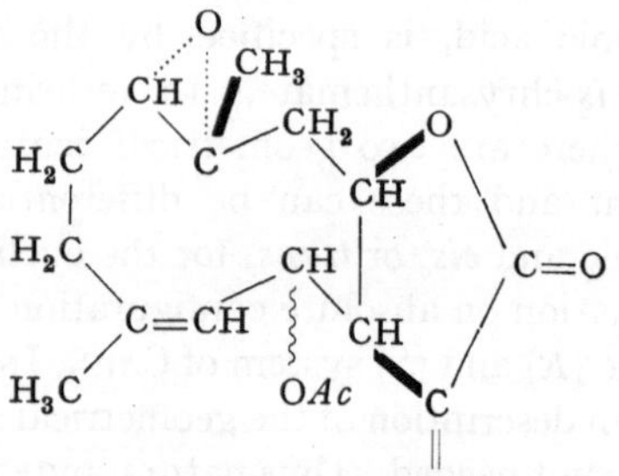

(XIII.) Pyrethrosin.

Extraction of a petrol solution of pyrethrum extract with nitromethane, and passage of the nitromethane solution through a short column of activated charcoal, gives pyrethrins concentrate of 90% purity or above (4, 5, 162). Molecular distillation can be successfully used to purify and decolourise pyrethrins concentrate (55).

Separation of the four esters (I–IV) presents formidable difficulty and, generally, chemical work has been concentrated on the derived acids and alcohols (V–VIII) obtained from the pyrethrins mixture. It is of interest that until WARD (187) was able to separate the natural esters by displacement chromatography on alumina contained in a tapered column, there was no rigorous proof that all four insecticidal esters (I–IV) were in fact present. Pyrethrin I with cinerin I is fairly easily separated from pyrethrin II with cinerin II on alumina or silica gel (*22, 131, 170*), but further separation is difficult. This situation applies also to paper chromatographic methods (*131, 152*) and to thin-layer chromatography (*170*). Amounts of the four esters adequate for most bioassay purposes have been obtained by displacement chromatography (*161, 180, 187*) and these have been compared with the much larger samples of natural rethrins available by re-esterifying the appropriate pure natural ketol and acid (*49, 50, 67, 120*).

The latest re-esterified materials are identical in biological activity with chromatographically derived materials (*161*), though earlier distilled specimens of pyrethrin I and especially pyrethrin II probably had reduced biological activity because of thermal isomerisation (p. 153).

IV. Structure and Chemistry of Chrysanthemic and Pyrethric Acids.

STAUDINGER and RUZICKA (*175*) found that hydrolysis of natural pyrethrins concentrate of Dalmatian origin gave (+)-chrysanthemic acid (XIV), $C_{10}H_{16}O_2$, and (+)-chrysanthemum dicarboxylic acid (XV), $C_{10}H_{14}O_4$. These monoterpenic acids* are separable by steam distillation and have m. p. 17–21°, $[\alpha]_D + 14.6°$ (in ethanol) (*17*), and m. p. 164°, $[\alpha]_D^{17} + 72.8°$ (in methanol) (*175*), respectively. (+)-Chrysanthemic acid absorbs bromine and takes up 1 mol. of hydrogen over a platinum or palladium catalyst in ethyl acetate to give a dihydro derivative in which the cyclopropane ring is intact (*86*). Hydrogenolytic ring cleavage is common in the cyclopropane series and hindrance by the substituents to the catalyst surface may play a part in the survival of the ring.

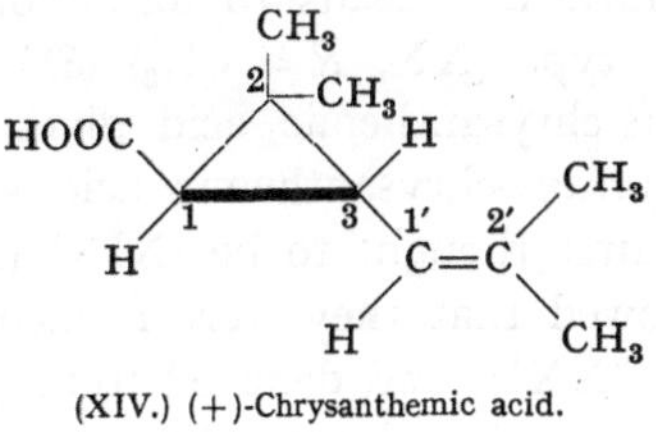

(XIV.) (+)-Chrysanthemic acid.

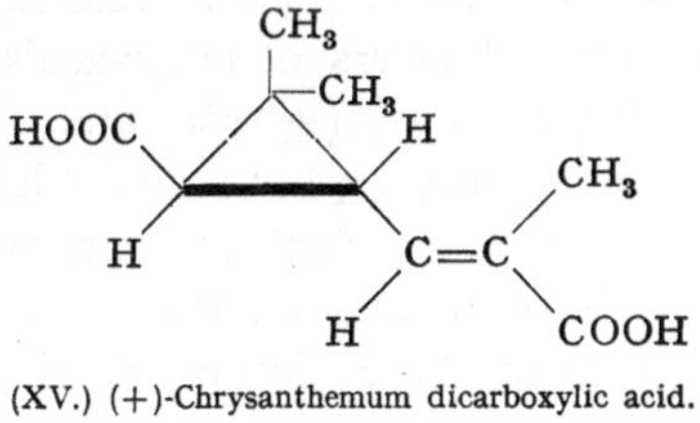

(XV.) (+)-Chrysanthemum dicarboxylic acid.

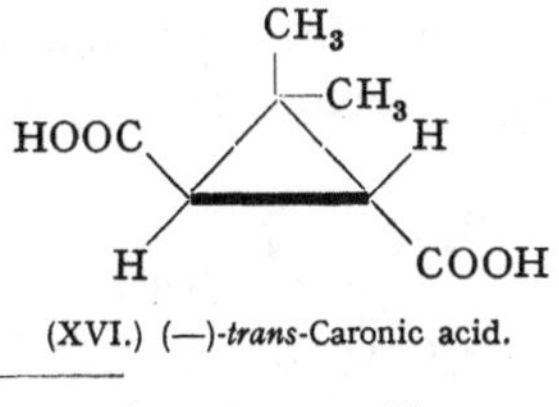

(XVI.) (—)-*trans*-Caronic acid.

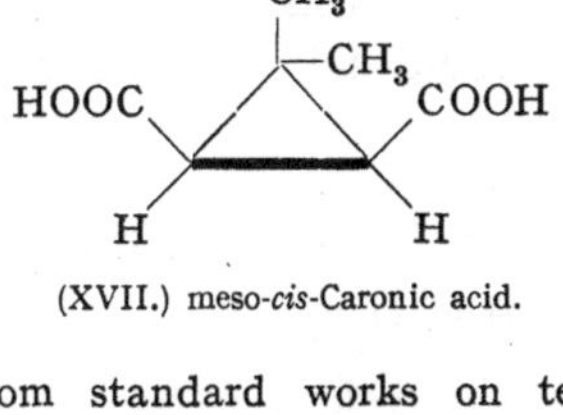

(XVII.) meso-*cis*-Caronic acid.

---

* Curiously, these acids are omitted from standard works on terpenes e. g. (*44, 167*), although they were RUZICKA's first introduction to terpene chemistry (*156*).

Ozonolysis of (+)-chrysanthemic acid (*175*) gave acetone and (—)-*trans*-caronic acid (XVI), recognised by epimerisation to meso-*cis*-caronic acid (XVII), thus establishing structure (XIV) apart from its absolute configuration. Both *cis*- and *trans*-chrysanthemic acids have been synthesised and optically resolved (see p. 134). In agreement with the above geometry the *cis* acids are converted by refluxing with dilute mineral acid into a lacto-enoic (60 : 40) equilibrium mixture, e. g. of (XVIII)

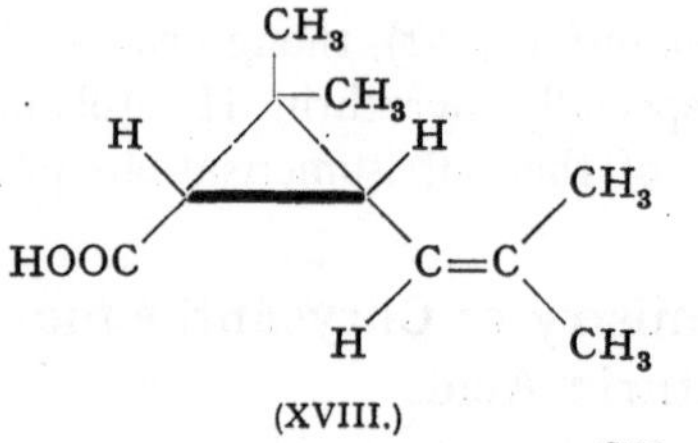

(XVIII.) 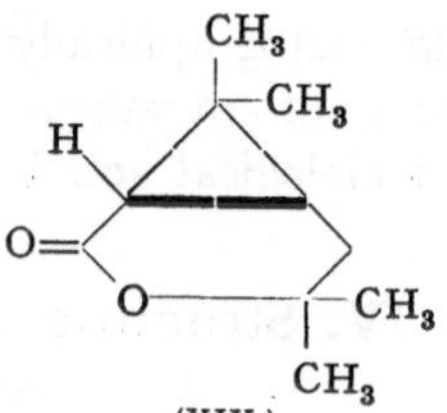   (XIX.)

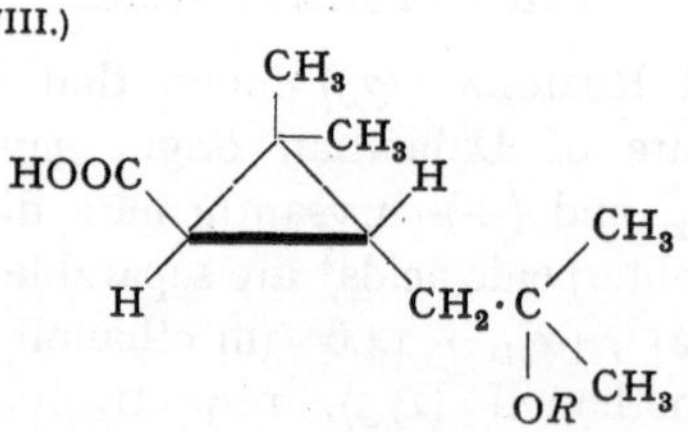

(XX.)

and (XIX) whilst the *trans*-acids, including the natural (+)-isomer, are hydrated to compounds of type (XX, R = H) (*40, 92*). As expected, a carbonium ion is readily formed at position 2′ in acid solution, and esterification of chrysanthemic acids is best carried out with diazomethane because methanolic sulphuric acid leads to formation of the methyl esters of ether-acids of the type (XX, R = CH₃) (*88*).

Pyrolysis of (±)-*cis*- and (±)-*trans*-chrysanthemic acid gives the same lactone, (±)-pyrocin, whilst (+)-*trans*-chrysanthemic acid gives a (—)-pyrocin. Such lactones were at first thought to be (XIX) (*134*) but CROMBIE and HARPER (*31, 40*) showed that they were formed by rupture of the cyclopropane ring as in (XXI), and deduced from (—)-pyrocin the absolute configuration of natural (+)-*trans*-chrysanthemic acid. Ozonolysis of (—)-(XXI) gave (+)-terebic acid and acetone. (+)-Terebic acid can be configurationally related, via (—)-isopropyl-

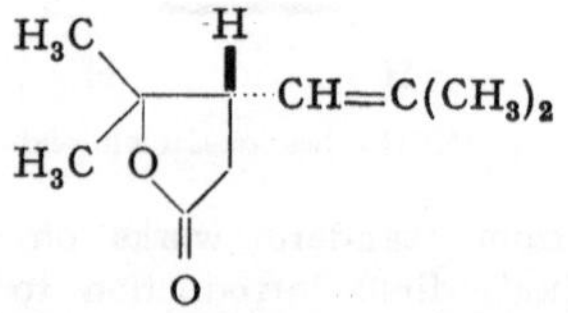

(XXI.) 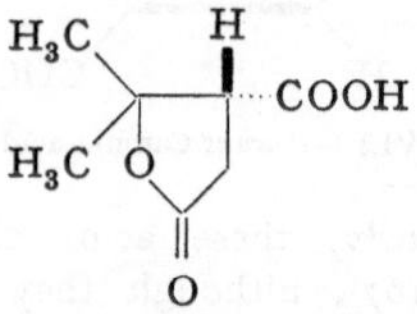   (XXII.)

succinic acid and (—)-methylsuccinic acid, to glyceraldehyde (*61*, *62*), and the absolute configuration at the unracemised position 2 in (—)-pyrocin and (+)-*trans*-chrysanthemic acid is (R) (*31*). The structure of (±)-pyrocin has also been confirmed by synthesis (*139*) from the acid chloride of terebic acid and isopropylzinc iodide. According to

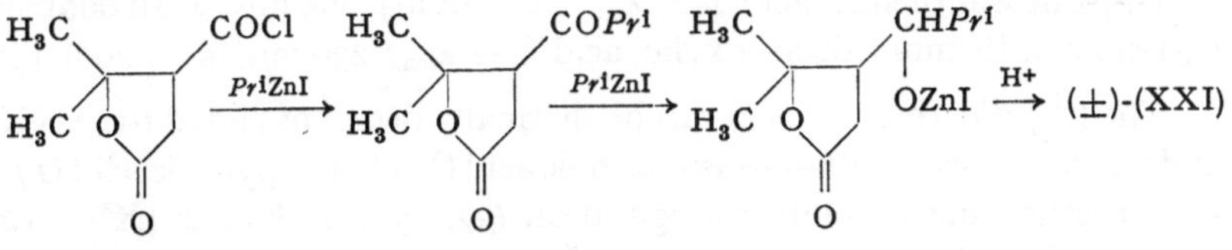

(XXIII.)   (XXIV.)   (XXV.)

MATSUI and his colleagues (*139*) the latter functions as a reducing agent in the second stage (XXIV → XXV) but other investigators (*140*) consider that the acid chloride is first reduced as in (XXVI) and that the product (XXVII or XXVIII) then reacts with more reagent to give pyrocin.

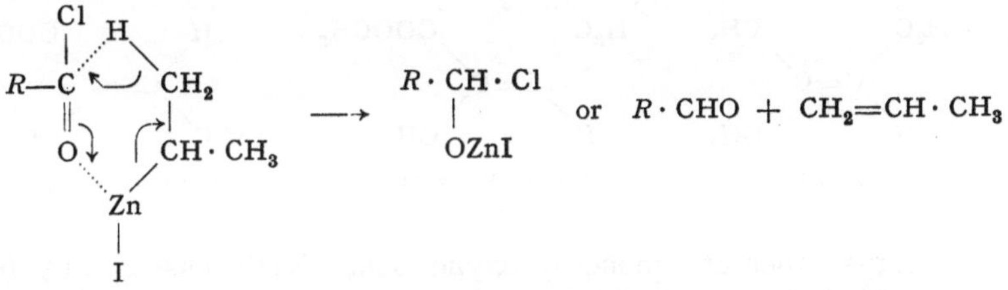

(XXVI.)   (XXVII.)   (XXVIII.)

That the 1-centre in natural (+)-chrysanthemic acid has the (R) configuration follows from the geometry of the acid but a direct determination of absolute configuration has been achieved by Arndt-

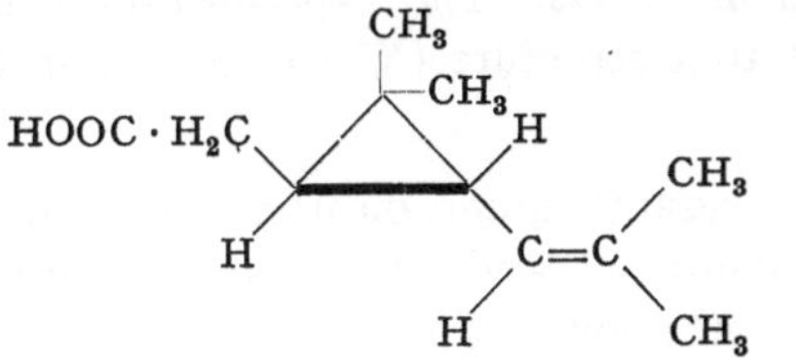

(XXIX.) 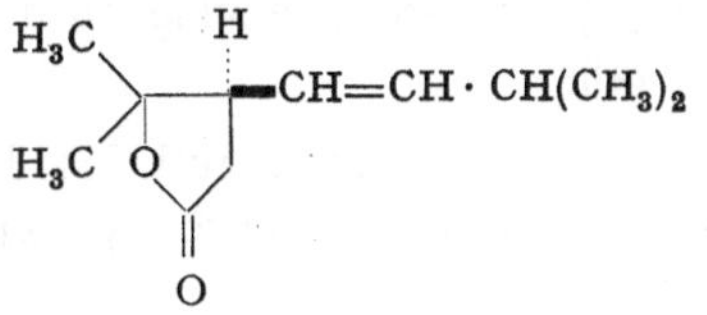   (XXX.)

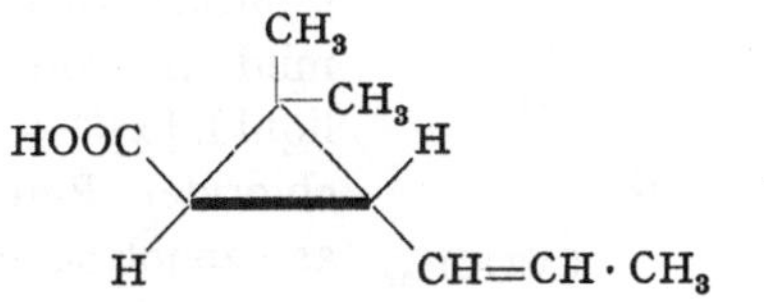

(XXXI.)

9*

Eistert homologation to give the homoacid (XXIX). When refluxed with acid, this lactonises to (XXX) and on ozonisation the lactone yields isobutyraldehyde and the enantiomorph of (XXII), i. e. (—)-terebic acid. The lactonisation preserves $C_{(1)}$ intact and the absolute configuration is thus (R) as in (XIV) (23).

(+)-Chrysanthemic acid has λ_{max} (apparent) 205 mμ (ε 10,000) but (+)-chrysanthemum dicarboxylic acid has λ_{max} 236 mμ (ε 14,200) (37).

The (+)-dicarboxylic acid can be thermally decarboxylated to (XXXI) and on ozonolysis gives (—)-trans-caronic acid (XVI) and pyruvic acid (175). Its structure and absolute configuration (37, 175) is thus as [XV, 1(R), 2(R)] with only the geometry of the side-chain requiring definition. Since (±)-trans-chrysanthemum dicarboxylic acid has been synthesised from trans-2-α,δ-dimethylsorbic acid (37), the geometry should be trans as in (XV). A more direct proof uses nuclear magnetic resonance spectroscopy as follows (23). The τ value for the olefinic hydrogen in (XXXII) is 4.79 p. p. m. but the 1'-hydrogen of the methyl ester of

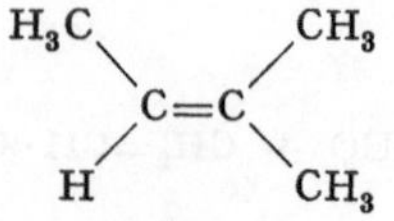

(XXXII.) 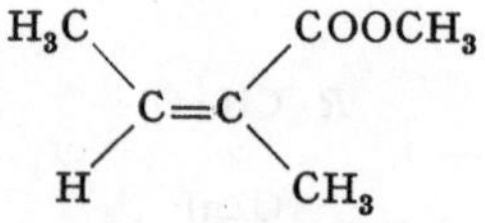   (XXXIII.) Methyl angelate. 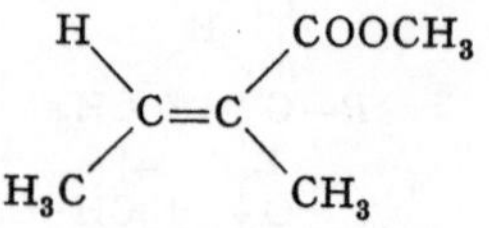   (XXXIV.) Methyl tiglate.

trans-chrysanthemum monocarboxylic acid (XIV) has τ 5.13 (mean value). Methyl tiglate (XXXIV) has τ 3.27 and methyl angelate (XXXIII) τ 4.02 for the β-hydrogen (109). Assuming the shifts relative to (XXXII), (XXXIII) and (XXXIV) are applicable to the chrysanthemic acid series, the calculated value for the 1'-hydrogen in the dimethyl ester of chrysanthemum dicarboxylic acid is 3.61 in the case of (XV) with a trans side-chain and 4.36 if the side-chain is cis. The experimental value is 3.58 (mean τ) leaving no doubt that structure (XV) is correct in all details.

In the natural rethrins II, chrysanthemum dicarboxylic acid is esterified at one carboxyl with methanol and at the other with a rethrolone. The orientation of the esters is known as pyrethrins II semicarbazone concentrate gives, on mild hydrolysis, pyrethric acid, a liquid, $[\alpha]_D^{18} + 103.9°$ (in carbon tetrachloride). Pyrethric acid is (XXXV) as ozonolysis gives (—)-trans-caronic acid and methyl pyruvate (175).

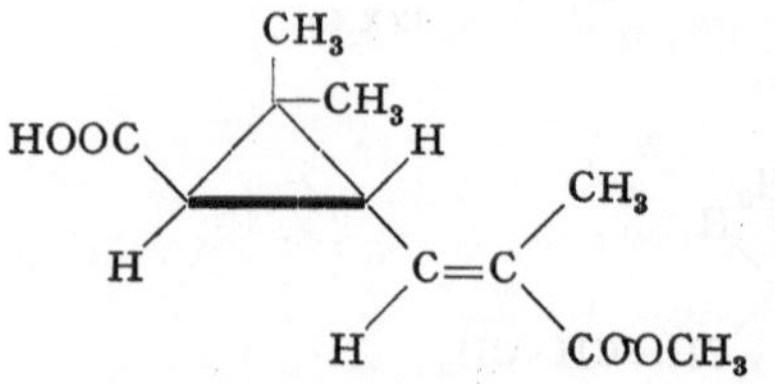

(XXXV.) Pyrethric acid.

V. Synthesis of Chrysanthemic and Pyrethric Acids.

The first synthetic work in this field was by STAUDINGER, RUZICKA and their collaborators (*173*) who added ethyl diazoacetate to 2,5-dimethyl-hexa-2,4-diene and, after hydrolysis, obtained (±)-*cis*-chrysanthemic acid in poor yield. The (±)-*trans*-acid was not isolated, though its presence was established. Such an approach has been greatly improved by CAMPBELL and HARPER (*17, 18*). 2,5-Dimethylhexa-2,4-diene (**XXXVII**) was made by self-coupling β-methylallyl chloride in the presence of magnesium and isomerising the product (**XXXVI**) to (**XXXVII**) in the vapour phase over alumina impregnated with chromic oxide (*96*),

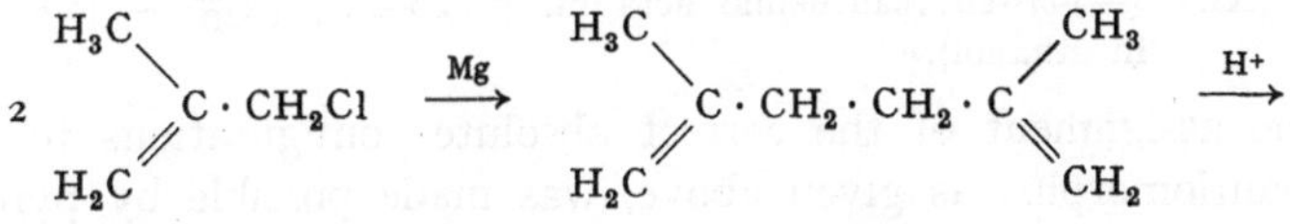

β-Methylallyl chloride.   (XXXVI.)

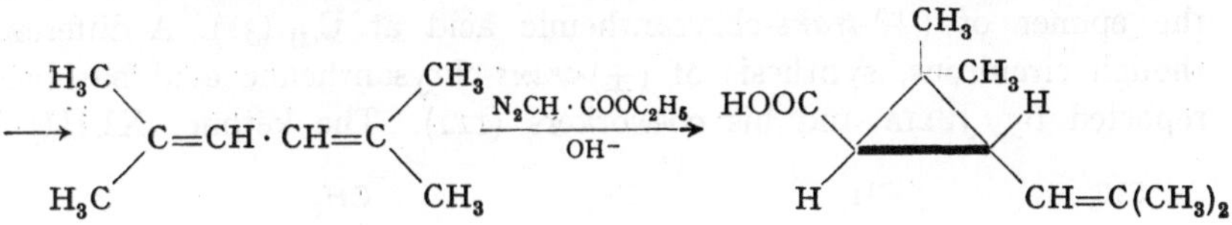

(XXXVII.) 2,5-Dimethylhexa-2,4-diene.   (XXXVIII.)

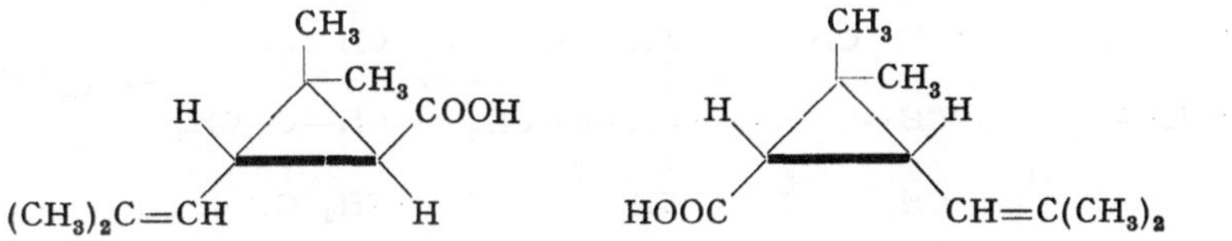

(XXXIX.)   (XL.)

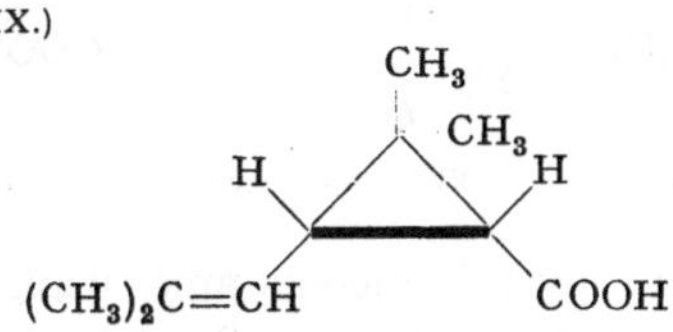

(XLI.)

or, better, in the liquid phase with toluene-*p*-sulphonic acid (*90*). Addition of ethyl diazoacetate as above gave ethyl (±)-*cis*- and (±)-*trans*-chrysanthemates and on hydrolysis a 33 : 77 mixture (*63*) of the corresponding acids was obtained. Improvements to the preparation (*90*) and a procedure for the safe handling of diazoacetic ester (*11*) have been reported: alternative ways to the inter-mediate (**XXXVII**) such as hydrogenation and dehydration of the readily obtainable diol (**XLII**) are available.

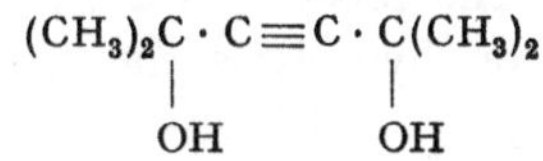

$$(CH_3)_2C \cdot C \equiv C \cdot C(CH_3)_2$$
$$\underset{OH}{|}\underset{OH}{|}$$

(XLII.)

The two racemates (XXXVIII–XXXIX) and (XL–XLI) are separable by crystallisation from ethyl acetate. Each racemate has been resolved using quinine and (—)- or (+)-α-phenylethylamine to give the four stereoisomers (XXXVIII) to (XLI), the first of these being the natural (+)-*trans*-acid. Their properties are as follows (*18*):

(XXXVIII), (+)-*trans*-chrysanthemic acid m. p. 17–21°, $[\alpha]_D^{20}$ + 14.4° (c. 2.77, in ethanol).

(XXXIX), (—)-*trans*-chrysanthemic acid m. p. 17–21°, $[\alpha]_D^{23}$ — 14.4° (c. 2.78, in ethanol).

(XLI), (+)-*cis*-chrysanthemic acid m. p. 40–42°, $[\alpha]_D^{22}$ + 40.8° (c. 1.775, in ethanol).

(XL), (—)-*cis*-chrysanthemic acid m. p. 40–42°, $[\alpha]_D^{20}$ — 40.8° (c. 1.52, in ethanol).

The assignment of the correct absolute configurations to the two *cis*-enantiomorphs, as given above, was made possible by pyrolysis of (+)-*cis*-chrysanthemic acid to (+)-pyrocin. The (—)-*cis*-acid is therefore the epimer of (+)-*trans*-chrysanthemic acid at $C_{(1)}$ (*31*). A different, though circuitous, synthesis of (±)-*trans*-chrysanthemic acid has been reported by JULIA and his co-workers (*111*). The ketone (XLIII) is

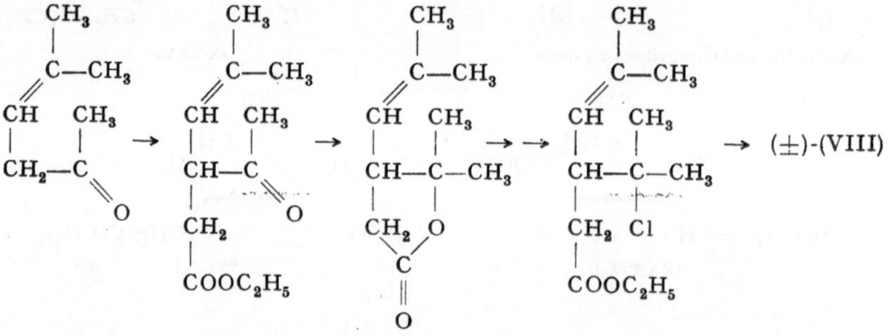

(XLIII.)   (XLIV.)   (XLV.) (±)-Pyrocin.   (XLVI.)

α-substituted via the pyrrolidine-enamine using ethyl bromoacetate. Treatment of the resulting keto-ester (XLIV) with methyl magnesium bromide yields (±)-pyrocin (XLV). This is treated with thionyl chloride in benzene and then with ethanolic hydrogen chloride to give the chloro-ester (XLVI) which cyclises under the influence of sodium *tert*-amylate to give ethyl (±)-*trans*-chrysanthemate. (±)-Dihydrochrysanthemic acid has been made similarly (*110*). (±)-*trans*-Chrysanthemic acid has also been made by adding diazoacetonitrile to tetramethylbutadiene (*91*). The product is a 73 : 27 mixture of (±)-*trans*- and (±)-*cis*-nitriles but on hydrolysis the *cis*-nitrile is epimerised and the product is almost pure (±)-*trans*-chrysanthemic acid. A disadvantage of this route is the hazardous nature of diazoacetonitrile.

Synthesis of chrysanthemum dicarboxylic acid has been effected by CROMBIE, HARPER and SLEEP (*36*, *37*). Reformatski reaction between β-methyl crotonaldehyde (conveniently obtained from acetoacetaldehyde dimethylacetal) and ethyl α-bromopropionate gives a mixture of retro-ester (XLVII) and fully conjugated ester (XLVIII). The former is converted into the latter by isomerisation with phosphorus oxychloride. When treated with diazoacetic ester and hydrolysed, (±)-*cis*$_c$- and (±)-*trans*$_c$-chrysanthemum dicarboxylic acids are obtained. Attack is at the γ,δ-double bond as with sorbic ester (*56*, *89*). The (±)-*cis*$_c$-racemate m. p. 208–210° is ozonised to (±)-*cis*-caronic acid and the (±)-*trans*$_c$-racemate m. p. 207–208° to (±)-*trans*-caronic acid. An earlier Japanese claim (*103*) to have synthesised chrysanthemum dicarboxylic acid is

$$(CH_3)_2C{=}CH\cdot CHO \xrightarrow[\text{Zn}]{CH_3CHBr\,.\,COOC_2H_5} \begin{array}{c} H_2C \\ {} \\ H_3C \end{array}\!\!\!\!\!\big> C\cdot CH{=}CH\cdot \overset{\overset{\displaystyle CH_3}{|}}{CH}\cdot COOC_2H_5$$

(XLVII.)

+

$$\begin{array}{c} H_3C \\ {} \\ H_3C \end{array}\!\!\!\!\!\big> \overset{\displaystyle C}{\underset{\displaystyle \underset{\displaystyle COOH}{|}}{\underset{CH}{\big|}}}{-}\!\!\overset{\displaystyle H}{\underset{}{C}}{-}C{=}\overset{\overset{}{}}{\underset{\underset{\displaystyle CH_3}{|}}{C}}\cdot COOH \xleftarrow[OH^-]{N_2CH\,.\,COOC_2H_5} \begin{array}{c} H_3C \\ {} \\ H_3C \end{array}\!\!\!\!\!\big> C{=}CH\cdot CH{=}\overset{\overset{\displaystyle CH_3}{|}}{C}\cdot COOC_2H_5$$

(XLIX.) (±)-*cis*- and (±)-*trans*-Chrysanthemum dicarboxylic acid.

(XLVIII.)

$$HOOC\cdot CH_2\cdot CH_2\cdot \overset{\overset{\displaystyle CH_3}{|}}{C}{=}CH\cdot CH{=}\overset{\overset{\displaystyle CH_3}{|}}{C}\cdot COOH$$

(L.)

erroneous, the product being the isomer (L) (*36*, *37*). The above synthesis has been confirmed (*107*, *108*) and the (±)-*trans*$_c$-racemate has been resolved with (—)-α-phenylethylamine to give the (+)-acid m. p. 163–164°, $[\alpha]_D^{11} + 70.9°$ (in ethanol), identical with the natural acid (*100*, *101*). The ester (XLVIII) has the *trans*$_c$-2-configuration since the corresponding acid is not lactonised by acid (*37*), and other evidence supports this (*104*, *105*). The synthetic (±)-*trans*- and (±)-*cis*-racemates are therefore considered to have *trans*-olefinic side-chains as in (XLIX) and in the case of the *trans*-racemate this is confirmed by the nuclear magnetic resonance work mentioned above.

Other routes to chrysanthemum dicarboxylic acids have since been reported. Thus oxidation of (±)- or (+)-methyl chrysanthemate with selenium dioxide gives the aldehyde (LI) (or its olefinic geometrical isomer), oxidised by silver oxide and hydrolysed to (±)- or (+)-

chrysanthemum dicarboxylic acid: the Willgerodt reaction has been used similarly (*138*). MATSUI and his colleagues (*138*) have also converted the aldehyde (LII) into chrysanthemum dicarboxylic acid by Perkin

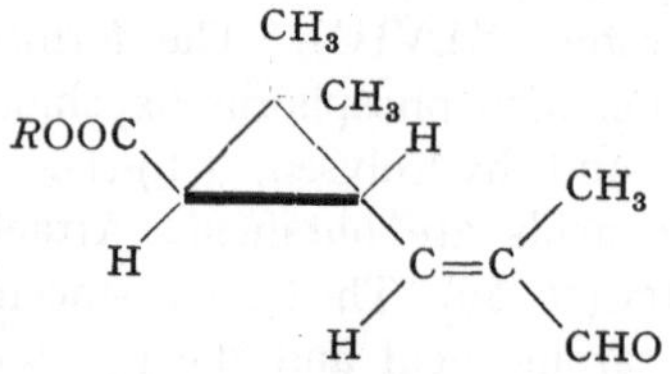

(LI.)

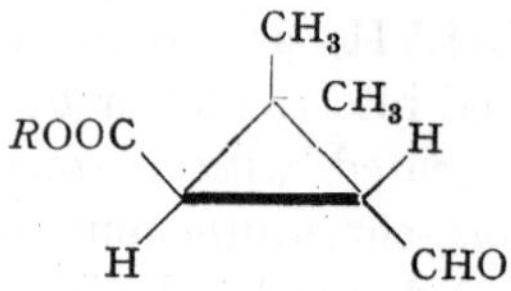

(LII.)

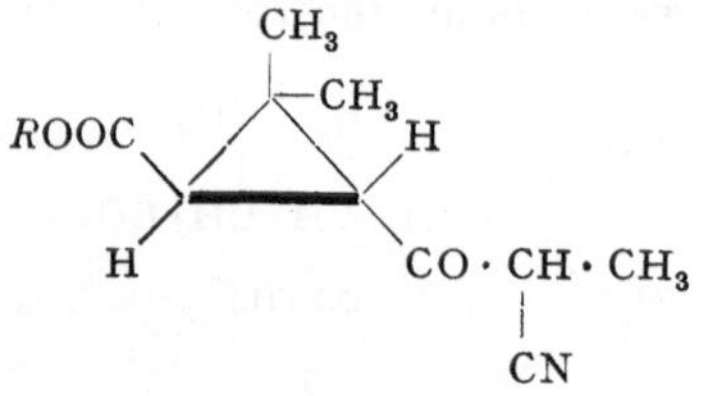

(LIII.)

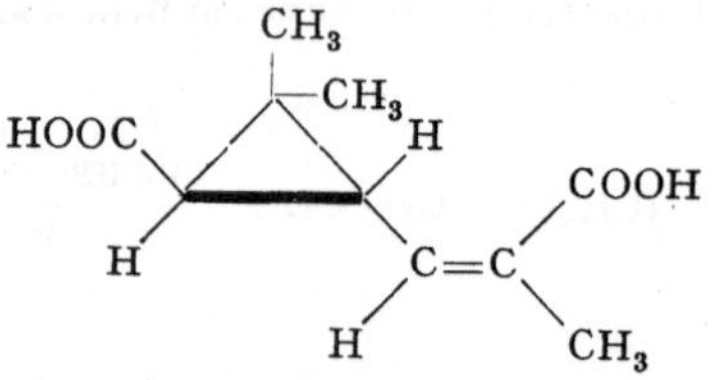

(LIV.)

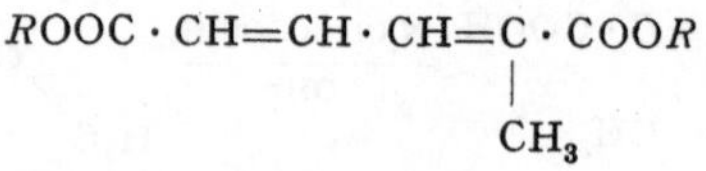

$$ROOC \cdot CH{=}CH \cdot CH{=}C \cdot COOR$$
$$| $$
$$CH_3$$

(LV.) α-Methyl-muconic esters.

reaction using potassium propionate and propionic anhydride: condensation of dimethyl *trans*-caronate with propionitrile and sodium ethoxide is stated to give the ketonitrile (LIII) which when reduced with boro-hydride, dehydrated and hydrolysed gives the *cis*-olefinic isomer (LIV), m. p. 175–177°. INOUE and co-workers (*106*) have approached the matter by adding dimethyldiazomethane to *trans*-2, *trans*-4 and *cis*-2, *trans*-4-α-methyl-muconic esters (LV): pyrazolines are formed which are pyrolysed. In the first case (±)-*trans*-chrysanthemum dicarboxylic acid with a *trans*-olefinic side-chain was obtained but the second case is said to give the same compound with a *cis*-side-chain (*183*). The m. p. 191° is not, however, the same as MATSUI's.

(+)-*trans*-Pyrethric acid (XXXV, p. 132), the acid from mild hydrolysis of rethrins II, was synthesised by complete esterification of (+)-*trans*-chrysanthemum dicarboxylic acid and then half-hydrolysis of the dimethyl ester (*37*). The product had $[\alpha]_D^{16} + 103.4°$ (in carbon tetrachloride) and when ozonised gave methyl pyruvate and (+)-*trans*-caronic acid. Its structure is also indirectly supported by biological evidence, for (±)-allylrethronyl (+)-*trans*-chrysanthemate was 4.1 times as toxic to house flies as the (±)-allylrethronyl (+)-*trans*-pyrethrate made from the synthetic specimen (*37*). Under the same conditions, using natural (+)-*trans*-pyrethrate, a ratio of 4.4 is reported (*121*).

VI. Structure and Chemistry of Pyrethrolone and Cinerolone.

Pyrethrolone.

Pyrethrolone (V, p. 126) was originally isolated from the pyrethrins semicarbazones by methanolysis followed by cleavage of the ketol semicarbazone with potassium hydrogen sulphate (*174*) or dilute sulphuric acid (*49*). It is then obtained in (+)- and (±)-forms though the latter is accepted as an artefact. The second alcoholic component of the pyrethrins mixture, cinerolone (VI, p. 126) was not recognised until much later (*118, 119*) and its isolation led to the realisation that earlier degradative work had been carried out on mixtures of pyrethrolone and cinerolone in the belief that a pure component was being handled. In the account which follows, adjustment for this and certain other early errors has been made and at times this may cause superficial discrepancy between statement and reference cited.

ELLIOTT (*49*) has recently shown that natural pyrethrolone forms a crystalline monohydrate m. p. 38.8–39.6° and crystallisation of this from ether at low temperature is an excellent method for obtaining pure (+)-pyrethrolone, a liquid, n_D^{20} 1.5475, λ_{max} (in ethanol), 226 mμ (ε 33,900), $[\alpha]_D^{20}$ + 15.1° (c. 12.7, in ether).

(+)-Pyrethrolone contains a keto-group (semicarbazone) and a hydroxyl group (acetate, methyl ether) and has three double bonds, two of which are easily hydrogenated (*79, 126, 175*). More vigorous catalytic hydrogenation, especially of the acetate, causes complete saturation and allylic cleavage to give hexahydropyrethrone (LIV) (*80*), on which important degradative work by STAUDINGER and RUZICKA pivots (*177*). When hexahydropyrethrone is oxidised with permanganate,

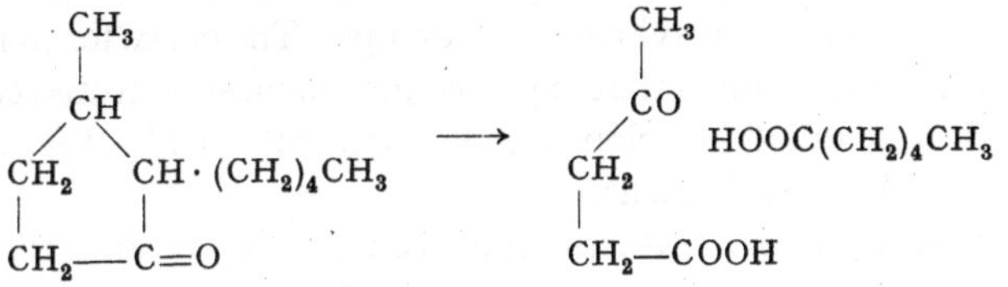

(LIV.) Hexahydropyrethrone.

laevulic acid and caproic acid are formed, allowing structures (LIV) or (LV). These were differentiated by Beckmann degradation. A lactam (LVI) hydrolysable to an amino acid, and not an amide, was isolated, so (LIV) is correct (*177*). Restrained catalytic hydrogenation of pyrethrolone yields tetrahydropyrethrolone (LVII, R = OH) (*127*); and similar hydrogenation of pyrethrone [made by reducing pyrethrolone with aluminium amalgam (*129*)] gives tetrahydropyrethrone (LVII,. $R = H$) (*127*), also obtainable from chlorotetrahydropyrethrone by

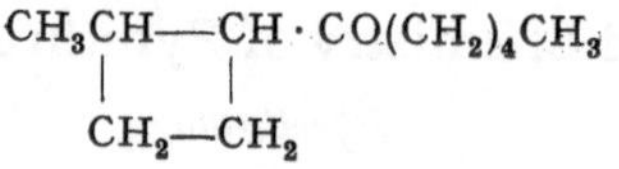

$$CH_3CH\!\!-\!\!CH \cdot CO(CH_2)_4CH_3$$
$$| \qquad |$$
$$CH_2\!\!-\!\!CH_2$$

(LV.)

$R = OH.$ Tetrahydropyrethrolone.
$R = H.$ Tetrahydropyrethrone.

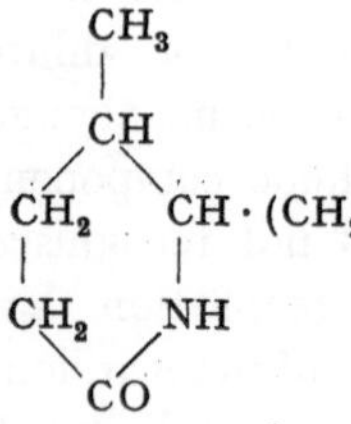

(LVI.) 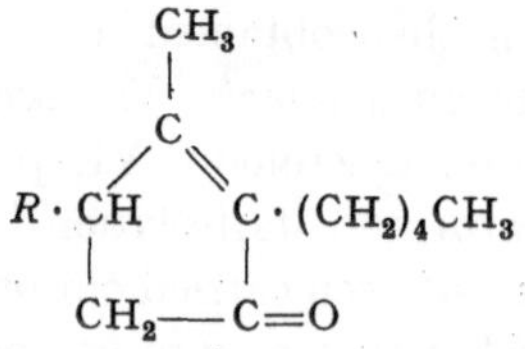   (LVII.) 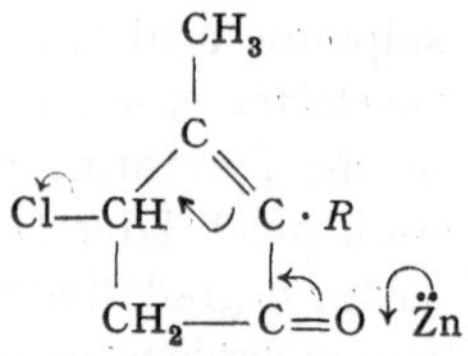   (LVIII.)

treatment with zinc dust (cf. LVIII) (*127*). Ultraviolet data for tetra-hydropyrethrone (*74*) are typical of an α-unsaturated cyclopentanone with a tetrasubstituted double bond and structures (LVII, $R = H$) and (LVII, $R = OH$) have been confirmed by synthesis (*178*).

STAUDINGER and RUZICKA (*176*) originally placed the hydroxyl at position 5 in pyrethrolone as α-ketol properties were thought detectable, but work on cinerolone (see below) makes this untenable. The placing on carbon 4 is rigorously established by synthesis and accords with its catalytic hydrogenolysis. Location of the side-chain unsaturation in pyrethrolone caused confusion for a long period. It is now apparent that this was because samples were contaminated with cinerolone and hence gave acetaldehyde on ozonolysis: pure pyrethrolone gives only formaldehyde (*39*). Because of the confusion, allenic (*115*, *129*, *176*) and other formulae (*75*, *157*, *189*) were at one time current, but the matter was settled by ultraviolet spectroscopy. The curve for pyrethrolone was shown to be the sum of that expected for a monosubstituted conjugated diene and a 2,3-dialkylcyclopentenone (*74—76*). Only the 2',4'-diene accomodates this requirement.

Two stereochemical features, the geometry at 2' and the absolute configuration at $C_{(4)}$, remained to be decided. Failure of pyrethrolone or its methyl ether to give a Diels-Alder adduct with various dienophiles (*189*) suggested a *cis*-2'-linkage and this was rigorously established by synthesis of *cis*- and *trans*-pyrethrolone. Infrared data are in agreement. Work on the absolute configuration at carbon 4 has been reported by INOUE (*102*, *113*). By exhaustive ozonolysis, the asymmetric centre of pyrethrolone methyl ether (LIX) was extracted as the keto acid (LX) which was oxidised to (—)-α-methoxysuccinic

References, pp. 155—164.

acid (LXI). The latter has been related to (—)-malic acid and to (—)-glyceraldehyde, so (+)-pyrethrolone is written as in (LIX).

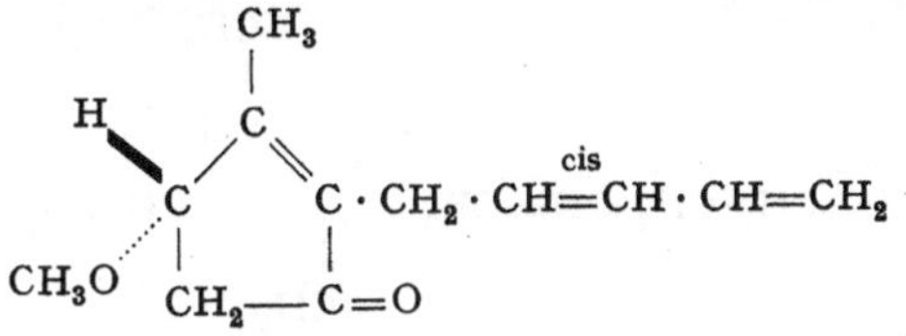

(LIX.) Pyrethrolone methyl ether.

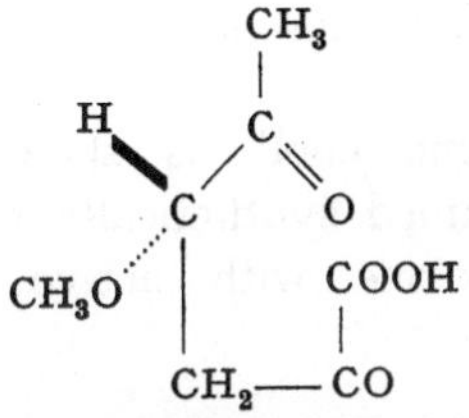

(LX.)

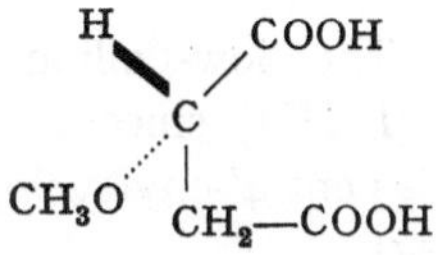

(LXI.) (—)-α-Methoxysuccinic acid.

Experiments by PHIPERS and BROWN (*12*) have indicated that pyrethrins I and II can be thermally isomerised and ELLIOTT (*50*) has recently shown that (+)-pyrethrolone is isomerised rapidly at 200° to the cross-conjugated (—)-isopyrethrolone (LXII), $[\alpha]_D^{20} - 5.6°$, n_D^{20} 1.5894, λ_{max} 260, 270 and 280 mμ. This finding helps to explain why LaFORGE and BARTHEL (*118*) reported five so-called "pyrethrolone" semicarbazones whilst WEST (*191*) found no less than six (A 1, A 2, B 1, B 2, C 1, C 2). Four of the latter were identified as follows: A 1, (+)-cinerolone semicarbazone; A 2, (±)-cinerolone semicarbazone; B 1, (+)-pyrethrolone semicarbazone; and B 2, (±)-pyrethrolone semicarbazone. If C 1 and C 2 represented the active and racemic forms of another alcohol, six natural rethrins should be found, but displacement chromatography gives no support for this (*162*). Scrutiny of the experimental work makes it likely that C 1 and C 2 are semicarbazones from (+)-*cis*- and (±)-*cis*-pyrethrolone contaminated with thermally isomerised material (*50*).

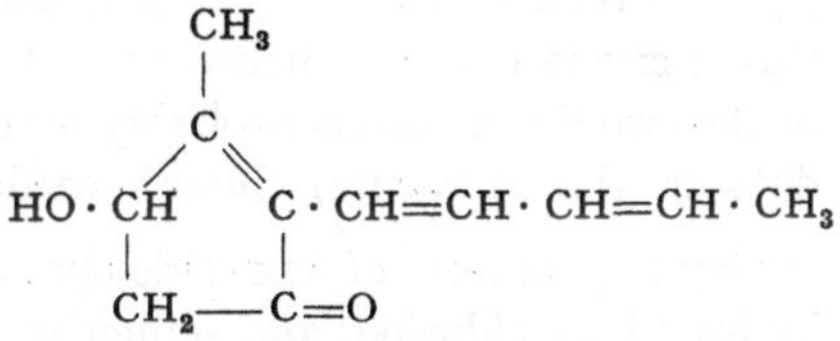

(LXII.) (—)-Isopyrethrolone.

When pyrethrolone is boiled with sodium methoxide for 7–20 hours two enols, "pyrethrolone enol", b. p. 82°/0.05 mm., and "isopyrethrolone enol", b. p. 150°/0.05 mm., are obtained (*75, 82, 176*). ELLIOTT (*51*) has shown the high-boiling enol to be (LXIV) and its absorption maximum is concentration dependent (243.5–256.0 mμ). The enol forms an acetate which can be hydrogenated to 2-methyl-3-*n*-amylcyclopentanone (*cis-trans* mixture): infrared evidence agrees with the cyclopentane-dione

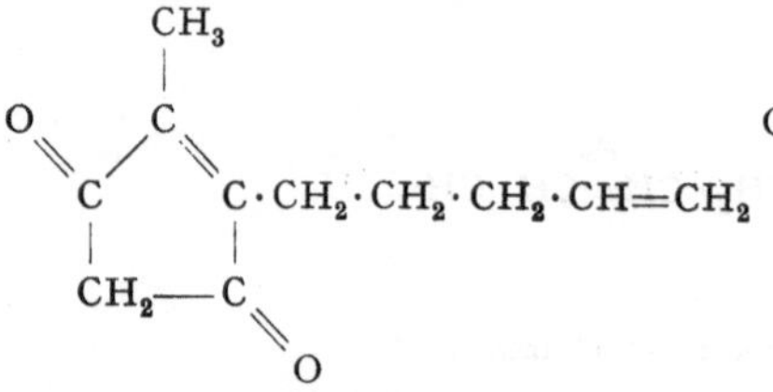

(LXIII.) "Pyrethrolone enol" (low boiling). 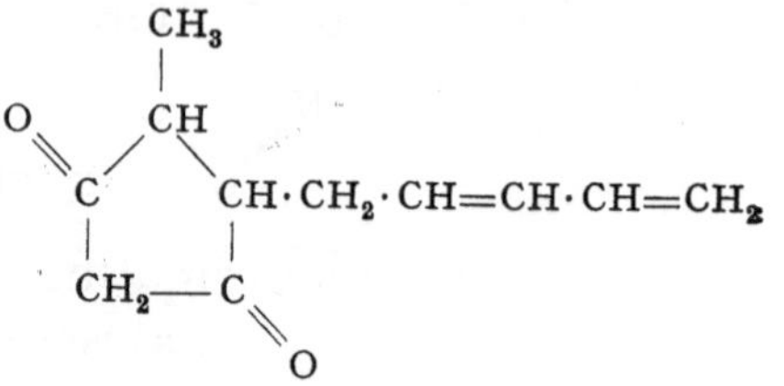   (LXIV.) "Pyrethrolone enol" (high boiling).

formula. The low-boiling "pyrethrolone enol" is shown to have structure (LXIII) since it can be obtained synthetically by oxidising 3-methyl-2-pent-4'-enylcyclopent-2-en-4-olone with active manganese dioxide (*51*).

Cinerolone.

LaForge and Barthel (*118*) showed that the ketol portion derived from natural pyrethrins mixture could be distilled as the acetate and split into low- and high-boiling fractions. The former yielded a semicarbazone mixture which on separation gave (+)-cinerolone and (±)-cinerolone, the latter being an artefact. Structural investigation led initially to the proposal (LXV) (*119*), though cinerolone is now known to be (VI, p. 126). (+)-Cinerolone, like (+)-pyrethrolone, can form a monohydrate but this is crystalline only below room temperature. The hygroscopic nature of the two liquid ketols probably accounts for some of the poor analytical data in the literature, though oxidation too may play a part (*50*).

Hydrogenation of the side-chain double bond, replacement of the hydroxyl by chlorine, and reductive elimination led to dihydrocinerone, shown to be (LXVI) by synthesis (*119*). The ultraviolet spectrum agrees with that expected for a five-membered α-unsaturated ketone with 2,3-disubstitution. As terminal methyl values tended to two for cinerolone, unsaturation was placed at 2' and as expected ozonolysis of pure samples gives acetaldehyde only: the formaldehyde found in early work probably arose from pyrethrolone contaminant.

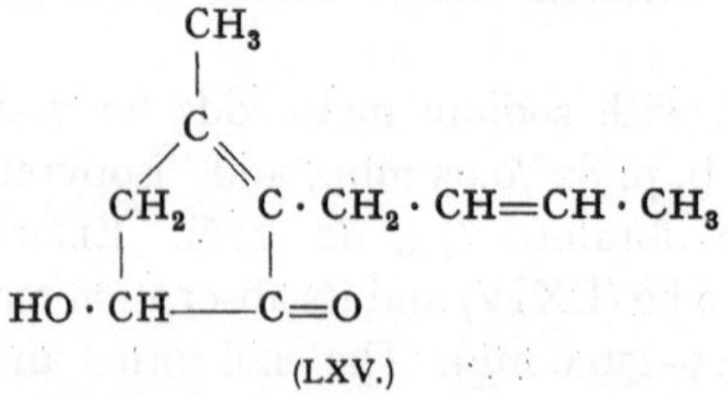

(LXV.) 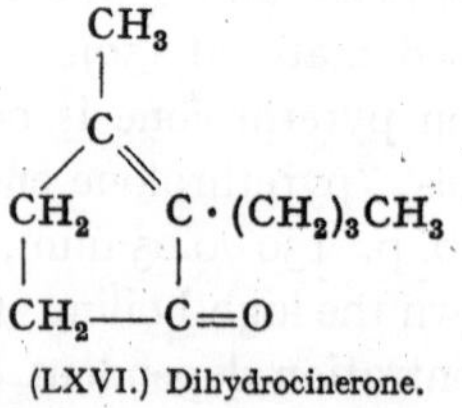   (LXVI.) Dihydrocinerone.

The incorrect placing of the hydroxyl as represented by (LXV) was shown by the synthetic experiments of LaForge and Soloway (*130*). Dihydrocinerone (LXVI) was ethoxycarbonylated at carbon 5 by diethyl carbonate and sodium hydride and the product (LXVII) was acetoxylated with lead tetraacetate and converted by treatment with aqueous ammonia and then acid into the 5-hydroxyketone (LXX). The latter was also made by treating the sodio-derivative of (LXVII) with toluene-*p*-sulphonyl-chloride to give (LXXI), hydrolysing and decarboxylating the product, and converting the chloro-compound (LXXII) into the ketol via the acetate. The ketol (LXX) differed from ($\pm$)-dihydro-

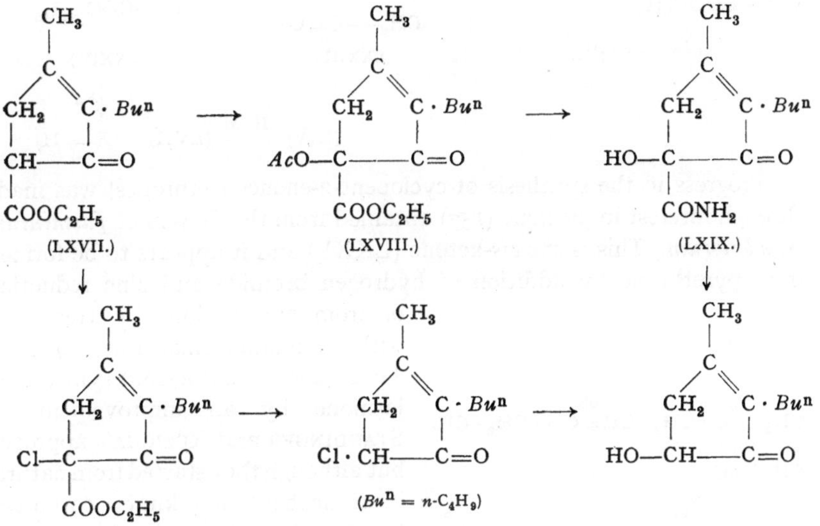

cinerolone from natural sources, having much more powerful reducing properties. Carbon 4 became the obvious location for the hydroxyl and this was later confirmed by synthesis. The *cis*-geometry of the side-chain was established by synthesis and later infrared work agrees (*42*). The absolute configuration at $C_{(4)}$ is reported to be the same as in (+)-pyrethrolone (*102*, *113*).

VII. Synthesis of *cis*-Pyrethrolone and *cis*-Cinerolone.

The first synthetic work directed towards the cyclopentenone part of the pyrethrins was carried out by Staudinger and Ruzicka (*178*). They prepared hexahydropyrethrone (their "tetrahydropyrethrone") by Reformatski reaction between ethyl laevulate and ethyl α-bromo-heptoate. This gives a mixture of ester (LXXIV) and lactone (LXXIII):

under Dieckmann cyclisation the former gives true tetrahydro-pyrethrone (LVII, $R = H$) and on hydrogenation hexahydropyrethrone (LIV) identical with material of natural origin is obtained. Further synthetic work at this period was sterile as it was thought that pyrethrolone was a 5-hydroxycyclopentanone.

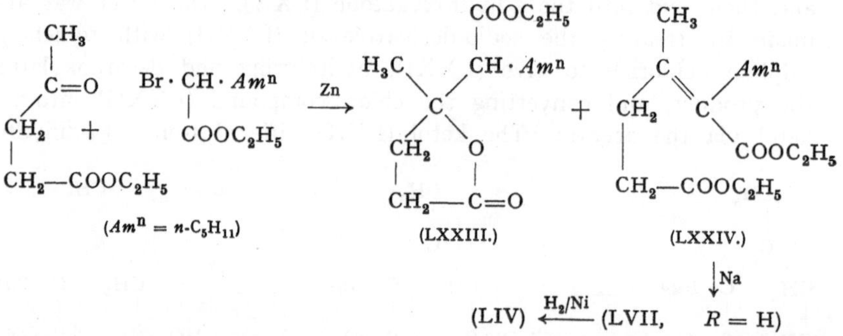

$(Am^n = n\text{-}C_5H_{11})$ (LXXIII.)   (LXXIV.)

$$(LIV) \xleftarrow{H_2/Ni} (LVII, \quad R = H)$$

Progress in the synthesis of cyclopent-2-enones (rethrones) was made through interest in jasmone (*157*) obtained from the flowers of *Jasminium grandiflorum*. This is the *cis*-ketone (LXXV) and it appears to be formed from pyrethrone by addition of hydrogen bromide and zinc reduction or from pyrethrolone by treatment with aluminium amalgam (*129, 190*). TREFF and WERNER (*184*) synthesised jasmone by an improvement of STAUDINGER and RUZICKA's approach but although they started from natural "leaf alcohol", now known to be *cis*-hex-3-enol (*28*), bromination and de-bromination of the *cis*-linkage during synthesis (for protection purposes) led to some stereochemical uncertainty about the product. Dehydration of lactones (LXXVI) prepared as below (*60, 153*), or via Stobbe condensation (*48*), has also been used for synthesis of cyclopentenones related to hydro-pyrethrones and hydro-cinerones.

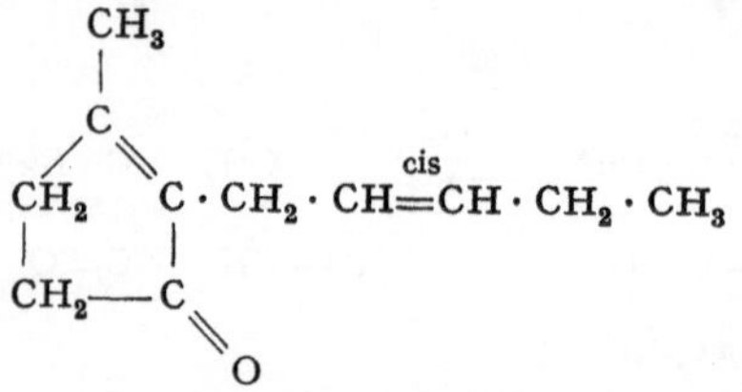

(LXXV.) Jasmone.

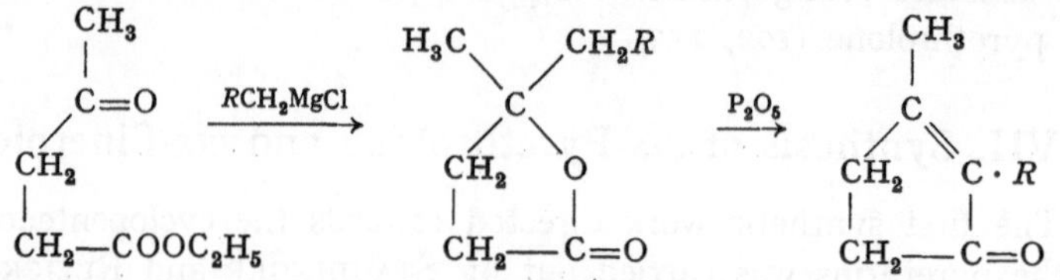

(LXXVI.)   (LXXVII.)

The most useful route to rethrones is the base-catalysed cyclisation of 2,5-diketones or α-acyl laevulic esters (LXXVIII), studied by

HUNDSIECKER (97, 98). The latter esters are accessible by treating an appropriately substituted sodio-acetoacetic ester with bromoacetone. By this route CROMBIE, HARPER and their colleagues have made cis-pyrethrone, cis-cinerone, and cis-jasmone, all identical with materials from natural sources (27, 29, 30, 34); the corresponding trans-isomers have been made (24, 30, 39) as well as various related compounds.

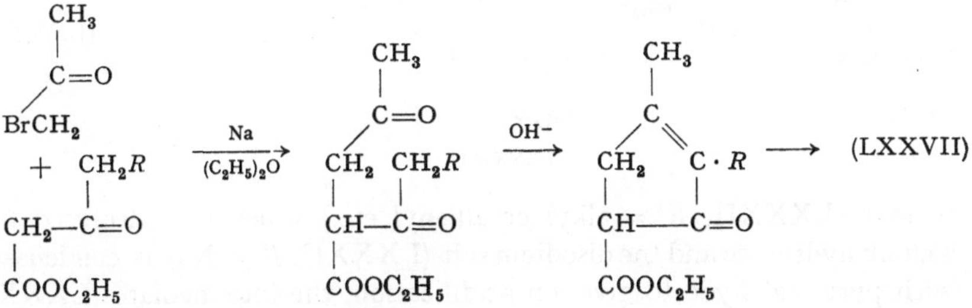

(LXXVIII.)

By treating dihydrocinerone or tetrahydropyrethrone with N-bromo-succinimide, allylic substitution at $C_{(4)}$ is effected and the bromo-derivatives can be converted either by direct hydrolysis or via the acetates prepared by reaction with silver acetate, into (±)-dihydrocinerolone (169, 25) or (±)-tetrahydropyrethrolone (25, 43). This approach, devised almost simultaneously by SOLOWAY and LAFORGE, and DAUBEN and WENKERT, is limited in scope because it cannot be successfully applied to rethrolones with unsaturated side-chains (25).

An important advance in rethrolone synthesis was the development of M. HENZE's early observations into an efficient and flexible synthesis by SCHECHTER, GREEN and LAFORGE (163). The appropriate substituted acetoacetate is hydrolysed by cold alkali and the sodium salt (LXXIX) is condensed with pyruvaldehyde at pH 8 to give a hydroxy-diketone (LXXX) with alkaline decarboxylation. The hydroxy-diketone is cyclised by dilute sodium hydroxide in the usual way to give the ketol (LXXXI) in which R may be saturated or unsaturated. Diethyl-amine, piperidine (136) and Amberlite IRA 400 (58) may also be used for the cyclisation.

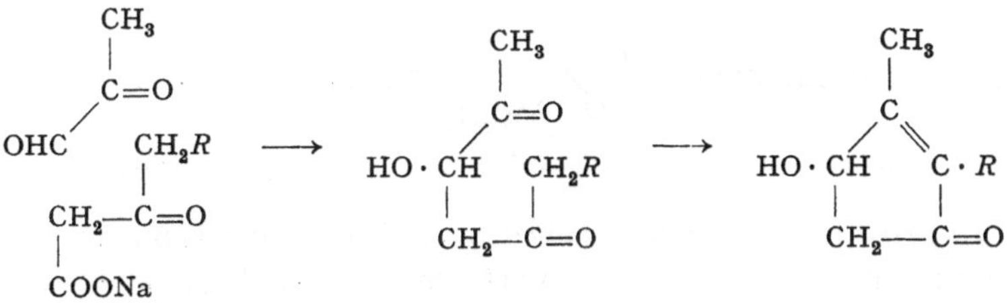

(LXXIX.)   (LXXX.)   (LXXXI.)

The general synthesis given above requires the construction of an appropriate γ-substituted acetoacetate and a later patent by BAVLEY and SCHREIBER (7) describes a modified general procedure. In this acetone dicarboxylic ester is alkylated in the presence of sodium alkoxide

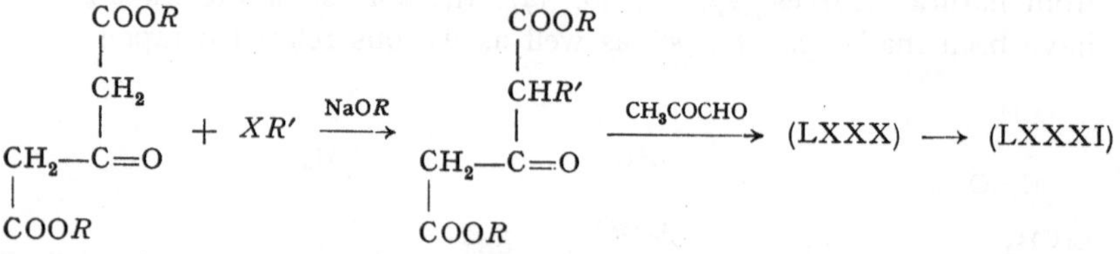

$$
\begin{array}{c}
COOR \\
| \\
CH_2 \\
| \\
CH_2{-}C{=}O \\
| \\
COOR
\end{array}
\quad + \ XR' \ \xrightarrow{NaOR} \quad
\begin{array}{c}
COOR \\
| \\
CHR' \\
| \\
CH_2{-}C{=}O \\
| \\
COOR
\end{array}
\quad \xrightarrow{CH_3COCHO} \ (LXXX) \ \longrightarrow \ (LXXXI)
$$

(LXXXII.)

to give (LXXXII, $R' = $ alkyl or alkenyl etc.) which is hydrolysed by sodium hydroxide and the disodium salt (LXXXII, $R = $ Na) is condensed with pyruvaldehyde to give, on acidification, the intermediate (LXXX) which is cyclised to (LXXXI).

By elaborating suitable γ-substituted acetoacetates of the correct geometry CROMBIE, HARPER and their collaborators have synthesised ($\pm$)-*cis*- and ($\pm$)-*trans*-cinerolone and ($\pm$)-*cis*- and ($\pm$)-*trans*-pyrethrolone (24, 27, 29, 34, 39). The two *cis*-isomers are identical with the ($\pm$)-ketols of natural origin. The side-chain of ($\pm$)-*cis*-cinerolone was introduced via the acetylenic alcohol (LXXXIII) derived from a β-halogeno-ether synthesis (29) which was semi-hydrogenated over a palladium catalyst to the *cis*-alcohol and elaborated first to hex-*cis*-4-enoic acid and then to the γ-substituted acetoacetate (LXXXIV, $R = CH_3$). The latter was condensed with pyruvaldehyde and cyclised to ($\pm$)-*cis*-cinerolone. But-2-ynol has also been employed. It is converted into the ketone (LXXXV), semi-hydrogenated, and then ethoxycarbonylated with sodium hydride and diethyl carbonate (38) to give the acetoacetate (LXXXIV, $R = C_2H_5$). As an alternative, semi-hydrogenation can be carried out as the last stage on the ketol (LXXXI, $R = CH_2 \cdot C{\equiv}C \cdot CH_3$) (164).

$$CH_3 \cdot C{\equiv}C \cdot (CH_2)_2OH \qquad\qquad \overset{cis}{CH_3 \cdot CH{=}CH \cdot (CH_2)_2 \cdot CO \cdot CH_2 \cdot COOR}$$

(LXXXIII.)   (LXXXIV.)

$$CH_3 \cdot C{\equiv}C \cdot (CH_2)_2 \cdot COCH_3 \qquad\qquad \overset{tr}{CH_2{=}CH \cdot CH{=}CH \cdot (CH_2)_2 \cdot COCH_3}$$

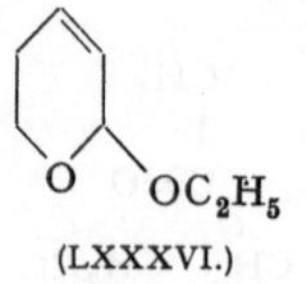

(LXXXV.)   (LXXXVI.)   (LXXXVII.)

($\pm$)-*cis*-Cinerolone was resolved by LaFORGE and GREEN (122) who esterified it with (+)-*trans*-chrysanthemic acid and separated the pair of diastereoisomers as their semicarbazones.

(±)-*trans*-Pyrethrolone (*34, 39*) was built up from pent-*trans*-2-en-4-ynol or penta-*trans*-2,4-dienol: the latter is accessible from penta-2,4-dienal formed by treating the dihydropyran derivative (LXXXVI) with phosphoric acid (*193*). Synthesis proceeded via ethoxycarbonylation of the ketone (LXXXVII) (made from 1-chloropenta-*trans*-2,4-diene)

$$CH_2{=}CH \cdot C{\equiv}C \cdot CH_2OH$$

(LXXXVIII.)

$$CH_2{=}CH \cdot C{\equiv}C \cdot CH_2 \cdot CH_2 \cdot CO \cdot CH_3$$

(LXXXIX.)

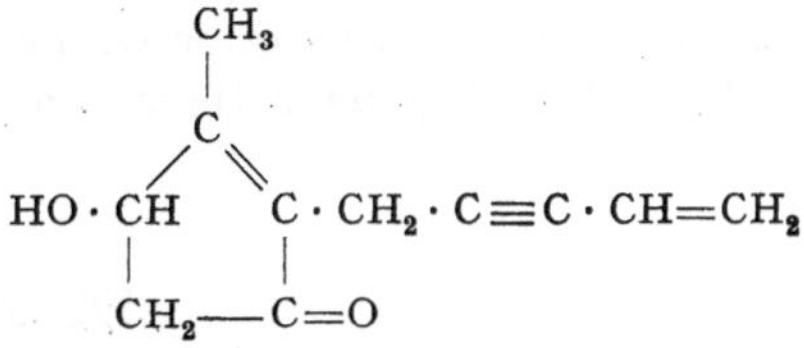

CH₃ structure:

HO·CH   C·CH₂·C≡C·CH=CH₂

CH₂——C=O

(XC.)

to the required γ-substituted acetoacetate. (±)-*cis*-Pyrethrolone, experimentally the most difficult proposition, was made (*34*), after testing various approaches (*35*), from vinylacetylene which was converted into the Grignard reagent and, by treatment with formaldehyde, into the alcohol (LXXXVIII). Elaboration via the chloride and acetoacetate synthesis gave the ketone (LXXXIX) which was ethoxycarbonylated and converted by condensation with pyruvaldehyde into the ketol (XC). Partial hydrogenation of the ketol gave (±)-*cis*-pyrethrolone. Ketones of the type $R \cdot CH_2COCH_3$ which are needed in this kind of work are sometimes conveniently made by reacting a Grignard reagent $R \cdot CH_2MgX$ with acetic anhydride at $-78°$ (*38, 147*).

VIII. Synthesis of Pyrethrins and Cinerins.

The first total synthesis of rethrins close in structure to the natural ones was the preparation of "dihydrocinerin I" [(±)-*n*-butylrethronyl (±)-*trans*-chrysanthemate] and "tetrahydropyrethrin I" [(±)-*n*-amyl-rethronyl (±)-*trans*-chrysanthemate] by CROMBIE, ELLIOTT and HARPER (*25, 26*). These were made by treating silver (±)-*trans*-chrysanthemate with (±)-4-bromo-*n*-butyl- or 4-bromo-*n*-amylrethrone and each product consists of a pair of racemic diastereoisomers. Partial synthesis of natural rethrins by re-esterification of the rethrolone with chrysanthemic or pyrethric acid was early established by STAUDINGER and RUZICKA (*49, 81, 120, 179*). Normally the acid chloride is treated with the ketol in the presence of pyridine and this method has been used on many occasions to make rethrins as pure optical isomers, or as mixtures of diastereoisomers, using the components synthesised above, e. g. for (+)-*cis*-cineronyl (+)-*trans*-chrysanthemate [cinerin I (III, p. 122)] (*29, 122*) or (±)-*cis*-pyrethronyl (+)-*trans*-chrysanthemate (a pair of diastereoisomers, one of which is the natural rethrin) (*34*). Other methods of esterification such as the use of chrysanthemic anhydride (*171*), or vacuum distillation of ethyl chrysanthemate with the rethrolone (*135*) have been used. Rethrins II are usually made from the acid chloride of pyrethric acid. Synthetic rethrins I derived from *cis*-chrysanthemic acid can be rapidly distinguished from those derived from the *trans*-acid as the former possess three strong bands near 1169, 1130 and 1075 cm^{-1} whilst the latter have three near 1193, 1151 and 1113 cm^{-1} (*32*).

IX. Allethrin.

The methods developed for making natural pyrethrins and their degradation products have been used to synthesise a considerable number of compounds resembling, to a greater or lesser extent, the natural esters. As much of this work presents little of new chemical interest, such compounds will be further mentioned only where they have bearing on those structural features in the natural rethrins which are responsible for high insecticidal activity (*46, 143, 146*). A limited number of such analogues have high activity, examples being allethrin (XCI a) (*154, 155*), furethrin (XCI b) (*137*) and cyclethrin (XCI c) (*94*). The first is produced commercially and as it has been well investigated chemically and is very close in structure to cinerin I (p. 122), it is discussed briefly below.

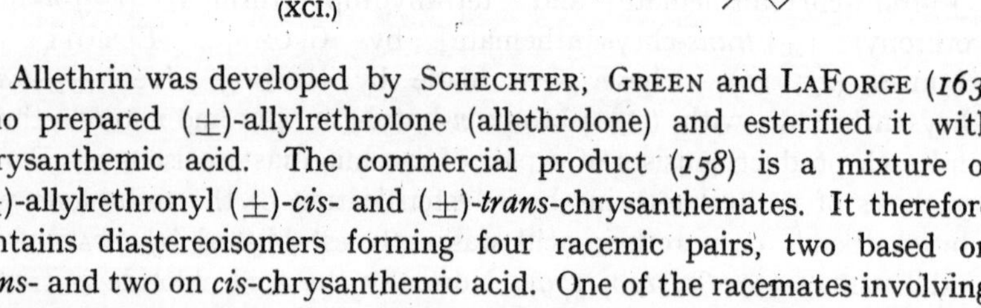

(a) $R = $ —CH$_2$CH=CH$_2$. Allethrin.

(b) $R = $. Furethrin.

(c) $R = $. Cyclethrin.

(XCI.)

Allethrin was developed by SCHECHTER, GREEN and LAFORGE (*163*) who prepared (±)-allylrethrolone (allethrolone) and esterified it with chrysanthemic acid. The commercial product (*158*) is a mixture of (±)-allylrethronyl (±)-*cis*- and (±)-*trans*-chrysanthemates. It therefore contains diastereoisomers forming four racemic pairs, two based on *trans*- and two on *cis*-chrysanthemic acid. One of the racemates involving *trans*-chrysanthemic acid, "α-*dl*-*trans*-allethrin", is crystalline, m. p. 51° (*165*), and consists of (+)-allylrethronyl (—)-*trans*-chrysanthemate and (—)-allylrethronyl (+)-*trans*-chrysanthemate (*47*). It is, however, less insecticidal than the liquid "β-*dl*-*trans*" racemate [(+) (+) with (—) (—)]. Allethrolone has been resolved and all eight stereoisomeric allethrins have been obtained by esterification of the two enantiomorphic alcohols with (+)- or (—)-*trans* or (+)- or (—)-*cis*-chrysanthemates (*124, 125*). For biological studies, allethrin labelled with ^{14}C in the ketol (*192*) or acid (*2*) portion is available. Synthesis of the latter starts from 2-^{14}C-glycine and proceeds via labelled diazoacetic ester to give chrysanthemic acid labelled at $C_{(1)}$ in the cyclopropane ring.

Treatment of allethrolone esters (XCII) with base yields considerable amounts of the cyclopentadienone (XCIII) but as expected (*3*) this dimerises at once. Two orientations are possible for the dimer and the main product is thought to be (XCIV and enantiomorph) and to be *exo* (*123*). On heating, the dimer decarbonylates, probably first to (XCV), finally giving the indanone (XCVI) (*123*). Dimers similar to (XCIV) result when 4-bromodihydrocinerone or 4-bromotetrahydropyrethrone are dehydrobrominated with trimethylamine or sodium methoxide (*25*). It is likely that some of the "altered" semicarbazones, encountered by STAUDINGER and RUZICKA (*174, 175*) and thought to have been derived through heterocyclisation with elimination of water, are in fact cyclopentadienone dimer derivatives.

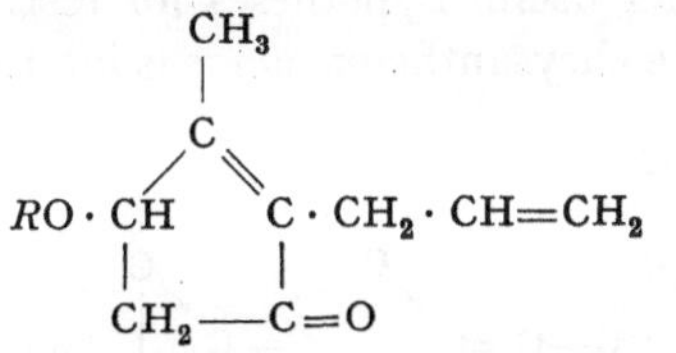

(XCII.) Allethrolone esters.

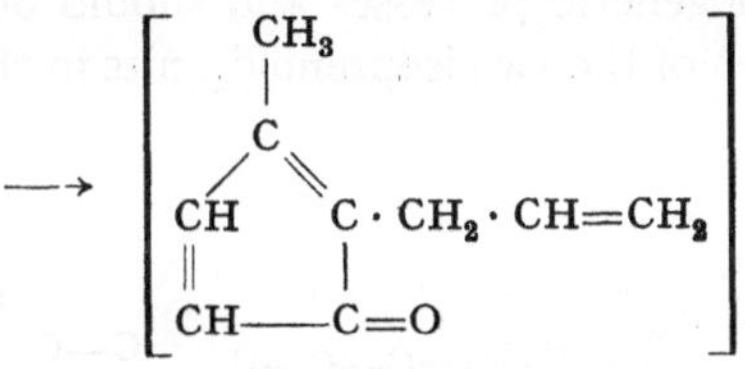

(XCIII.)

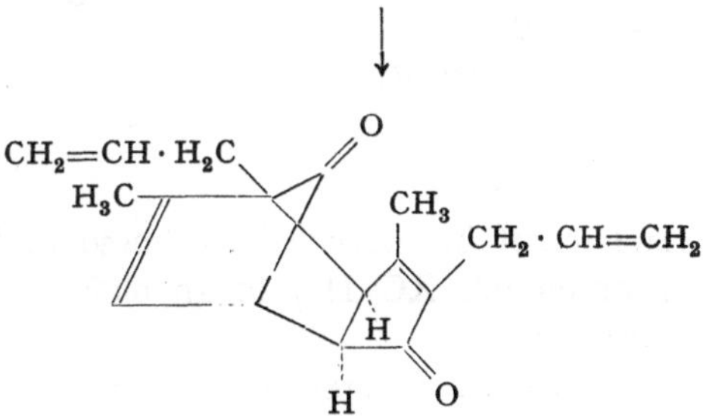

(XCIV.)

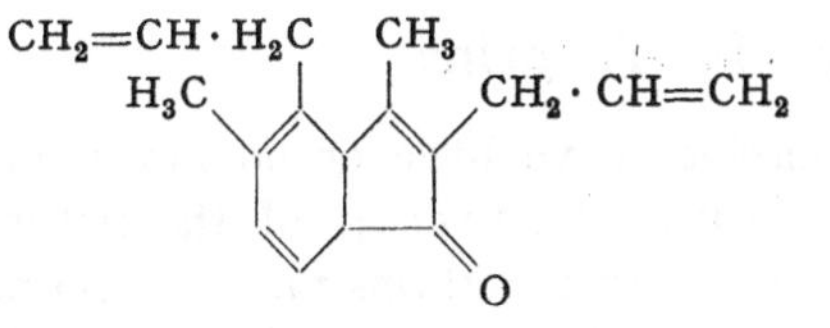

(XCV.)

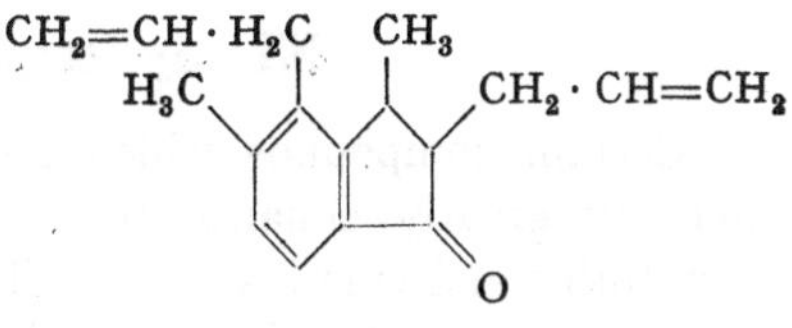

(XCVI.)

X. Biosynthesis of the Pyrethrins and Cinerins.

Knowledge of biosynthesis in this group is at present limited. Chrysanthemic acid and chrysanthemum dicarboxylic acid are each formally made up of two isoprene units whilst one isoprene unit can be formally dissected from cinerolone or pyrethrolone in more than one way. Recent work by THAIN and his colleagues (41) shows that $(\pm)$-2-^{14}C-mevalonic acid is incorporated into the two chrysanthemic acids when fed to the achenes of pyrethrum flowers, and thus supports the isoprenoid derivation. There is no evidence that $(\pm)$-2-^{14}C-mevalonic acid is directly involved in the formation of the ketols, which do not seem to be built up on a partial isoprenoid basis. The latter conclusion is of interest in connection with circumstantial evidence which suggests that the structurally similar compound jasmone (p. 142) is also not of partial isoprenoid derivation. Thus cis-hex-3-en-1-ol occurs in Mentha species, like jasmone (LXXV) (166) and since the alcohol accounts for the five carbon atoms of the chain and one of the ring (cf. XCVII), if it is directly involved in the biosynthesis, an isoprenoid unit would not be involved. Indeed, the in vitro synthetic pathways described earlier are suggestive of biogenetic processes and should offer useful hypotheses for test. The fusion of the two isoprenoid units in the chrysanthemic acids is interesting

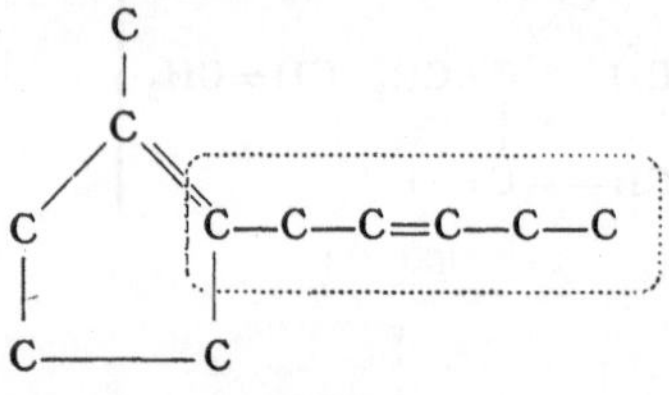

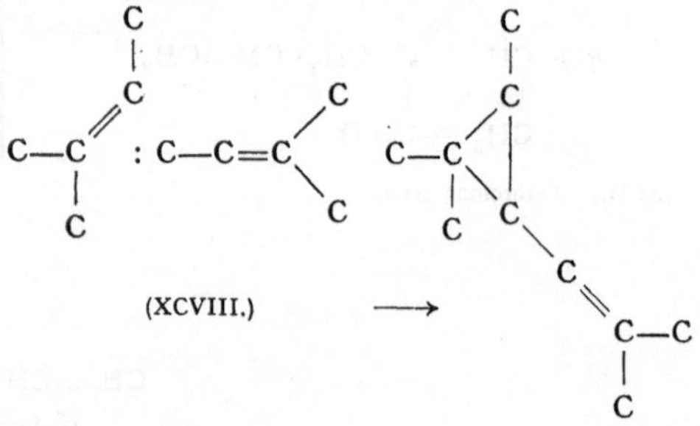

and might be described as "meso to tail" fusion. A carbene type of addition (cf. XCVIII) to an olefinic unit is an interesting biogenetic possibility.

[14]C-Labelled pyrethrins-cinerins mixture is available for insecticidal studies from pyrethrum flowers grown in $^{14}CO_2$ (148).

XI. Synergists for Rethrins.

Certain compounds which themselves have little or no insecticidal activity greatly enhance the insecticidal effectiveness of the natural pyrethrins and cinerins. Generally, the natural rethrins mixture appears to be capable of being synergized to a higher degree than allethrin.

Because of its economic importance synergism has been extensively studied but it is intended here only to give an indication of the types of compound which are active.

Most of the synergists contain a methylenedioxyphenyl group and/or an amide linkage (10). Some are of natural origin, some synthetic. Among those in the former class are piperine (93, 182), its vinylogue piperettine (XCIX) (71), and various lignans such as hibbalactone (133), hinokinin (132), sesamolin (8, 9), and the group of stereoisomers (+)-sesamin (C), (+)-asarinin (CI), (+)-epiasarinin (CII) and their enantiomorphs (64, 83).

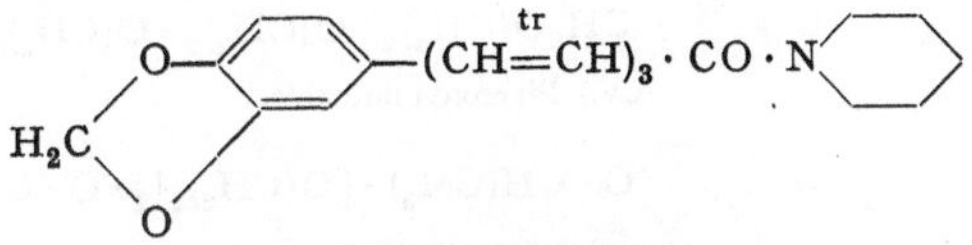

(XCIX.) Piperettine.

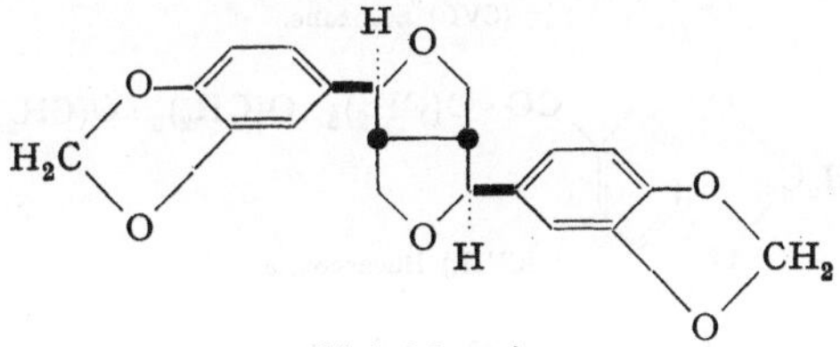

(C.) (+)-Sesamin.

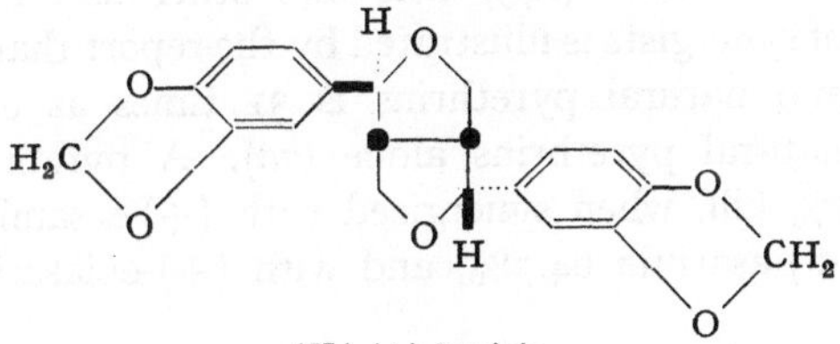

(CI.) (+)-Asarinin.

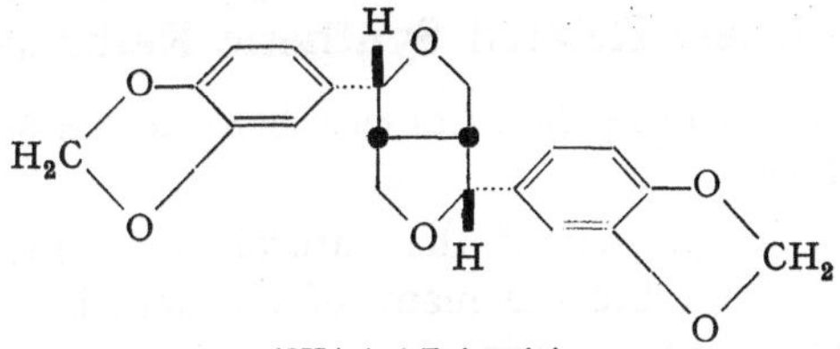

(CII.) (+)-Epiasarinin.

The amide N-isobutylundecylenamide has long been used as a synthetic synergist and other suitable synthetic compounds are piperonylcyclonene (95, 185), a mixture of (CIII) and (CIV), piperonyl butoxide (CV) (185), sesoxane (CVI) (57) and bucarpolate (CVII) (144).

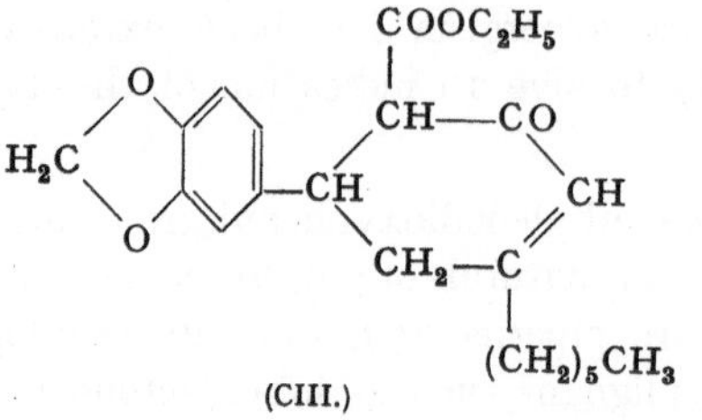

(CIII.)

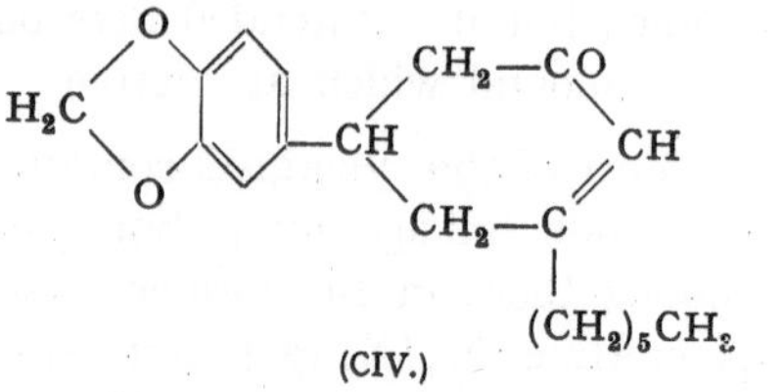

(CIV.)

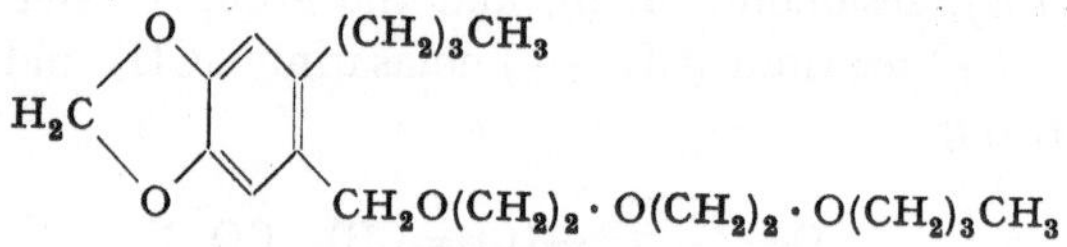

(CV.) Piperonyl butoxide.

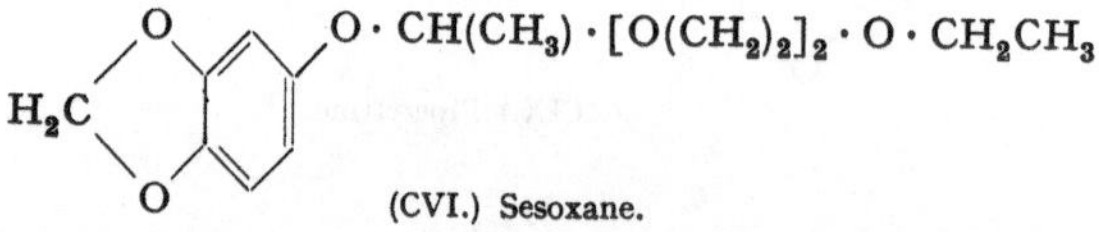

(CVI.) Sesoxane.

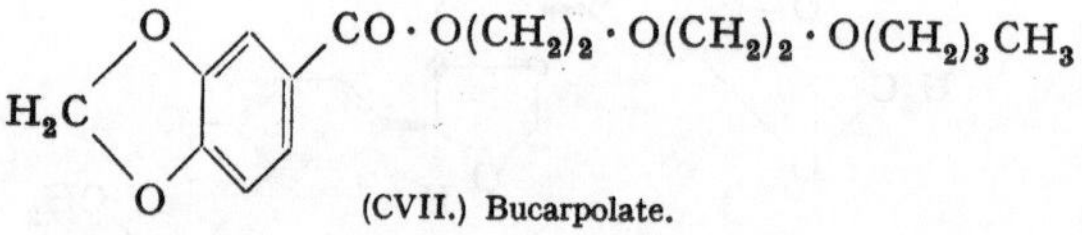

(CVII.) Bucarpolate.

Features essential for synergistic activity have been studied recently by Moore and Hewlett (145) who cite other useful references. The striking effect of synergists is illustrated by the report that an 8 : 1 mixture of sesamolin and natural pyrethrins is 31 times as effective towards houseflies as natural pyrethrins alone (70). A pyrethrins preparation giving an 18.4% kill, when synergized with (+)-sesamin gave a kill of 82.3%, with (+)-asarinin 64.7%, and with (+)-epiasarinin 53.6% (73).

XII. The Insecticidal Activity of Natural and Closely Related Synthetic Rethrins.

An earlier survey of this subject was contributed to this Series by Feinstein and Jacobson (58a).

Quantitative comparison of the natural and synthetic rethrins as insecticides is difficult because many of the data in the literature are not on a common testing basis.

The absolute and relative toxicity of the rethrins may vary with the insect species, with the age, sex, and storage environment before and after treatment, and with the medium and method by which the insecticide is applied (151): even moderately complete quantitative knowledge of the general effectiveness of one insecticide thus requires a large programme of study. In comparing two compounds, the method of application must be such that each insect receives the same dose and this is best done by treating individual insects (by topical application or injection) with measured volumes of insecticide dissolved in a suitable solvent. On the basis of the weight required to affect a given weight of one insect species, the natural pyrethrins compare well with many insecticides in use to-day (DDT, lindane, rotenone, aldrin, dieldrin, paraoxon, parathion) (15, 142). Because they are unstable in air and light, few films of diminishing potency persist and this factor may have hindered insects from acquiring resistance to them. There is no evidence of synergistic or antagonistic activity towards the pyrethrins by other compounds occurring in pyrethrum extract (162).

The toxicity of the natural rethrins mixture relative to commercial allethrin or to (±)-allylrethronyl (+)-trans-chrysanthemate varies widely within four species of insect (54) but recent work has shown (159) that the action of the four natural rethrins is complex, so that interpretation of results using mixtures becomes difficult. Now that pure specimens of the four natural rethrins are more accessible, scientific information is better based on these.

Table 1. Insecticidal Activity of the Natural Pyrethrins
(Molecule-molecule basis).

	Pyrethrin I*	Cinerin I	Pyrethrin II	Cinerin II
Relative Toxicity to *Phaedon cochleariae* Fab. (Mustard Beetles)[a]	10	4	3	2
Relative Toxicity to *Musca domestica* L. (Houseflies)[b]	10	7	3	2
Relative Toxicity to *Musca domestica* L.[c]	10	4	18	7
Relative Toxicity to *Musca domestica* when synergized with piperonyl butoxide at 8 : 1[d]	10	10	6	5
Relative Susceptibility to Synergism by piperonyl butoxide (8 : 1) against *Musca domestica*[d]	10	17	4	9
Relative Knock-down Potency after 15 minutes to *Musca domestica* L.[e]	10	8	26	10

* Used as Standard and given potency 10.

[a] Topical application of acetone solutions to adult males and females (53, 187).

[b] Kerosene sprays in Campbell turntable to 2 day old adult males and females (67, 99).

[c] Topical application of acetone solutions to 5–6 day old females (161, 162).

[d] Topical application of acetone solutions to 5–6 day old females (159).

[e] Topical application of dodecane solutions (160).

Table 1 (p. 151) summarises data on the relative activities of the natural rethrins and illustrates the wide variation in relative toxicity, especially between pyrethrins I and II, with species and test method. In topical application to houseflies, pyrethrin II is especially effective for kill and knock-down but the differences in potency are considerably altered when the natural rethrins are synergized by piperonyl butoxide [SAWICKI et al. (*159*)]. Esters of pyrethric acid [pyrethrin II, cinerin II, (CXVI) and (CXVII)] are less toxic than those of (+)-chrysanthemic acid towards Mustard Beetles, and, according to work using a spraying technique, to houseflies (*Tables 1, 3*, pp. 151, 154). These housefly data were, however, obtained with distilled pyrethrins I and II before the lability of pyrethrin II to thermal isomerisation, and its extreme instability, had been recognised. SAWICKI's data apply to fresh undistilled pyrethrin II.

The insecticidal effectiveness of the natural rethrins is highly responsive to structure and stereochemical alterations. Thus (±)-allyl-rethronyl methyl (±)-*trans*-caronate (CVIII) (*86*) and (±)-H-rethronyl (±)-*trans*-chrysanthemate (CIX) (*87*) are not toxic to Mustard Beetles (*53*), so appropriate side-chains in both acid and alcohol are essential for activity. For highest potency these side-chains should be unsaturated, since when the double bond in both *cis*- and *trans*-chrysanthemic acid is hydrogenated [(CX) and (CXI), *Table 3*, p. 154], the new dihydroesters are less toxic. Similarly rethrins from rethrolones with saturated side-chains (CXII–CXV) have reduced activity.

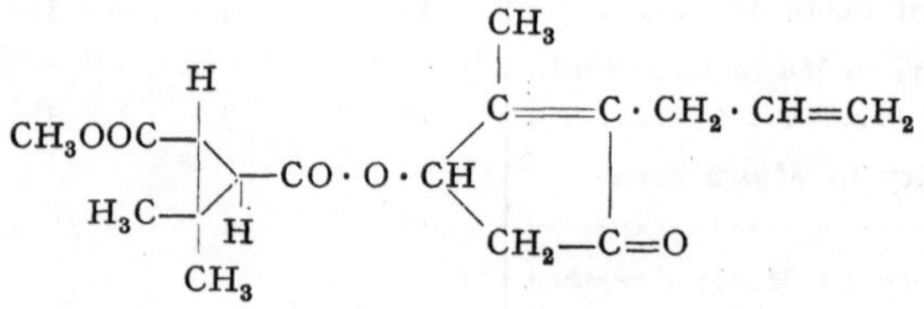

(CVIII.) (±)-Allylrethronyl methyl (±)-*trans*-caronate.

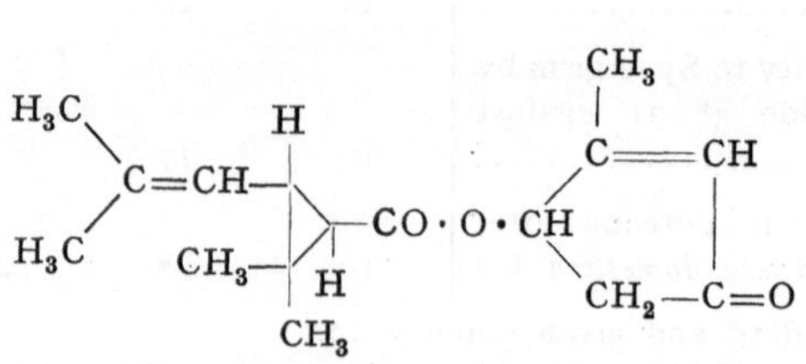

(CIX.) (±)-H-rethronyl (±)-*trans*-chrysanthemate.

The highest potency is usually associated with the *cis*-penta-2′,4′-dienyl side-chain (pyrethrins I and II) and the effect of the position of unsaturation in the side-chain has been examined (Tables 3 and 4). If a double bond is separated from the cyclopentenolone ring by two or more methylene groups, it has little effect in raising the toxicity above that of a rethrin with a saturated side-chain [see (CXXII, p. 154)]. Similarly the but-3-enylrethrin (CXX) is less toxic than the allyl-rethrin (CXXI) to Mustard Beetles, and much less to houseflies. The change from a methyl (CXXIII) to an ethyl (CXXIV) on the terminus of a 2' double bond makes little difference: nor does change from a cis- (CXXV) to a trans- (CXXVI) configuration, but unless a terminal double bond is present (CXXVII) chain branching (CXXVIII) produces a substantial reduction in activity. The penta-1',3'-dienylrethrolene (LXII, p. 139), produced by thermal isomerisation, gives an ester only $1/_{16}$th as toxic to Mustard Beetles as that with a penta-2',4'-dienyl group (pyrethrin I) (53), again stressing that it is those rethrolones with a 1'-methylene group activated by a side-chain 2'-double bond or bonds which give highly toxic esters.

For details of the effects of other changes in molecular structure see ELLIOTT (46, 47).

Table 2. Insecticidal Activity of Isomers of Allethrin.

( )-Allethronyl ( )-	-chrysanthemate		Phaedon[a] cochleariae FAB.	Musca[b] domestica L.
+	+ trans		100*	100*
+	— trans			4
—	+ trans			17
—	— trans		0.4	0.7
+	+ cis			53
+	— cis			4
—	+ trans			10
—	— cis			1.7
+	+ trans } "β-dl-trans"		54	50
—	— trans }			
+	— trans } "α-dl-trans"		4	9
—	+ trans }			
±	+ trans		56	[58]**
±	— trans		2	[2]**
±	± trans		28	35
±	+ cis		27	[27]**
±	± cis		13	16

* Used as Standard and given potency 100.

** Calculated values: arithmetical mean of potencies of stereoisomers.

[a] Topical application of measured drops in acetone to adult male and female (53).

[b] Kerosene spray in Campbell turntable to 2–3 day old adult male and female (72).

Stereochemical factors, especially absolute stereochemistry, are of great importance in relation to insecticidal activity. Thus rethrins made from (+)-*trans*-chrysanthemic acid (*Table 2*, p. 153) are more toxic than those from the (+)-*cis*-acid, and both (+)-acids give rethrins much more toxic (40–50 times) than those from their (—)- counterparts. The esters from (+)-cinerolone and (+)-allethrolone are at least five times as toxic as those from the (—)-alcohols.

Table 3. Insecticidal Activity of Various Rethrins[a]
(Molecule-molecule basis).

No.	Ref.	Rethrin		Potency
	(24)	(±)-Allylrethronyl	(±)-*trans*-chrysanthemate	100*
	(24)	(±)-Allylrethronyl	(±)-*cis*-chrysanthemate ..	45
(CXII)	(69, 87)	(±)-Methylrethronyl	(±)-*trans*-chrysanthemate	5
(CXIII)	(69, 87)	(±)-Ethylrethronyl	(±)-*trans*-chrysanthemate	15
(CXIV)	(86)	(±)-Propylrethronyl	(±)-*trans*-chrysanthemate	22
(CXV)	(24)	(±)-Butylrethronyl	(±)-*trans*-chrysanthemate	12
(CXXV)	(29)	(±)-*cis*-Cineronyl	(±)-*trans*-chrysanthemate	93
(CXXVI)	(29)	(±)-*trans*-Cineronyl	(±)-*trans*-chrysanthemate	105
(CXXII)	(30)	(±)-Pent-4'-enylrethronyl	(±)-*trans*-chrysanthemate	7
(CXXIX)	(34)	(±)-*trans*-Sorbylrethronyl	(±)-*trans*-chrysanthemate	41
(CXXVIII)	(33)	(±)-*trans*-1'-Methylcrotonylrethronyl		
			(±)-*trans*-chrysanthemate	25
(CXVIII)	(33)	(±)-Furfurylrethronyl	(±)-*trans*-chrysanthemate	122
(CXIX)	(33)	(±)-Phenylrethronyl	(±)-*trans*-chrysanthemate	22
(CX)	(86)	(±)-Allylrethronyl (±)-*trans*-dihydrochrysanthemate		35
(CXI)	(86)	(±)-Allylrethronyl (±)-*cis*-dihydrochrysanthemate..		11
(CXVI)	(37)	(±)-Allylrethronyl	(±)-*trans*-pyrethrate	24
(CXVII)	(37)	(±)-Allylrethronyl	(±)-*cis*-pyrethrate	6

* Used as Standard and given potency 100.
[a] Topical application of acetone solutions to adult male and female *Phaedon cochleariae* FAB. (53).

Table 4. Insecticidal Activity of Esters of
(+)-*trans*-Chrysanthemic Acid.

No.	Ref.	Rethrin	Mustard Beetles[a]	Houseflies[b]
(CXXI)	(24)	(±)-Allylrethronyl..............	100*	100*
(CXXIII)	(29)	(±)-*cis*-Cineronyl................	64	22
(CXXIV)	(38)	(±)-*cis*-*n*-Pent-2'-enylrethronyl....	62	
(CXXVII)	(24)	(±)-2'-Methylallylrethronyl.......	72	52
(CXX)	(24)	(±)-But-3'-enylrethronyl	49	9

* Used as Standard and given potency 100.
[a] Topical application of acetone solutions to adult males and females (53).
[b] Kerosene sprays on Campbell turntable to 2–3 day old males and females (69).

By Campbell turntable method GERSDORFF and MITLIN (*68*) found (±)-allylrethronyl (+)-*trans*-chrysanthemate to be six times as toxic as natural pyrethrins mixture to 2–3 day old houseflies. In later tests on other species, the best synthetic analogues of the pyrethrins [e. g. (CXVIII) and (CXIX), Table 3, and (XCI c, p. 146)] have always shown high insecticidal activity but in many cases lower than that of the natural esters (*54, 146, 154, 155*).

Although the synthetic rethrins have a useful role in commerce, earlier expectations that natural pyrethrins would be displaced by synthetic substitutes have not so far been realised.

Present theories of toxic action [cited by NEGHERBON (*146*)] have limited use in explaining the physiological action of the pyrethrins. As is so common with drugs and toxins, absolute configuration is a highly important factor indicating specific interaction with other asymmetric molecules and surfaces. But the understanding which can come from study of structure and stereochemical variation alone on biological activity is strictly limited in compass, and there is an urgent need for deeper penetration into the biochemical and biophysical aspects.

References.

1. ACREE, F., Jr. and F. B. LaFORGE: Constituents of Pyrethrum Flowers. X. Identification of the Fatty Acids Combined with Pyrethrolone. J. Organ. Chem. (USA) **2**, 308 (1937).

2. ACREE, F., Jr., C. C. ROAN and F. H. BABERS: The Synthesis and Chromatographic Purification of Radioactive Allethrin. J. econ. Entomol. **47**, 1066 (1954).

3. ALLEN, C. F. H.: Carbonyl Bridge Compounds and Related Substances. Chem. Rev. **37**, 209 (1945).

4. BARTHEL, W. F. and H. L. HALLER: Purification of Pyrethrum Extract with Nitromethane. U. S. Patent 2 372 183 (1945).

5. BARTHEL, W. F., H. L. HALLER and F. B. LaFORGE: Pyrethrins for Aerosols. Soap **20**, (7), 121 (1944).

6. BARTON, D. H. R., O. C. BÖCKMAN and P. DE MAYO: Sesquiterpenoids. Part XII. Further Investigations on the Chemistry of Pyrethrosin. J. Chem. Soc. (London) **1960**, 2263.

7. BAVLEY, A. and E. C. SCHREIBER: Process for the Manufacture of 2,5-Diketononen-3-ol. U. S. Patent 2 768 967 (1956).

8. BEROZA, M.: Pyrethrum Synergists in Sesame Oil. Sesamolin, a Potent Synergist. J. Amer. Oil Chem. Soc. **31**, 302 (1954).

9. — Sesamolin and Related Compounds as Synergists for Pyrethrum. Soap **32**, (7), 128 (1956).

10. BEROZA, M. and W. F. BARTHEL: Chemical Structure and Activity of Pyrethrin and Allethrin Synergists for Control of the Housefly. J. Agric. Food Chem. **5**, 855 (1957).

11. BOON, W. R. and F. HALL: Ethyl Chrysanthemate. Brit. Patent 740014 (1955).

12. Brown, N. C., D. T. Hollinshead, R. F. Phipers and M. C. Wood: New Isomers of the Pyrethrins Formed by the Action of Heat. Pyrethrum Post 4, No. 2, 13 (1957).

13. — — — — Application of Chromatography to Analysis of Pyrethrins. Soap 33, (9), 87, (10), 91 (1957).

14. Brown, N. C. and R. F. Phipers: The Analysis of Pyrethrins. Errors Arising During the Examination of Partially Degraded Materials. Pyrethrum Post 3, No. 4, 23 (1955).

15. Busvine, J. R. and R. Nash: The Potency and Persistence of Some New Synthetic Insecticides. Bull. entom. Res. 44, 371 (1953).

16. Cahn, R. S., C. K. Ingold and V. Prelog: The Specification of Asymmetric Configuration in Organic Chemistry. Experientia 12, 81 (1956).

17. Campbell, I. G. M. and S. H. Harper: Experiments on the Synthesis of the Pyrethrins. Part I. Synthesis of Chrysanthemum Monocarboxylic Acid. J. Chem. Soc. (London) 1945, 283.

18. — — The Chrysanthemumcarboxylic Acids. IV. Optical Resolution of the Chrysanthemic Acids. J. Sci. Food Agric. 3, 189 (1952).

19. Chandler, S. E.: Botanical Aspects of Pyrethrum. General Considerations; the Seat of the Active Principles. Pyrethrum Post 2, No. 3, 1 (1951).

20. — Botanical Aspects of Pyrethrum. II. Further Observations. Pyrethrum Post 3, No. 3, 6 (1954).

21. — Botanical Aspects of Pyrethrum. III. The Natural History of the Secretory Organs; the Pyrethrins Content of the Fertile Achenes ("Seed"). Pyrethrum Post 4, No. 1, 10 (1956).

22. Cornelius, J. A.: A Chromatographic Procedure for the Determination of Pyrethrins in Pyrethrum Extracts. Analyst 79, 458 (1954).

23. Crombie, L. and J. Crossley: Unpublished work.

24. Crombie, L., A. J. B. Edgar, S. H. Harper, M. W. Lowe and D. Thompson: Experiments on the Synthesis of the Pyrethrins. Part V. Synthesis of Side-chain Isomers and Analogues of Cinerone, Cinerolone, and Cinerin-I. J. Chem. Soc. (London) 1950, 3552.

25. Crombie, L., M. Elliott and S. H. Harper: Experiments on the Synthesis of the Pyrethrins. Part III. Synthesis of Dihydrocinerin-I and Tetrahydro-pyrethrin-I; a Study of the Action of N-Bromosuccinimide on 3-Methyl-2-n-alkyl (and alkenyl)-cyclopent-2-en-1-ones. J. Chem. Soc. (London) 1950, 971.

26. Crombie, L., M. Elliott, S. H. Harper and H. W. B. Reed: Total Synthesis of Some Pyrethrins. Nature (London) 162, 222 (1948).

27. Crombie, L. and S. H. Harper: Synthesis of Cinerone, Cinerolone and Cinerin-I. Nature (London) 164, 534 (1949).

28. — — "Leaf Alcohol" and the Stereochemistry of the cis- and the trans-n-Hex-3-en-1-ols and n-Pent-3-en-1-ols. J. Chem. Soc. (London) 1950, 873.

29. — — Experiments on the Synthesis of the Pyrethrins. Part IV. Synthesis of Cinerone, Cinerolone, and Cinerin-I. J. Chem. Soc. (London) 1950, 1152.

30. — — Experiments on the Synthesis of the Pyrethrins. Part VIII. Stereo-chemistry of Jasmone and Identity of Dihydropyrethrone. J. Chem. Soc. (London) 1952, 869.

31. — — The Chrysanthemumcarboxylic Acids. Part VI. The Configurations of the Chrysanthemic Acids. J. Chem. Soc. (London) 1954, 470.

32. — — Spectroscopic Assignment of Geometrical Configuration to Rethrins-I. Chem. and Ind. 1958, 1001.

33. Crombie, L., S. H. Harper and K. Mackenzie: Unpublished work.

34. CROMBIE, L., S. H. HARPER and F. C. NEWMAN: Experiments on the Synthesis of the Pyrethrins. Part XI. Synthesis of *cis*-Pyrethrolone and Pyrethrin I: Introduction of the *cis*-Penta-2,4-dienyl System by Selective Hydrogenation. J. Chem. Soc. (London) **1956**, 3963.

35. CROMBIE, L., S. H. HARPER, F. C. NEWMAN, D. THOMPSON and R. J. D. SMITH: Experiments on the Synthesis of the Pyrethrins. Part X. Intermediates for the Synthesis of *cis*-Pyrethrolone. J. Chem. Soc. (London) **1956**, 126.

36. CROMBIE, L., S. H. HARPER and K. C. SLEEP: Synthesis of the Naturally Derived Geometrical Isomer of Chrysanthemum Dicarboxylic Acid. Chem. and Ind. **1954**, 1538.

37. — — — Experiments on the Synthesis of the Pyrethrins. Part XIII. Total Synthesis of $(\pm)$-*cis*- and *trans*-Chrysanthemumdicarboxylic Acid, $(\pm)$-*cis*- and *trans*-Pyrethric Acid, and Rethrins II. J. Chem. Soc. (London) **1957**, 2743.

38. CROMBIE, L., S. H. HARPER, R. E. STEDMAN and D. THOMPSON: Experiments on the Synthesis of the Pyrethrins. Part VI. New Syntheses of the Cinerolones. J. Chem. Soc. (London) **1951**, 2445.

39. CROMBIE, L., S. H. HARPER and D. THOMPSON: Experiments on the Synthesis of the Pyrethrins. Part VII. Synthesis of *trans*-Pyrethrone, *trans*-Pyrethrolone and a Pyrethrin-I. J. Chem. Soc. (London) **1951**, 2906.

40. CROMBIE, L., S. H. HARPER and R. A. THOMPSON: The Chrysanthemum-carboxylic Acids. III. Lactonisation of the Chrysanthemic Acids. J. Sci. Food Agric. **2**, 421 (1951).

41. CROWLEY, M. P., H. S. INGLIS, M. SNAREY and E. M. THAIN: Personal communication.

42. CUPPLES, H. L.: Infrared Spectra of Cinerolone and Synthetic *trans*-(2-Butenyl)-4-hydroxy-3-methyl-2-cyclopenten-1-one. J. Amer. Chem. Soc. **72**, 4522 (1950).

43. DAUBEN, H. J., Jr. and E. WENKERT: Synthesis and Structure of Tetra-hydropyrethrolone. J. Amer. Chem. Soc. **69**, 2074 (1947).

44. DE MAYO, P.: The Chemistry of Natural Products, Vol. II. Mono- and Sesquiterpenoids. New York: Interscience Publ. 1959.

45. ELLIOTT, M.: Unpublished work.

46. — The Insecticidal Activity of the Pyrethrins and Related Compounds. Pyrethrum Post **2** (3), 18 (1951).

47. — Allethrin. J. Sci. Food Agric. **5**, 505 (1954).

48. — The Preparation of *cyclo*Pentenones from the Products of Stobbe Condensations with Aliphatic Ketones. J. Chem. Soc. (London) **1956**, 2231.

49. — Isolation and Purification of $(+)$-Pyrethrolone from Pyrethrum Extract: Reconstitution of Pyrethrins I and II. Chem. and Ind. **1958**, 685.

50. — Pyrethrolone and Related Compounds. Chem. and Ind. **1960**, 1142.

51. — The Structures of the Enols of Pyrethrolone. Proc. Chem. Soc. (London) **1960**, 406.

52. — The Pyrethrins and Related Compounds. Part II. Infra-red Spectra of the Pyrethrins and of Other Constituents of Pyrethrum Extract. J. Applied Chem. (London) **11**, 19 (1961).

53. ELLIOTT, M. and P. H. NEEDHAM: Unpublished work.

54. ELLIOTT, M., P. H. NEEDHAM and C. POTTER: The Insecticidal Activity of Substances Related to the Pyrethrins. I. Toxicities of Two Synthetic Pyrethrin-like Esters Relative to that of the Natural Pyrethrins and the Significance of the Results in the Bioassay of Closely Related Compounds. Ann. appl. Biol. **37**, 490 (1950).

55. ELLIOTT, M., J. S. OLEJNICZAK and J. J. GARNER: Laboratory-Scale Molecular Distillation of the Pyrethrins. Pyrethrum Post **5** (2), 8 (1959).

56. Ettlinger, M. G., S. H. Harper and F. Kennedy: The Addition of Ethyl Diazoacetate to Sorbic Esters: A Correction. J. Chem. Soc. (London) 1957, 922.

57. Fales, J. H., O. F. Bodenstein and M. Beroza: New Pyrethrum Synergist. Soap 33, (2) 79 (1957).

58. Farkaš, J., H. Komrsová, J. Krupička und J. J. K. Novák: Beziehung zwischen chemischem Bau und Insektizid-Aktivität in der Reihe der Pyrethrum-Verbindungen. IV. Über den Einfluß der Substituenten in der Seitenkette auf den Verlauf der LaForge-Cyclisierung. Coll. Czech. Chem. Comm. 25, 1824 (1960).

58a. Feinstein, L. and M. Jacobson: Insecticides Occurring in Higher Plants. Fortschr. Chem. organ. Naturstoffe 10, 423 (1953).

59. Feldman, A.: Stereo Numbers: A Short Designation for Stereoisomers. J. Organ. Chem. (USA) 24, 1556 (1959).

60. Frank, R. L., R. Armstrong, J. Kwiatek and H. A. Price: The Preparation of Cyclopentenones from Lactones. J. Amer. Chem. Soc. 70, 1379 (1948).

61. Fredga, A.: Optically Active Forms of Terebic Acid. Svensk Papperstidn. 50, No. 11 B, 91 (1947).

62. Fredga, A. und E. Leskinen: Zur Konfiguration einiger Terpene. Ark. Kemi, Mineral. Geol. 19 B, No. 1 (1945).

63. Freeman, S. K.: Infra-red Spectrophotometric Determination of Allethrin. Analyt. Chemistry 27, 1268 (1955).

64. Freudenberg, K. und G. S. Sidhu: Die absolute Konfiguration des Sesamins und Pinoresinols. Tetrahedron Letters 1960, No. 20, 3.

65. Fujitani, J.: Chemistry and Pharmacology of Insect Powder. Arch. exp. Pathol. Pharmakol. 61, 47 (1909).

66. Fukushi, S.: The Components of the Unsaponifiable Matter of the Wax of Chrysanthemum cinerariaefolium. J. Agric. Chem. Soc. Japan 26, 1 (1952).

67. Gersdorff, W. A.: Toxicity to Houseflies of the Pyrethrins and Cinerins, and Derivatives, in Relation to Chemical Structure. J. econ. Entomol. 40, 878 (1947).

68. Gersdorff, W. A. and N. Mitlin: A Bioassay of some Stereoisomeric Constituents of Allethrin. J. Washington Acad. Sci. 42, 313 (1952).

69. — — The Relative Toxicity to House Flies of the Methyl and Ethyl Analogues of Allethrin. J. econ. Entomol. 46, 945 (1953).

70. Gersdorff, W. A., N. Mitlin and M. Beroza: Comparative Effects of Sesamolin, Sesamin and Sesamol in Pyrethrum and Allethrin Mixtures as House-Fly Sprays. J. econ. Entomol. 47, 839 (1954).

71. Gersdorff, W. A. and P. G. Piquett: Comparative Effect of Piperetine in Pyrethrum and Allethrin Mixtures as Housefly Sprays. J. econ. Entomol. 50, 164 (1957).

72. — — Effect of Molecular Configuration on Relative Toxicity to House Flies as Demonstrated with the Four cis-Isomers of Allethrin. J. econ. Entomol. 51, 181 (1958).

73. Gersdorff, W. A., P. G. Piquett and M. Beroza: Comparative Effects of the Optical Forms of Epiasarinin, Asarinin and Sesamin in Pyrethrum Mists as Housefly Sprays. J. econ. Entomol. 50, 409 (1957).

74. Gillam, A. E. and T. F. West: Observations on the Absorption Spectra of Terpenoid Compounds. Part IV. Five-Atom-Ring Unsaturated Ketones. J. Chem. Soc. (London) 1942, 486.

75. — — Absorption Spectra and the Structure of Pyrethrins I and II. J. Chem. Soc. (London) 1942, 671.

76. GILLAM, A. E. and T. F. WEST: Absorption Spectra and the Structure of Pyrethrins I and II. Part II. J. Chem. Soc. (London) 1944, 49.

77. GNADINGER, C. B.: Pyrethrum Flowers, 2nd Ed. Minneapolis, Minn.: McLaughlin Gormley King Co. 1936.

78. — Pyrethrum Flowers. Suppl. to 2nd Ed. (1936–1945). Minneapolis, Minn.: McLaughlin Gormley King Co. 1945.

79. HALLER, H. L. and F. B. LaFORGE: Constituents of Pyrethrum Flowers. IV. The Semicarbazones of Pyrethrins I and II and of Pyrethrolone. J. Organ. Chem. (USA) 1, 38 (1936).

80. — — Constituents of Pyrethrum Flowers. VII. The Behavior of the Pyrethrins on Hydrogenation. J. Organ. Chem. (USA) 2, 49 (1937).

81. — — Constituents of Pyrethrum Flowers. IX. The Optical Rotation of Pyrethrolone and the Partial Synthesis of Pyrethrins. J. Amer. Chem. Soc. 59, 1678 (1937).

82. — — Constituents of Pyrethrum Flowers. XIV. The Structures of the Enols of Pyrethrolone. J. Organ. Chem. (USA) 3, 543 (1939).

83. HALLER, H. L., F. B. LaFORGE and W. N. SULLIVAN: Some Compounds Related to Sesamin: Their Structures and Their Synergistic Effect with Pyrethrum Insecticides. J. Organ. Chem. (USA) 7, 185 (1942).

84. HARPER, S. H.: Experiments on the Synthesis of the Pyrethrins. Part II. The Structure of Cinerone. J. Chem. Soc. (London) 1946, 892.

85. — A Nomenclature for the Pyrethrins. Chem. and Ind. 1949, 636; Pyrethrum Post 2, (1) 20 (1950).

86. — The Chrysanthemumcarboxylic Acids. VII. Catalytic Hydrogenation of the Chrysanthemic Acids. J. Sci. Food Agric. 5, 529 (1954).

87. HARPER, S. H. and M. A. KAZI: Unpublished work.

88. HARPER, S. H. and H. W. B. REED: The Chrysanthemumcarboxylic Acids. II. Esterification of the Chrysanthemic Acids. J. Sci. Food Agric. 2, 414 (1951).

89. — — Experiments on the Synthesis of the Pyrethrins. Part IX. The Addition of Ethyl Diazoacetate to Sorbic Esters. J. Chem. Soc. (London) 1955, 779.

90. HARPER, S. H., H. W. B. REED and R. A. THOMPSON: The Chrysanthemumcarboxylic Acids. I. Preparation of the Chrysanthemic Acids. J. Sci. Food Agric. 2, 94 (1951).

91. HARPER, S. H. and K. C. SLEEP: The Chrysanthemumcarboxylic Acids. VIII. A Modified Route to the Chrysanthemic Acids. J. Sci. Food Agric. 6, 116 (1955).

92. HARPER, S. H. and R. A. THOMPSON: The Chrysanthemumcarboxylic Acids. V. Hydration of the Chrysanthemic Acids. J. Sci. Food Agric. 3, 230 (1952).

93. HARVILL, E. K., A. HARTZELL and J. M. ARTHUR: Toxicity of Piperine Solutions to Houseflies. Contrib. Boyce Thompson Inst. 13, 87 (1943).

94. HAYNES, H. L., H. R. GUEST and H. A. STANSBURY et al.: Cyclethrin. Soap 31, (2) 141 (1955).

95. HEDENBURG, O. F. and H. WACHS: Methylenedioxyphenyl Cyclohexenones. J. Amer. Chem. Soc. 70, 2216 (1948).

96. HENNE, A. L. and H. H. CHANAN: Conjugated Diolefins by Double Bond Displacement. II. J. Amer. Chem. Soc. 66, 395 (1944).

97. HUNSDIECKER, H.: Über das Verhalten der γ-Diketone. II. Mitt. Der Cyclopentenon-Ringschluß der γ-Diketone vom Typus $CH_3 \cdot CO \cdot CH_2 \cdot CH_2 \cdot CO \cdot CH_2 \cdot R$. Ber. dtsch. chem. Ges. 75, 455 (1942).

98. HUNSDIECKER, H. und E. WIRTH: Über das Verhalten der γ-Diketone. III. Mitt. Die Synthese des Jasmons. Ber. dtsch. chem. Ges. 75, 460 (1942).

99. Incho, H. H. and H. Greenberg: Synergistic Effect of Piperonyl Butoxide with the Active Principles of Pyrethrum and with Allethrolone Esters of Chrysanthemum Acids. J. econ. Entomol. 45, 794 (1952).

100. Inoue, Y.: Synthesis and Stereochemistry of Chrysanthemum Dicarboxylic Acid. Bull. Inst. Chem. Res. Kyoto Univ. 35, 49 (1957).

101. Inoue, Y. and M. Ohno: Resolution of (±)-trans-3-(trans-2-carboxypropenyl)-2,2-Dimethylcyclopropane-1-carboxylic Acid. Bull. Inst. Chem. Res., Kyoto Univ. 34, 90 (1956).

102. — — Absolute Configuration of Natural Pyrethrins. Kagaku (Tokyo) 28, 636 (1958).

103. Inoue, Y., T. Shinohara and M. Ohno: An Approach to the Synthesis of Pyrethric Acid. Botyu Kagaku 19, 35 (1954).

104. Inoue, Y., Y. Sugita and M. Ohno: Synthesis of Pyrethroids. IX. Assignment of Geometrical Configuration to $\alpha\delta$-Dimethylsorbic Acid. Bull. Agric. Chem. Soc. Japan 21, 5 (1957).

105. — — — Synthesis of Pyrethroids. XI. Another Piece of Evidence for the trans-Configuration of $\alpha\delta$-Dimethylsorbic Acid. Bull. Agric. Chem. Soc. Japan 21, 222 (1957).

106. — — — A Novel Route of Synthesis to Chrysanthemumdicarboxylic Acid. Bull. Agric. Chem. Soc. Japan 22, 269 (1958).

107. Inoue, Y., Y. Takeshita and M. Ohno: Studies on Synthetic Pyrethroids. V. Synthesis of Geometrical Isomers of Chrysanthemumdicarboxylic Acid. Botyu-Kagaku 20, 102 (1955).

108. — — — Geometrical Isomers of Chrysanthemum Dicarboxylic Acid. Bull. Inst. Chem. Res., Kyoto Univ. 33, 73 (1955).

109. Jackman, L. M. and R. H. Wiley: Studies in Nuclear Magnetic Resonance. Part III. Assignment of Configurations of $\alpha\beta$-Unsaturated Esters and the Isolation of Pure trans-β-Methylglutaconic Acid. J. Chem. Soc. (London) 1960, 2886.

110. Julia, M., S. Julia et C. Jeanmart: Synthèses de l'acide trans-dihydro-chrysanthémique. C. R. hebd. Séances Acad. Sci. 250, 4003 (1960).

111. — — — Synthèses de la pyrocine, de l'acide trans-(±)-chrysanthémique et de quelques acides cyclopropaniques apparantés. C. R. hebd. Séances Acad. Sci. 251, 249 (1960).

112. Kageyama, I.: Extraction of Active Principles from Pyrethrum, Derris, Tobacco etc. Japanese Patent 3649 (1952) [Chem. Abstr. 47, 8328 (1953)].

113. Katsuda, Y., T. Chikamoto and Y. Inoue: The Absolute Configuration of Naturally Derived Pyrethrolone. Bull. Agric. Chem. Soc. Japan 22, 427 (1958); Relationship between Stereoisomers and Biological Activity of Pyrethroids. Part V. The Absolute Configuration of (+)-Pyrethrolone and (+)-Cinerolone. Bull. Agric. Chem. Soc. Japan 23, 174 (1959).

114. Katsuda, Y. and T. Chikamoto: Studies on the Degradation of Pyrethrins. IV. Botyu-Kagaku 23, 60 (1958).

115. LaForge, F. B. and F. Acree, Jr.: Constituents of Pyrethrum Flowers. XV. Presence of the Cumulated System in the Pyrethrolone Side-Chain. J. Organ. Chem. (USA) 7, 416 (1942).

116. LaForge, F. B. and W. F. Barthel: Constituents of Pyrethrum Flowers. XVI. Heterogeneous Nature of Pyrethrolone. J. Organ. Chem. (USA) 9, 242 (1944).

117. — — Constituents of Pyrethrum Flowers. XVII. The Isolation of Five Pyrethrolone Semicarbazones. J. Organ. Chem. (USA) 10, 106 (1945).

118. LaForge, F. B. and W. F. Barthel: Constituents of Pyrethrum Flowers. XVIII. The Structure and Isomerism of Pyrethrolone and Cinerolone. J. Organ. Chem. (USA) 10, 114 (1945).

119. — — Constituents of Pyrethrum Flowers. XIX. The Structure of Cinerolone. J. Organ. Chem. (USA) 10, 222 (1945).

120. — — Constituents of Pyrethrum Flowers. XX. The Partial Synthesis of Pyrethrins and Cinerins and their Relative Toxicities. J. Organ Chem. (USA) 12, 199 (1947).

121. LaForge, F. B., W. A. Gersdorff, N. Green and M. S. Schechter: Allethrin-Type Esters of Cyclopropanecarboxylic Acids and their Relative Toxicities to House Flies. J. Organ. Chem. (USA) 17, 381 (1952).

122. LaForge, F. B. and N. Green: Constituents of Pyrethrum Flowers. XXV. The Synthesis of d-Cinerolone, Cinerin I, and its Optical Isomers. J. Organ. Chem. (USA) 17, 1635 (1952).

123. LaForge, F. B., N. Green and M. S. Schechter: Dimerized Cyclopentadienones from Esters of Allethrolone. J. Amer. Chem. Soc. 74, 5392 (1952).

124. — — — Allethrin. Resolution of dl-Allethrolone and Synthesis of the Four Optical Isomers of trans-Allethrin. J. Organ. Chem. (USA) 19, 457 (1954).

125. — — — Allethrin. Synthesis of Four Isomers of cis-Allethrin. J. Organ. Chem. (USA) 21, 455 (1956).

126. LaForge, F. B. and H. L. Haller: Constituents of Pyrethrum Flowers. V. Concerning the Structure of Pyrethrolone. J. Amer. Chem. Soc. 58, 1061 (1936).

127. — — Constituents of Pyrethrum Flowers. VI. The Structure of Pyrethrolone. J. Amer. Chem. Soc. 58, 1777 (1936).

128. — — Constituents of Pyrethrum Flowers. VIII. The Presence of a New Ester of Pyrethrolone. J. Organ. Chem. (USA) 2, 56 (1937).

129. — — Constituents of Pyrethrum Flowers. XII. The Nature of the Side-Chain of Pyrethrolone. J. Organ. Chem. (USA) 2, 546 (1938).

130. LaForge, F. B. and S. B. Soloway: Constituents of Pyrethrum Flowers. XXI. Revision of the Structure of Dihydrocinerolone. J. Amer. Chem. Soc. 69, 2932 (1947).

131. Lord, K. A., J. Ward, J. A. Cornelius and M. W. Jarvis: Chromatographic Separation of the Pyrethrins. J. Sci. Food Agric. 3, 419 (1952).

132. Matsubara, H.: A New Synergist for Pyrethrins (Hinokinin). Science (Japan) 20, 183 (1950).

133. — Hibalactone (Savinin) as a Synergist with Pyrethrins and Allethrin. Bull. Agric. Chem. Soc. Japan 21, 132 (1957).

134. Matsui, M.: Pyrocin (a New Insecticide). Botyu-Kagaku 15, 1 (1950).

135. — Synthetic Insecticide. Japanese Patent 5626 (1955) [Chem. Abstr. 52, 1218 (1958)].

136. Matsui, M., S. Kitamura, T. Kato and S. Sugihara: Synthesis of Cyclopentenolones of the Cinerolone Type. J. Chem. Soc. Japan 71, 235 (1950).

137. Matsui, M., F. B. LaForge, N. Green and M. S. Schechter: Furethrin. J. Amer. Chem. Soc. 74, 2181 (1952).

138. Matsui, M., M. Miyano, K. Yamashita, H. Kubo and K. Tomita: Syntheses of Pyrethridic (Chrysanthemumdicarboxylic) Acid. Bull. Agric. Chem. Soc. Japan 21, 22 (1957).

139. Matsui, M., T. Ohno, S. Kitamura and M. Tayao: The Lactones Derived from Chrysanthemic Acids. Bull. Chem. Soc. Japan 25, 210 (1952).

140. Matsumoto, T. and A. Suzuki: Reaction of Isopropyl Zinc Iodide with Terebic Acid Chloride. A Suggested New Mode of Action of the Blaise Reagent. J. Organ. Chem. (USA) 25, 1666 (1960).

141. Merritt, R. P. and T. F. West: Notes on the Oil Distilled from Pyrethrum Flowers. J. Soc. Chem. and Ind. 57, 321 (1938).

142. Metcalf, R. L.: Methods of Topical Application and Injection. In: H. H. Shepard, Methods of Testing Chemicals on Insects, Vol. 1, p. 92. Minneapolis, Minn.: Burgess Publ. Co. 1958.

143. — Organic Insecticides. Their Chemistry and Mode of Action. New York and London: Interscience Publ. 1955.

144. Mitchell, W.: "Bucarpolate," Pyrethrum Synergist. Pyrethrum Post 5, No. 1, 19 (1959).

145. Moore, B. P. and P. S. Hewlett: Insecticidal Synergism with the Pyrethrins: Studies on the Relationship between Chemical Structure and Synergistic Activity in 3,4-Methylenedioxyphenyl Compounds. J. Sci. Food Agric. 9, 666 (1958).

146. Negherbon, W. O.: Handbook of Toxicology. Vol. III, Insecticides. A Compendium. Philadelphia and London: W. B. Saunders Co. 1959.

147. Newman, M. S. and W. T. Booth, Jr.: The Preparation of Ketones from Grignard Reagents. J. Amer. Chem. Soc. 67, 154 (1945).

148. Pellegrini, J. P., A. C. Miller and R. V. Sharpless: Biosynthesis of Radioactive Pyrethrins using $^{14}CO_2$. J. econ. Entomol. 45, 532 (1952).

149. Phipers, R. F.: Pyrethrins and Allied Compounds. In: K. Paech and M. V. Tracey, Modern Methods of Plant Analysis, Vol. III, p. 43. Berlin: Springer-Verlag. 1955.

150. — The Analysis of Pyrethrins. A Review of Recent Publications. Pyrethrum Post 4 (3), 3 (1958).

151. Potter, C. and M. J. Way: Precision Spraying. In: H. H. Shepard, Methods of Testing Chemicals on Insects, Vol. I, p. 154. Minneapolis, Minn.: Burgess Publ. Co. 1958.

152. Quayle, J. R.: Paper Chromatography of Pyrethrins and their Derivatives. Nature (London) 178, 375 (1956).

153. Rai, C. and S. Dev: Organic Reactions with Polyphosphoric Acid. IV. Intramolecular Acylation with Lactones: cycloPentenones from γ-Lactones. J. Indian Chem. Soc. 34, 178 (1957).

154. Roark, R. C.: A Digest of Information on Allethrin. U. S. Dept. Agric., Bur. Entomology and Plant Quarantine, Agric. Res. Admin., E 846, Sept. 1952.

155. Roark, R. C. and R. H. Nelson: A Second Digest of Information on Allethrin and Related Compounds. U. S. Dept. Agric., Agric. Res. Serv. ARS-33-12 (1955).

156. Ruzicka, L.: History of the Isoprene Rule. Proc. Chem. Soc. (London) 1959, 341.

157. Ruzicka, L. und M. Pfeiffer: Über Jasminriechstoffe. I. Die Konstitution des Jasmons. Helv. Chim. Acta 16, 1208 (1933).

158. Sanders, H. J. and A. W. Taff: Allethrin. Ind. Eng. Chem. 46, 414 (1954).

159. Sawicki, R. M.: Insecticidal Activity of the Four Constituents of Pyrethrum. Rothamsted Exp. Stat. Report for 1960 (in press).

160. — A Technique for the Knockdown Assessment of Topically Treated Normally Active Houseflies. Bull. entom. Res. 51, 715 (1961).

161. SAWICKI, R. M., M. ELLIOTT, J. C. GOWER, M. SNAREY and E. M. THAIN: Insecticidal Activity of Pyrethrum Extract and its Active Constituents against Houseflies. I. Preparation and Relative Toxicity of the Pure Constituents; Statistical Analysis of Mixtures of these Compounds. J. Sci. Food Agric. (1961) (in press).

162. SAWICKI, R. M. and E. M. THAIN: The Chemical and Biological Examination of Commercial Pyrethrum Extracts for Insecticidal Constituents. J. Sci. Food Agric. **12,** 137 (1961).

163. SCHECHTER, M. S., N. GREEN and F. B. LaFORGE: Constituents of Pyrethrum Flowers. XXIII. Cinerolone and the Synthesis of Related Cyclopentenolones. J. Amer. Chem. Soc. **71,** 3165 (1949).

164. — — — Constituents of Pyrethrum Flowers. XXIV. Synthetic *dl-cis*-Cinerolone and Other Cyclopentenolones. J. Amer. Chem. Soc. **74,** 4902 (1952).

165. SCHECHTER, M. S., F. B. LaFORGE, A. ZIMMERLI and J. M. THOMAS: Crystalline Allethrin Isomer. J. Amer. Chem. Soc. **73,** 3541 (1951).

166. SCHMIDT, H.: Zur Kenntnis des Pfefferminzöls. Vorkommen von Jasmon im ätherischen Öl von *Mentha piperita* L. Chem. Ber. **80,** 538 (1947).

167. SIMONSEN, J. L. and L. N. OWEN: The Terpenes, Vol. I. Cambridge: Univ. Press. 1947.

168. "Socophar": Extraction of Pyrethrin with Isopropyl Ether. Belgian Patent 509803 (1952) [Chem. Abstr. **52,** 6708 (1958)].

169. SOLOWAY, S. B. and F. B. LaFORGE: The Synthesis of Dihydrocinerolone. J. Amer. Chem. Soc. **69,** 979 (1947).

170. SPICKETT, R. G. W.: A Method for the Separation of the Constituents of Pyrethrum Extract. Chem. and Ind. **1957,** 561.

171. STANSBURY, H. A., Jr. and H. R. GUEST: Derivatives of Chrysanthemum Monocarboxylic Acid. British Patent 744268 (1956).

172. — — [A] Substituted *cyclo*-Pentenones. [B] An Ester of Chrysanthemum-monocarboxylic Acid and Insecticidal Constituents Therefrom. British Patent 790841 (1958).

173. STAUDINGER, H., O. MUNTWYLER, L. RUZICKA und S. SEIBT: Insektentötende Stoffe. VII. Synthesen der Chrysanthemumsäure und anderer Trimethylen-carbonsäuren mit ungesättigter Seitenkette. Helv. Chim. Acta **7,** 390 (1924).

174. STAUDINGER, H. und L. RUZICKA: Insektentötende Stoffe. I. Über Isolierung und Konstitution des wirksamen Teiles des dalmatinischen Insektenpulvers. Helv. Chim. Acta **7,** 177 (1924).

175. — — Insektentötende Stoffe. II. Zur Konstitution der Chrysanthemum-monocarbonsäure und -dicarbonsäure. Helv. Chim. Acta **7,** 201 (1924).

176. — — Insektentötende Stoffe. III. Konstitution des Pyrethrolons. Helv. Chim. Acta **7,** 212 (1924).

177. — — Insektentötende Stoffe. IV. Konstitution des Tetrahydropyrethrons. Helv. Chim. Acta **7,** 236 (1924).

178. — — Insektentötende Stoffe. V. Synthese des Tetrahydropyrethrons, des Reduktionsproduktes des Pyrethrolons. Helv. Chim. Acta **7,** 245 (1924).

179. — — Insektentötende Stoffe. X. Über die Synthese von Pyrethrinen. Helv. Chim. Acta **7,** 448 (1924).

180. STEPHENSON, H.: The Separation and Estimation of the Four Insecticidal Constituents of *Chrysanthemum cinerariaefolium* by Elution Chromatography on a Column of Adsorptive Charcoal. Pyrethrum Post **5,** No. 4, 22 (1960).

181. SUBBARATNAM, A. V. and S. SIDDIQUI: The Irritant Factor of Pyrethrum Flowers. J. Sci. Industr. Res. (India) **15 B,** 243 (1956).

182. SYNERHOLM, M. E., A. HARTZELL and J. M. ARTHUR: Derivatives of Piperic Acid and their Toxicities towards Houseflies. Contrib. Boyce Thompson Inst. **13**, 433 (1945).

183. TAKEI, S., T. SUGITA und Y. INOUYE: Eine neue Synthese der Chrysanthemum-dicarbonsäure. Liebigs Ann. Chem. **618**, 105 (1958).

184. TREFF, W. und H. WERNER: Über die Synthese des Jasmons. Ber. dtsch. chem. Ges. **68**, 640 (1935).

185. WACHS, H.: Synergistic Insecticides. Science (Washington) **105**, 530 (1947).

186. WANLESS, G. G., W. H. KING and J. J. RITTER: Hydrocarbons in Pyrethrum Cuticle Wax. Biochemic. J. **59**, 684 (1955).

187. WARD, J.: Separation of the Pyrethrins by Displacement Chromatography. Chem. and Ind. **1953**, 586.

188. WEED, A.: New Insecticide Compound. Soap **14**, No. 6, 133 (1938).

189. WEST, T. F.: The Structure of Pyrethrolone and Related Compounds. Part II. J. Chem. Soc. (London) **1944**, 239.

190. — The Structure of Pyrethrolone and Related Compounds. Part IV. J. Chem. Soc. (London) **1945**, 412.

191. — The Structure of Pyrethrolone and Related Compounds. Part V. J. Chem. Soc. (London) **1946**, 463.

192. WINTERINGHAM, F. P. W., A. HARRISON and P. M. BRIDGES: Absorption and Metabolism of ^{14}C-Pyrethroids by the Adult Housefly, *Musca domestica*, in vivo. Biochemic. J. **61**, 359 (1955).

193. WOODS, G. F. and H. SANDERS: Studies in Pyrane Chemistry. J. Amer. Chem. Soc. **68**, 2483 (1946).

194. YAMAMOTO, R.: The Insecticidal Principle in *Chrysanthemum cinerariaefolium*. I. J. Tokyo Chem. Soc. **40**, 126 (1919).

195. — Studies on the Insecticidal Principle in *Chrysanthemum cinerariaefolium*. Parts II and III. On the Constitution of Pyrethronic Acid. J. Chem. Soc. Japan **44**, 311 (1923).

196. YAMAMOTO, R. and M. SUMI: Studies on the Insecticidal Principle in *Chrysanthemum cinerariaefolium*. J. Chem. Soc. Japan **44**, 1080 (1923).

(Received, January 27, 1961.)

Conformational Analysis of Steroids and Related Natural Products.

By D. H. R. BARTON and G. A. MORRISON, London.

Contents.

I. Introduction.

During the past decade, the principles of conformational analysis (*31, 32, 37, 125, 223*) have been widely employed as an aid in the elucidation of the stereochemistry of steroids (*163*), terpenoids (*32, 44, 223*), alkaloids (*97*), and carbohydrates. The conformational analysis of carbohydrates has already been adequately treated elsewhere (*162, 260, 279*), and the use of this approach in the study of sesquiterpenoids has also been surveyed (*265*; see also *91*). The present review will be concerned mainly with the application of conformational principles in the steroid and triterpenoid fields. The more important papers on this topic published up to the end of 1960 have been surveyed.

II. Definition of Conformation.

It is important that configuration and conformation should be clearly distinguished. The possible *configurations* of an organic molecule are limited by the principles of free rotation about single bonds and restricted rotation about double bonds. Applying these principles, it follows that a molecule containing n dissimilar asymmetric centres can exist as 2^n stereoisomers. Similarly, there are 2^m possible geometrical isomers of a structure containing m suitably substituted double bonds. Each of these isomers is a stable molecular species, and together they represent the possible configurations of the molecule.

The first indication that the principle of free rotation about single bonds required modification came with the discovery that certain 2,2',6,6'-tetrasubstituted diphenyl derivatives could be resolved into optical isomers (*299*). In these cases, the bulk of the substituents prevents free rotation about single bonds. Following on this it has been shown that even in a simple molecule such as ethane there is some hindrance to free rotation about a carbon-carbon single bond (*251, 262*). Although the magnitude of the barrier to rotation is low—it has been calculated to be $2,875 \pm 125$ cal. per mole (*276*)—it is sufficient to ensure that most of the molecules of ethane will exist in one preferred stereochemical arrangement. In principle, of course, there is an infinite number of these non-superposable molecular arrangements, and the term used to describe each of them is *conformation*. The conformations of a molecule may be defined as those steric arrangements of the atoms of the molecule, which, although they may be represented as one configurational species, are non-superposable on each other. Such a definition includes arrangements of atoms in which angle strain has been introduced, but these are not normally considered.

III. The Existence of Preferred Conformations.

1. Acyclic Compounds.

It has already been noted that, although the ethane molecule can exist theoretically in an infinite number of conformations, one of these is energetically preferred over all the others. In the absence of intra-molecular forces due to hydrogen-bonding or electrostatic effects, the most stable conformation of a molecule is the one in which non-bonded interactions are at a minimum. Thus, for ethane, the most stable conformation is the one in which the hydrogen atoms attached to the two methyl groups are fully staggered, and the least stable conformation is that in which the hydrogen atoms are totally eclipsed. These may be represented respectively by (I) and (II).

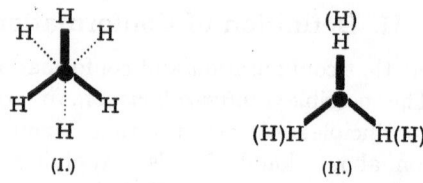

(I.)   (II.)

Owing to the symmetry of the molecule, the methyl group rotation in ethane has a potential barrier with three equal maxima and minima. In aliphatic compounds, in general, the preferred conformation is that with a maximum staggering of bonds. The carbon skeleton adopts a preferred zig-zag arrangement.

2. Cyclohexane Derivatives.

(For a full discussion, see *270*.)

a) Boat and Chair Conformations. SACHSE (*288*) and MOHR (*263*) first drew attention to the fact that cyclohexane could exist in two conformations free from angle strain. These are known as the chair (III) and boat (IV) conformations.

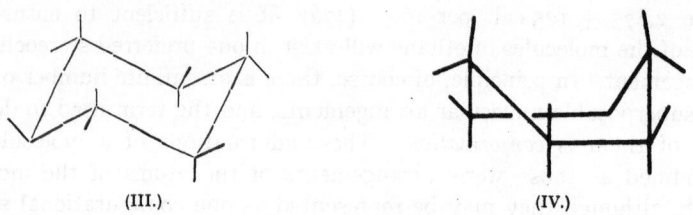

(III.)   (IV.)

The energy difference between the boat and chair conformations has been estimated by empirical and semi-empirical procedures (*29, 63, 273, 315*) to be between 1.3 and 10.6 kcal. per mole. All such calculations have indicated that the chair conformation is more stable than the boat. This is as would be predicted from an examination of the two conformers. In the chair conformation (III) all the C—H bonds on adjacent carbon atoms are staggered, while in the boat form the C—H bonds along the sides of the boat are eclipsed. The boat conformation is further destabilised by a 1,4-interaction between the two "flagpole" hydrogen atoms.

Two recent experimental determinations of the enthalpy difference between boat and chair conformations are in excellent agreement. JOHNSON and his co-workers (*216*) found a value for ΔH of 5.3 $\pm$ 0.3 kcal. per mole by measuring the heats of combustion, of (V), which can adopt an all-chair conformation, and (VI) which must have its central ring in a boat conformation. Using a quite different approach, ALLINGER and FREIBERG (*16*) found a value for ΔH of 5.9 $\pm$ 0.6 kcal. per mole.

H H   H H

=O   =O

(V.)   (VI.)

It has been pointed out (9, 195, 279) that the regular boat conformation of cyclohexane (VIIa) is less favoured than the skewed form (VIII), which may be regarded as intermediate between the boat forms (VIIa) and (VIIb). All these variants of the flexible form of cyclohexane are readily interconvertible by rotation about single bonds, without distorting the bond angles. In the skewed form (VIII) the severe 1,4-transannular

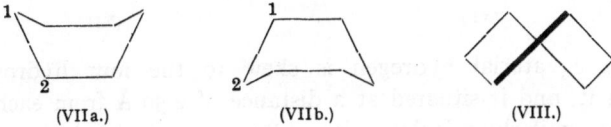

(VIIa.)   (VIIb.)   (VIII.)

interaction present in the regular boat conformation is somewhat relieved. The energy of the skewed form has been calculated (9) as 5.1 kcal. per mole greater than that of the chair conformation.

Until recently, it was thought that, wherever it was configurationally possible, derivatives of cyclohexane would adopt the chair conformation, and in most instances this is indeed true. However, under certain circumstances this rule does not hold (245). These special cases will be considered later in a separate section (p. 212).

b) Axial and Equatorial Bonds. The chair conformation of cyclohexane has a six-fold axis of alternating symmetry. As a consequence of this the twelve carbon-hydrogen bonds are of two distinct geometrical classes. Six lie parallel to the axis (three on each side of the ring) (IX) and are designated "axial" (a); the remaining six bonds radiate out from the ring (X) and are called "equatorial" (e) (40).

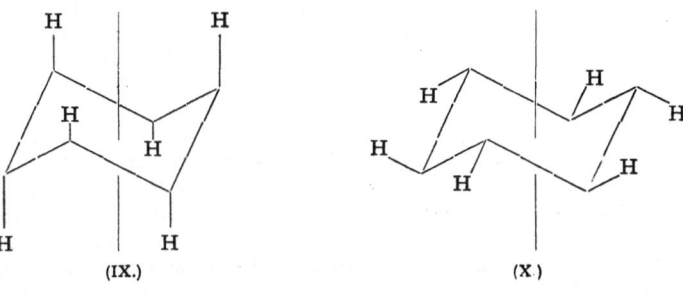

(IX.)   (X.)

It is to be noted that, by a flexing of bond angles, the chair conformation (XI) of cyclohexane can be converted into an alternative chair conformation (XII) in which all the bonds which were formerly equatorial have become axial, and those which were axial have become equatorial. At ordinary temperatures the energy barrier separating the two forms is only a few kcal. per mole (see *213*), so that in a simple cyclohexane derivative, the ring can readily flip from one conformation to the other.

(XI.)   (XII.)

Each equatorial hydrogen is skew to the four hydrogen atoms flanking it, and is situated at a distance of 2.50 Å from each (*109*). It has been similarly calculated by vector analysis (*109*) that each axial

Table 1. Free Energy Differences Between Equatorial and Axial Substituents in Cyclohexane.*

Group	$-\Delta F$ (kcal./mole)	Temp. (°)	Solvent	Reference
—OH	0.8	40	75% HO*Ac*	(*335*)
—OH	0.5	25	pyridine	(*154*)
—OH	0.9	25	water	(*22*)
—OH	0.96	90	isopropanol	(*156*)
—OH	0.4	20	carbon disulphide	(*275*)
—O*Ac*...............	0.36	25	*Ac*OH—*Ac*$_2$O	(*241*)
p-NO$_2$ · C$_6$H$_4$ · COO.....	0.9	25	80% acetone	(*202*)
—O*Ts*...............	1.7	50	C$_2$H$_5$OH, *Ac*OH, HCOOH	(*335*)
—O*Ts*...............	0.7	25	87% ethanol	(*157*)
—Br	0.7	25	87% ethanol	(*153*)
—Br	0.48	—81	carbon disulphide	(*65*)
—CH$_3$	1.5–1.8	35	ether	(*155*)
C$_6$H$_5$	ca. 2.6	35	ether	(*155*)
—COOH...............	1.6 ± 0.3	20	80% methyl cellosolve	(*314*)
—COO$^{\ominus}$	2.2 ± 0.3	20	80% methyl cellosolve	(*314*)
—COOC$_2$H$_5$	1.2–1.4	25	70–100% ethanol	(*151*)
—HgBr...............	ca. 0	95	pyridine	(*211*)

* It has been suggested (*151*) that the wide variation in the values found for the hydroxyl group is due to hydrogen-bonding between the hydroxyl group and the solvent, the equatorial form being relatively more favoured in proton-donor solvents. The small preference for an equatorial conformation exhibited by the bromo- and mercuri-bromo-groups is thought to be due to the long C—Br and C—HgBr bond lengths, and to the operation of London attractive forces in these cases (*151, 211*).

References, pp. 225—241.

hydrogen is 2.53 Å from the other two axial hydrogen atoms on the same side of the ring.

When a substituent larger than hydrogen is in the axial conformation it is closer to the two axial hydrogens on the same side of the ring than it would be to the adjacent axial and equatorial hydrogen atoms if it were in an equatorial conformation. Hence, a substituent is subjected to greater non-bonded interactions when it is in an axial conformation than when it is in an equatorial conformation. For this reason the stable conformation of a substituted cyclohexane is, in general, that with the greatest number of substituents in equatorial positions (*114*). This has been borne out by electron diffraction studies on a number of cyclohexane derivatives in the vapour phase (*187–189*).

Various methods have been devised to determine the free energy difference between the equatorial and axial conformations of substituents attached to a cyclohexane ring. These have recently been reviewed by ELIEL (*151*), and some of the results so far obtained are listed in *Table 1*. (See, however, *206*.)

c) Auwers-Skita Rules. An interesting consequence of the recognition that substituents on a cyclohexane ring are generally more stable in the equatorial conformation is that it has been found possible to restate the AUWERS (*26*) and SKITA (*303*) rules more accurately in conformational terms.

The original rules, which state that the *cis* compound of a pair of *cis-trans* isomers has the higher refractive index, density, and boiling point, fail when applied to 1,3-disubstituted cyclohexanes (*114*). Various reformed versions of the rules, couched in conformational terms, have been proposed (*6, 7, 221*). The most generally applicable form of the rules has been put forward by ALLINGER (*8*) who proposes that for stereoisomers in cyclic systems, not differing in dipole moment, the isomer which has the smaller molecular volume (i. e. higher refractive index, density, and boiling point) is the one which has the higher heat content and is the less stable isomer as judged by conformational analytical considerations. ELIEL and HABER (*152*) have recently found that the epimeric 2-, 3-, and 4-methylcyclohexanols present an exception, in their boiling points, to this conformational rule. This was explained as being due to a lesser degree of intermolecular hydrogen-bonding in the (*e, a*) isomers, because of reduced accessibility of the hydroxyl group, so leading to a decreased boiling point on the part of these isomers. This explanation is valid only if the less stable epimers adopt a conformation in which the hydroxyl group is axial and the methyl group equatorial, and not vice versa. Reference to *Table 1*, p. 170 will indicate the reasonableness of this assumption.

d) Exceptions to the Usual Stability Relationships. Where strong electrostatic forces are in operation the usual order of stability for axial and equatorial substituents on a cyclohexane ring may be reversed. Thus, for instance, in 2-bromo-cyclohexanone the conformation in which the bromine atom is equatorial (XIII) is destabilised by repulsion between the C $\xrightarrow{\;\;+\;\;}$ O and C $\xrightarrow{\;\;+\;\;}$ Br dipoles, and by a non-bonded interaction between the carbonyl oxygen atom and the bromine atom (*224, 282, 283*). If the molecule undergoes a conformational inversion the species (XIV) is obtained, in which the bromine atom is axial. Both the above interactions are relieved in this conformation, but of course, (XIV) is destabilised by two 1:3-diaxial (H:Br) non-bonded interactions.

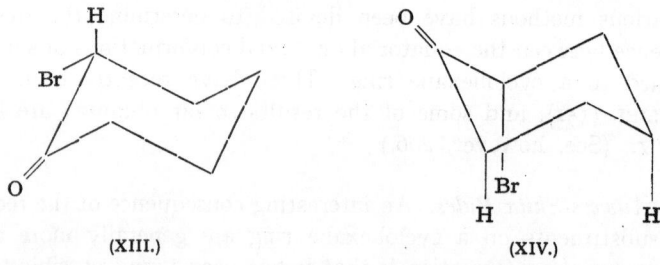

(XIII.)   (XIV.)

It is thus seen that in this case the stable conformation will be determined by a balance of steric and electrostatic effects. In fact, in solution, (XIV) predominates, the exact point of equilibrium depending on the polarity of the solvent; in solvents of high dielectric constant, (XIII) is relatively more favoured (*10, 103, 231*). A similar situation has been found (*11*) in the equilibrium between the *cis-* and *trans*-2-bromo-4-*t*-butylcyclohexanones, where the bromine atom is respectively equatorial (XV) and axial (XVI), since in both compounds the bulky tertiary butyl group will be in the equatorial conformation (*147, 335*). Other examples of the dependence of the conformation of 2-halocyclohexanones on the polarity of the solvent are known (*12, 13, 136, 137*).

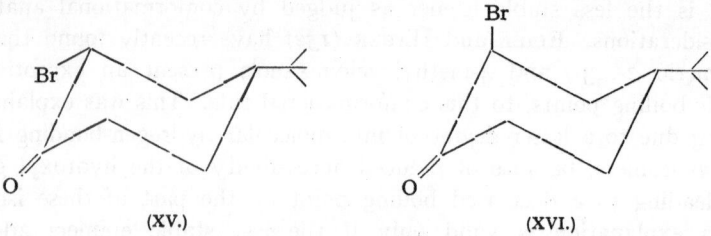

(XV.)   (XVI.)

Integral charge interaction, and the formation of intramolecular hydrogen bonds can also result in a modification or reversal of the normal conformational preferences (*37*).

Even in systems where neither electrostatic forces nor hydrogen-bonding operate it is possible to conceive of circumstances in which a substituent on a cyclohexane ring will be more stable in the axial than in the equatorial conformation. Let us consider two conformationally rigid molecules (XVII) and (XVIII); if the substituents R_1, R_2 and R_3 are bulky, it is possible that the $1:2$-(a, e)-interactions between R_1 and R_3 and between R_2 and R_3 in (XVII) will be more severe than the $1:3$-interactions between R_3 and the two relatively small axial hydrogen atoms on the same side of the ring in (XVIII).

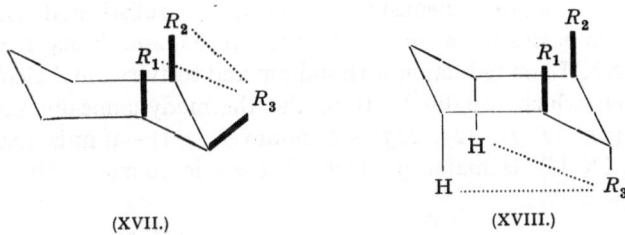

(XVII.)   (XVIII.)

Such a situation, resulting in a reversal of the usual conformational stability relationship, has been encountered on several occasions during work in the terpenoid series. For example, base treatment of acetyl-oleanolic lactone-11,12-seco-dicarboxylic acid methyl ester (XIX) epimerises the $C_{(9)}$-carboxymethyl group from the equatorial to the more stable axial conformation (287). Similarly, BROWNLIE and SPRING (86) found that the axial carboxymethyl group attached at $C_{(9)}$ in (XX), a degradation product of methyl glycyrrhetate, could not be epimerised with base.

(XIX.) Acetyloleanolic lactone-11,12-seco-
dicarboxylic acid methyl ester.

(XX.)

COREY and URSPRUNG (113) have described another example of the same phenomenon. They obtained a degradation product of friedelin, possessing structure (XXI), in which the carboxyl group was found to be stable in the quasi-axial conformation (see p. 176 for a definition of this term).

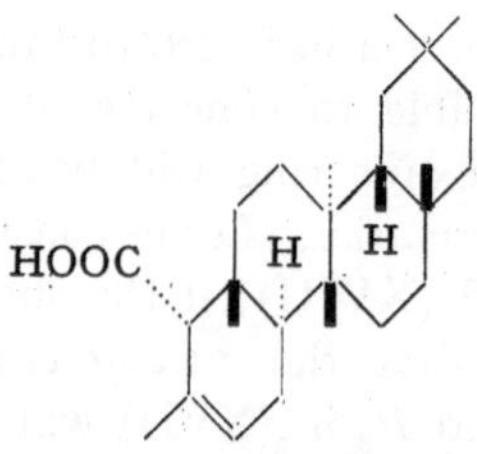

(XXI.)

Stability inversions related to those described above have been encountered in the chemistry of certain 19-substituted triterpenoid compounds possessing a *trans D/E* junction (*20*, *46*). Thus, 18α-oleanan-19-one (XXII), on reduction with sodium and tertiary amyl alcohol (*20*)— a reaction which usually leads to the thermodynamically more stable alcohol (*31*, *32*, *37*, *125*, *223*; see, however, *271*)—affords 18α-oleanan-19β-ol (XXIII) as major product. The steric course of this reduction

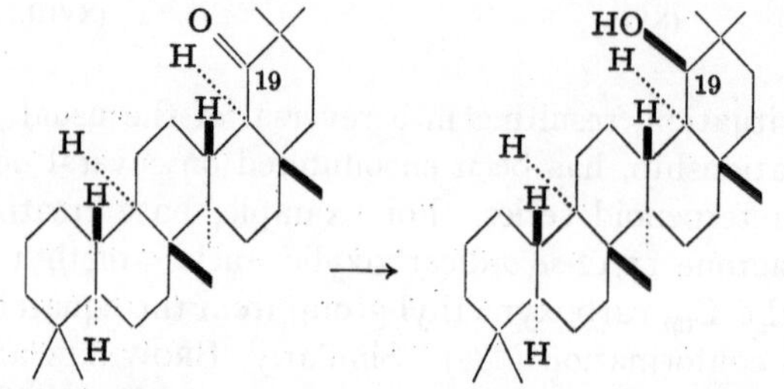

(XXII.) 18α-Oleanan-19-one.   (XXIII.) 18α-Oleanan-19β-ol.

implies that an axial hydroxyl group at $C_{(19)}$ is more stable than its equatorial epimer. The instability of the equatorial hydroxyl group has been ascribed (*20*) to its participation in a non-bonded interaction with the $C_{(12)}$-methylene group (see XXIV).

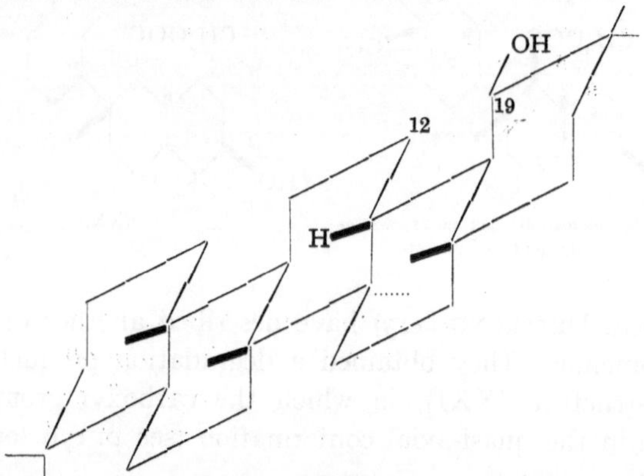

(XXIV.)

It is also known (20) that (XXV) is more stable than its $C_{(19)}$-epimer (XXVI). However, this may not be a strictly analogous case: since ring E contains a trigonal carbon atom, the possibility that it assumes a boat conformation (XXVII) in the 19β-isomer cannot be disregarded

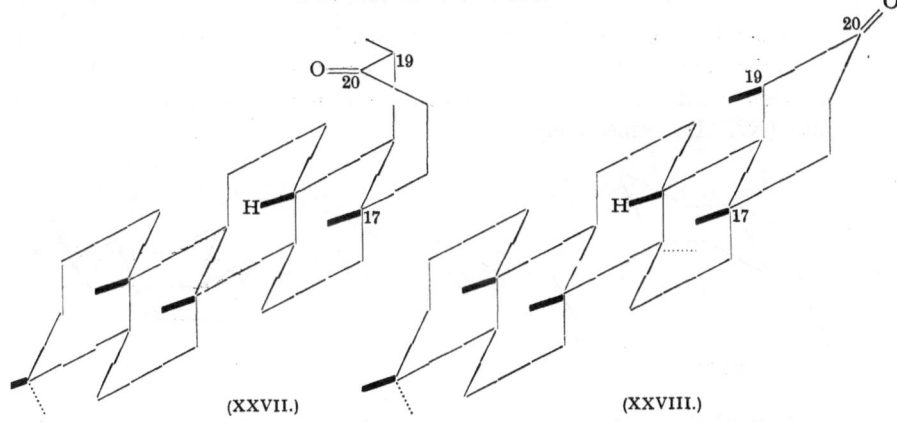

(XXV.)   (XXVI.)

(cf. Chapter VII, p. 212). In this way, the 19β-methyl group would occupy an equatorial position, thus eliminating the $1:3$-diaxial interaction between the β-methyl groups at $C_{(17)}$ and $C_{(19)}$, present in the alternative, all-chair, conformation (XXVIII).

(XXVII.)   (XXVIII.)

e) Six-membered Rings Containing Trigonal Carbon Atoms. This topic has been discussed in an earlier review (37). Introduction of a single trigonal carbon atom, as in cyclohexanone, produces only a small distortion of the ring.

Cyclohexene, which contains two trigonal carbon atoms, can exist in two conformations, (XXIX) and (XXX), known respectively as the "half-chair" and "half-boat" forms (38) which are free from appreciable Baeyer strain. It has been calculated (62) from thermodynamic data that the half-chair conformation is more stable than the half-boat by

2.7 kcal. per mole, and this is the conformation normally adopted by cyclohexene derivatives, where it is configurationally possible.

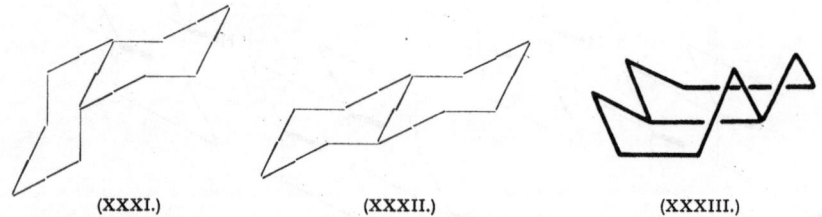

(XXIX.)   (XXX.)

In the half-chair conformation, the two trigonal and the two allylic carbon atoms are approximately coplanar. The remaining two carbon atoms are staggered above and below this plane, and the bonds attached to them are essentially normal equatorial and axial bonds. The bonds attached to the allylic carbon atoms are designated quasi-axial (a') and quasi-equatorial (e'). In the absence of complicating electrostatic forces or hydrogen bonding, substituents are generally more stable in a quasi-equatorial conformation than in a quasi-axial conformation.

3. Decalin Derivatives.

Since the boat form of cyclohexane is less stable than the chair, it would be expected that the preferred conformations of cis- and trans-decalin would each involve the fusion of two chairs, as shown in (XXXI) and (XXXII) respectively.

(XXXI.)   (XXXII.)   (XXXIII.)

Nevertheless, for nearly thirty years it was accepted that cis-decalin existed in the two-boat conformation (XXXIII) (263). The two chair form was finally established as a result of electron diffraction measurements (57, 58, 186, 190), and this is now supported by a great deal of chemical evidence (31, 32, 37, 223) and by dipole moment measurements (240).

There are three skew interactions in cis-decalin which do not occur in trans-decalin, and, by assigning to each of these an interaction energy of 0.8 kcal. per mole, as found (276) in butane, Turner (315) has estimated that the heat content of the cis-isomer is 2.4 kcal. per mole greater than that of the *trans*-isomer. This value is in very good agreement with the values of —2.12 kcal. per mole (*128*) and —2.69 kcal. per mole (*261*) obtained for the heat of isomerisation (*cis → trans*) from combustion measurements, and with the value of —2.72 kcal. per mole obtained (*14*) from equilibration data.

Application of TURNER's empirical method of calculating ΔH to the 9-methyldecalins indicates that the *trans*-isomer is 0.8 kcal. per mole more stable than the *cis*-isomer. In this series, combustion measurements have not yet been carried out, but the difference in heat contents, as revealed by a comparison of the physical constants of the two isomers (*7*) is certainly small. In fact, it has been found that in the case of 9-methyldecal-1-one (XXXIV) the *cis* isomer predominates at equilibrium (*286*), thus indicating that at the temperature of equilibration it had the lower free energy.

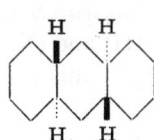

(XXXIV.)
9-Methyldecal-
1-one.

4. Perhydroanthracenes and Perhydrophenanthrenes.

The empirical approach employed by TURNER (*315*) in estimating the relative stabilities of cyclohexane and decalin derivatives has been applied by JOHNSON (*215*; see also *214*) to the perhydroanthracenes and perhydrophenanthrenes. The results thus obtained have already been reviewed in detail (*223*). It is of interest to note that, because it is impossible for two six-membered rings to be fused through *trans*-diaxial bonds, *trans-anti-trans*-perhydroanthracene (XXXV) and *trans-syn-trans*-perhydrophenanthrene (XXXVI) must each have one ring in a boat conformation.

(XXXV.) *trans-anti-trans*-Perhydroanthracene.   (XXXVI.) *trans-syn-trans*-Perhydrophenanthrene.

5. Steroids.

On the assumption that the preferred conformation of a system of fused cyclohexane rings will contain the maximum number of chair forms, unique conformations, (XXXVIII) and (XL), can be written for 5α-cholestane (XXXVII) and 5β-cholestane (XXXIX) respectively (*31*). FRIDRICHSONS and MATHIESON (*166*) have shown by X-ray analysis that the iodoacetate of lanost-8-en-3β-ol (XLI), which may be regarded as a trimethylsteroid exists in an all-chair and half-chair conformation in the crystal lattice.

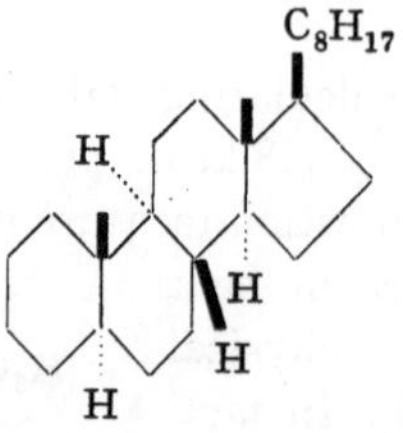

(XXXVII.) 5α-Cholestane.

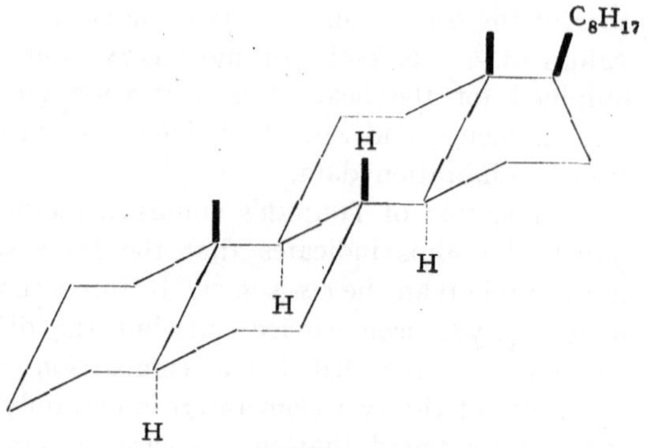

(XXXVIII.)

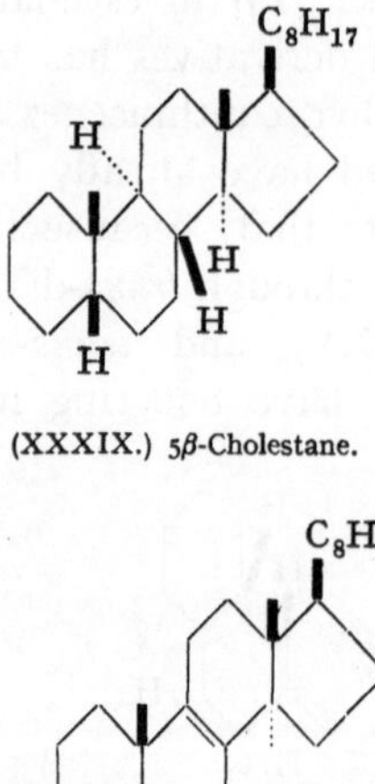

(XXXIX.) 5β-Cholestane.

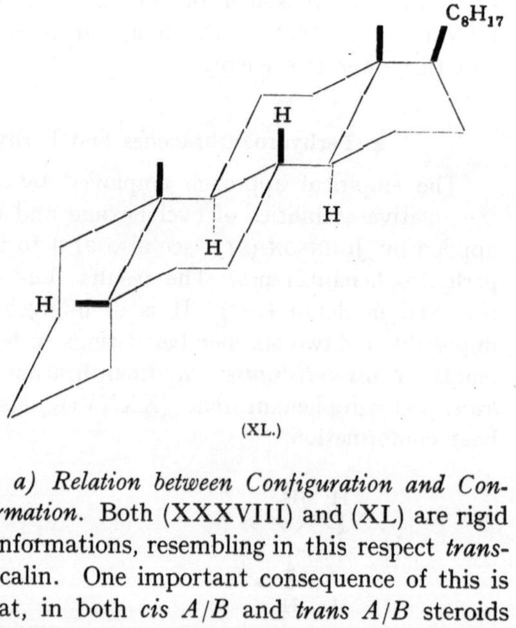

(XL.)

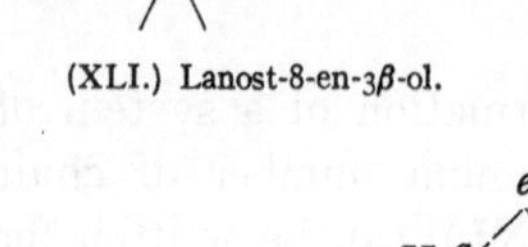

(XLI.) Lanost-8-en-3β-ol.

a) *Relation between Configuration and Conformation.* Both (XXXVIII) and (XL) are rigid conformations, resembling in this respect *trans*-decalin. One important consequence of this is that, in both *cis A/B* and *trans A/B* steroids there is a fixed relationship between the configuration and the conformation of a substituent group attached at any position on the nucleus.

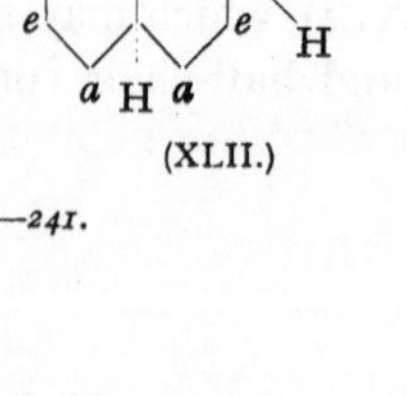

(XLII.)

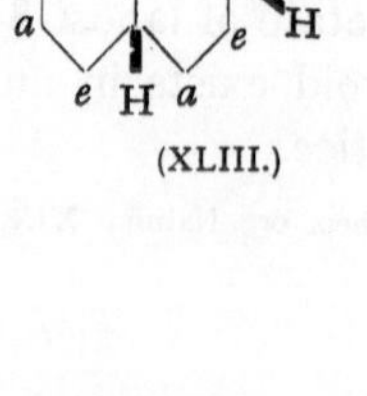

(XLIII.)

This is summarised in the 5α- and 5β-series by the expressions
(XLII) and (XLIII) respectively, in which the conformation of a
β-substituent at each position is indicated. Conformations at positions 15
and 17 in ring D are quoted relative to ring C. It is because of this
relationship between conformation and configuration that conformational
analysis has found such wide application in the determination of the
stereochemistry of steroids and triterpenoids.

b) Stability of Ring Junction in Hydrindanones Incorporated into
Fused Ring Systems. In recent years the factors governing the relative
stabilities of the cis and trans isomers of various hydrindane systems have attracted a great deal of attention.
For hydrindane (XLIV) itself the thermodynamic quan-
tities for the isomerisation (cis → trans) are (15):
$\Delta H^{522} = -1,070 \pm 90$ kcal. per mole and $\Delta S^{522} =
= -2.30 \pm 0.1$ e. u. Hence, at room temperature, the
trans isomer is more stable, but above 466° K. the
cis isomer predominates at equilibrium. In the hydrindanones the
configuration of the more stable isomer varies, depending on the
position of the carbonyl group. Structures (XLV), (XLVI), and (XLVII)
represent in each case the stable isomer (247).

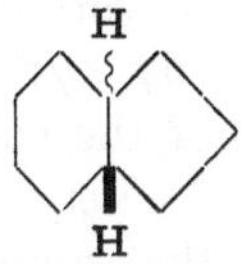

(XLIV.)

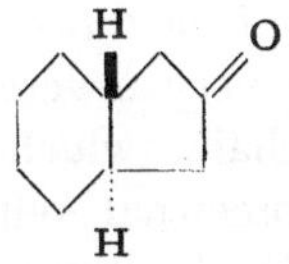

(XLV.) 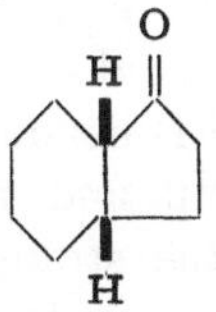   (XLVI.) 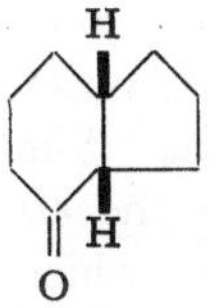   (XLVII.)

The question of deciding, by conformational analysis, which isomer
of a hydrindanone will be the more stable, is, of course, of importance
in steroid chemistry. The complexity of this problem may be illustrated
by reference to structures (XLVIII) (140), (XLIX) (138), (L) (239, 246),

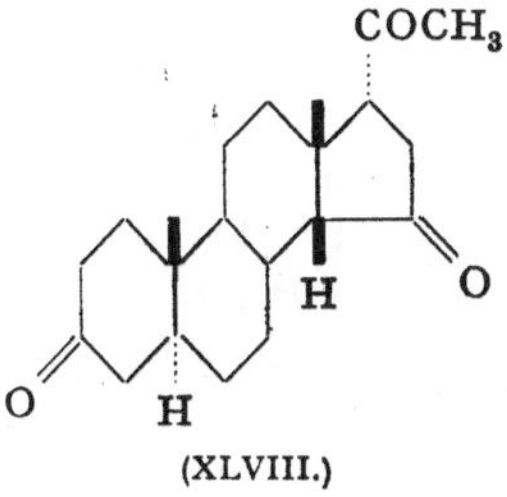

(XLVIII.) 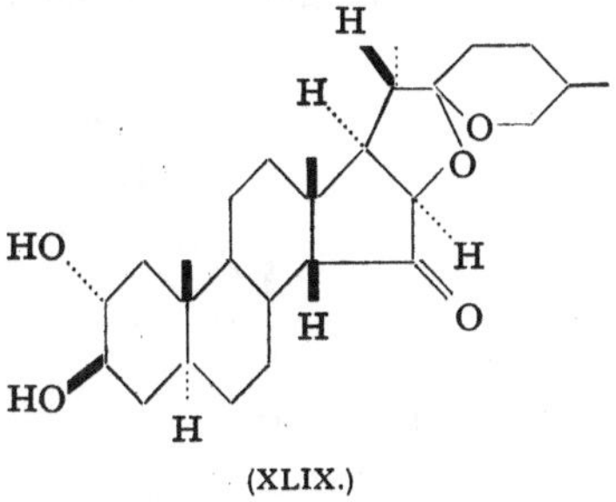   (XLIX.)

12 *

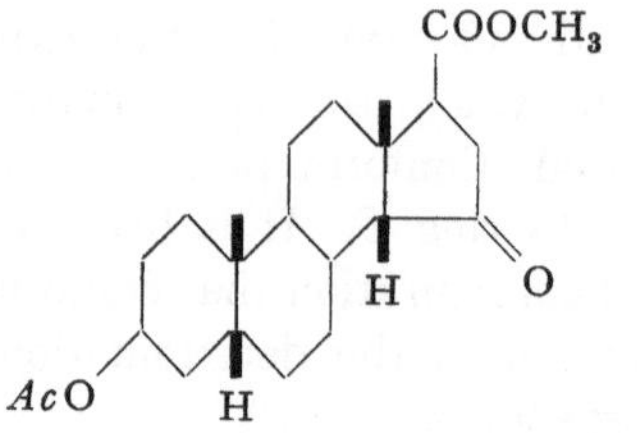

(L.) 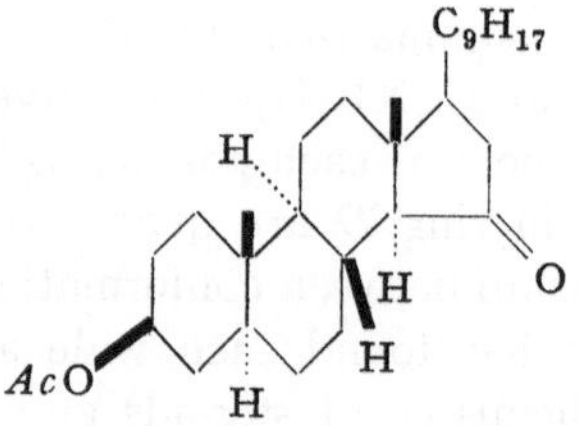   (LI.)

and (LI) (*47*). In each case, the more stable fusion of the *C/D* ring
junction is indicated. In the past, various attempts have been made
to correlate structure and stability in hydrindanone systems (*7, 150,
164, 277*), but none of these appears to be adequate to account for all
of the cases studied.

Very recently, a new and more promising empirical approach to
the problem has been described (*17*). Basically, this consists of
calculating the relative heat contents of the corresponding decalin
isomers (which, neglecting entropy, are assumed to approximate to the
free energy differences) and applying to these values a correction derived
from the known energy relationships in the parent decalin (*14, 261*)
and hydrindane (*15*) systems. The results obtained in this way for a
number of steroidal and triterpenoid hydrindanones are in reasonable
accord with those obtained by equilibration experiments *(Table 2)*. It
is suggested (*17*) that the observed stability (*47*) of the *trans C/D* junction
in 15-ketostanols of type (LI) is due to the very severe interaction
between the 18-methyl group and the side-chain, which are almost
eclipsed in the *cis* isomer, the strain being somewhat relieved in the
trans isomer which has a more puckered ring *D*. In compound (L) the
planar methoxycarbonyl can adopt a conformation in which its interaction
with the 18-methyl group is much smaller, and so, in this case the
equilibrium is displaced towards the *cis* configuration.

BIELLMANN, FRANCETIĆ and OURISSON (*68*) have carried out equilibration studies on some 3-oxo-*A*-norsteroids and triterpenoids (and
molecules of related structure) and have succeeded in relating qualitatively
the relative energies of the conformations of the *cis* and *trans* isomers

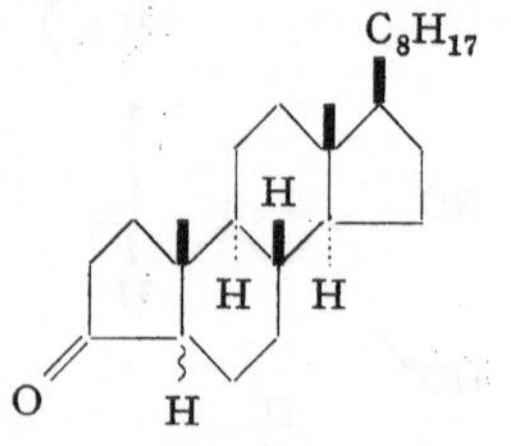

(LII.) 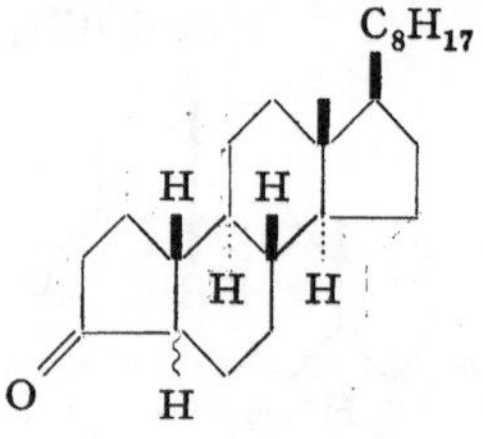   (LIII.)

Table 2. Calculated and Observed Free Energy Differences between *cis*- and *trans*-Ring Junctions of Hydrindanones Incorporated into Steroid and Triterpenoid Molecules (17).

	% *cis* isomer at equilibrium	— ΔF calculated (*trans* → *cis*)	— ΔF found (*trans* → *cis*)
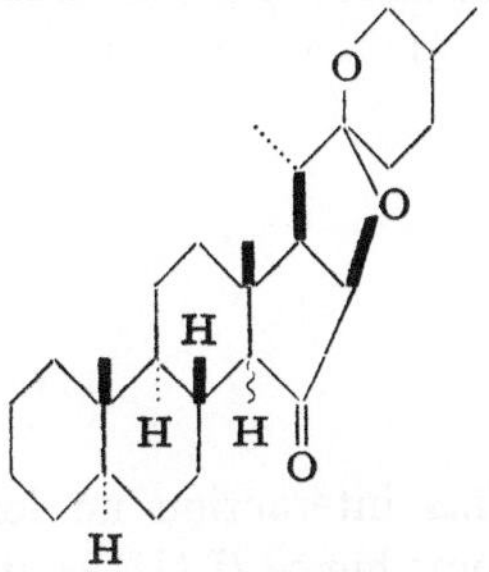	> 98		
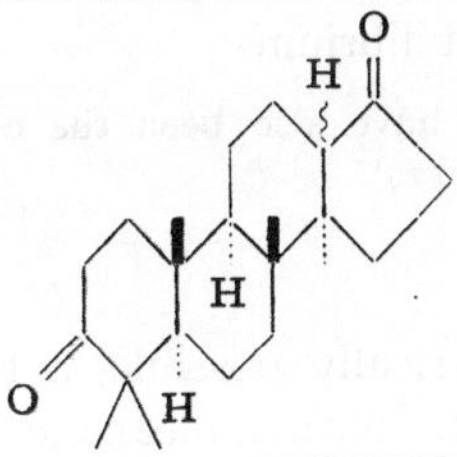	61.4	— 0.4	0.4
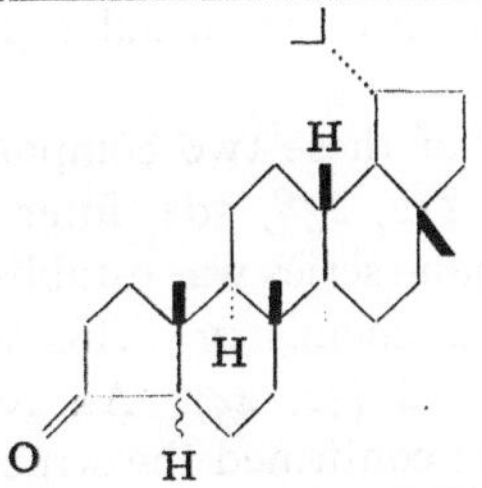	> 99	1.4	2.7
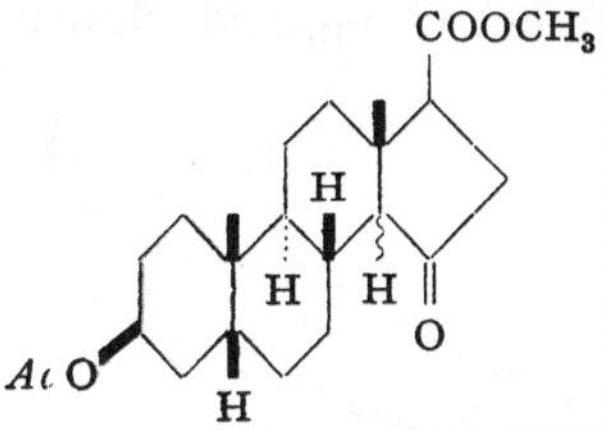	87	1.4	1.0
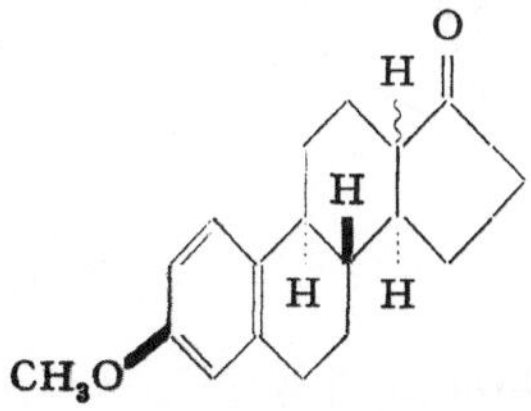	55	0.0	0.1

with their point of equilibrium. For example, the equilibrium mixture of (LII) contains 97% of the *cis* isomer, while that of (LIII) contains only 61%. The conformation of the *trans* isomer of (LII) may be represented by (LIV). This conformation is destabilised by the gauche

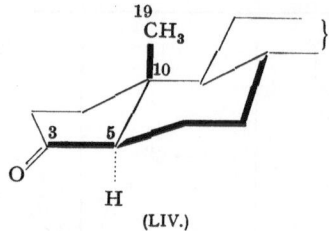

(LIV.)

arrangement of the 3,5 and 10,19 bonds. This interaction no longer exists when $C_{(19)}$ is replaced by a hydrogen atom: hence (LIII) contains a greater percentage of the *trans* isomer at equilibrium.

Hydrindanone systems in the *B*-norsteroid series have also been the object of recent conformational studies (*117, 119, 121, 173, 174*).

6. Triterpenoids.

Assuming once again that, wherever geometrically possible, a fused carbocyclic molecule will exist in an all-chair conformation, oleanane (LV) and ursane (LVII) may be represented by the perspective formulae (LVI) and (LVIII) respectively.

Both the structure and the stereochemistry of these two compounds have been firmly established (*5, 61, 106, 111, 112, 258, 302*, inter al.). Indeed, much of the stereochemistry of the oleanane series was established by application of conformational analysis to its chemistry. This work has already been discussed in considerable detail (*32, 44*). An X-ray investigation of methyl oleanolate iodoacetate (*1*) confirmed the structure and stereochemistry (excluding the configuration at $C_{(13)}$) of oleanane denoted by (LV).

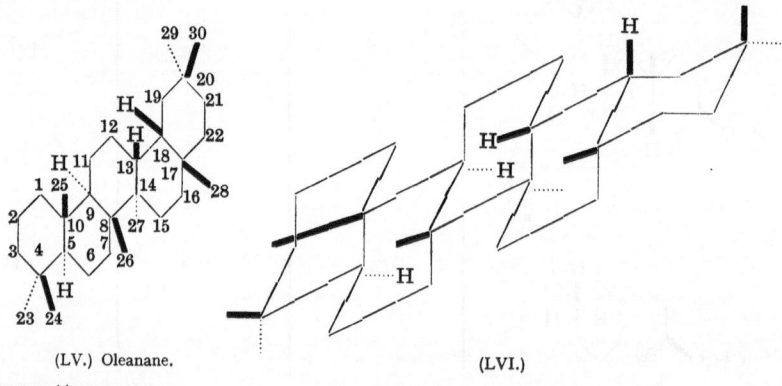

(LV.) Oleanane.   (LVI.)

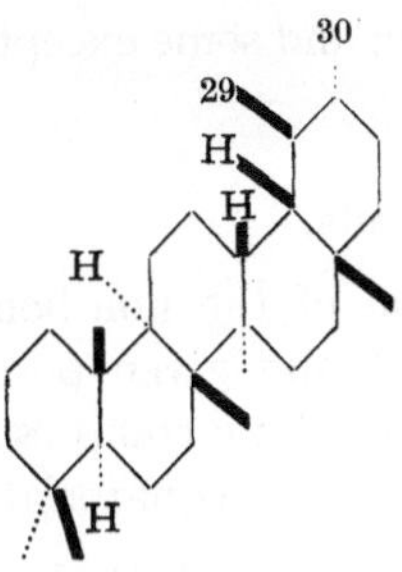

(LVII.) Ursane. 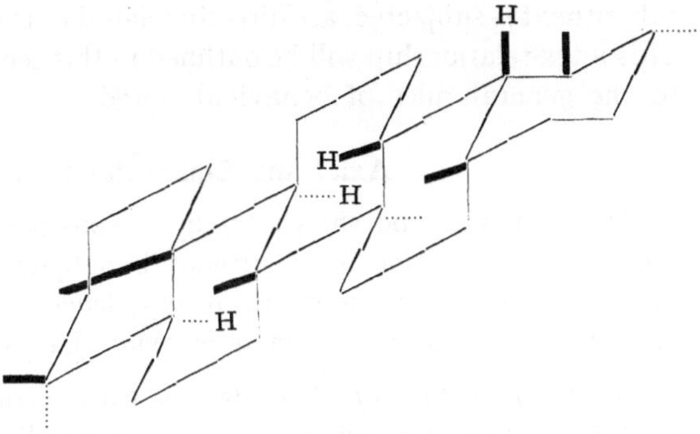   (LVIII.)

Many of the differences observed in the chemistry of α-amyrin (LIX: $R = CH_3$) and of β-amyrin (LX: $R = CH_3$) are readily explained in terms of conformational analysis. For example, it has been established that, while the D/E ring junction of methyl oleanolate (LX: $R = COOCH_3$) is isomerised from *cis* to *trans* by base treatment of its 11-oxo derivative, that of the 11-oxo derivative of methyl ursolate (LIX: $R = COOCH_3$)

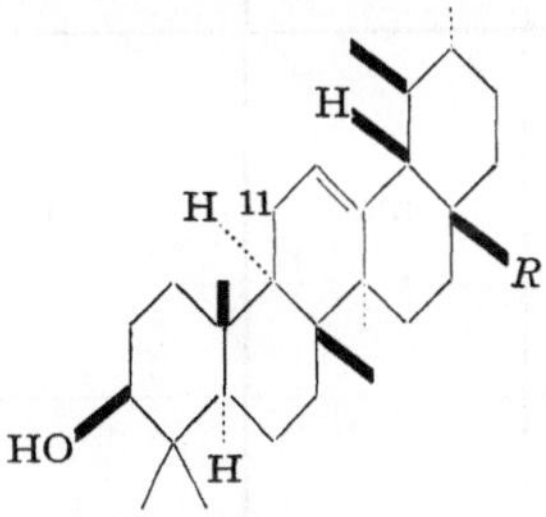

(LIX.) α-Amyrin. ($R = CH_3$). 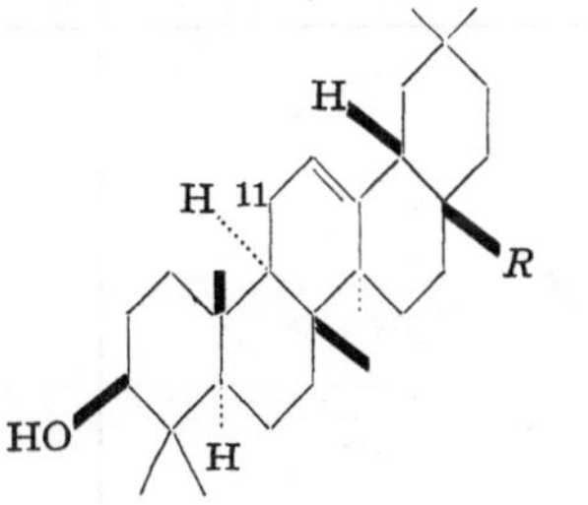   (LX.) β-Amyrin. ($R = CH_3$).

is stable in the *cis* configuration (44). This difference in behaviour may be attributed to the fact that isomerisation of the D/E ring junction in the α-amyrin series would involve changing the conformation of the $C_{(29)}$- and $C_{(30)}$-methyl groups from equatorial to the energetically unfavoured axial form (112).

Preferred all-chair and half-chair conformations can be written for most of the other triterpenoids.

IV. Conformational Analysis of Reactions Controlled by Steric Hindrance and Steric Compression.

In conformationally rigid molecules of the steroid and triterpenoid type, the steric hindrance exhibited by the molecule towards any particular reaction, and the steric compression to which any functional substituent is subjected, are directly related to the molecular conformation. This inter-relationship will be outlined in this section, and some exceptions to the general rules of behaviour noted.

1. Axial and Equatorial Substituents.

It can readily be shown (2) by a consideration of the non-bonded interactions to which a substituent is subjected at any given position in a steroid or triterpenoid that it experiences greater steric compression, and is more hindered, when it is axial than when it is equatorial.

a) Esterification of Alcohols and Carboxylic Acids, and Hydrolysis of Esters. One consequence of this generalisation is that equatorial hydroxyl groups should be more readily esterified, and their esters more readily saponified than their epimers. This has been amply demonstrated

Table 3. Relative Ease of Hydrolysis of Steroidal 3-Acetates with and without Participation of a 5-Hydroxyl Group.

Structure	Conformations relative to ring A		% Hydrolysis under standard conditions
	— OAc	— OH	
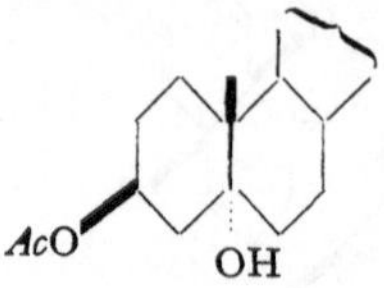	eq.	ax.	18
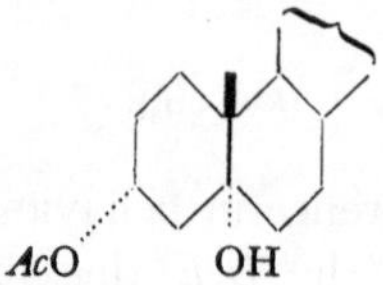	ax.	ax.	70
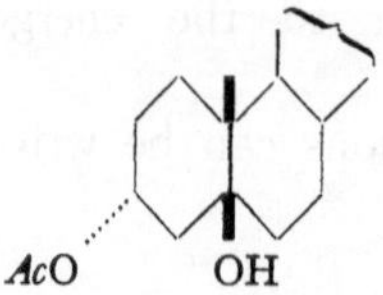	eq.	ax.	13
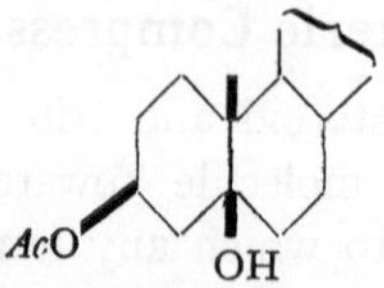	ax.	ax.	78

by work in the steroid series (*32*), and has been frequently employed in determining the configuration of hydroxyl groups. The esterification of axial carboxyl groups and the hydrolysis of the resulting esters are also slower than for their epimers. Recent work in the cyclohexane series (*88*) has indicated that, at 30°, an axial methoxycarbonyl group is hydrolysed at least seventeen times slower than an equatorial one under the same conditions. This difference in reactivity has proved of diagnostic value in configurational studies on triterpenoid acids; see (*60*) for an example.

It has been found, however, that the monoacetates of $1:3$-diaxial diols are hydrolysed much faster than is usual for axial acetates. Indeed, in such cases, the axial acetate may sometimes be more readily hydrolysed than its epimer. This is illustrated by the results obtained by HENBEST and LOVELL (*196*) in the steroid series. These are set out in *Table 3*.

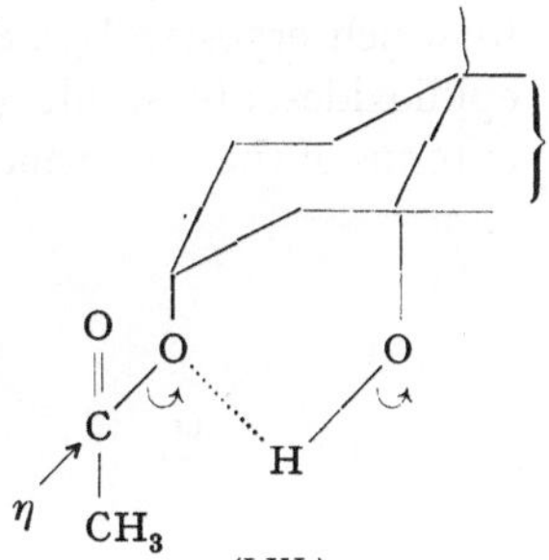

The explanation of this reversal of the usual order of hydrolysis of epimeric acetates is that in the saponification of the diaxial isomers the hydroxyl group is suitably situated for hydrogen-bonding to the ester (*196, 327*) and so can facilitate its hydrolysis (LXI: η = nucleophile).

(LXI.)

This effect has been invoked to account for the ready methanolysis of the axial $C_{(16)}$-acetate in certain derivatives of cevine (LXII) (*233*), germine (part structure LXIII) (*234*) and protoverine (part structure LXIII) (*232*). In these cases, the axial hydroxyl group at $C_{(20)}$ facilitates

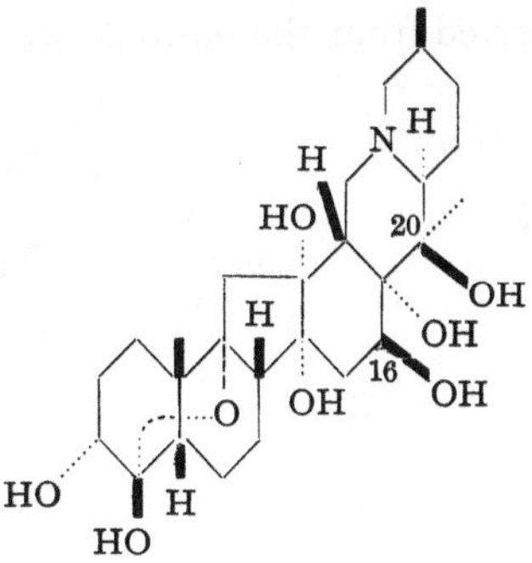

(LXII.) Cevine.

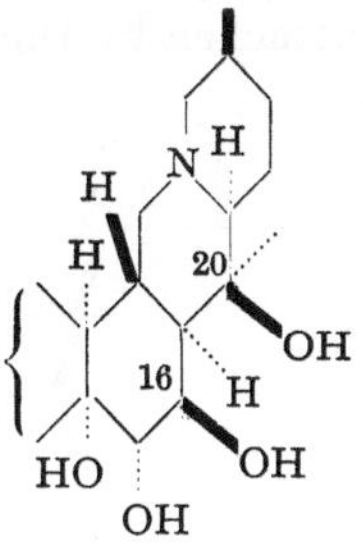

(LXIII.) Germine and protoverine (part structures).

the solvolysis. An axial acetate at $C_{(7)}$ in the germine series (part structure LXIV) is also easily methanolysed, the participating hydroxyl group being axially situated at $C_{(14)}$ (*234*).

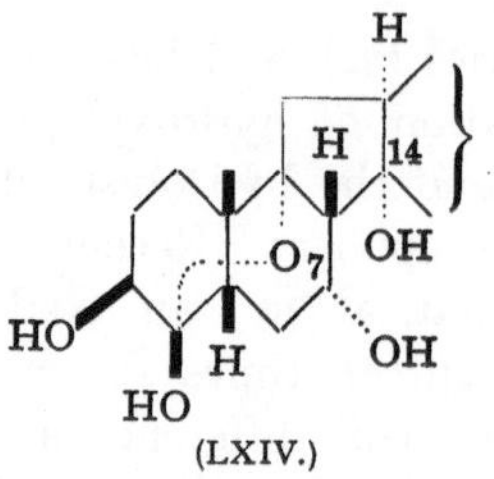

(LXIV.)

A preliminary account has been given (*235*) of experiments which indicate that the base-catalysed hydrolysis of an alicyclic acetate is also facilitated by the presence of an adjacent *cis* hydroxyl group.

A somewhat related reaction which may be noted here is the base-catalysed cleavage of lumisterol-5β,8β-epidioxide (LXV) under conditions to which ergosterol-5α,8α-epidioxide (LXVI) and most other di-tertiary epidioxides are stable (*75*). This difference finds a ready explanation in terms of the conformations of the molecules involved. The conformation

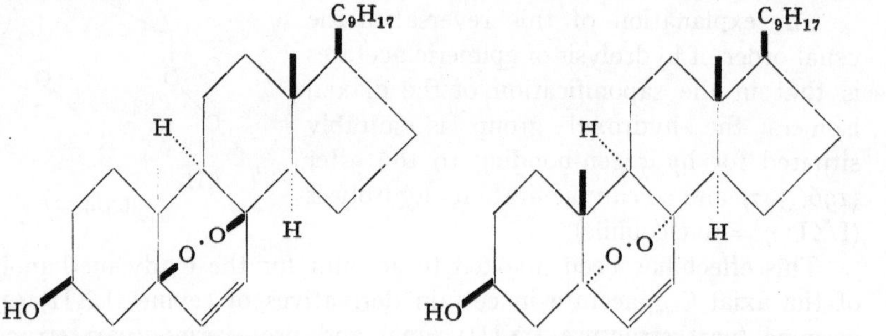

(LXV.) Lumisterol-5β,8β-epidioxide.   (LXVI.) Ergosterol-5α,8α-epidioxide.

of lumisterol-5β,8β-epidioxide (LXVII) is such that the dioxide bridge can be readily attacked by the oxanion derived from the 3β-hydroxyl group:

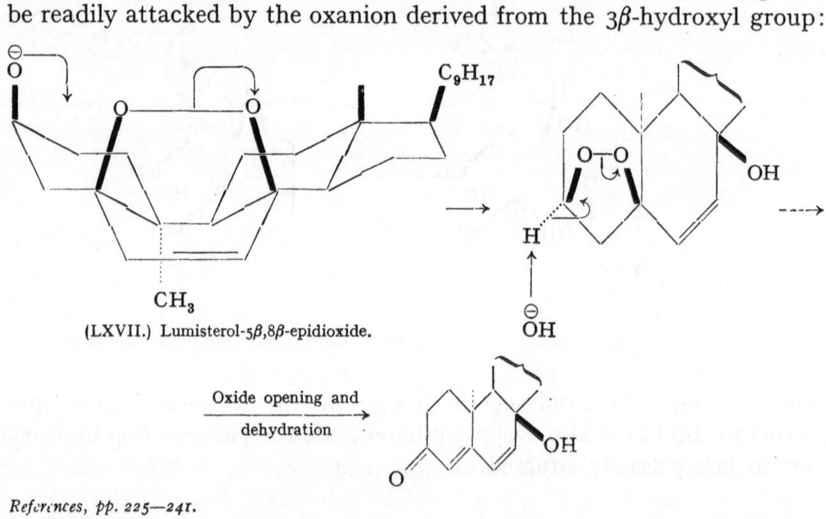

(LXVII.) Lumisterol-5β,8β-epidioxide.

Oxide opening and
dehydration

In contrast, the equatorial 3β-hydroxyl group in ergosterol-5α,8α-epi-dioxide is not suitably oriented for intramolecular attack on the dioxide bridge (see LXVIII).

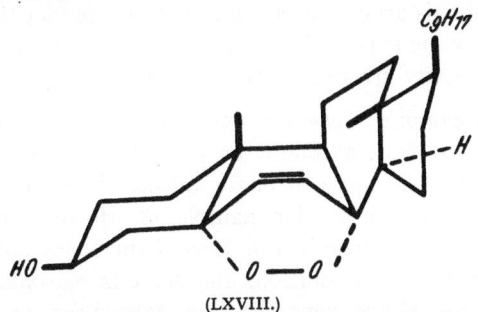

(LXVIII.)

b) Strengths of Acids and Amines. Since the ionisation of acids is determined to some extent by the degree of solvation of the corresponding ions (*181*), it is reasonable to expect that an equatorial carboxyl group would be a stronger acid than its epimer, since there is less steric hindrance to the approach of solvent molecules to an equatorial group. Experimental evidence that this is so has been presented (*130, 310, 314*).

SOMMER, ARYA, and SIMON (*307*) have developed a semi-empirical method for calculating the apparent pK_a of a carboxyl group attached to an alicyclic system, in 80% methylcellosolve/20% water (w : w) as solvent.

In this procedure an allowance of 0.25 pK units is made for each 1 : 3-diaxial interaction in which the carboxyl group participates, including (H: COOH) interactions, and a correction of 0.22 pK units is applied for substituents in the α-position. Hence, $pK_a = 7.44 + 0.25\,a + 0.22\,b$

where a = number of 1 : 3-diaxial interactions
and b = 1 if there is an α-substituent,
            0 if there is an α-hydrogen atom.

The results obtained for three decalin carboxylic acids have been reported in a preliminary communication (*307*), and they are in very good agreement with the observed values.

The basic strengths of aminosteroids have also been investigated (*72, 73*), and here again it appears that the equatorial amine is usually a stronger base, presumably because the corresponding ammonium ion is more easily solvated. The difference in pK_a values increases with increasing hindrance of the position on the steroid nucleus (cf. *32*). An interesting anomaly is observed with the epimeric 7-dimethylaminocholestanes. Here the axial epimer is the stronger base (by 0.79 pK units). BIRD and COOKSON (*73*) have suggested that this is because the most stable conformation of the cyclohexyldimethylammonium ion is probably that in which the N⊕—H bond is parallel to the axial bonds, with the hydrogen atom strongly bonded to the solvent. This conformation is denied to the ion derived from 7β-dimethylaminocholestane because of interference by the 15-methylene group, and so the hydrogen atom is less accessible to the solvent. This argument is supported by the fact that the 7-amino-cholestanes show the expected order of basic strength, with the equatorial amine 0.34 pK units stronger than its epimer.

c) *Chromatographic Behaviour of Alcohols.* It has already been noted (*31, 32, 37*) that, as might be predicted, an equatorial hydroxyl group is, in general, more strongly adsorbed on a chromatographic column than is its epimer (see, for example, *28*). However, the conformation adopted by a molecule adsorbed on a solid surface is not necessarily the one in which it exists in solution, and so it is reasonable that various exceptions to the above generalisation have been observed (*84*). In partition chromatography, a molecule is more likely to exist in the conformation predicted by conformational analysis, and so a closer adherence to the rule would be expected. It has, in fact, been reported (*291, 292*; see also *274*) that axial steroidal alcohols have higher R_f values than their equatorial epimers. Even in paper chromatography a few anomalies have been noted (see, for example, *85, 250, 290*), possibly attributable in some cases to preferential solvation of the equatorial hydroxyl group, with a consequent reduction in its polarity.

d) *Oxidation of Secondary Alcohols with Chromium Trioxide.* SCHREIBER and ESCHENMOSER (*293*) have measured the rates of oxidation by chromic acid of a series of epimeric pairs of steroidal secondary alcohols. In every case, they found that the axial epimer reacted more rapidly. This was attributed to steric acceleration: The more compressed the alcohol the closer will its energy be to that of the transition state, and the faster it will be oxidised. If the reaction proceeds by way of an intermediate chromate ester (*182*), then these results imply that its formation must be a fast step. Other detailed mechanisms have been proposed for chromic acid oxidation (*236, 284*), but these mainly concern the way in which the secondary hydrogen atom is removed from the carbon atom becoming trigonal, and in no way require any alteration of the principle expressed above.

GRIMMER (*177*) has utilised the different rates of oxidation of equatorial and axial alcohols to develop a procedure for determining the position and configuration of hydroxyl groups attached to the steroid nucleus.

e) *Solvolysis.* (For a detailed discussion, see *312*.) Since the approach of a reagent to the back of an equatorial substituent is hindered by the axial groups (see LXIX), while its approach to the rear of an axial substituent is hindered only by the 2-substituents (see LXX), it might be expected that the latter would be more rapidly replaced in an $S_N 2$

process. This has recently been shown to be the case for a pair of epimeric cyclohexyl tosylates (*157*). *Cis*-4-*t*-butylcyclohexyl tosylate,

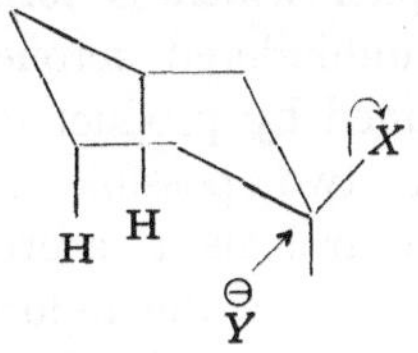

(LXIX.) 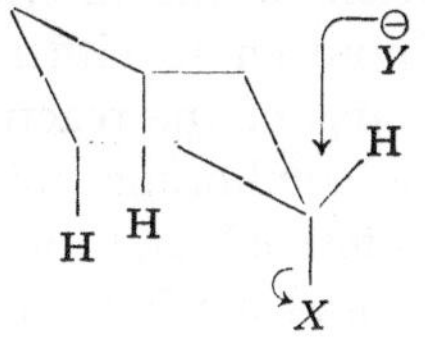   (LXX.)

in which the tosyloxy group is axial, undergoes an S_N2 replacement reaction with thiophenolate ion nineteen times faster than the *trans*-isomer, in which the tosyloxy group is equatorial. Similarly, in the steroid series, it has been found (*171*) that an axial 11β-bromine atom undergoes S_N2 displacement by a hydroxide ion more rapidly than does its equatorial epimer.

In S_N1 replacement reactions (and in the related E_1 eliminations), the greater relief of steric compression associated with the departure of an axial substituent, would suggest that the axial epimer should react more rapidly. This expectation has been borne out by studies of the rates of acetolyses of a number of decalyl (*264*) and cholestanyl (*268*) tosylates.

2. Reactivity of Carbonyl Groups.

a) Formation of Carbonyl Derivatives. Since the formation of most of the common carbonyl derivatives involves an intermediate such

as $\diagdown\mathrm{C}\diagup$ with OH above and NH— below, it is possible to estimate the relative hindrance of carbonyl groups at different positions in a steroid or triterpenoid molecule by summing for each position the non-bonded interactions in which the epimeric hydroxyl groups would participate (assuming that —NH— and —OH groups have approximately the same steric requirements) (*32*). The predictions made from conformational analysis are in reasonable agreement with the experimental findings (*163, 191–193, 248*).

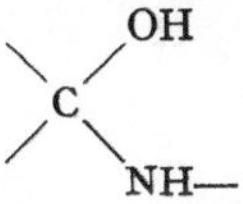

b) Hydride Reduction of Carbonyl Groups. It was early recognised (*32*) that reduction of unhindered ketones with sodium borohydride or lithium aluminium hydride tended to give the equatorial alcohol, while with hindered ketones the axial epimer was obtained. DAUBEN and his co-workers have accounted for this observation by postulating that there are two controlling factors in these reductions (*118, 122*; see also *253, 330, 331*). One of these—*"steric approach control"*—takes account of the relative hindrance to approach of the reagent from either side of the carbonyl group. The other—*"product development control"*—determines the orientation of the intermediate complex which is formed. Thus, since the approach to either side of an unhindered ketone is equally easy, the course of the reaction is determined by product development control; the equilibrium mixture of the two possible intermediate complexes is formed, and this leads, after hydrolysis, to a preponderance of the equatorial alcohol. An example of this is the reduction of 5α-cholestan-3-one with lithium aluminium hydride to afford 5α-cholestan-3β-ol in 90% yield (*124*). With more hindered ketones, steric approach control becomes more important and the intermediate complex, formed by approach of the reagent predominantly from the more accessible side of the carbonyl group, will afford more than the equilibrium concentration of the less stable alcohol. An extreme example of the operation of steric approach control is provided by the reduction of 11-oxosteroids with lithium aluminium hydride. In these compounds, approach of the reagent from the β-side of the molecule is severely hindered by the $C_{(18)}$- and $C_{(19)}$-methyl groups, and by the axial hydrogen at $C_{(8)}$ (see LXXI); consequently, the axial 11β-alcohol is

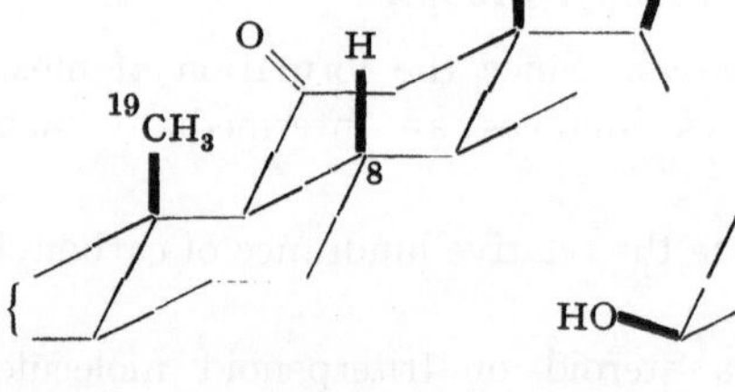

(LXXI.) 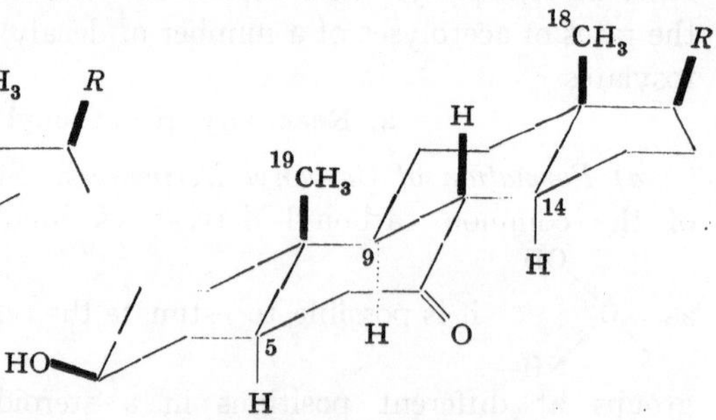   (LXXII.) 5α-Cholestan-3β-ol-7-one.

obtained in high yield as the sole isolable product (*66*). 5α-Cholestan-3β-ol-7-one (LXXII) is an intermediate case. Attack from the β-side is hindered only slightly by the $C_{(18)}$- and $C_{(19)}$-methyl groups, while approach of the α-side of the molecule is severely inhibited by the three axial hydrogen atoms at $C_{(5)}$, $C_{(9)}$, and $C_{(14)}$. Accordingly, the mixture of products obtained should contain more of the axial 7α-hydroxy compound than the 20% found in the equilibrium mixture of epimers (*55*). In fact, lithium aluminium hydride reduction of 3β-acetoxy-5α-cholestan-7-one afforded 55% of the 7α-epimer, and 45% of the 7β-epimer (*118*).

The same principles may be applied in predicting the ratio of isomers obtained in reductions with sodium borohydride: in these cases, however, the greater steric requirements of the reagent render steric approach control more important. In line with this, reduction of 3β-acetoxy-5α-cholestan-7-one with sodium borohydride gives 73% of the 7α-hydroxy-compound, and 27% of its epimer (*118*).

The theory outlined above has been criticised by HARDY and WICKER (*185*; cf. *120*) on the basis of experiments performed with mono-cyclic ketones. However, in the conformationally rigid steroid series, the results obtained experimentally accord well with prediction (see, however, *328*).

3. *Cis* Addition to Double Bonds.

Reagents which add in a *cis*-fashion to double bonds will obviously more readily attack from the less hindered side. In the steroid series, where a great many data have been accumulated, a good correlation exists between the stereochemistry of addition to an unsaturated molecule, and its conformation. The axial $C_{(18)}$- and $C_{(19)}$-methyl groups hinder the approach of reagents to the β-face of 5α-steroids of natural configuration, and in this series there is a general tendency for attack to occur from the α-side (*163*).

Hydroxylations with potassium permanganate and osmium tetroxide, hydroboration reactions, and epoxidations with peracids are subject to the "*Rule of α attack*" in natural steroids (*163*). The hindrance to approach of the α-side of a molecule imposed by the "folding back" of ring *A* in *cis* A/B steroids may sometimes inhibit reaction completely. Thus, only the Δ^{17}-double bond of (LXXIII) reacts with perbenzoic acid (*228*). Both double bonds in the corresponding 5α-compound react under the same conditions to give α-epoxides (*39*). Parallel results

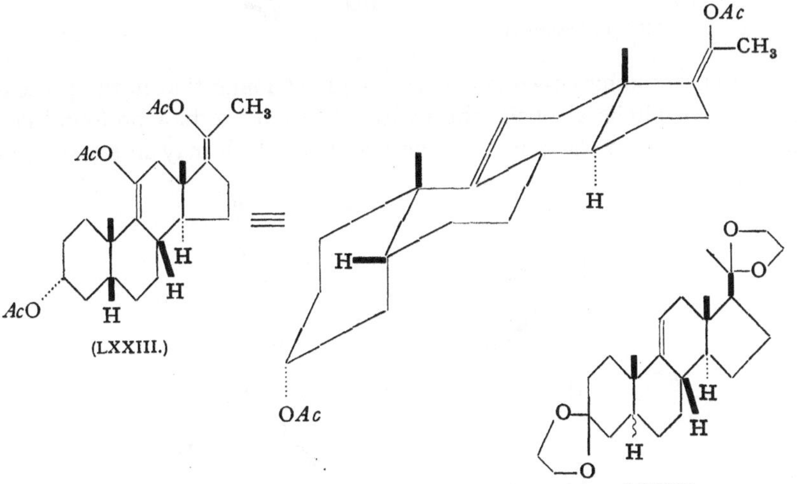

(LXXIII.)

(LXXIV.)

were obtained (*319*) in the hydroboration of the pair of $C_{(5)}$ epimers represented by (LXXIV).

The stereochemistry of epoxidation of steroids of unnatural configuration is, like that of their hydrogenation (see, for example, *76, 87*), controlled by their conformations. The Δ^7-5α,9β-steroid (LXXV) affords a β-epoxide with perbenzoic acid, while its 9α-epimer gives an α-epoxide

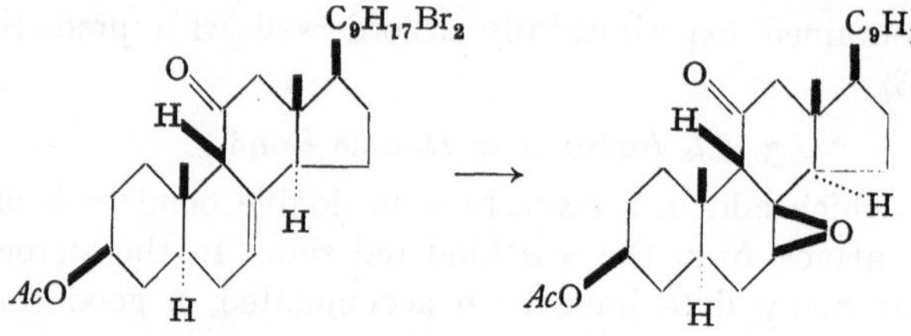

(LXXV.)

under the same conditions (*176*). This difference in behaviour is due to the fact that, in the 9β-compound, ring C is obliged to adopt a distorted boat conformation, in which the angular methyl groups are directed away from each other, so allowing easier access to the β-face of the molecule. The peracid oxidation of lumisterol (LXXVI) is another example of β-attack.

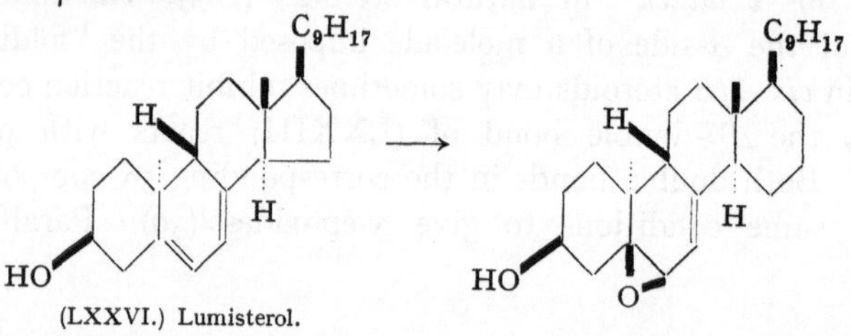

(LXXVI.) Lumisterol.

Henbest and his co-workers (*198, 199*) have found that in the peracid oxidation of allylic alcohols, the hydroxyl group exerts a profound *cis*-directing effect on the course of the reaction, which may in some cases

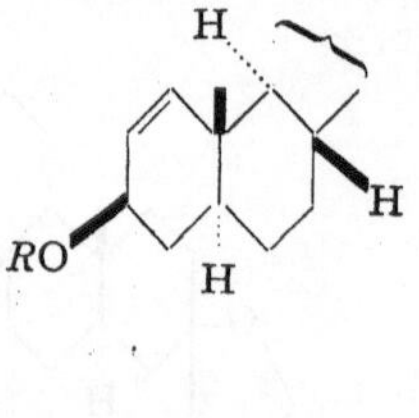

(LXXVII.) 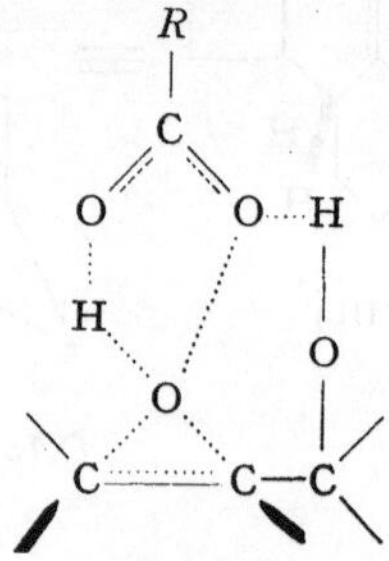   (LXXVIII.)

completely outweigh steric effects acting in the opposite sense. ALBRECHT and TAMM (3) have reported a good example of this effect in the steroid series. Perbenzoic acid oxidation of (LXXVII: $R = Ac$) affords, following the usual α-attack, the corresponding 1α,2α-epoxide; oxidation of (LXXVII: $R = H$) under the same conditions gives the β-epoxide. The influence of an allylic hydroxyl group on such oxidations, which is attributed (199) to H-bonding with the reagent (see LXXVIII), has been the subject of detailed study (198).

The influence of molecular conformation on the properties of double bonds is, of course, as important in triterpenoids as it is in the steroid series. An important instance of this is the inertness of the double bond in α-amyrin (LXXIX) as compared with that in β-amyrin (LXXX) (112).

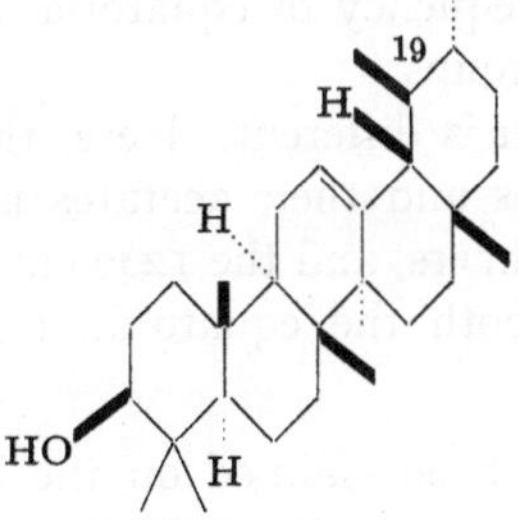

(LXXIX.) α-Amyrin. 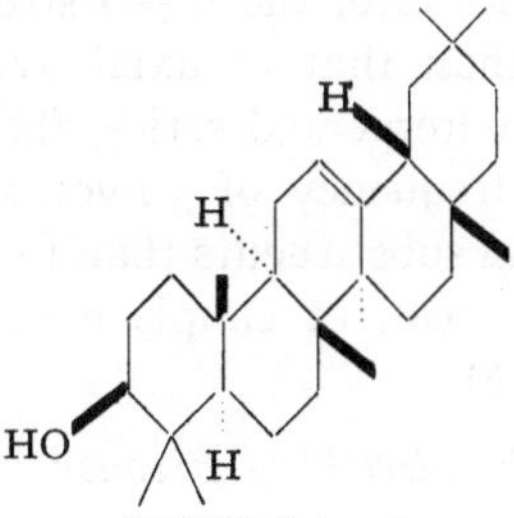   (LXXX.) β-Amyrin.

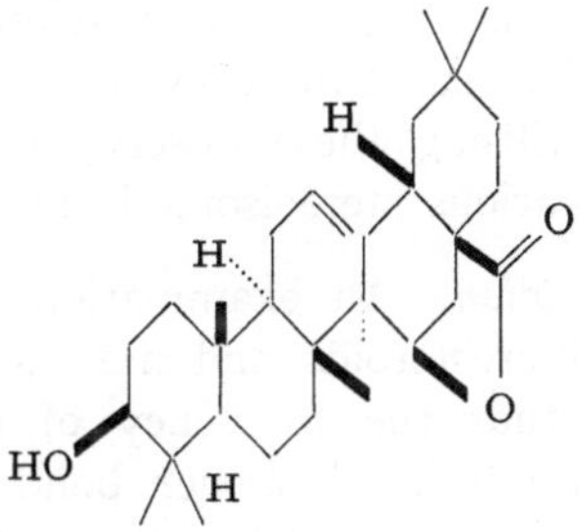

(LXXXI.) Dumortierigenin.

Doubtless, the equatorial methyl group at $C_{(19)}$ in α-amyrin plays a major role in hindering the double bond. Similarly, owing to the shielding effect of the lactone ring, the ethylenic linkage in dumortierigenin (LXXXI), a triterpenoid of the β-amyrin series, has a reactivity comparable with that found in the α-amyrins (145).

V. Spectroscopic Correlations.

1. Infrared Spectra.

Many useful correlations exist between the conformation of a substituent, and its characteristic infrared absorption. A previous review in this series by COLE (92) surveys the literature on this topic up to about 1956.

a) C—O Vibrations. The C—O stretching frequency of steroidal alcohols is higher for equatorial groups (ca. 1040 cm.$^{-1}$) than for axial ones (ca. 1000 cm.$^{-1}$) (*92*), though in the case of *cis A/B*-steroids the axial 3β-epimer has an absorption frequency of 1030–1036 cm.$^{-1}$ (*92, 220*), the higher value being attributed to the greater restoring forces operating in the L-shaped 5β-steroids. These differences between epimers are paralleled in the corresponding acetates and methyl ethers. Equatorial epimers absorb at higher frequencies (in the 1000–1150 cm.$^{-1}$ range) than the axial ones (*272*). Also, the band at 1200–1250 cm.$^{-1}$ in the spectra of steroidal 3-acetates is simple for the equatorial epimers, but consists of two or three peaks for the axial substituents (*218*).

Hückel and Riad (*207*) have found that in the cyclohexane and decalin series also, the C—O stretching frequency of equatorial alcohols is higher than that of axial hydroxyl groups.

In the triterpenoid series, the situation is different. Here, the C—O stretching frequency of 3-hydroxyl groups and their acetates is higher for the axial substituents than for their epimers, and the 1240 cm.$^{-1}$ band is a single peak of simple contour for both the equatorial and axial acetates (*18*).

b) O—H Stretching Frequency. It has been claimed on the basis of high resolution spectral measurements, that axial secondary hydroxyl groups exhibit a somewhat higher O—H stretching frequency in triterpenoids (*18, 96*) and in substituted cyclohexanols (*93*) than do their equatorial epimers. Utilising these observations, conformations have been assigned to the various stereoisomeric menthols (*93*).

c) C-Halogen Absorption. An examination of the infrared spectra of 2-, 3-, 5-, 6-, and 7-chlorosteroids, and of 2-, 3-, 4-, 5-, 6-, and 7-bromosteroids has revealed that the frequency of the "halogen-sensitive" band (cf. *36*) for an equatorial C-halogen bond is greater than for the corresponding axial linkage (*52*). The observed frequencies lie in the range 580–780 cm.$^{-1}$, the difference between epimers being 39–56 cm.$^{-1}$ for the chlorides and 17–37 cm.$^{-1}$ for the bromides. Similarly it has been found (*116*) that the halogen-sensitive band of α-halo-oxosteroids and α-halo-oxotriterpenoids (in the range 850–670 cm.$^{-1}$) is of higher frequency for equatorial than for axial halogen atoms.

d) C—D Stretching Frequencies. The high frequency absorption band of an equatorial C—D linkage (in the range 2100–2200 cm.$^{-1}$) is higher than that of the corresponding axial bond in several deuterosteroids and deuterocyclohexanes (*107*).

e) C=O Stretching Frequencies. The carbonyl stretching frequency of a keto-group in a six-membered ring is increased by about 20 cm.$^{-1}$

by the introduction of an equatorial bromine atom, but an axial bromine atom has virtually no effect (*219*). Similar effects are produced by α-substitution of chlorine (*116*; see also *105*). This difference in behaviour has been attributed to the greater dipole interaction which exists in the equatorial epimer (in which the $C \overset{\rightarrow}{=\!=\!=} Br$ and $C \overset{\rightarrow}{=\!=\!=} O$ dipoles are almost coplanar) (*219*). The less regular effect of an α-acetoxyl group on the stretching frequency of a ketone (see, e. g., *159*) is in agreement with such an explanation, since here the relative orientations of the two $C \overset{\rightarrow}{=\!=\!=} O$ dipoles are not fixed.

f) Intramolecular Hydrogen Bonding. The ease with which a hydrogen bond is formed between groups in a molecule varies inversely with their separation distance (*229, 230*). In steroids and triterpenoids it has been established (*95, 267*) that only weak or negligible hydrogen bonds are formed between neighbouring diaxial groups, while neighbouring *cis* or diequatorially *trans* groups will form an intramolecular bond, the strength of which varies with the nature of the groups. COLE and MICHELL (*94*) took advantage of this relationship to confirm the stereochemistry of ring *A* of β-boswellic acid (LXXXII). The O—H stretching frequency indicated the 3-hydroxyl group to be in the axial conformation, and the absence of absorption due to intramolecular hydrogen-bonding established that the 4-carboxyl group must also be axial (as in LXXXIII).

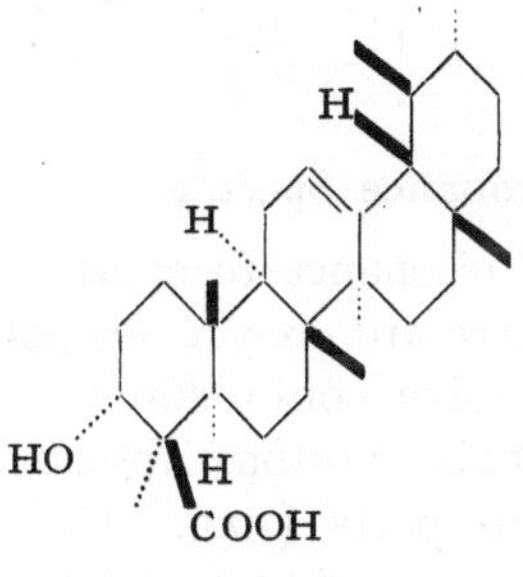

(LXXXII.) β-Boswellic acid. 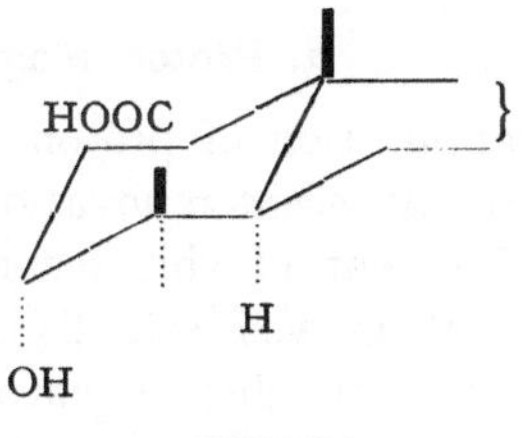   (LXXXIII.)

From a study of their infrared spectra, it is believed that 1 : 3-diaxial cyclohexanic diols are almost completely intramolecularly hydrogen-bonded (*327*). The ready solvolysis of monoesters of these compounds is attributed to intramolecular hydrogen-bonding of this type (see Chapter IV, p. 183).

2. Ultraviolet Spectra.

It has been found (*98*) that the introduction of an equatorial halogen atom into the α-position of a cyclohexanone derivative results in a slight hypsochromic shift of the weak absorption band exhibited by saturated ketones at about 280 mμ. In contrast, an axial α-halogen atom produces a much larger bathochromic displacement. Similar results were obtained with α-hydroxy- and α-acetoxyketones (99; see also 59). The approximate

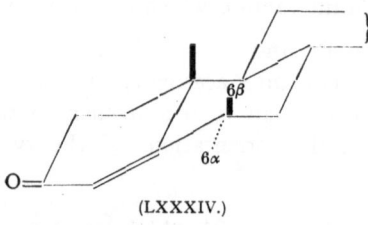

(LXXXIV.)

wavelength shifts are summarised in the second column of *Table 4*. In order to investigate the effect of γ-substituents on the long wavelength R band of α,β-unsaturated ketones, the ultraviolet spectra of 6α- and 6β-substituted cholest-4-en-3-ones (see LXXXIV) were measured (74). In these cases, both axial and equatorial substituents produced a bathochromic shift, but the displacement was much more marked for the axial substituents. These results are summarised in the third column of Table 4.

Table 4. Shifts Produced in the Wavelengths of the R Bands in the Ultraviolet Spectra of Saturated, and α,β-Unsaturated Ketones, by the Introduction of α- and γ-Substituents Respectively. $\Delta\lambda$ (mμ).

Substituent	α-substituted saturated ketone		γ-substituted α,β-unsaturated ketone	
	equatorial	axial	equatorial	axial
Cl..............	− 5	+ 15	+ 3	+ 14
Br	− 5	+ 28	+ 5	+ 20
OH	− 12	+ 17	+ 2	+ 7
OAc...........	− 5	+ 10	+ 2	+ 10

3. Proton Magnetic Resonance Spectra.

The application of proton magnetic resonance to organic chemistry is a relatively recent innovation, and there are, as yet, few examples of the employment of this technique to solve conformational problems. However, it is well established that axial protons absorb at higher frequencies than their equatorial counterparts (209). The frequency difference between the signals from axial and equatorial protons in acetylated carbohydrates has been found (242) to be about 8 c. p. s. (This, and other frequency differences quoted, refer to spectra measured at 40 megacycles per second.)

In large molecules, such as steroids, it is only possible to measure the difference in chemical shift between epimeric protons attached to a carbon atom if their signals can be distinguished from those of the other skeletal protons. A proton attached to a carbon atom bearing a hydroxyl group gives a readily discernible signal, since the hydroxyl group induces a large paramagnetic shift in the frequency of the proton (i. e. a shift toward lower fields). Because of this effect, SHOOLERY and

ROGERS (*295*) were able to study the difference in chemical shift between axial and equatorial protons attached to hydroxyl-substituted carbon atoms. They found that, for an axial proton attached to $C_{(3)}$, the chemical shift is 22–25 c. p. s. higher than for the equatorial. At the $C_{(11)}$ position the difference is 18–25 c. p. s. Similarly, an axial proton attached to a carbon atom bearing an acetoxyl group has a resonance frequency some 25 c. p. s. higher than that of an equatorial proton in the same environment (*306*).

A difference in the frequency of the signals from the methyl protons of axial and equatorial acetoxyl groups in the cyclohexane series has been reported (*242, 243*), but no corresponding difference has been found with steroids (*295*). However, it appears that in a series of 6-methyl-steroids, the axial 6β-methyl group absorbs at a lower frequency than its epimer (*304*). It has been suggested (*304*) that any structural changes which increase the crowding of a methyl group will result in the methyl signal being observed at a lower field.

The solvent shifts observed (*305*) in the N. M. R. spectra of steroids on changing from deuteriochloroform to pyridine as solvent are different for different protons: axial methyl groups exhibit larger solvent shifts than those which are equatorial.

It is evident that nuclear magnetic resonance is of great potential value for conformational studies. SHOOLERY and his colleagues (*285*) have recently used this technique to confirm the side-chain stereochemistry of various steroidal sapogenins.

4. Optical Rotatory Dispersion.

In the past five years or so, optical rotatory dispersion has been developed into a valuable technique for tackling conformational problems. Full accounts of this approach have been given in reviews by DJERASSI (*131*) and KLYNE (*226*).

Table 5. Shifts Produced in the Wavelengths of the First Extrema in the O. R. D. Curves of Steroidal Ketones by the Introduction of α-Substituents.

α-Substituent	$\Delta\lambda$ (mμ) of first extremum	
	equatorial	axial
F	+ 2	+ 15 to + 18
Cl.............................	+ 3	+ 15 to + 20
Br	+ 3	+ 20 to + 25
I...............................	+ 8	+ 32
OH	− 5 to − 10	+ 12 to + 23
OAc.........................	− 5	+ 10

Because of the close relationship between the positions of the peaks and troughs in the O. R. D. curve of a ketone, and the wavelength at which it absorbs light, it might be expected that axial and equatorial substituents α- to the carbonyl group would produce changes paralleling those found in its U. V. spectrum. This expectation has been borne out by measurements on a number of steroidal ketones (*135, 139, 144*). *Table 5* summarises the shifts in the wavelength of the first extremum in the O. R. D. curve of a steroidal ketone, caused by the introduction of various α-substituents (cf. Table 4, p. 196).

Introduction of an α-hydroxyl or α-acetoxyl group into a ketone does not, in general, affect the sign of its Cotton effect (see *139* for some exceptions). This is also true of an equatorial α-halogen atom; however, an axial α-halogen atom may or may not invert the sign of the Cotton effect. It is possible to predict the sign in such cases by means of a useful generalisation known as the *Axial Haloketone Rule* (*141, 144*). This states that if a cyclohexanone ring containing an α-chlorine or an α-bromine atom in the axial conformation is viewed along the O=C linkage, as indicated in (LXXXV), then if the halogen substituent lies on the left, as in (LXXXVI), the ketone will exhibit a negative

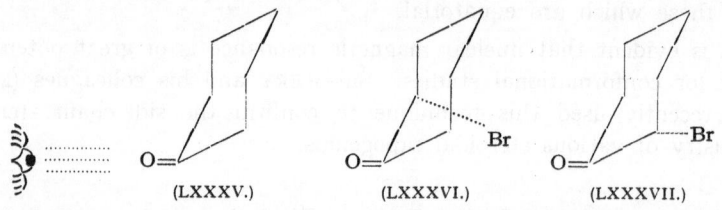

(LXXXV.)   (LXXXVI.)   (LXXXVII.)

Cotton effect, while if it lies on the right, as in (LXXXVII), the observed Cotton effect will be positive. Axial α-fluoroketones, however, exhibit Cotton effects of opposite sign to those of the corresponding bromo- or chloro-ketones.

Apart from the obvious value of this rule in determining the conformation of α-haloketones (or, conversely, the position of an axial α-haloketone) in compounds of known absolute configuration (such as steroids and terpenoids), it has proved well suited for employment in studies on the conformational equilibria of simple α-halocyclohexanones in various solvents (*13, 136, 137, 147*). The discovery (*134*) that the rule also holds for cyclohexanones in which the ring adopts a boat conformation has further extended its range of usefulness.

The Axial Haloketone Rule is now considered to be a particular case of a more widely applicable generalisation known as the *Octant Rule* (*131, 225, 226*). Publication of the details of this rule is awaited.

VI. Effect of Conformation on Reactions with Stereochemically Demanding Transition States.

1. Reactions Involving Four Coplanar Centres.

a) Bimolecular Elimination. The most stable transition state for 1,2-bimolecular elimination is that in which the two groups which are eliminated, X and Y, and the two carbon atoms to which they are attached, lie in one plane (*51*, *55*, *208*), as in (LXXXVIII).

Neighbouring substituents on a cyclohexane ring satisfy this requirement only if they are *trans*-diaxial. The debromination of steroidal dibromides with iodide ion illustrates this point very well (*19*, *51*, *55*). Diaxial dibromides react very much faster than those which are diequatorial. For example, it was found (*55*) that methyl 11β,12α-dibromo-3α,9α-epoxy-5β-cholanate (part formula LXXXIX) was

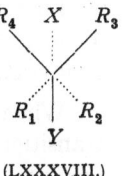

(LXXXVIII.)

more than 90% debrominated at the end of 12 days, while under the same conditions the isomeric 11α,12β-dibromo-compound (part formula XC) was completely unreacted even after six months.

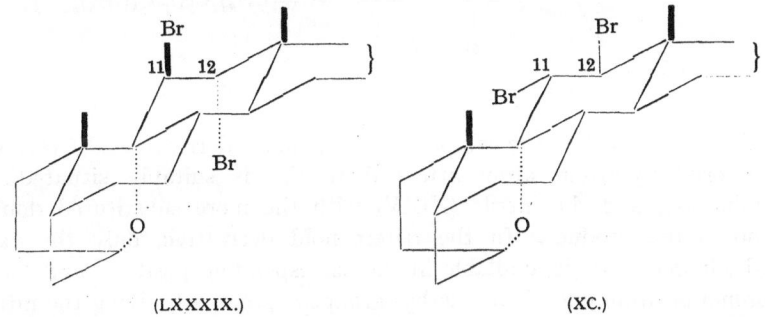

(LXXXIX.)   (XC.)

Similar preference for diaxial elimination is observed in the Hofmann degradation of steroidal amines (*172*, *194*) and in the bimolecular dehydration of steroidal alcohols. Thus, treatment of 3α-hydroxy-3β-

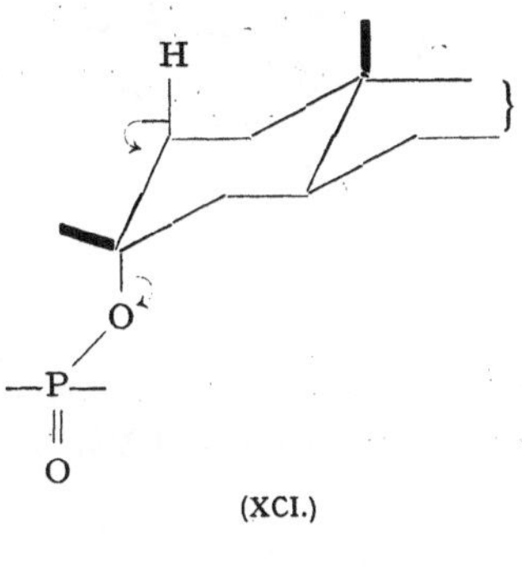

(XCI.) 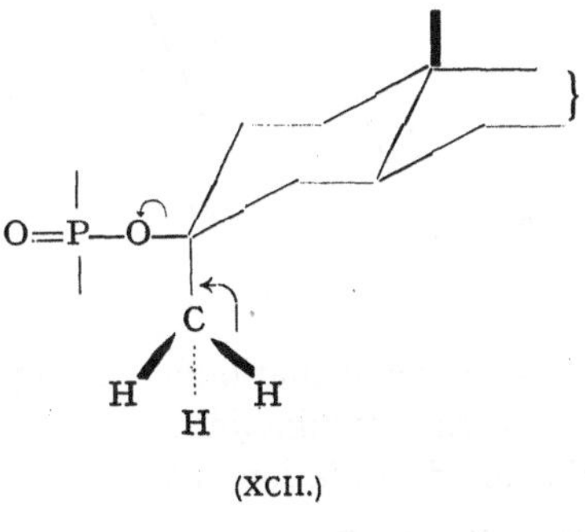   (XCII.)

methyl-5α-cholestane with phosphorus oxychloride and pyridine readily gives 3-methyl-5α-cholest-2-ene (see XCI); the 3-epimeric alcohol however, can achieve the necessary coplanarity only with one of the hydrogen atoms attached to the methyl group, and so in this case the exocyclic olefin results (see XCII) (36). (For other examples, see 67; and for an exception, 244.)

As a further illustration of the preference for a coplanar transition state in elimination reactions, it may be mentioned that in the Wolff-Kishner reduction of α-ketols, equal amounts of olefin and saturated alcohol are obtained when the hydroxyl group is equatorial, but olefin formation predomiantes when the hydroxyl group is axial (316).

When a cyclohexane ring is cleaved in an E 2 process a coplanar transition state can be achieved only if an *equatorial* proton is lost. This being so, the different courses of the abnormal Beckmann rearrangement in 2,2-dimethylcyclohexanone oxime and in β-amyranone oxime are easily

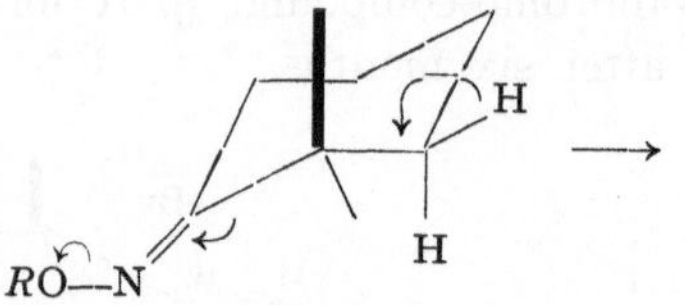

$$\longrightarrow \quad NC-(CH_2)_3CH=C(CH_3)_2$$

(XCIII.)   (XCIV.)

explained (332). In the simple cyclohexanone derivative (XCIII) the equatorial hydrogen atom attached to $C_{(3)}$ is suitably situated for elimination, and the nitrile (XCIV) with the more substituted double bond is the product. In the triterpenoid derivative, only the axial 5α-hydrogen atom is available in the corresponding position, and *trans*-elimination from one of the methyl groups is preferred, giving the nitrile with the less substituted double bond (see XCV).

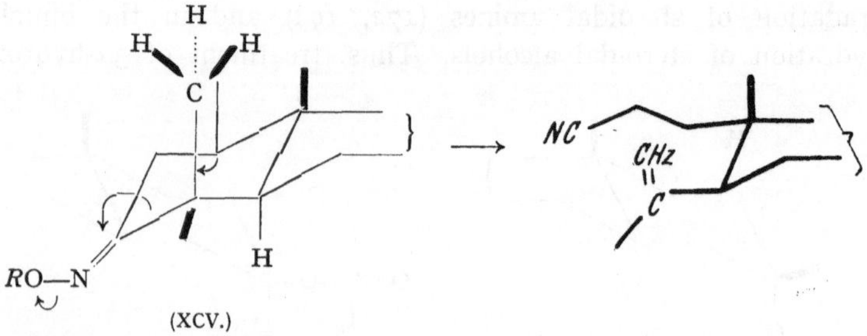

(XCV.)

A similar explanation has been proposed for the difference in behaviour of *cis*-perhydroquinoline and the related azasteroid (XCVI) in the Hofmann degradation (252).

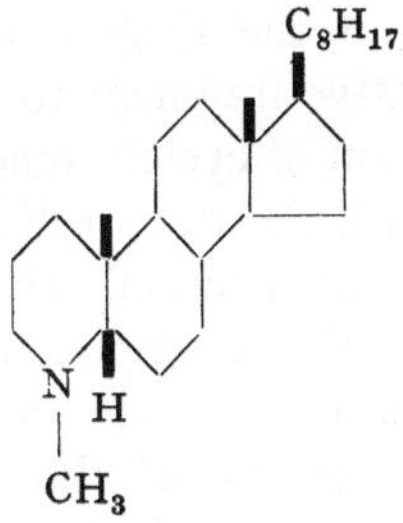

(XCVI.)

b) Diaxial Electrophilic Addition. It has been found (*19*) in conformationally unambiguous compounds, such as steroids, that bromine adds to a double bond to give the *trans*-diaxial dibromide. In agreement with this, 5α-cholest-2-ene affords, as main product, 2β,3α-dibromo-5α-cholestane, and 5α-cholest-3-ene gives principally 3α,4β-dibromo-5α-cholestane, presumably through the bromonium ions (XCVII) and (XCVIII) respectively (*280*). There is some evidence that the addition of dinitrogen tetroxide to steroidal double bonds also proceeds diaxially (*21*).

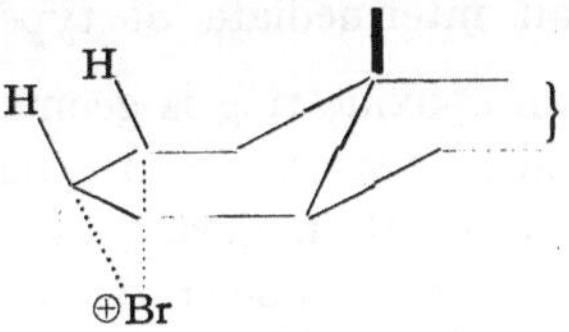

(XCVII.)

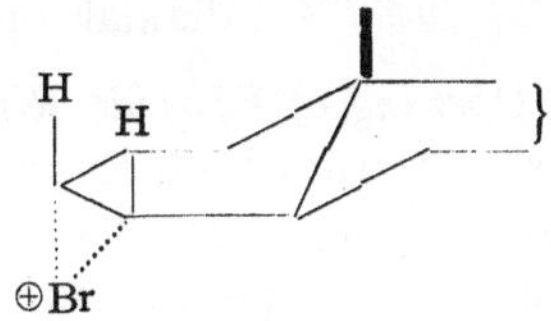

(XCVIII.)

It is well known that in aliphatic compounds, electrophilic addition to double bonds obeys the generalised Markownikoff Rule (*208*), that

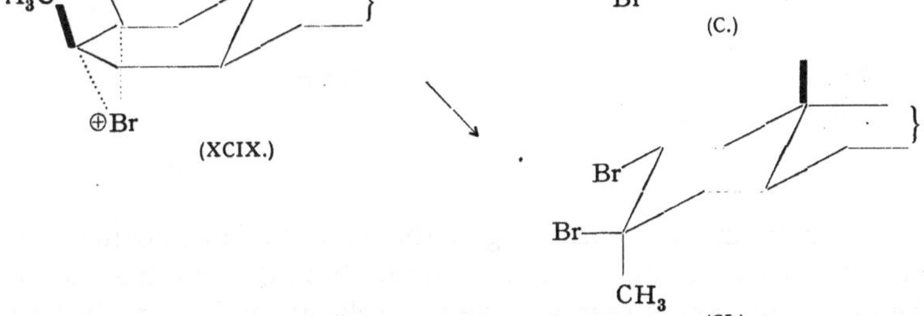

(C.)

(XCIX.)

(CI.)

is, the positive fragment of the reagent adds to the less substituted carbon atom, and the negative fragment to the more substituted carbon. However, in the bromonation of cyclohexene derivatives, it is found that where this rule conflicts with the rule of diaxial addition, it is the latter which directs the course of the reaction. Thus, 3-methyl-5α-cholest-2-ene gives the bromonium ion (XCIX) (by attack of bromine on the less hindered side of the molecule), which could give either the diaxial dibromide (C), with infringement of the Markownikoff Rule, or the diequatorial dibromide (CI), in compliance with the Markownikoff Rule, but contrary to the rule of diaxial addition. In fact, the diaxial compound (C) is obtained (36).

The preferred diaxial opening of the bromonium ion (XCIX) is no doubt a consequence of the fact that in this way the four participating centres are most nearly coplanar in the transition state.

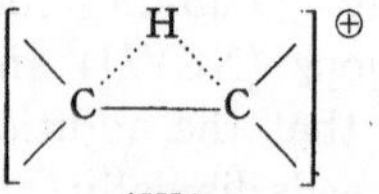

(CII.)

The addition of hydrogen halides to steroidal double bonds appears to proceed (36) by addition of halide ion to a classical carbonium ion rather than by diaxial opening of an intermediate of type (CII).

c) Opening of Epoxide Rings. Since an epoxide ring is geometrically very similar to a bromonium ion it would be reasonable to expect that it also should open diaxially. For the great majority of epoxide openings—nucleophilic and electrophilic—this is indeed the case (*32, 168, 169, 170, 179*; see also *180*). Normal epoxide cleavage is well documented (see, for example, *32*). In this Section, some of the few exceptions which have been encountered will be discussed.

Cookson and Hudec (*100*) have found that the product of perchloric acid cleavage of 2α,3α-epoxy-3β-phenyl-5α-cholestane (CIII) is 2α,3α-dihydroxy-3β-phenyl-5α-cholestane (CIV). Clearly this is a violation of the rule of diaxial opening. It would appear that in this case, the protonated epoxide is very highly polarised (owing to the presence of

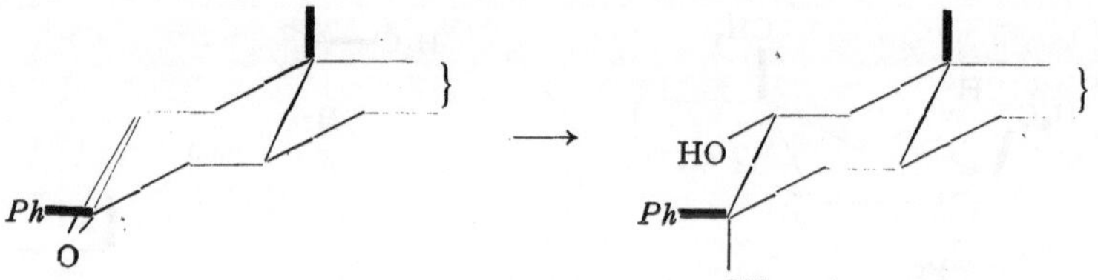

(CIII.)   (CIV.)

the phenyl group) and opens to give the more stable carbonium ion (CV) without any substantial amount of "push" from the entering nucleophile. The carbonium ion is then attacked by solvent predominantly from the less hindered side to give $2\alpha,3\alpha$-dihydroxy-3β-phenyl-5α-cholestane (cf. *36, 83*).

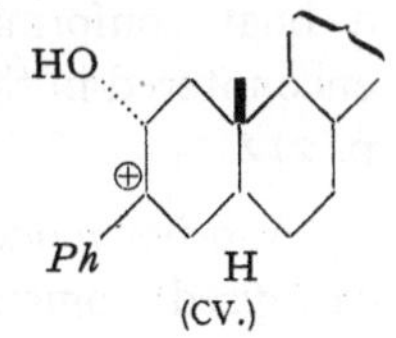

Other similar cases are known. The cleavage of $5\beta,6\beta$-epoxydihydro-lumisterol (CVI) with benzoic acid to afford 5β-hydroxy-6β-benzoyl-oxydihydrolumisterol (CVII) (*256*) is presumably another case of uni-molecular epoxide opening to afford a resonance-stabilised carbonium ion, followed by anion attack from the less hindered side of the molecule.

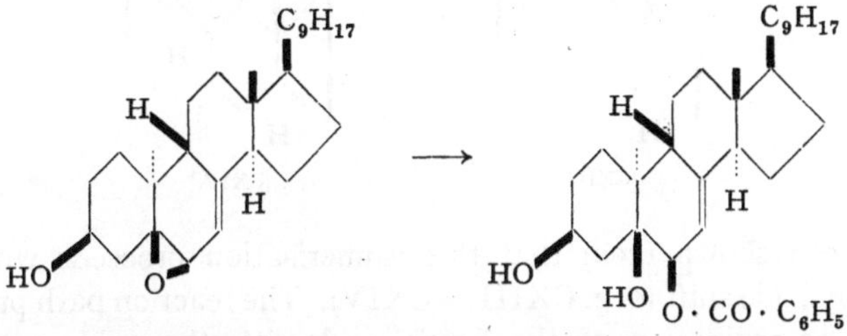

(CVI.) $5\beta,6\beta$-Epoxydihydrolumisterol.   (CVII.) 5β-Hydroxy-6β-benzoyloxydihydrolumisterol.

The oxirane ring in 3α-acetoxy-$16\alpha,17\alpha$-epoxy-$11,17$a-dioxo-*D*-homo-5β-androstane (CVIII) is cleaved with hydrobromic acid to give the diequatorial bromohydrin (CIX) (*325*). This was at first attributed (*325*) to the influence of the electron-withdrawing 17a-carbonyl group in inhibiting polarisation of the $C_{(17)}$—O bond. However, it is difficult to reconcile this explanation with the report that $4\beta,5\beta$-epoxycholestan-3-one (CX) reacts with hydrochloric or hydrobromic acid to give (CXII; $X = $ Cl or Br) (*294*), presumably by dehydration of the initially formed diaxial halohydrin (CXI). It may be that ring *D* of (CVIII) reacts in

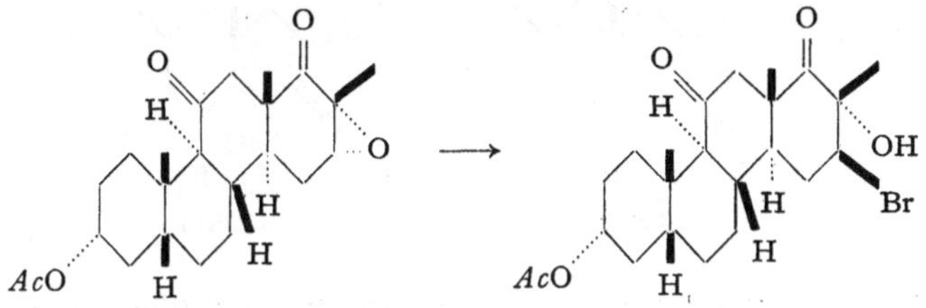

(CVIII.) 3α-Acetoxy-$16\alpha,17\alpha$-epoxy-$11,17$a-dioxo-*D*-homo-5β-androstane.   (CIX.)

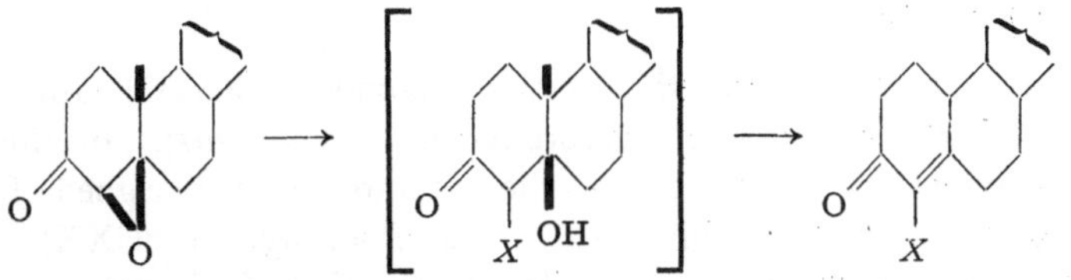

(CX.) $4\beta,5\beta$-Epoxycholestan-3-one.   (CXI.)   (CXII.)

a boat conformation (48, 323); this, together with similar reactions encountered in the lanostane series (48) will be considered in Chapter VII, p. 212.

Another mode of epoxide cleavage is observed in the boron trifluoride catalysed isomerisation of epoxides to ketones (see, e. g., 76, 203, 204).

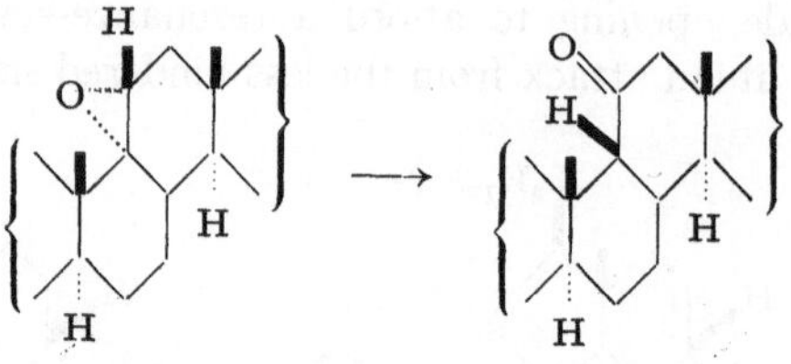

(CXIII.)   (CXIV.)

It has been shown (200) that this isomerisation proceeds with stereospecific hydride shift (e. g. CXIII → CXIV). The reaction path presumably involves co-ordination of the Lewis acid with the oxide oxygen; the powerfully electrophilic boron trifluoride promotes considerable polarisation of the bond to the more alkylated carbon centre, and the ketone is formed by 1 : 2-hydride migration to that centre (CXV).

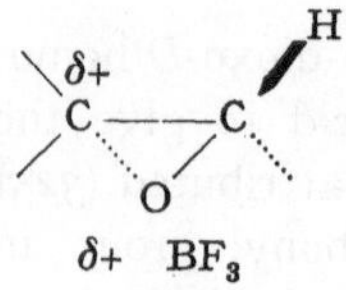

(CXV.)

Investigations of this reaction (200, 201; see also 81, 82) have shown that its course is controlled by a nice balance of electronic and conformational factors. The -I effect of the 3-carbonyl group in (CXVI) inhibits boron trifluoride ionisation of the $C_{(5)}$—O bond; hence, an alternative reaction involving

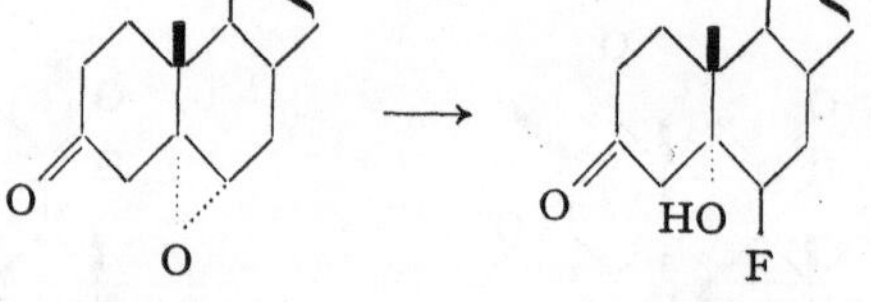

(CXVI.)   (CXVII.)

dual attack of the Lewis acid and an external fluoride nucleophile prevails to yield the fluorohydrin (CXVII) by diaxial opening of the epoxide ring (82).

Similar treatment of the acetoxy-epoxide (CXVIII) affords the ketone (CXIX) and not the fluorohydrin (CXX) (201). In this case, the unfavourable -I effect of the acetate is outweighed by the conformational factors involved. The fluorohydrin (CXX), which contains a 1 : 3-diaxial (OAc : OH) interaction (although somewhat mitigated by hydrogen-bonding of the type described in Chapter IV), is of higher energy than the ketone (CXIX), with its equatorial acetoxyl group.

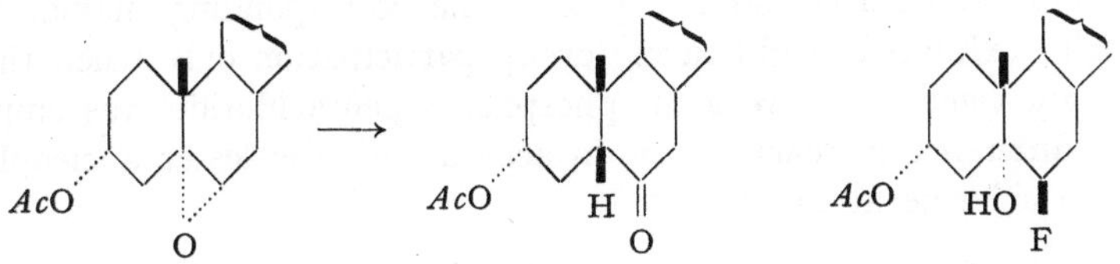

(CXVIII.)   (CXIX.)   (CXX.)

d) Formation of Epoxide Rings, and Neighbouring Group Participation. As would be expected from the reactions already discussed, diaxial halohydrins form epoxides much faster than their diequatorial isomers (*48*). The geometrical preference for four coplanar centres is well illustrated by the data accumulated on the stereochemistry of neighbouring group participation (*333*; for a recent review, see *249*). It has been found (*19*) that replacement reactions of steroidal halohydrins proceed much more readily when the halogen atom and the hydroxyl group are both axial, than when they are both equatorial. Thus, the 2β,3α-bromohydrin (CXXI) with thionyl chloride gave mainly the 2β,3α-bromochloride (CXXIII),

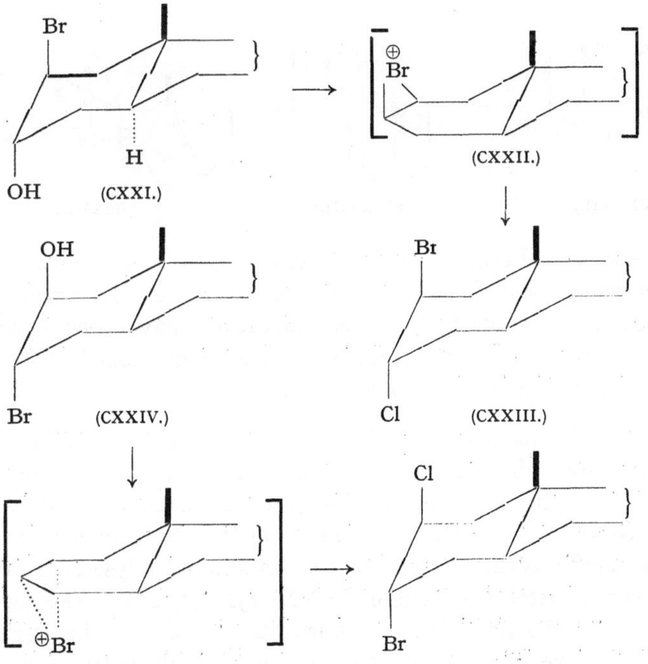

(CXXII.)

(CXXI.)

(CXXIV.)   (CXXIII.)

(CXXV.)   (CXXVI.)

presumably via the bromonium ion (CXXII). Similarly, the $3\alpha,2\beta$-bromohydrin (CXXIV) gave mainly the $3\alpha,2\beta$-bromochloride (CXXVI), through the intermediate (CXXV). Under the same conditions, the diequatorial isomers did not react. The corresponding diaxial chloro-hydrins exhibited neighbouring group participation only when the very strongly electrophilic reagent, phosphorus pentachloride was employed. This difference in reactivity is in accord with the lesser nucleophilicity of the chlorine atom (334).

e) 1:2-Diaxial Rearrangements. In cyclohexane derivatives, Wagner-Meerwein type rearrangements which do not involve a change in ring size take place most readily when the participating centres are coplanar, a condition which is fulfilled only when both the leaving group and the migrating carbon atom are axial. As a typical example of such a re-arrangement, the conversion of (CXXVII) to (CXXVIII) by treatment with phosphorus oxychloride and pyridine (217), may be mentioned. The stereochemistry of the reaction is indicated in the perspective formula (CXXIX).

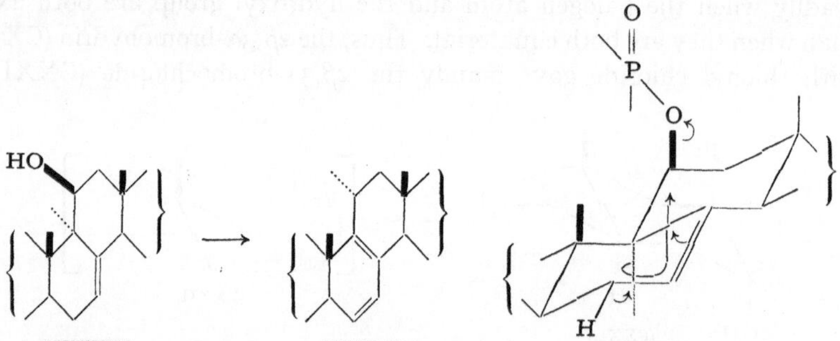

(CXXVII.)   (CXXVIII.)   (CXXIX.)

The Westphalen-Lettré Rearrangement, which has been recently re-investigated (2), also falls into the category of 1:2-diaxial rearrangements. Other examples of such concerted migrations have been described in earlier reviews (32, 37), and a few more will be mentioned in the Chapter on conformational driving forces. (p. 216).

f) Ring Contraction and Ring Expansion Reactions. Like the other transformations discussed above, ring *contractions* take place most readily when the four centres involved are coplanar, i. e. when the eliminated group is equatorial. Probably the best-known example of ring contraction is the phosphorus pentachloride catalysed rearrangement of 3β-hydroxytriterpenoids (see CXXX) (31, 37, 223). The mechanism illustrated implies retention of configuration at $C_{(5)}$ and inversion at $C_{(3)}$. In most of the ring A contractions of 3β-hydroxytriterpenoids, the stereochemistry at $C_{(3)}$ is, of course, lost by proton elimination. However, in a recent re-examination of the reaction, BIELLMANN and OURISSON (70)

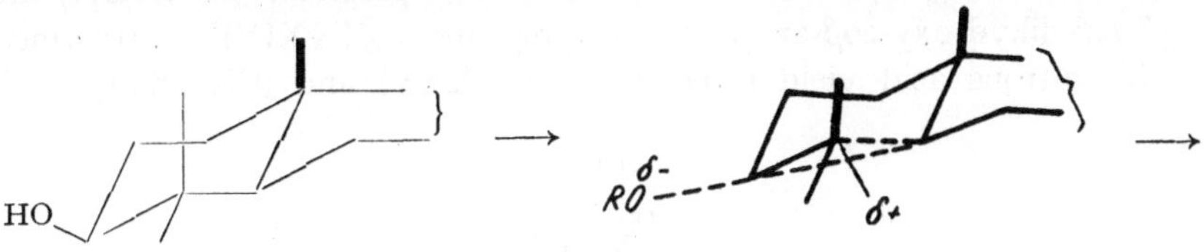

(CXXX.)

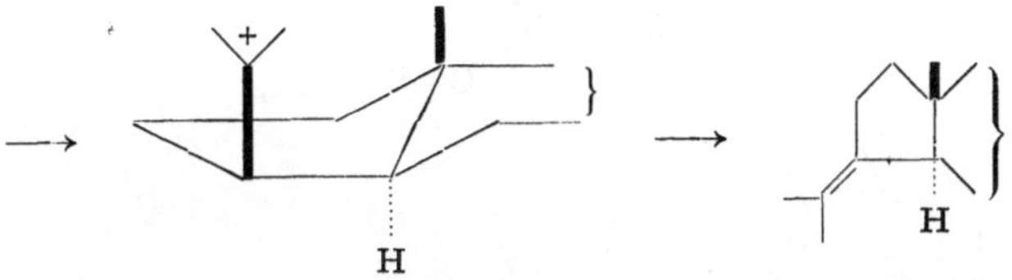

solvolysed the 3β-sulphonic esters of some triterpenoids in aqueous acetone in the presence of calcium carbonate, and were able to shown that a Walden inversion does indeed occur at $C_{(3)}$.

An interesting illustration of the effect of conformation on reaction course is provided by the contrasting behaviour of the epimeric 17-amino-17a-β-hydroxy-17-a-α-methyl-D-homo steroids when treated with nitrous acid (115). The 17β-epimer (CXXXI) in which both the 17a-methyl group and the 17-amino group are axial undergoes a Wagner-Meerwein re-arrangement of the type discussed above, while the 17α-epimer (CXXXII), with an equatorial amino group undergoes ring contraction.

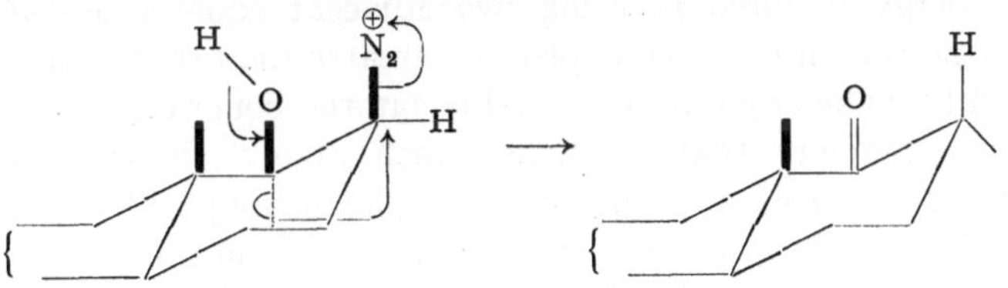

(CXXXI.)

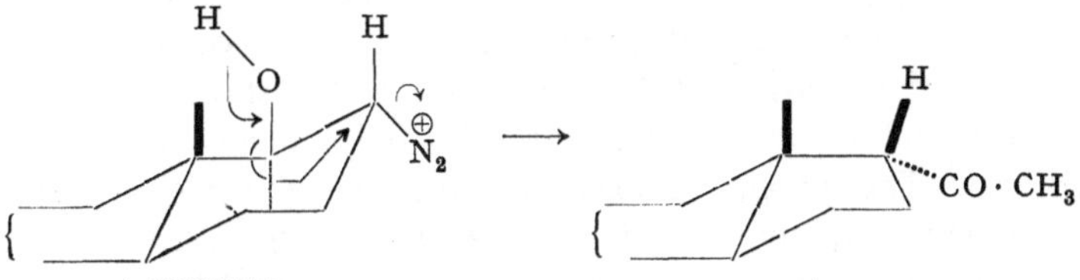

(CXXXII.)

Many other stereospecific ring contractions are known (see, for example, 24, 158, 205, 278, 324, 326).

The stereochemistry of ring *expansion* reactions is also determined by the conformational preference for a four-centre coplanar transition state. Thus, 3β,17α-dihydroxy-20α-amino-5α-pregnane (CXXXIII) and 3β,17β-dihydroxy-20β-amino-5α,17α-pregnane (CXXXIV) on treatment with nitrous acid yield respectively (CXXXV) and (CXXXVI) (*278*).

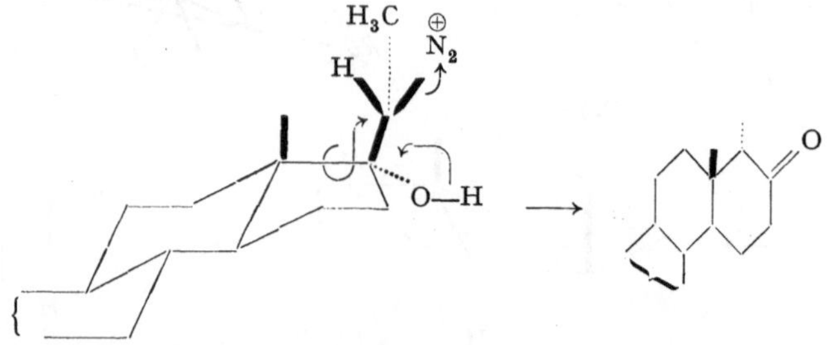

(CXXXIII.) 3β,17α-Dihydroxy-20α-amino-5α-pregnane.   (CXXXV.)

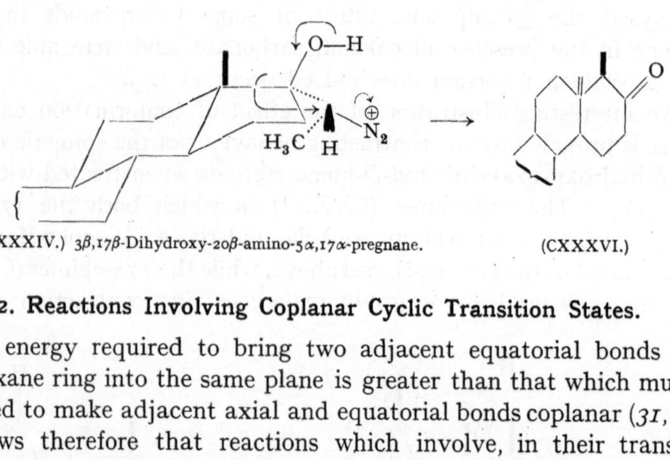

(CXXXIV.) 3β,17β-Dihydroxy-20β-amino-5α,17α-pregnane.   (CXXXVI.)

2. Reactions Involving Coplanar Cyclic Transition States.

The energy required to bring two adjacent equatorial bonds on a cyclohexane ring into the same plane is greater than that which must be expended to make adjacent axial and equatorial bonds coplanar (*31, 189*). It follows therefore that reactions which involve, in their transition state, a 1,2-fusion of an approximately planar ring will proceed more easily with *cis* (*e, a*) compounds than with *trans* (*e, e*) compounds. *Trans* (*a, a*) compounds should be the least reactive in such conversions. The cleavage of glycols with periodic acid or lead tetraacetate are reactions of the type under consideration. Angyal and Young (*23*) measured the rates of oxidation of the isomeric 5α-cholestane-3β,6,7-triols with lead tetraacetate, and obtained results fully in agreement with those predicted. The 3β,6β,7α-isomer in which the glycol grouping is *trans*-diaxial remained unattacked (cf. *132, 165, 175*). Configurations have been assigned to triterpenoid glycols on the basis of their rates of cleavage with lead tetraacetate (*146*).

Other reactions involving an approximately coplanar transition state—and which are therefore *cis*-stereospecific—are the thermal elimination of esters (*30, 43, 54, 55, 64, 129, 167*), the decarboxylation of β,γ-unsaturated carboxylic acids (*25, 35*) and the pyrolysis of amine oxides (*101*).

An interesting exception to the general *cis* elimination of esters by pyrolysis has been reported by BORDWELL and LANDIS (*78*) in the cyclohexane series. They have shown that *trans* elimination is preferred in the pyrolysis of (CXXXVII), presumably owing to the rather acidic nature of the proton attached to the carbon atom bearing the sulphonyl group.

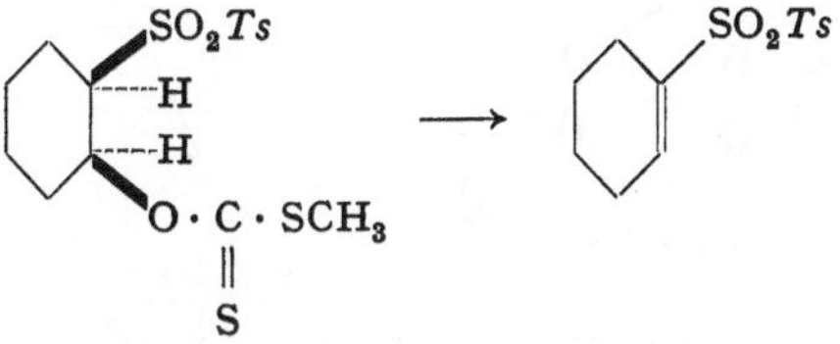

(CXXXVII.)

This reaction may be regarded as complementary to the examples of base-catalysed bimolecular elimination described by BORDWELL and his colleagues (*77, 79, 320*) in which a *cis*-hydrogen atom, rendered acidic by virtue of a strongly electron-attracting group, is eliminated instead of an alternative *trans* one. Typical of these cases is the second-order base-catalysed elimination of *p*-toluenesulphonic acid from (CXXXVIII).

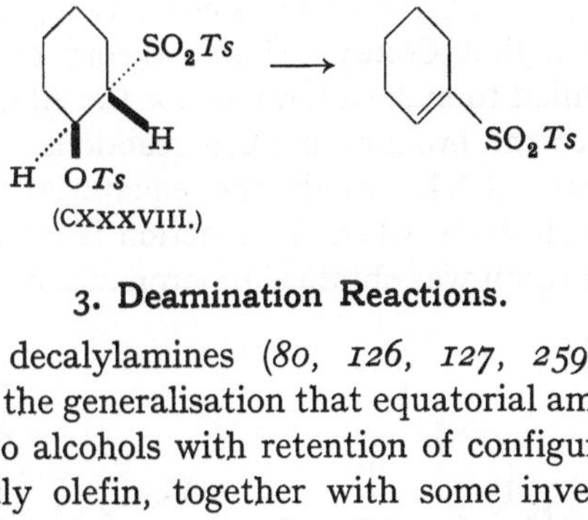

(CXXXVIII.)

3. Deamination Reactions.

Work on the decalylamines (*80, 126, 127, 259*) and hydrindylamines (*123*) led to the generalisation that equatorial amines are converted by nitrous acid into alcohols with retention of configuration, while axial amines yield mostly olefin, together with some inverted alcohol; and mechanistic theories to account for this stereospecificity were put forward (*259, 313*). However, later investigations on steroidal (*160, 296–298*) and triterpenoid (*148, 149*) amines would suggest that these series differ from the simpler systems in that the small amount of alcohol obtained from an axial amine is of the same configuration as the starting material.

4. Bromination and Protonation of Enols.

On the basis of spectroscopic studies on the products of bromination of cyclohexanone derivatives, COREY (*102–104*) proposed that the product of bromination under kinetically-controlled conditions is always the axial α-bromoketone. This was explained (*102*) as being due to the stereochemistry of the π orbitals of the enol, which are more favourably arranged for overlap with the vacant orbital of the bromine when it adds axially, than when it approaches equatorially (CXXXIX).

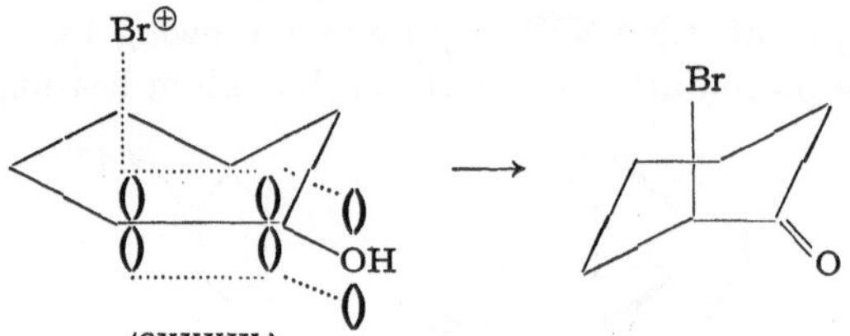

(CXXXIX.)

Deuterium isotope studies on the ketonisation of enols have shown that, in the absence of steric complications, axial protonation is preferred to equatorial attack (*110*; see also *159*). In ketonisation reactions it is well known that steric hindrance to approach of the proton donor also plays a major rôle in determining the orientation of the incoming proton (*339*), and in the protonation of steroidal enols the proportions of equatorial and axial attack are influenced by both the stereoelectronic and the steric effects (*110*).

Recent work (*133, 255, 318*) on steroidal ring *A* ketones and enol acetates has shown that Corey's simple theory of ketone bromination must also be modified to make allowance for the effect of steric hindrance to the approach of the bromine in such reactions. Thus, 5α-androstan-17β-ol-3-one acetate (CXL) yields the equatorial 2α-bromo-derivative (CXLI) almost exclusively when the reaction is carried out with kinetic control. A similar result was obtained by bromination of the corresponding

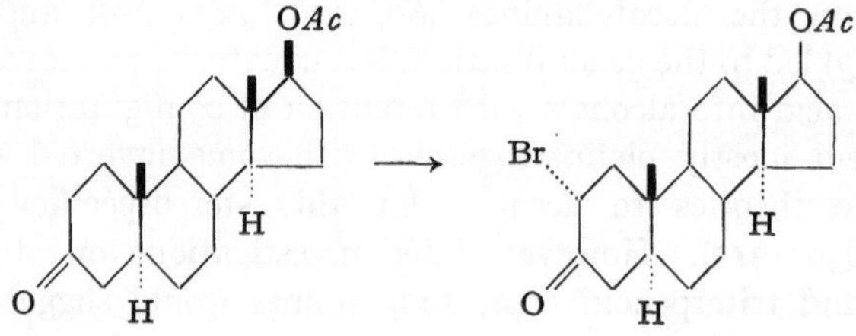

(CXL.) 5α-Androstan-17β-ol-3-one acetate.   (CXLI.)

enol acetate under the same conditions. Similar treatment of the 19-nor analogues of these compounds, (CXLII) and (CXLIII), in which there is no steric hindrance of the 2,3-enol by an axial methyl group at $C_{(10)}$, gave in each case a mixture of the 2β-bromo- and 4β-bromo-derivatives,

(CXLIV) and (CXLV), in both of which the bromine atom is axially disposed.

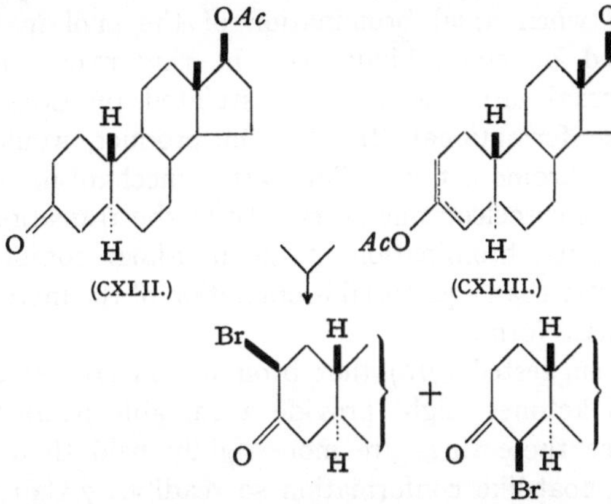

(CXLII.)   (CXLIII.)

(CXLIV.)   (CXLV.)

On the basis of these and other results, DJERASSI and his colleagues (*318*) have proposed that where axial attack by bromine in sterically hindered, as in the bromination of (CXL), the equatorial bromoketone is more rapidly formed. It has also been found (*56*) that 3-acetoxy-5α-cholest-2-ene can be chlorinated, in the presence of 2β-chloro-5α-cholestanone labelled with radioactive chlorine, to afford, as main product, 2α-chloro-5α-cholestanone containing only a fraction of the radioactivity present in the starting material. Hence, in this reaction, most of the equatorial 2α-chloro-compound has been formed directly, and not through the 2β-isomer as intermediate.

Even in the absence of such a steric effect, it has been shown (*318*) that a substantial amount of equatorial bromoketone may be formed. Thus, 2α-methyl-19-nor-5α-androstan-17β-ol-3-one acetate (CXLVI), when brominated with kinetic control gives a mixture of approximately equal amounts of the 2α- and 2β-bromo-compounds.

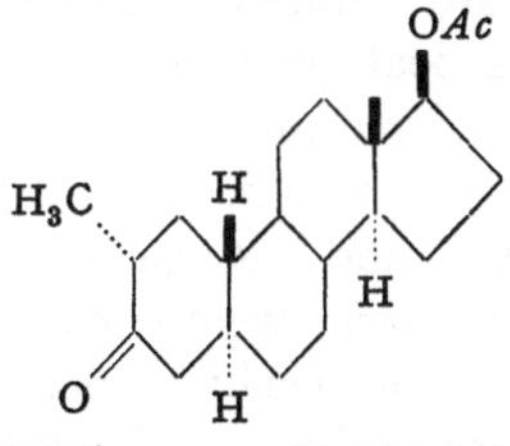

(CXLVI.) 2α-Methyl-19-nor-5α-androstan-17β-ol-3-one acetate.

Two possible hypotheses to account for these observations present themselves (*318*). Most simply, it can be admitted that direct equatorial attack on the enol can occur, and that this is relatively more important when axial approach is strongly hindered. Alternatively, it might be postulated that when axial bromination of the enol in its half-chair form is inhibited by steric hindrance, the ring reacts in a boat-like conformation; axial bromination, as postulated by Corey (*102–104*), followed by a conformational "flip" of the product, would then result in an equatorial bromo-ketone. This latter mechanism would require a high degree of stereoelectronic specificity in the transition state, since it implies that axial bromination of the half-boat conformation must be very much faster than equatorial bromination of the more energetically favoured half-chair form.

It has been suggested (*318*) that bromination studies on ring *B* or ring *C* steroidal ketones might provide a valuable pointer to the true mechanism, since these rings are more rigidly held than ring *A*, and cannot adopt a boat-like conformation so readily. 7-Oxo- and 12-oxo-steroids, which would both afford sterically hindered enols, are reported to give axial bromo-derivatives (see *318* for leading references). If these results are valid, they support the theory involving conformational change in ring *A*. However, it has been pointed out (*318*) that these ketones required special bromination conditions, which may have caused some change in the initial product composition.

VII. Preferred Boat Conformations.

(For a detailed account, see *245*.)

Many steroids and terpenoids are known in which it is configurationally impossible for the molecule to exist in an all-chair conformation, and some examples of these have already been cited in this review.

In recent years it has become evident that even when this geometrical restriction does not apply, a molecule may adopt a conformation in which one ring is in the boat form (or, more precisely, in one of the flexible forms) if, by so doing, the non-bonded interactions present in the alternative chair form are reduced sufficiently to compensate for the inherent instability of a boat relative to a chair.

This situation was first recognised (*48*) in one of the two products, (CXLVII) and (CXLVIII) obtained by bromination of lanost-8-en-3-one (CXLIX). The UV and IR spectra of the two bromoketones revealed that the halogen atoms were equatorial. The configurations of the bromine atoms were established by reducing the ketones to the bromo-hydrins (CL) and (CLI) respectively, and treating with base to afford, in the case of the 2α-bromo-3β-hydroxy compound (CL) the β-epoxide (CLII), and, from the *cis*-bromohydrin (CLI), the starting ketone (CXLIX). (CL) and (CLI) could be reduced to lanost-8-en-3β-ol. This reaction

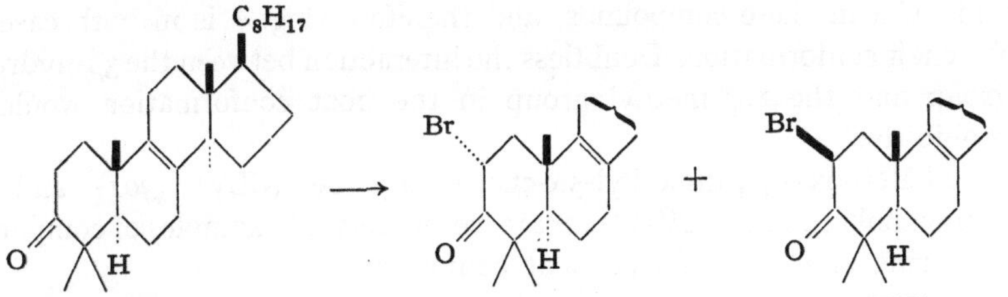

(CXLIX.) Lanost-8-en-3-one.
$\lambda_{max.}$ 288 mμ
$\nu_{max.}$ 1703 cm.$^{-1}$

(CXLVII.)
$\lambda_{max.}$ 291 mμ
$\nu_{max.}$ 1728 cm.$^{-1}$

(CXLVIII.)
$\lambda_{max.}$ 282 mμ
$\nu_{max.}$ 1734 cm.$^{-1}$

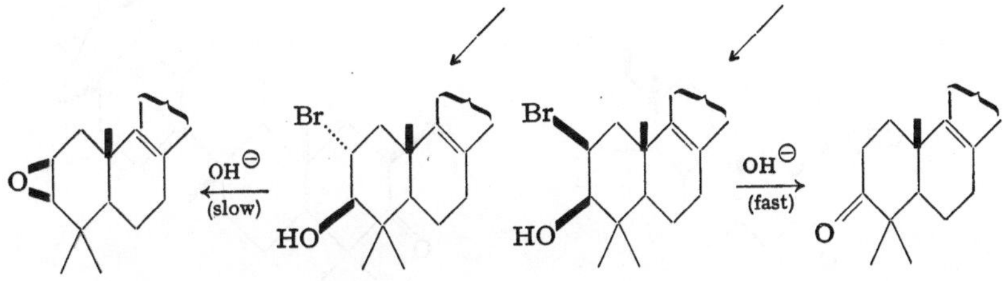

(CLII.)

(CL.)
$\nu_{max.}$ 724 cm.$^{-1}$

(CLI.)
$\nu_{max.}$ 515 cm.$^{-1}$

(CXLIX.)

sequence, which is the standard procedure for determining the configurations of steroidal α-haloketones, provides a good illustration of the application of the conformational principles described in Chapter VI (p. 199).

The establishment of the 2-bromine atom in (CXLVIII) as
conformationally equatorial and configurationally β means that ring A
must exist in the boat conformation (CLIII). Presumably this is preferred,

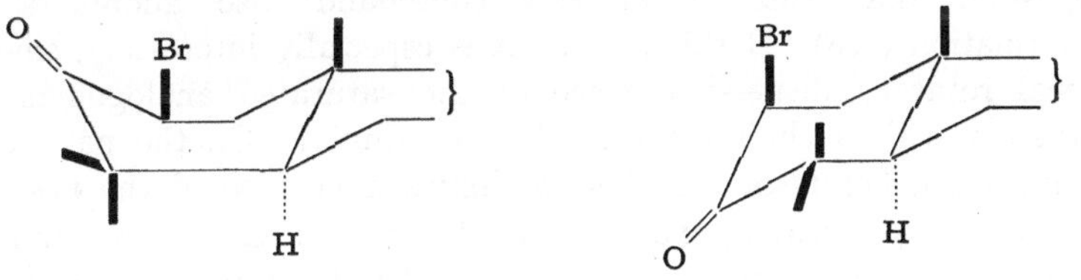

(CLIII.)

(CLIV.)

because of the severe 1 : 3-diaxial (CH$_3$: CH$_3$) and (Br : CH$_3$) interactions
which would be present in the alternative chair conformation (CLIV):
another important factor is that, since C$_{(3)}$ is trigonal, the energetically
unfavourable 1 : 4-transannular interaction which destabilises the boat
form of cyclohexane itself, is absent. Significantly, all the cyclohexane
derivatives for which it has been established that a boat conformation
is preferred to a chair, contain a trigonal carbon atom. It is worth noting
that the frequencies of the halogen-sensitive bands in the infrared spectra
of (CL) and (CLI) imply that the bromine atom is respectively equatorial

and axial in these compounds, and therefore ring A is in both cases in the chair conformation. Doubtless the interaction between the 3β-hydroxyl group and the 10β-methyl group in the boat conformation would be prohibitive.

2β-Methoxy-4,4-dimethyl-5α-cholestan-3-one (CLV) (*301*) and 2β-bromoallobetulone (CLVI) (*222*) are closely related examples of compounds with ring A in a preferred boat conformation.

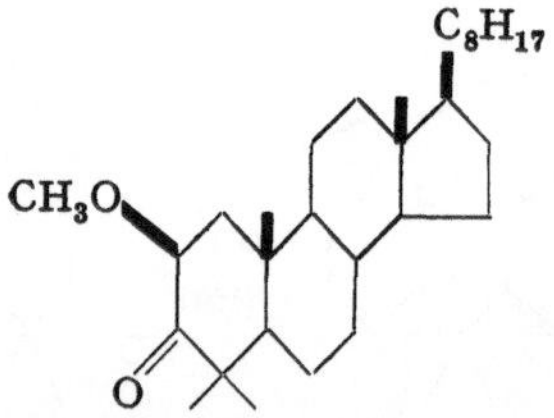

(CLV.) 2β-Methoxy-4,4-dimethyl-5α-cholestan-3-one. 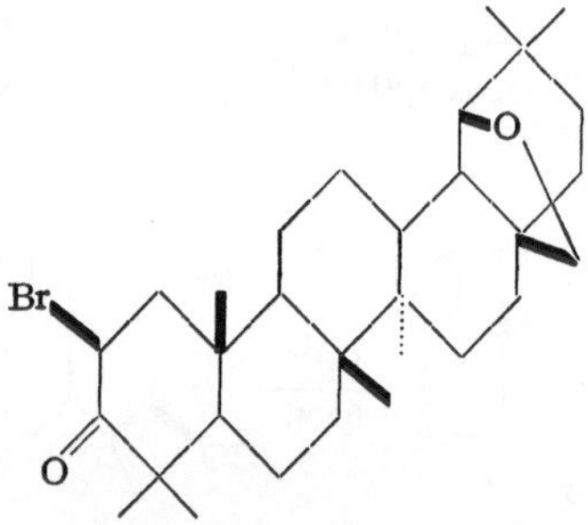   (CLVI.) 2β-Bromoallobetulone.

In the 5α-steroid series, a boat conformation for ring A has been established for certain 2α-bromo-2β-methyl-3-oxo-derivatives (*133, 255*). In these compounds, the chair conformation would be greatly destabilised by the 1:3-diaxial interaction between the methyl groups attached to $C_{(2)}$ and $C_{(10)}$.

Application of the Octant Rule (Chapter V, p. 193) to the optical rotatory dispersion curve of 2β-methyl-5α-cholest-6-en-3-one has led to the suggestion that ring A of this compound also adopts a boat conformation (*308*). If this is true, it is especially interesting, since the optical rotatory dispersion curve of the saturated analogue is quite consistent with a chair conformation for ring A. On the basis of the Octant Rule, Ourisson and his colleagues have mooted the possibility that in all 8β-methyl-3-oxotriterpenoids, ring A exists predominantly in the boat conformation (*183*; cf. *222*). In this way, the severe 1:3-diaxial interactions between the β-methyl groups attached at positions 4, 8, and 10 in the alternative all-chair conformation would be relieved. Further work is required, however, before the conformation of ring A in these compounds can be considered as established.

It has been suggested that the apparently abnormal ring opening observed when the 2β,3β-epoxides of lanost-8-ene and lanostane are treated with lithium aluminium hydride or hydrobromic acid may also be due to ring A reacting as a distorted boat (CLVII) (*48*). The same explanation may hold for the case of "diequatorial" epoxide opening reported by Wendler and his colleagues (*325*).

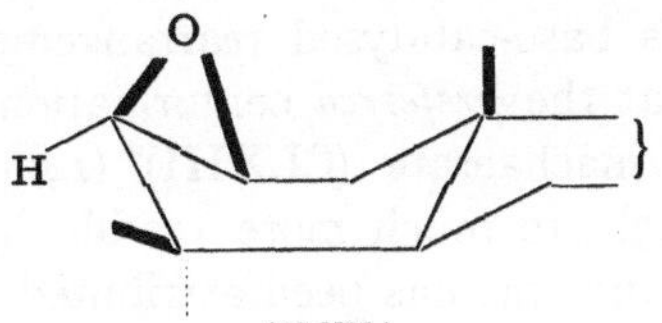

(CLVII.)

Treatment of the *D*-homosteroid (CLVIII) with base converts it to the methyl ketone (CLIX) rather than to the diosphenol (CLX), which would be predicted if the steroid were to react in its all chair conformation (as CLXI) (*322, 323*). To account for this, WENDLER (*322, 323*) has proposed that ring *D* has assumed a boat conformation for the re-

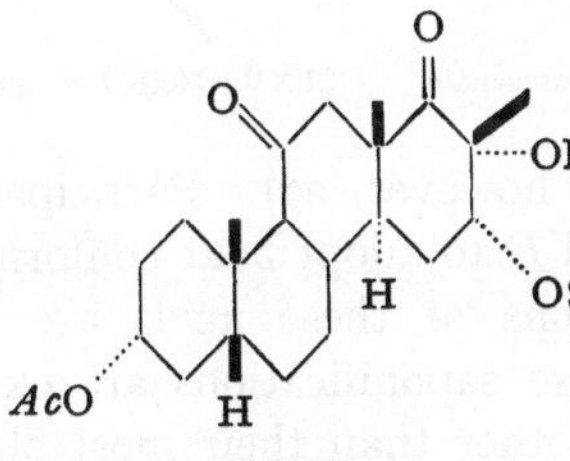

(CLVIII.)

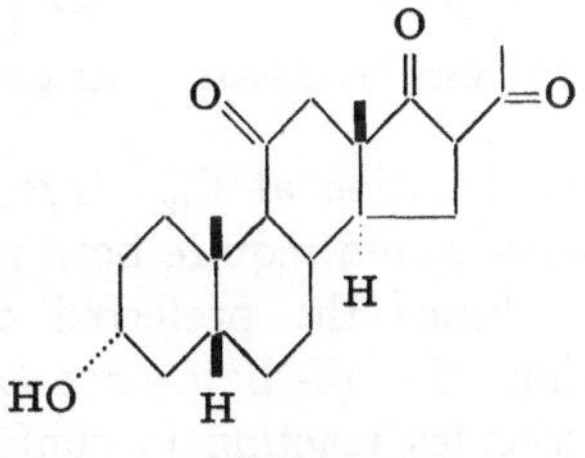

(CLIX.)

arrangement (as CLXII). In this conformation, the diaxial interaction between the methyl groups attached to $C_{(13)}$ and $C_{(17)}$ in (CLXI) has been eliminated, and the mesylate group is equatorially situated; it is probable that the energies of (CLXI) and (CLXII) are quite close (see *317* for an analogous pinacol rearrangement).

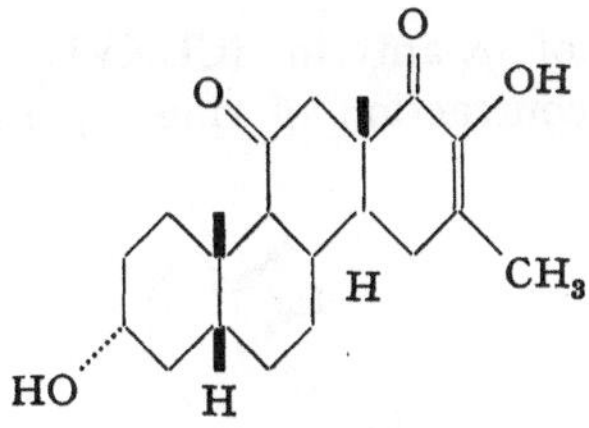

(CLX.)

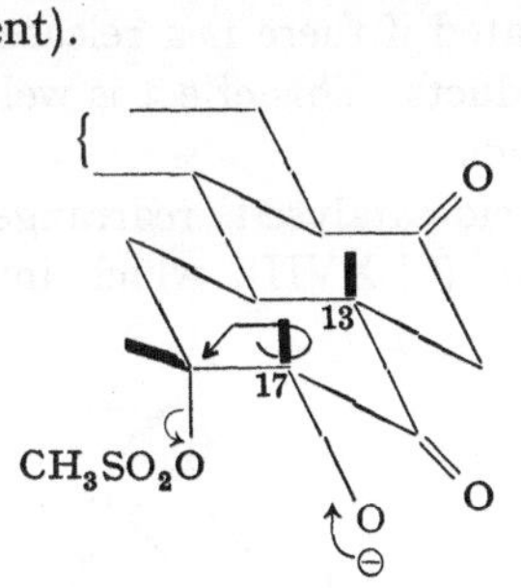

(CLXI.)

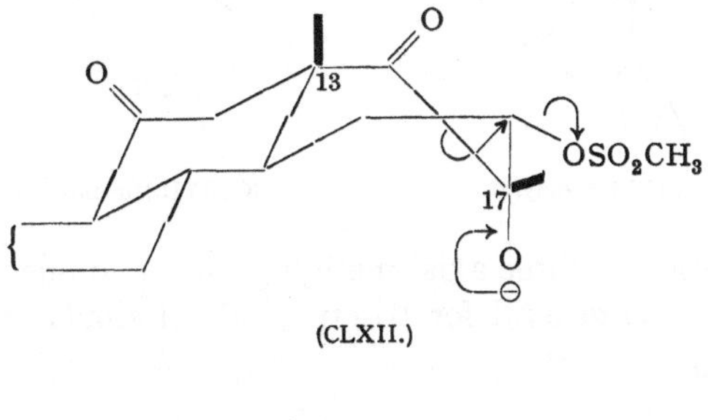

(CLXII.)

The course of its base-catalysed rearrangement, however, does not necessarily imply that the *preferred* conformation of ring D of (CLVIII) is a boat. Methyl machaerate (CLXIII) (*142*) and methyl treleasegenate (CLXIV) (*143*) are much more readily hydrolysed than methyl oleanolate (CLXV), and this has been attributed to participation by the

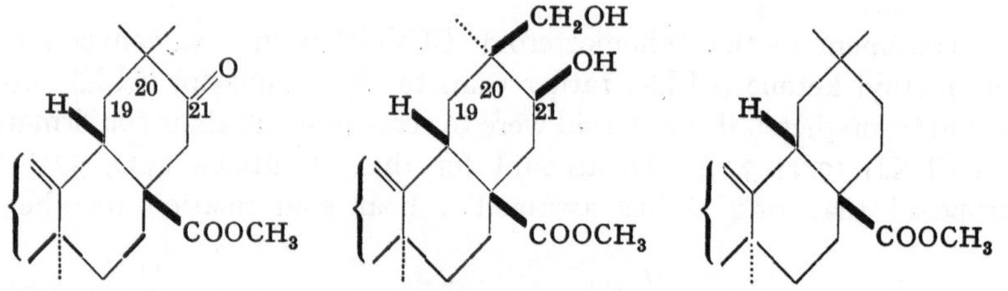

(CLXIII.) Methyl machaerate.   (CLXIV.) Methyl treleasegenate.   (CLXV.) Methyl oleanolate.

oxygen function at $C_{(21)}$ (*142, 143*; see, however, *27*). Participation of this kind would require both rings C and D to adopt boat conformations (*245*). Since the preferred conformations of these acids are almost certainly the all-chair ones (*1, 44*), these saponifications are examples of molecules reacting in conformations other than their most stable.

For recent examples of the same phenomenon in the cyclohexane series, see (*227, 300*).

VIII. Conformational Driving Forces.

It has been suggested (*37, 50*) that carbonium ion rearrangements are facilitated if there is a release of steric compression in the formation of the products. This effect is well illustrated by the chemistry of certain triterpenoids.

The acid-catalysed rearrangement of α-amyrin (CLXVI) to α-amyradiene (CLXVII), which involves contraction of ring A, and the

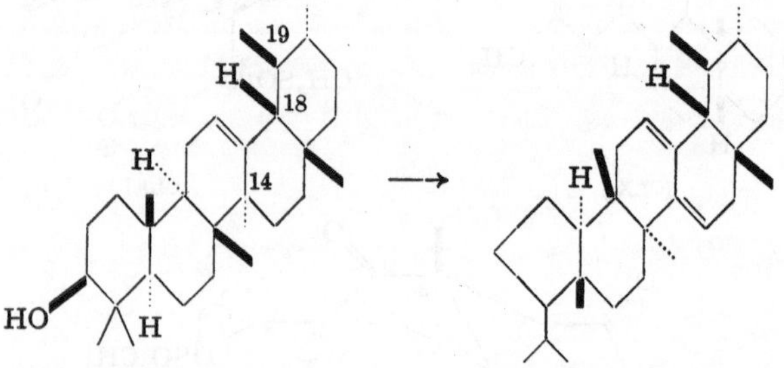

(CLXVI.) α-Amyrin.   (CLXVII.) α-Amyradiene.

concerted migration of three axial methyl groups and one axial hydrogen atom, appears to be general for triterpenoids of similar structure, with rings *D* and *E* *cis* fused: those with *trans*-fused *D* and *E* rings do not undergo the rearrangement (*4, 161*; see also *238*). It is probable that part of the driving force in this rearrangement is provided by the 1 : 3 di-axial interaction between the $C_{(19)}$-methine group and the $C_{(14)}$-methyl group, which is present in α-amyrin, but absent from α-amyradiene. In the *trans* fused triterpenoids, the corresponding interaction is between the $C_{(14)}$-methyl group, and the α-hydrogen atom attached to $C_{(18)}$.

The rearrangement of diaxial *vic*-dibromides to the diequatorial isomers, which is a general phenomenon in the steroid series (*33*), is another example of the operation of a conformational driving force—in this case it has its origin in the higher energy of an axial bromine atom as compared with its equatorial epimer. The mechanism of this reaction has been studied (*178, 237*), and it is generally accepted that it proceeds intramolecularly through a dipolar transition state (CLXVIII), without

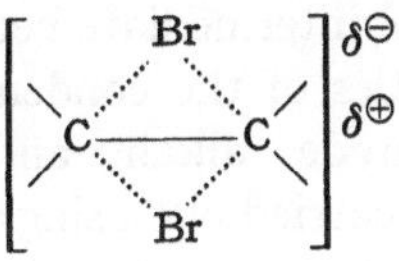

$$\left[\overset{\text{Br}}{\underset{\text{Br}}{\diagup C\!\!-\!\!C\diagup}} \right] \overset{\delta\ominus}{\underset{\delta\oplus}{}} \qquad \left[\overset{\text{Br } \delta\ominus}{\underset{\underset{\text{CH}_3}{\overset{|}{\underset{\delta\oplus}{C}}}}{\overset{C\!\!-\!\!C}{O \diagdown \; \diagup O}}} \right]$$

(CLXVIII.) 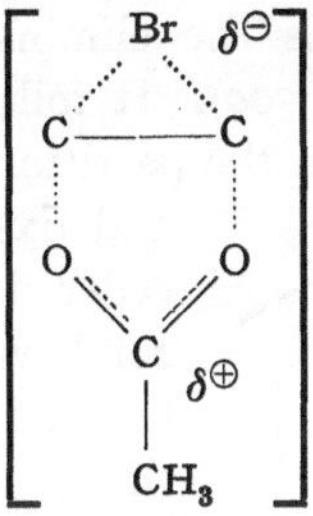   (CLXIX.)

the formation of free ions. It has recently been established (*34, 45*) that the dibromide rearrangement is only one example of a whole family of such diaxial-diequatorial rearrangements. Kinetic and equilibrium studies on the rearrangements of a series of 2β-halo-3α-acyloxy-5α-cholestanes—for which a transition state of type (CLXIX) may be written—have shown that the rate of rearrangement depends on the polarisability of the C-halogen bond and the electronegativity of the ester group (*45*).

IX. Conformational Transmission.

1. Long Range Effects.

More subtle aspects of conformational analysis, concerning the effect on the reactivity of a steroid or triterpenoid molecule of substitution or unsaturation at a carbon atom remote from the reacting site, have recently been the object of a considerable amount of research (*34, 41, 42, 49*; see also *254, 329*).

It has been found (*42, 49*) that the rate of condensation of steroidal or triterpenoid 3-ketones is markedly influenced by structural modifications in other rings. The mechanism of benzaldehyde condensation with a ketone may be expressed by *Chart 1* (*269, 309*) ($Ph = C_6H_5$):

$$-CH_2-CO- + OH^{\ominus} \underset{k_2}{\overset{k_1}{\rightleftarrows}} -\overset{\ominus}{CH}-CO- + H_2O \qquad\qquad \text{(i, ii)}$$

$$-\overset{\ominus}{CH}-CO- + PhCHO \underset{k_4}{\overset{k_3}{\rightleftarrows}} Ph\cdot CH(-\overset{\ominus}{O})-\overset{|}{CH}\cdot CO- \qquad\qquad \text{(iii, iv)}$$

$$PhCH(-\overset{\ominus}{O})-\overset{|}{CH}-CO- + H_2O \underset{k_6}{\overset{k_5}{\rightleftarrows}} Ph\cdot CH(OH)-\overset{|}{CH}-CO- + OH^{\ominus} \qquad \text{(v, vi)}$$

$$PhCH(OH)-\overset{|}{CH}-CO- + OH^{\ominus} \underset{k_8}{\overset{k_7}{\rightleftarrows}} Ph\cdot CH(OH)-\overset{\ominus}{\underset{|}{C}}-CO- + H_2O \qquad \text{(vii, viii)}$$

$$PhCH(OH)-\overset{\ominus}{\underset{|}{C}}-CO- \overset{k_9}{\rightarrow} Ph\cdot CH=\overset{|}{C}-CO- + OH^{\ominus} \qquad\qquad \text{(ix)}$$

Chart 1. Mechanism of the Condensation of Benzaldehyde with Ketones.

The rate determining step for the formation of the benzylidene ketone (like CLXX) is the elimination of $OH^{\ominus}$ from the aldol intermediate (step ix) (*269, 309*). It follows that the formation of the final product is dependent on the partition of the aldol intermediate between steps (iv) and (ix). The kinetics of the condensation are first order in benzaldehyde, alkali, and ketone. The condensations were carried out using a large excess of benzaldehyde, and, since the alkali is not consumed, the reaction was then first order in ketone, and could be followed by measuring the development of a band in the ultraviolet spectrum at 292 mμ, which is typical of benzylidene ketones. The rates of condensation,

(CLXX.)

relative to an arbitrary value of 100 for lanost-8-en-3-one (CLXXI) are appended to the formulae of some of the many 3-ketones which have been investigated (Chart 2). In those molecules possessing an unsubstituted 4-position, or a second carbonyl group it was demonstrated that, under the conditions employed, condensation occurred only at $C_{(2)}$.

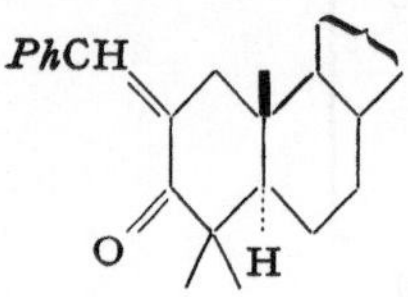

It is clear, from an examination of the data, that the rate differences are not explicable simply on the basis of electrostatic or of bond induction effects.

Large variations in the rates of condensation of steroidal 3-ketones are observed: it follows, therefore, that the part played by the axial $C_{(4)}$-methyl group in triterpenoids is, at most, of minor importance. This argues against any explanation of long range effects in which a major rôle is assigned to buttressing.

The simplest rationalisation of the data available is that long range effects have their origin in a distortion of bond angles caused by the introduction of unsaturated linkages, particularly ethylenic linkages. This is supported by the fact that the results obtained with steroids

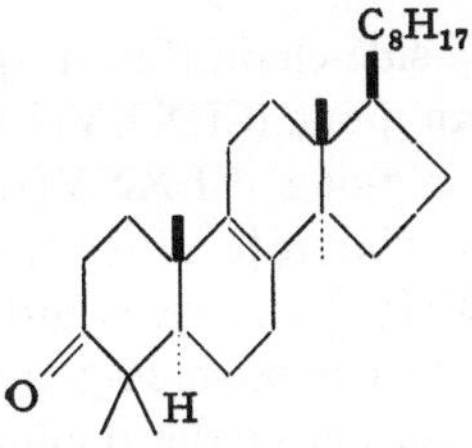

(CLXXI; 100.) Lanost-8-en-3-one.

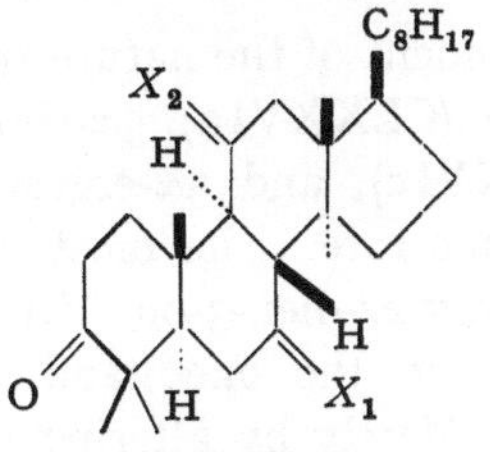

(CLXXII.)

(a) $X_1 = H_2$, $X_2 = H_2$; 55.
(b) $X_1 = O$, $X_2 = H_2$; 200.
(c) $X_1 = H_2$, $X_2 = O$; 43.
(d) $X_1 = O$, $X_2 = O$; 92.

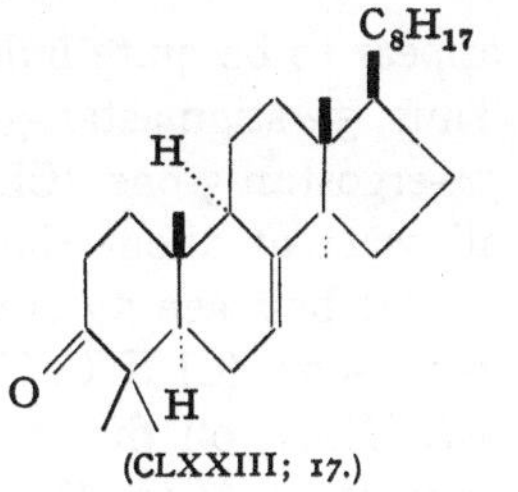

(CLXXIII; 17.)

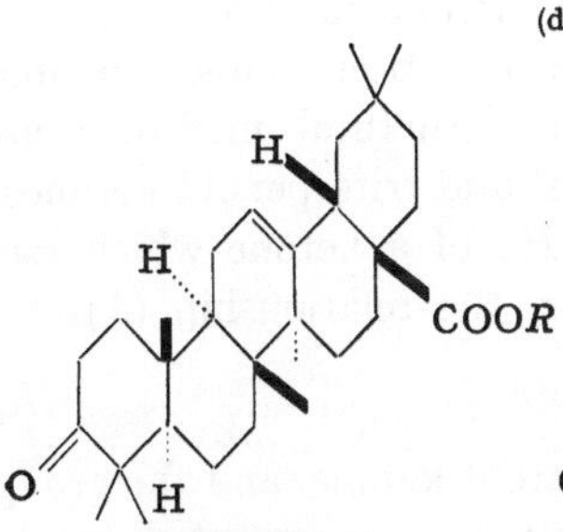

(CLXXIV.)

(a) $R = H$; 97.
(b) $R = CH_3$; 113.

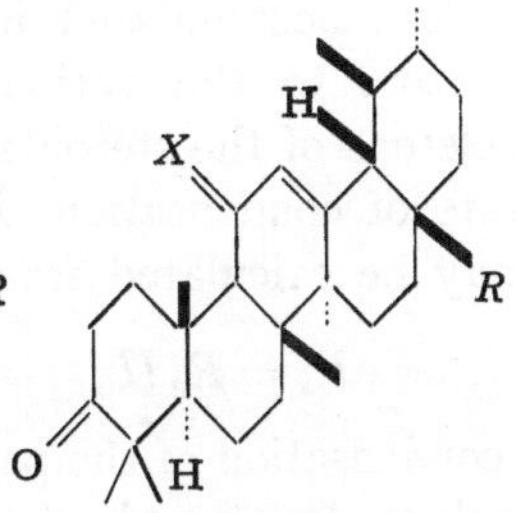

(CLXXV.)

(a) $R = COOH$, $X = H_2$; 91.
(b) $R = COOCH_3$, $X = H_2$; 111.
(c) $R = CH_3$, $X = H_2$; 100.
(d) $R = CH_3$, $X = O$; 75.
(e) $R = CH_3$; $X = CH_2$; 66.

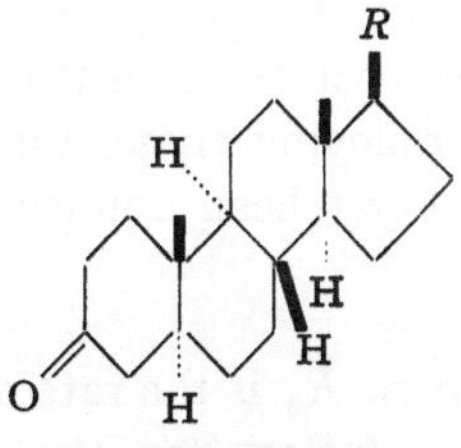

(CLXXVI.)

(a) $R = C_{10}H_{21}$; 180.
(b) $R = C_8H_{17}$; 182.
(c) $R = C_9H_{19}$; 188.
(d) $R = C_9H_{17}$; 188.

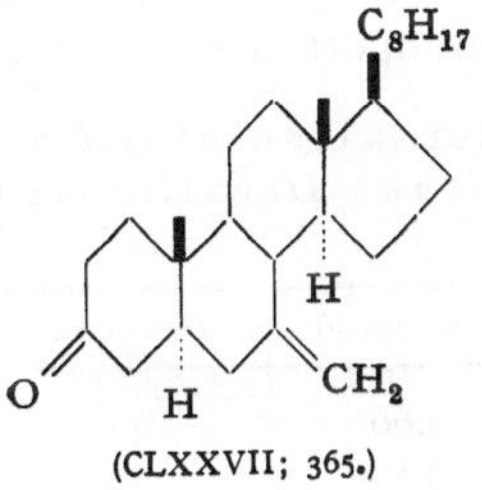

(CLXXVII; 365.)

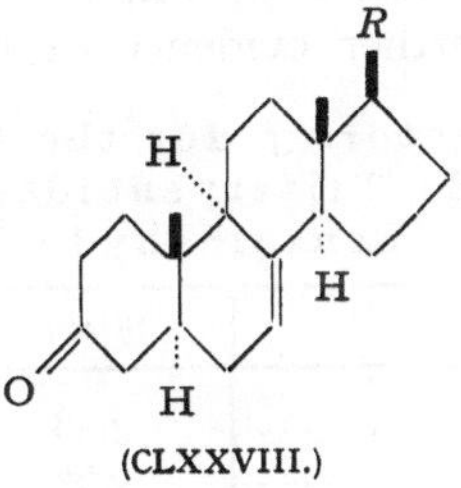

(CLXXVIII.)

(a) $R = C_9H_{17}$; 43.
(b) $R = C_9H_{19}$; 47.

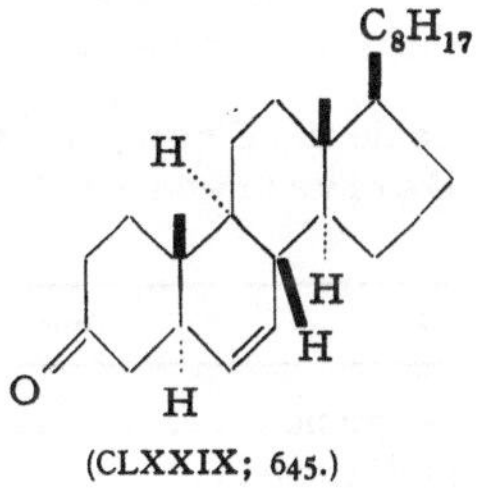

(CLXXIX; 645.)

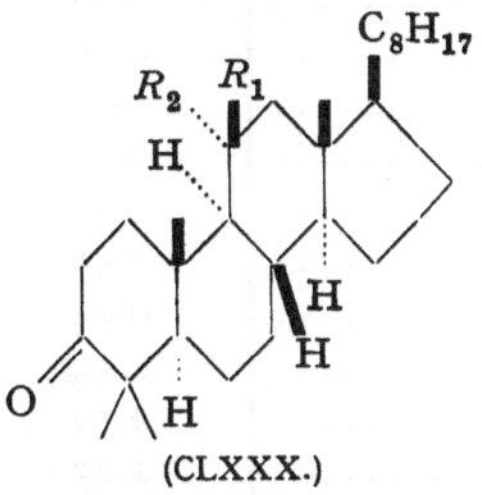

(CLXXX.)

(a) $R_1 = H$, $R_2 = OH$; 35.
(b) $R_1 = OH$, $R_2 = H$; 17.

Chart 2. Relative Rates of Condensation of Benzaldehyde with 3-Oxo Derivatives of Steroids and Triterpenoids.

appear to be quite independent of the nature of the side-chain (Chart 2). Thus, 5α-stigmastan-3-one (CLXXVIa), 5α-cholestan-3-one (CLXXVIb), 5α-ergostan-3-one (CLXXVIc), and 5α-ergost-22-en-3-one (CLXXVId) all react at about the same rate. The most spectacular rate difference is that between 5α-ergosta-7,22-dien-3-one (CLXXVIIIa) and 5α-ergost-7-en-3-one (CLXXVIIIb) on the one hand and 5α-cholest-6-en-3-one (CLXXIX) on the other. Merely by altering the position of the double bond from Δ^7 to Δ^6, the rate of condensation is increased by a factor of 15. Clearly, neither buttressing nor electrostatic forces could be responsible for the differing effects of these olefinic linkages.

From the extensive data accumulated it has been found possible to assign group rate factors to the various structural modifications incorporated into the skeletons of the steroidal and triterpenoid ketones. Employing these, the rate of condensation, R_s, of a ketone which has not yet been studied may be calculated from the relationship (A),

$$R_s = R_0 \mathop{\Pi}_{1}^{i} f_i \qquad\qquad (A)$$

where R_0 is the rate of condensation of the parent ketone, and the group rate factors for the various functional groups are represented by f_1, $f_2 \ldots f_i$. The group rate factors were determined by measurements made on monosubstituted ketones; the values obtained are set out in *Table 6*. The close correspondence between the f values observed for steroids and triterpenoids and, where structurally appropriate, for decalin derivatives, is further evidence of the unimportance of axial

Table 6. Group Rate Factors, f, for the Condensation of 3-Oxo Derivatives of Steroids, Triterpenoids and *trans*-Decalin with Benzaldehyde.

Group	Steroid	Triterpenoid	Decalinoid
7-Ketone	3.38	3.66	—
11-Ketone	0.62	0.78	—
12-Ketone	1.89	1.93	—
11α-Hydroxyl	0.67	0.64	—
11β-Hydroxyl	0.36	0.31	—
10-Methyl	—	3.40	4.20
4,4-Dimethyl	0.31	—	0.23
5(6)-Ethylenic linkage	1.0	1.0	—
6(7)-Ethylenic linkage	3.55	—	—
7(8)-Ethylenic linkage	0.24	0.31	0.23
8(9)-Ethylenic linkage	1.49	1.82	—
8(14)-Ethylenic linkage	0.50	—	—
9(11)-Ethylenic linkage	1.27	1.33	—
11(12)-Ethylenic linkage	2.18	—	—
14(15)-Ethylenic linkage	1.19	—	—
16(17)-Ethylenic linkage	1.14	—	—
7(8), 9(11)-Diene	0.53	0.80	—

buttressing of a $C_{(4)}$-methyl group. *Table 7* presents a comparison between the observed and calculated rates for three representative ketones. The agreement obtained, although good, is not perfect, nor would this be expected, since functional groups must interact with each other as well as with $C_{(2)}$.

Table 7. Observed and Calculated Rates of Condensation of 3-Oxo Steroids with Benzaldehyde
(Relative to lanost-8-en-3-one = 100).

Parent compound and its rate	Derivative	Appropriate group rate factors	Calculated rate	Observed rate
5α-Ergost-22-en-3-one (CLXXVI d) (188)	7,11-Dioxo-	7-(C=O); 3.38 11-(C=O); 0.62	393	360
5α-Ergost-22-en-3-one (CLXXVI d) (188)	7(8), 14(15)-Dehydro-	7(8)-(C=C); 0.24 14(15)-(C=C); 1.19	54	62
5α-Cholestan-3-one (CLXXVI b) (182)	4,4-Dimethyl- 5(6)-Dehydro-	4,4-Dimethyl; 0.31 5(6)-(C=C); 1.0	56	51

It seems probable that, as a deeper understanding of these long-range conformational effects is achieved, it will be possible to exert some control over chemical reactions by introducing substituents or unsaturation at points remote from the reacting centre. The large increase in the rate of condensation produced by the introduction of a 6(7)-double bond into 5α-cholestan-3-one suggests that the 2(3),6(7)-diene system in the corresponding enolate is geometrically favourable; it might therefore be predicted that the 3(4),7(8)-diene system, which is geometrically analogous would be similarly favoured. The discovery (*266, 321*) that methylation of 5α-cholest-7-en-3-one occurs preferentially at $C_{(4)}$ rather than at $C_{(2)}$, as is usual with *trans A/B* steroids can be construed as evidence that this is indeed the case.

The fact that benzaldehyde condenses with 5α-ergosta-7,22-dien-3-one (CLXXVIII a, p. 219) at position 2 (*49*) is easily reconciled with the methylation result if it is assumed that the benzaldehyde does condense at $C_{(4)}$ initially, but that the reverse step (iv, p. 218) is more favourable than at $C_{(2)}$, because of extra hindrance by the 6-methylene group. The final product is that in which condensation has occurred at $C_{(2)}$, only because at this position the elimination step (ix) is relatively favoured over the reversed aldol step (iv).

2. Reflex Interactions.

Ourisson and his colleagues (*69, 184, 289*) have recently investigated some of the consequences attendant upon the distortion of a hexa-carbocycle to accommodate 1 : 3-diaxial interactions. They have pointed out that in a molecule like 3,3,5,5-tetramethylcyclohexanone (CLXXX), the non-bonded interaction between the two axial methyl groups (see CLXXXI) can be relieved by a distortion of bond angles (as in CLXXXII).

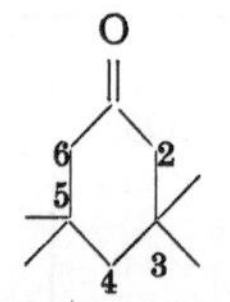

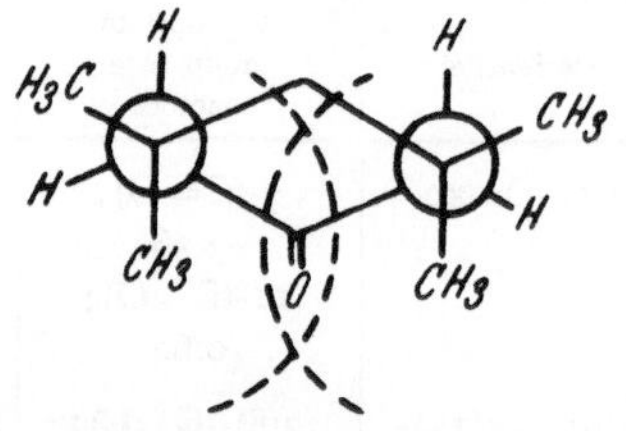

(CLXXX.) 3,3,5,5-Tetramethyl-   (CLXXXI.)   (CLXXXII.)
cyclohexanone.

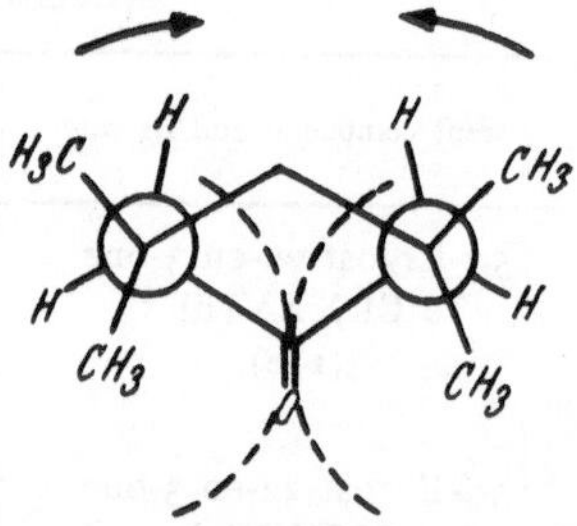

Such a distortion brings the axial hydrogen atoms at $C_{(2)}$ and $C_{(6)}$ closer to each other, and it is obvious that if one of these hydrogen atoms were to be replaced by a bulkier group, a serious non-bonded interaction would arise on the side of the ring opposite to the two axial methyl groups. The term *"reflex effect"* has been coined (*69, 289*) to describe such an enhancement of a 1 : 3-interaction by bond angle distortion which has its origin in the steric situation on the other side of the molecule.

This effect has been invoked (*289*) to explain the point of equilibrium of certain heavily substituted bromocyclohexanones. It has also been employed to account for the fact that, while friedelanone (CLXXXIII) contains none of its 4α-epimer at equilibrium (*289*), the *D*-homosteroid (CLXXXIV) is converted into its 17a-epimer to the extent of about 30% under equilibration conditions. Reference to the perspective formulae of 4-epifriedelanone (CLXXXV) and the 17a-epi-*D*-homosteroid (CLXXXVI) makes it clear that the repulsion between the axial methyl groups at $C_{(5)}$ and $C_{(9)}$ in the former gives rise to a destabilising reflex interaction between the axial methyl group at $C_{(4)}$ and the hydrogen attached to $C_{(10)}$; in the latter the 1 : 3-diaxial ($CH_3 : CH_3$) interaction on the β-face of the molecule has been replaced by a ($CH_3 : H$) interaction, and the reflex effect is correspondingly diminished. The operation of a reflex effect in certain steroidal hydrindanones has been suggested by Biellmann, Francetić and Ourisson (*68*) as a possible rationale for the relative stabilities of the *cis* and *trans* isomers of these compounds. The fact that a 3 : 5-oxide is obtained by base treatment of a 5α-hydroxy-3β-tosyl-

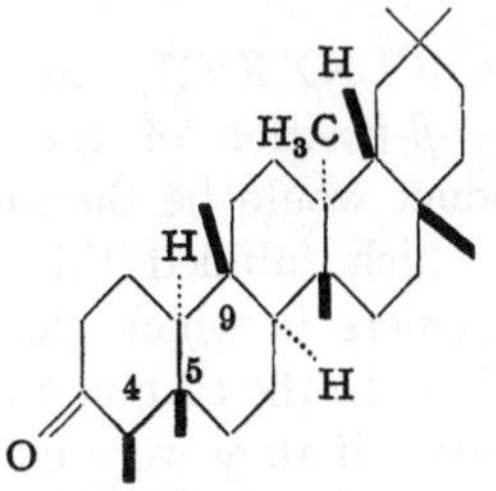

(CLXXXIII.) Friedelanone.

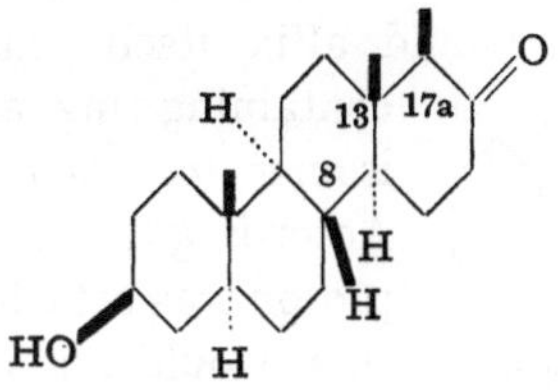

(CLXXXIV.)

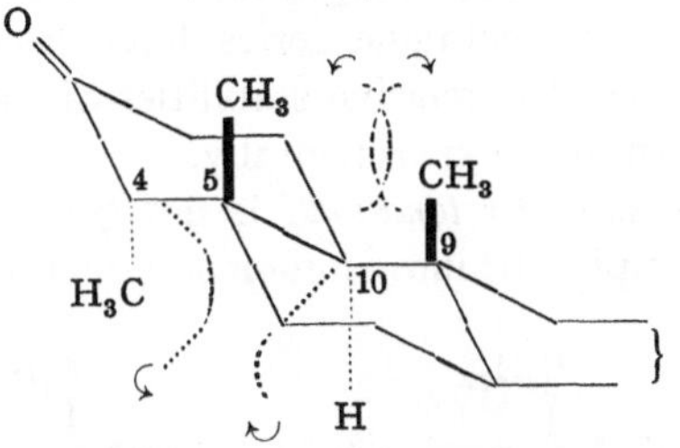

(CLXXXV.) 4-Epifriedelanone.

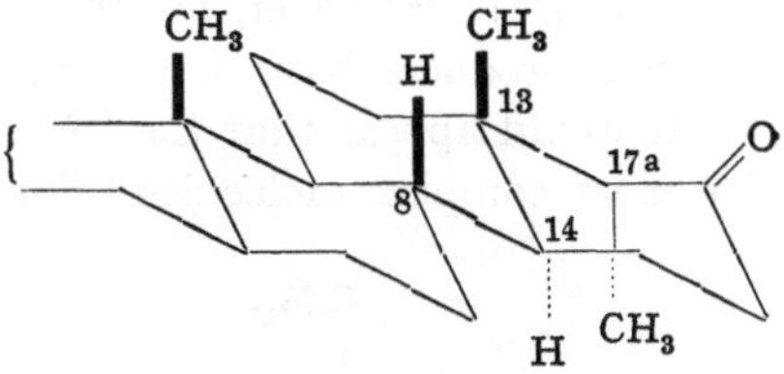

(CLXXXVI.)

oxysteroid, while a 5β-hydroxy-3α-tosyloxysteroid, does not yield an analogous compound under the same conditions (*90*; see also *197*) may be similarly explained.

X. Carbanion Reduction Processes.

In general, reductions of ketones or oximes with sodium and alcohol give mixtures of alcohols or amines, in which the stable epimer predominates (*31, 32, 37, 125, 223*). OURISSON and RASSAT (*271*), however, have pointed out that in the [2,2,1]-bicycloheptan-2-one series the *endo*-alcohol is invariably the major product, whether or not it is the more thermodynamically stable one.

Work in the steroid and triterpenoid fields led to the generalisation (*34, 53*; see also *71, 108, 281*) that all reductions by alkali metals and proton donors proceeding through carbanions—including reductions of conjugated dienes, α,β-unsaturated ketones, aromatic systems, and halides—give mainly the more stable product. This is most simply explained if it is assumed that a carbanion has a definite, though easily inverted tetrahedral configuration, and that the steric requirements of an electron pair are intermediate between those of a C—H bond and those of a C—C bond.

Recently, it has been proposed (*311*) that the stereochemical course of the lithium-ammonia reduction of α,β-unsaturated ketones in the octalone series is determined by the necessity for the proton-donor to attack the intermediate carbanion in such a way as to maintain maximum orbital overlap in the transition state. A consequence of this would be that, in the reduction of enones like (CLXXXVII), the incoming proton would affix itself axially at the β-position of the carbonyl-containing ring, and the product would be the more stable isomer (cis- or trans-fused) which fulfilled this condition. According to this theory, isomers in which the incoming proton was attached equatorially to the carbonyl-containing ring, would not be formed, even if they were energetically very favoured. In support of these proposals, the results of some metal ammonia reductions in the octalone series have been quoted (311). However, in the cases cited, the relative stabilities of the possible products have not been determined experimentally.

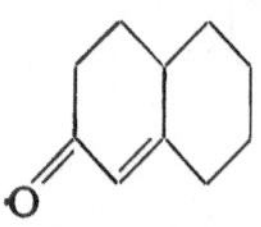

(CLXXXVII.)

It would appear that the "*Rule of Axial Protonation*" is not general for more complex molecules. For example, lithium-ammonia reduction

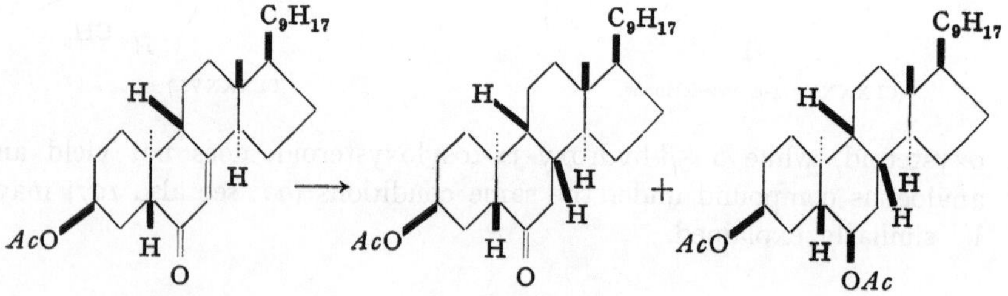

(CLXXXVIII.)   (CLXXXIX.)   (CXC.)

of (CLXXXVIII), followed by re-acetylation, gave as sole ketonic product (CLXXXIX) in a yield of 45%, together with 23% of the corresponding diacetate (CXC) (257); the structure of the latter compound, however, does not seem to have been firmly established. In this case, the attacking proton has approached the carbonyl-containing ring

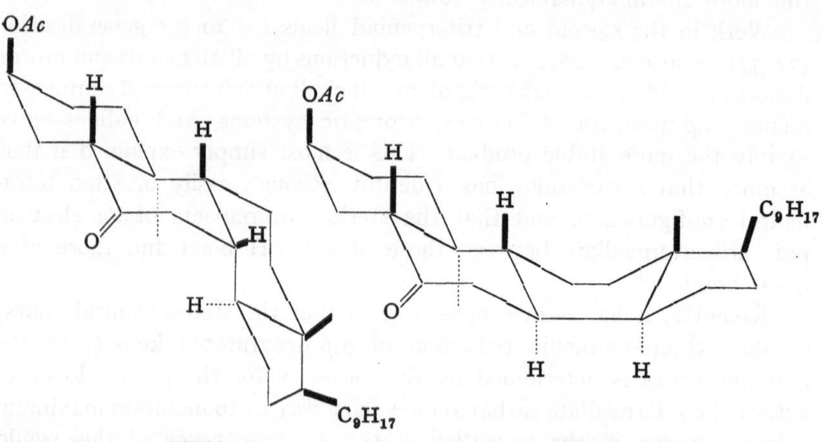

(CXCI.)   (CXCII.)

equatorially (see CXCI). Axial attack would have led to an 8α-configuration in the product, and necessitated a boat conformation for ring C (see CXCII).

It is probable that the mechanism of alkali metal—ammonia reductions is more complex than has been supposed in the past. In particular, the intermediacy, in certain reductions, of species containing C-metal bonds, which might then suffer electrophilic displacement by a proton with retention of configuration (cf. *89, 210, 212, 336–338*), merits consideration. Such a concept would account for the fact that, in the reduction of camphor with an alkali metal and ammonia (*271*), the proportion of isoborneol (the unstable, *exo*-isomer) which is obtained, increases with increasing size of the metal atoms.

References.

1. ABD EL REHIM, A. M. and C. H. CARLISLE: The Structure of Methyl Oleanolate Iodoacetate. An X-Ray Determination. Chem. and Ind. **1954**, 279.

2. AEBLI, H., C. A. GROB und E. SCHUMACHER: Die Isomerisierung von 5-Methyl-19-nor-koprosten-Derivaten. I. Teil. Die Irreversibilität der Westphalen-Lettré-Umlagerung. Helv. Chim. Acta **41**, 774 (1958).

3. ALBRECHT, R. und CH. TAMM: Reduktion von Cholesten-(1)-on-(3) mit Metall-hydriden und mit Aluminiumisopropylat. Helv. Chim. Acta **40**, 2216 (1957).

4. ALLAN, G. G., M. B. E. FAYEZ, F. S. SPRING and R. STEVENSON: Triterpenoids. Part XLVII. The Constitution of Some Compounds Obtained by the Dehydration of β-Amyrin and Related Alcohols. Further Observations on the Stereochemistry of α-Amyrin. J. Chem. Soc. (London) **1956**, 456.

5. ALLAN, G. G. and F. S. SPRING: Triterpenoids. Part XXXVI. Reactions of 18α-Olean-12-en-3β-ol Derivatives and Observations on the Stereochemistry of α-Amyrin. J. Chem. Soc. (London) **1955**, 2125.

6. ALLINGER, N. L.: The Relative Stabilities of *Cis* and *Trans* Isomers. Experientia **10**, 328 (1954).

7. — The Relative Stabilities of *cis* and *trans* Isomers. II. The Decalin and Hydrindan Ring Systems. J. Organ. Chem. (USA) **21**, 915 (1956).

8. — The Relative Stabilities of *cis* and *trans* Isomers. III. The Cyclodecenes. J. Amer. Chem. Soc. **79**, 3443 (1957).

9. — Conformational Analysis. III. Applications to Some Medium Ring Compounds. J. Amer. Chem. Soc. **81**, 5727 (1959).

10. ALLINGER, N. L. and J. ALLINGER: The Conformers of 2-Bromo*cyclo*hexanone. Tetrahedron **2**, 64 (1958).

11. — — Conformational Analysis. II. The 2-Bromo-4-*t*-butylcyclohexanones. J. Amer. Chem. Soc. **80**, 5476 (1958).

12. ALLINGER, N. L., J. ALLINGER, L. A. FREIBERG, R. F. CZAJA and N. A. LeBEL: Conformational Analysis. XI. The Conformers of 2-Chlorocyclohexanone. J. Amer. Chem. Soc. **82**, 5876 (1960).

13. ALLINGER, N. L., J. ALLINGER, L. E. GELLER and C. DJERASSI: Conformational Analysis. VI. Optical Rotatory Dispersion Studies. XXVII. Quantitative Studies of an α-Haloketone by the Rotatory Dispersion Method. J. Organ. Chem. (USA) **25**, 6 (1960).

14. ALLINGER, N. L. and J. L. COKE: The Relative Stabilities of *cis* and *trans* Isomers. VI. The Decalins. J. Amer. Chem. Soc. **81**, 4080 (1959).

15. Allinger, N. L. and J. L. Coke: The Relative Stabilities of *cis* and *trans* Isomers. VII. The Hydrindanes. J. Amer. Chem. Soc. 82, 2553 (1960).
16. Allinger, N. L. and L. A. Freiberg: Conformational Analysis. X. The Energy of the Boat Form of the Cyclohexane Ring. J. Amer. Chem. Soc. 82, 2393 (1960).
17. Allinger, N. L., R. B. Hermann and C. Djerassi: Relative Stabilities of *cis* and *trans* Isomers. VIII. Optical Rotatory Dispersion Studies. XXXIV. Kinetic and Equilibrium Measurements on Some Steroidal Hydrindanones. J. Organ. Chem. (USA) 25, 922 (1960).
18. Allsop, I. L., A. R. H. Cole, D. E. White and R. L. S. Willix: Infrared Spectra of Natural Products. Part VI. The Characterization of Equatorial and Axial 3-Hydroxyl Groups in Triterpenoids. J. Chem. Soc. (London) 1956, 4868.
19. Alt, G. H. and D. H. R. Barton: Some Conformational Aspects of Neighbouring-group Participation. J. Chem. Soc. (London) 1954, 4284.
20. Ames, T. R., J. L. Beton, A. Bowers, T. G. Halsall and E. R. H. Jones: The Chemistry of the Triterpenes and Related Compounds. Part XXIII. The Structure of Taraxasterol, ψ-Taraxasterol (Heterolupeol), and Lupenol-I. J. Chem. Soc. (London) 1954, 1905.
21. Anagnostopoulos, C. E. and L. F. Fieser: Nitration of Unsaturated Steroids. J. Amer. Chem. Soc. 76, 532 (1954).
22. Angyal, S. J. and D. J. McHugh: Interaction Energies of Axial Hydroxyl Groups. Chem. and Ind. 1956, 1147.
23. Angyal, S. J. and R. J. Young: Glycol Fission in Rigid Systems. II. The Cholestane-3β,6,7-triols. Existence of a Cyclic Intermediate. J. Amer. Chem. Soc. 82, 5251 (1959).
24. Anliker, R., O. Rohr und H. Heusser: Über Steroide und Sexualhormone. 205. Mitt. Über weitere Umlagerungen in den Ringen C und D der Steroide. Helv. Chim. Acta 38, 1171 (1955).
25. Arnold, R. T. and M. J. Danzig: Thermal Decarboxylation of Unsaturated Acids. II. J. Amer. Chem. Soc. 79, 892 (1957).
26. Auwers, K. v.: Über Beziehungen zwischen Konstitution und physikalischen Eigenschaften hydroaromatischer Verbindungen. Liebigs Ann. Chem. 420, 84 (1920).
27. Barnes, C. S.: Methylsteroids. V. The Facilitation of Hydrolysis of Sterically Hindered Acetoxy Groups by Carbonyl Groups in the Lanosterol Series. Austral. J. Chem. 11, 546 (1958).
28. Barnes, C. S. and A. Palmer: Methylsteroids. I. Some Observations on 3-Methyl-substituted Steroids Derived from Cholesterol. Austral. J. Chem. 9, 105 (1956).
29. Barton, D. H. R.: Interactions between Non-bonded Atoms, and the Structure of *cis*-Decalin. J. Chem. Soc. (London) 1948, 340.
30. — *cis*-Elimination in Thermal Decompositions. J. Chem. Soc. (London) 1949, 2174.
31. — The Conformation of the Steroid Nucleus. Experientia 6, 316 (1950).
32. — The Stereochemistry of *cyclo*Hexane Derivatives. J. Chem. Soc. (London) 1953, 1027.
33. — Conformation et réactivité des structures. Bull. soc. chim. France 1956, 973.
34. — Some Recent Progress in Conformational Analysis. In: Theoretical Organic Chemistry, p. 127. London: Butterworths Sci. Publ. 1959.
35. Barton, D. H. R. and C. J. W. Brooks: Triterpenoids. Part I. Morolic Acid, A New Triterpenoid Sapogenin. J. Chem. Soc. (London) 1951, 257.

36. BARTON, D. H. R., A. DA S. CAMPOS-NEVES and R. C. COOKSON: The 3-Methyl-cholestanols and their Derivatives. J. Chem. Soc. (London) **1956**, 3500.

37. BARTON, D. H. R. and R. C. COOKSON: The Principles of Conformational Analysis. Quart. Rev. Chem. Soc. (London) **10**, 44 (1956).

38. BARTON, D. H. R., R. C. COOKSON, W. KLYNE and C. W. SHOPPEE: The Conformation of cycloHexene. Chem. and Ind. **1954**, 21.

39. BARTON, D. H. R., R. M. EVANS, J. C. HAMLET, P. G. JONES and T. WALKER: Studies in the Synthesis of Cortisone. Part VII. The Preparation of $3\beta : 17\alpha$-Dihydroxyallopregnane-11 : 20-dione. J. Chem. Soc. (London) **1954**, 747.

40. BARTON, D. H. R., O. HASSEL, K. S. PITZER and V. PRELOG: Nomenclature of cycloHexane Bonds. Nature (London) **172**, 1096 (1953).

41. BARTON, D. H. R. and A. J. HEAD: Long-range Effects in Alicyclic Systems. Part I. The Rates of Rearrangement of Some Steroidal Dibromides. J. Chem. Soc. (London) **1956**, 932.

42. BARTON, D. H. R., A. J. HEAD and P. J. MAY: Long-range Effects in Alicyclic Systems. Part II. The Rates of Condensation of Some Triterpenoid Ketones with Benzaldehyde. J. Chem. Soc. (London) **1957**, 935.

43. BARTON, D. H. R., A. J. HEAD and R. J. WILLIAMS: Stereospecificity in Thermal Elimination Reactions. Part III. The Pyrolysis of (—)-Menthyl Benzoate. J. Chem. Soc. (London) **1953**, 1715.

44. BARTON, D. H. R. and N. J. HOLNESS: Triterpenoids. Part V. Some Relative Configurations in Rings C, D and E of the β-Amyrin and the Lupeol Group of Triterpenoids. J. Chem. Soc. (London) **1952**, 78.

45. BARTON, D. H. R. and J. F. KING: The Generalised Diaxial → Diequatorial Rearrangement. J. Chem. Soc. (London) **1958**, 4398.

46. BARTON, D. H. R. and W. KLYNE: Abstract of Chemical Society Meeting. Chem. and Ind. **1953**, 1386.

47. BARTON, D. H. R. and G. F. LAWS: Some Oxidation Products of Ergosta-7 : 14 : 22-trien-3β-yl Acetate (Ergosterol B_3 Acetate). J. Chem. Soc. (London) **1954**, 52.

48. BARTON, D. H. R., D. A. LEWIS and J. F. McGHIE: Conformational Anomalies in Some Triterpenoid Bromo-ketones. J. Chem. Soc. (London) **1957**, 2907.

49. BARTON, D. H. R., F. McCAPRA, P. J. MAY and F. THUDIUM: Long-range Effects in Alicyclic Systems. Part III. The Relative Rates of Condensation of Some Steroid and Triterpenoid Ketones with Benzaldehyde. J. Chem. Soc. (London) **1960**, 1297.

50. BARTON, D. H. R., J. F. McGHIE, M. K. PRADHAN and S. A. KNIGHT: The Constitution and Stereochemistry of Euphol. J. Chem. Soc. (London) **1955**, 876.

51. BARTON, D. H. R. and E. MILLER: Stereochemistry of the Cholesterol Dibromides. J. Amer. Chem. Soc. **72**, 1066 (1950).

52. BARTON, D. H. R., J. E. PAGE and C. W. SHOPPEE: Infrared Absorption of Halogeno-steroids. J. Chem. Soc. (London) **1956**, 331.

53. BARTON, D. H. R. and C. H. ROBINSON: The Stereospecificity of Carbanion Reduction Processes. J. Chem. Soc. (London) **1954**, 3045.

54. BARTON, D. H. R. and W. J. ROSENFELDER: The Application of the Method of Molecular-rotation Differences to Steroids. Part XII. Cholest-6-en-3β-ol. J. Chem. Soc. (London) **1949**, 2459.

55. — — The Stereochemistry of Steroids. Part IV. The Concept of Equatorial and Polar Bonds. J. Chem. Soc. (London) **1951**, 1048.

56. BARTON, D. H. R. and C. SCHUERCH: Unpublished results.

57. BASTIANSEN, O. and O. HASSEL: Structure of the So-called cis-Decalin. Nature (London) **157**, 765 (1946).

58. Bastiansen, O. and O. Hassel: The Molecular Structure of the So-called "cis" Decalin. Tidsskr. Kjemi, Bergvesen Metall. 6, 70 (1946).
59. Baumgartner, G. und Ch. Tamm: 3α,12α-Dioxy-11-ketocholansäure. Helv. Chim. Acta 38, 441 (1955).
60. Beaton, J. M. and F. S. Spring: Triterpenoids. Part XLII. The Configuration of the Carboxyl Group in Glycyrrhetic Acid. J. Chem. Soc. (London) 1955, 3126.
61. Beaton, J. M., F. S. Spring, R. Stevenson and W. S. Strachan: Triterpenoids. Part XXXIX. The Constitution and Stereochemistry of the Ursane Group of Triterpenoids. J. Chem. Soc. (London) 1955, 2610.
62. Beckett, C. W., N. K. Freeman and K. S. Pitzer: The Thermodynamic Properties and Molecular Structure of Cyclopentene and Cyclohexene. J. Amer. Chem. Soc. 70, 4227 (1948).
63. Beckett, C. W., K. S. Pitzer and R. Spitzer: The Thermodynamic Properties and Molecular Structure of Cyclohexane, Methylcyclohexane, Ethylcyclohexane and the Seven Dimethylcyclohexanes. J. Amer. Chem. Soc. 69, 2488 (1947).
64. Benkeser, R. A. and J. J. Hazdra: Factors Influencing the Direction of Elimination in the Chugaev Reaction. J. Amer. Chem. Soc. 81, 228 (1959).
65. Berlin, A. J. and F. R. Jensen: Variation of Conformational Interactions with Radii and Polarisabilities. The Conformational Equilibria of the Cyclohexanes. Chem. and Ind. 1960, 998.
66. Bernstein, S., R. H. Lenhard and J. H. Williams: Steroidal Cyclic Ketals. V. Transformation Products of Adrenosterone. The Synthesis of Related $C_{19}O_3$-Steroids. J. Organ. Chem. (USA) 18, 1166 (1953).
67. Beton, J. L., T. G. Halsall, E. R. H. Jones and P. C. Phillips: The Chemistry of the Triterpenes and Related Compounds. Part XXX. The Relative Stabilities of Ring-A Unsaturated Hydrocarbons Derived from 3:4-Dimethylcholestane and 3-Methyl-24-norurs-12-ene. J. Chem. Soc. (London) 1957, 753.
68. Biellmann, J. F., D. Francetić et G. Ourisson: Effets conformationnels sur l'équilibre cis-trans d'α-hydrindanones. Tetrahedron Letters 1960, No. 18, 4.
69. Biellmann, J. F., R. Hanna, G. Ourisson, C. Sandris et B. Waegell: Étude d'interactions 1—3. Bull. soc. chim. France 1960, 1429.
70. Biellmann, J. F. et G. Ourisson: Stéréochimie de la contraction du cycle A de triterpènes et de corps apparentés. Bull. soc. chim. France 1960, 348.
71. Birch, A. J., H. Smith and R. E. Thornton: Reduction by Dissolving Metals. Part XIV. Some Stereochemical Aspects of the Reduction of αβ-Unsaturated Ketones. J. Chem. Soc. (London) 1957, 1339.
72. Bird, C. W. and R. C. Cookson: Control of Basic Strength by Steric Hindrance to Solvation of Ammonium Ions. Chem. and Ind. 1955, 1479.
73. — — Linear Free-Energy Relations in the Steroid Series. Basic Strengths of Aminocholestanes. J. Chem. Soc. (London) 1960, 2343.
74. Bird, C. W., R. C. Cookson and S. H. Dandegaonker: Absorption Spectra of Ketones. Part V. γ-Substituted αβ-Unsaturated Ketones. J. Chem. Soc. (London) 1956, 3675.
75. Bladon, P.: Studies in the Steroid Group. Part LXVIII. Epidioxides Derived from Lumisterol and Related Compounds. J. Chem. Soc. (London) 1955, 2176.
76. Bladon, P., H. B. Henbest, E. R. H. Jones, B. J. Lovell, G. W. Wood, G. F. Woods, J. Elks, R. M. Evans, D. E. Hathway, J. F. Oughton and G. H. Thomas: Studies in the Steroid Group. Part LXII. Studies in the Synthesis of Cortisone. Part I. A Novel Route to 11-Ketosteroids. J. Chem. Soc. (London) 1953, 2921.

77. BORDWELL, F. G. and R. J. KERN: Elimination Reactions in Cyclic Systems. I. *cis* Eliminations in the Cyclohexane and Cyclopentane Series. J. Amer. Chem. Soc. **77**, 1141 (1955).

78. BORDWELL, F. G. and P. S. LANDIS: Elimination Reactions. VIII. A *trans* Chugaev Elimination. J. Amer. Chem. Soc. **80**, 2450 (1958).

79. BORDWELL, F. G. and M. L. PETERSON: Elimination Reactions in Cyclic Systems. II. *cis* Eliminations Promoted by the Sulfonate Group. J. Amer. Chem. Soc. **77**, 1145 (1955).

80. BOSE, A. K.: The Stereochemistry of the Reaction of Nitrous Acid with Cyclohexylamines. Experientia **9**, 256 (1953).

81. BOWERS, A., E. DENOT, R. URQUIZA and L. M. SANCHEZ-HIDALGO: Steroids. CXXVI. Some Fission Reactions of Steroid 5,6-Epoxides Induced by Boron Trifluoride Etherate. Tetrahedron **8**, 116 (1960).

82. BOWERS, A., L. C. IBÁÑEZ and H. J. RINGOLD: Steroids. CXX. Synthesis of Halogenated Steroid Hormones. New Routes to 6α-Fluorotestosterone and the 6α- and 6β-Fluoro Analogs of Progesterone. The Synthesis of 6α- and 6β-Fluoro Reichstein's Compound "S" and 6α- and 6β-Fluorodesoxycorticosterone Acetate. Tetrahedron **7**, 138 (1959).

83. BREWSTER, J. H.: The Configuration of Atrolactic Acid. Retention of Configuration in the Acid-catalyzed Ring Opening of Stilbene Oxides. J. Amer. Chem. Soc. **78**, 4061 (1956).

84. BROOKS, R. V., W. KLYNE and E. MILLER: The Separation of Steroid Alcohols by Chromatography of their Benzoates on Alumina. Biochemic. J. **54**, 212 (1953).

85. BROOKS, S. G., J. S. HUNT, A. G. LONG and B. MOONEY: Studies in the Synthesis of Cortisone. Part XIX. Paper Chromatography of Some Steroidal 11 : 12-Diols and -Ketols. J. Chem. Soc. (London) **1957**, 1175.

86. BROWNLIE, G. and F. S. SPRING: Triterpenoids. Part XLIX. The Constitution of the Ester $C_{33}H_{46}O_7$ Obtained by Oxidation of Methyl Glycyrrhetate Acetate. J. Chem. Soc. (London) **1956**, 1949.

87. CASTELLS, J., G. A. FLETCHER, E. R. H. JONES, G. D. MEAKINS and R. SWINDELLS: Steroids of Unnatural Configuration. Part II. Reduction Products of Lumisterol: Hexahydrocompounds. J. Chem. Soc. (London) **1960**, 2627.

88. CAVELL, E. A. S., N. B. CHAPMAN and M. D. JOHNSON: Conformation and Reactivity. Part I. Kinetics of the Alkaline Hydrolysis of the Methyl Cyclohexane-mono- and -di-carboxylates and 4-t-Butylcyclohexanecarboxylates. J. Chem. Soc. (London) **1960**, 1413.

89. CHARMAN, H. B., E. D. HUGHES and Sir C. K. INGOLD: Mechanism of Electrophilic Substitution at a Saturated Carbon Atom. Part II. Kinetics, Stereochemistry, and Mechanism of the Two-alkyl Mercury-exchange Reaction. J. Chem. Soc. (London) **1959**, 2530.

90. CLAYTON, R. B., H. B. HENBEST and M. SMITH: Aspects of Stereochemistry. Part V. Reactions of *cyclo*Hexane-1 : 3-diol Monoarenesulphonates with Alkali. J. Chem. Soc. (London) **1957**, 1982.

91. COCKER, W. and T. B. H. McMURRY: Stereochemical Relationships in the Eudesmane (Selinane) Group of Sesquiterpenes. Tetrahedron **8**, 181 (1960).

92. COLE, A. R. H.: Infrared Spectra of Natural Products. Fortschr. Chem. organ. Naturstoffe **13**, 1 (1956).

93. COLE, A. R. H., P. R. JEFFERIES and G. T. A. MÜLLER: Infrared Spectra of Natural Products. Part X. Conformations and Infrared Spectra of Substituted *cyclo* Hexanols. J. Chem. Soc. (London) **1959**, 1222.

94. COLE, A. R. H. and A. J. MICHELL: Infrared Spectra of Natural Products. Part XII. Triterpenoid and Diterpenoid Carboxylic Acids. J. Chem. Soc. (London) **1959,** 2005.

95. COLE, A. R. H. and G. T. A. MÜLLER: Infrared Spectra of Natural Products. Part XI. Intramolecular Hydrogen-bonding and Stereochemistry of Triterpenoid Diols and Related Compounds. J. Chem. Soc. (London) **1959,** 1224.

96. COLE, A. R. H., G. T. A. MÜLLER, D. W. THORNTON and R. L. S. WILLIX: Infrared Spectra of Natural Products. Part IX. Frequencies and Intensities of Hydroxyl Absorption Bands in Triterpenoids and Similar Compounds. J. Chem. Soc. (London) **1959,** 1218.

97. COOKSON, R. C.: The Stereochemistry of Alkaloids. Chem. and Ind. **1953,** 337.

98. — Absorption Spectra of Ketones. Part I. α-Halogenoketones. J. Chem. Soc. (London) **1954,** 282.

99. COOKSON, R. C. and S. H. DANDEGAONKER: Absorption Spectra of Ketones. Part II. The Configuration of Some Bromoderivatives of 6-Oxocholestanyl Acetate. Absorption Spectra of α-Ketols. J. Chem. Soc. (London) **1955,** 352.

100. COOKSON, R. C. and J. HUDEC: The Stereochemistry of Acid-catalysed Opening of Styrene Oxides. Proc. Chem. Soc. (London) **1957,** 24.

101. COPE, A. C. and E. M. ACTON: Amine Oxides. V. Olefins from N,N-Dimethylmenthylamine and N,N-Dimethylneomenthylamine Oxides. J. Amer. Chem. Soc. **80,** 355 (1958).

102. COREY, E. J.: Prediction of the Stereochemistry of α-Brominated Ketosteroids. Experientia **9,** 329 (1953).

103. — The Stereochemistry of α-Haloketones. I. The Molecular Configurations of Some Monocyclic α-Halocyclanones. J. Amer. Chem. Soc. **75,** 2301 (1953).

104. — The Stereochemistry of α-Haloketones. V. Prediction of the Stereochemistry of α-Brominated Ketosteroids. J. Amer. Chem. Soc. **76,** 175 (1954).

105. COREY, E. J. and H. J. BURKE: The Stereochemistry of α-Haloketones. VII. The Stereochemistry and Spectra of Some α-Chlorocyclohexanones. J. Amer. Chem. Soc. **77,** 5418 (1955).

106. COREY, E. J. and E. W. CANTRALL: Proof of the Structure and Stereochemistry of α-Amyrin.by Synthesis from a β-Amyrin Derivative, Glycyrrhetic Acid. J. Amer. Chem. Soc. **81,** 1745 (1959).

107. COREY, E. J., M. G. HOWELL, A. BOSTON, R. L. YOUNG and R. A. SNEEN: Spectral and Stereochemical Studies with Deuterated Cyclohexanes. J. Amer. Chem. Soc. **78,** 5036 (1956).

108. COREY, E. J. and R. A. SNEEN: The Stereochemistry of the 3-Carboxy-, 3-Carboxymethyl- and 3-Acetylcholestanes and Δ⁵-Cholestenes. J. Amer. Chem. Soc. **75,** 6234 (1953).

109. — — Calculation of Molecular Geometry by Vector Analysis. Application to Six-membered Alicyclic Rings. J. Amer. Chem. Soc. **77,** 2505 (1955).

110. — — Stereoelectronic Control in Enolization-Ketonization Reactions. J. Amer. Chem. Soc. **78,** 6269 (1956).

111. COREY, E. J. and J. J. URSPRUNG: The Stereochemistry of the α-Amyrins. Chem. and Ind. **1954,** 1387.

112. — — The Stereochemistry of the α-Amyrins. J. Amer. Chem. Soc. **78,** 183 (1956).

113. — — The Structure of the Triterpenes Friedelin and Cerin. J. Amer. Chem. Soc. **78,** 5041 (1956).

114. CORNUBERT, R.: Le principe de la prééminence des liaisons équatoriales chez les dérivés cyclohexaniques. Bull. soc. chim. France **1956,** 996.

115. CREMLYN, R. J. W., D. L. GARMAISE and C. W. SHOPPEE: Steroids and Walden Inversion. Part X. The Reconversion of D-Homosteroids into Steroids. J. Chem. Soc. (London) 1953, 1847.

116. CUMMINS, E. G. and J. E. PAGE: Studies in the Synthesis of Cortisone. Part XX. The Infrared Absorption of α-Halogeno-oxo-steroids. J. Chem. Soc. (London) 1957, 3847.

117. DAUBEN, W. G.: Stereochemistry in Polycyclic Systems. Bull. soc. chim. France 1960, 1338.

118. DAUBEN, W. G., E. J. BLANZ, J. JIU and R. A. MICHELI: The Stereochemistry of the Hydride Reduction of Some Steroidal Ketones. J. Amer. Chem. Soc. 78, 3752 (1956).

119. DAUBEN, W. G., G. A. BOSWELL and G. H. BEREZIN: The Configuration of B-Norsteroid Derivatives. J. Amer. Chem. Soc. 81, 6082 (1959).

120. DAUBEN, W. G. and R. E. BOZAK: Lithium Aluminum Hydride Reduction of Methylcyclohexanones. J. Organ. Chem. (USA) 24, 1596 (1959).

121. DAUBEN, W. G. and G. J. FONKEN: Reactions of B-Norcholesterol. J. Amer. Chem. Soc. 78, 4736 (1956).

122. DAUBEN, W. G., G. J. FONKEN and D. S. NOYCE: The Stereochemistry of Hydride Reductions. J. Amer. Chem. Soc. 78, 2579 (1956).

123. DAUBEN, W. G. and J. JIU: The Stereochemistry and Reactivity of the cis-5-Hydrindanyl Derivatives. J. Amer. Chem. Soc. 76, 4426 (1954).

124. DAUBEN, W. G., R. A. MICHELI and J. F. EASTHAM: The Reduction of Steroidal Enol Acetates with Lithium Aluminum Hydride and Sodium Borohydride. J. Amer. Chem. Soc. 74, 3852 (1952).

125. DAUBEN, W. G. and K. S. PITZER: Conformational Analysis. In: M. S. NEWMAN, Steric Effects in Organic Chemistry, p. 1. New York: J. Wiley and Sons. 1956.

126. DAUBEN, W. G., R. C. TWEIT and R. L. MacLEAN: The Configuration and Reactivity of 9-Substituted Decalins. J. Amer. Chem. Soc. 77, 48 (1955).

127. DAUBEN, W. G., R. C. TWEIT and C. MANNERSKANTZ: Decahydronaphthoic Acids and their Relationship to the Decalols and Decalylamines. A Stereochemical Study of the Reaction of Nitrous Acid with Decalylamines. J. Amer. Chem. Soc. 76, 4420 (1954).

128. DAVIES, G. F. and E. C. GILBERT: The Heat of Combustion of cis- and trans-Decahydronaphthalene. J. Amer. Chem. Soc. 63, 1585 (1941).

129. DePUY, C. H. and R. W. KING: Pyrolytic Cis Eliminations. Chem. Rev. 60, 431 (1960).

130. DIPPY, J. F. J., S. R. C. HUGHES and J. W. LAXTON: Chemical Constitution and the Dissociation Constants of Monocarboxylic Acids. Part XIV. Mono-methylcyclohexanecarboxylic Acids. J. Chem. Soc. (London) 1954, 4102.

131. DJERASSI, C.: Optical Rotatory Dispersion. New York: McGraw-Hill. 1960.

132. DJERASSI, C. and R. EHRLICH: Lead Tetraacetate Oxidation of Steroidal Glycols. Some Observations on Ouabagenin. J. Organ. Chem. (USA) 19, 1351 (1954).

133. DJERASSI, C., N. FINCH, R. C. COOKSON and C. W. BIRD: Optical Rotatory Dispersion Studies. XXXVI. α-Haloketones (Part 7). Demonstration of Boat Form in the Bromination of 2α-Methylcholestan-3-one. J. Amer. Chem. Soc. 82, 5488 (1960).

134. DJERASSI, C., N. FINCH and R. MAULI: Optical Rotatory Dispersion Studies. XXX. Demonstration of Boat Form in a 3-Ketosteroid. J. Amer. Chem. Soc. 81, 4997 (1959).

135. DJERASSI, C., I. FORNAGUERA and O. MANCERA: Optical Rotatory Dispersion Studies. XXIII. α-Haloketones (Part 3). J. Amer. Chem. Soc. 81, 2383 (1959).

136. Djerassi, C. and L. E. Geller: Optical Rotatory Dispersion Studies. XVIII. Demonstration of Conformational Mobility in 2-Chloro-5-methyl-*cyclo*hexanone. Tetrahedron 3, 319 (1958).

137. Djerassi, C., L. E. Geller and E. J. Eisenbraun: Optical Rotatory Dispersion Studies. XXVI. α-Haloketones (Part 4). Demonstration of Conformational Mobility in α-Halocyclohexanones. J. Organ. Chem. (USA) 25, 1 (1960).

138. Djerassi, C., T. T. Grossnickle and L. B. High: The Constitution and Stereochemistry of Digitogenin. J. Amer. Chem. Soc. 78, 3166 (1956).

139. Djerassi, C., O. Halpern, V. Halpern, O. Schindler und Ch. Tamm: Untersuchungen der optischen Rotationsdispersion. XV. Anwendung auf Steroid-Ketole und herzwirksame Aglykone. Helv. Chim. Acta 41, 250 (1958).

140. Djerassi, C., L. B. High, J. Fried and E. F. Sabo: Correlation of Digitogenin with Progesterone. J. Amer. Chem. Soc. 77, 3673 (1955).

141. Djerassi, C. and W. Klyne: Optical Rotatory Dispersion Studies. X. Determination of Absolute Configuration of α-Halocyclohexanones. J. Amer. Chem. Soc. 79, 1506 (1957).

142. Djerassi, C. and A. E. Lippman: Terpenoids. XIII. The Structures of the Cactus Triterpenes Machaeric Acid and Machaerinic Acid. J. Amer. Chem. Soc. 77, 1825 (1955).

143. Djerassi, C. and J. S. Mills: Terpenoids. XXXII. The Structure of the Cactus Triterpene Treleasegenic Acid. Ring Conformational Alterations in a Pentacyclic Triterpene. J. Amer. Chem. Soc. 80, 1236 (1958).

144. Djerassi, C., J. Osiecki, R. Riniker and B. Riniker: Optical Rotatory Dispersion Studies. XIV. α-Haloketones (Part 2). J. Amer. Chem. Soc. 80, 1216 (1958).

145. Djerassi, C., C. H. Robinson and D. B. Thomas: Terpenoids. XXV. The Structure of the Cactus Triterpene Dumortierigenin. J. Amer. Chem. Soc. 78, 5685 (1956).

146. Djerassi, C., D. B. Thomas, A. L. Livingston and C. R. Thompson: Terpenoids. XXXI. The Structure and Stereochemistry of Medicagenic Acid. J. Amer. Chem. Soc. 79, 5292 (1957).

147. Djerassi, C., E. J. Warawa, R. E. Wolff and E. J. Eisenbraun: Optical Rotatory Dispersion Studies. XXXIII. α-Haloketones (Part 6). *trans*-2-Bromo-5-*t*-butylcyclohexanone. J. Organ. Chem. (USA) 25, 917 (1960).

148. Drefahl, G. und S. Huneck: Über die epimeren 3-Amino-friedelane. Chem. Ber. 93, 1961 (1960).

149. — — Über Reduktionsprodukte verschiedener Triterpenoxime und Triterpensäureamide. Chem. Ber. 93, 1967 (1960).

150. Dreiding, A. S.: The Conformations of Hydrindanes and the Relative Stabilities of the *cis*- and *trans*-Configurations at the C/D Ring Juncture in Steroids. Chem. and Ind. 1954, 992.

151. Eliel, E. L.: Conformational Analysis in Mobile Systems. J. Chem. Education 37, 126 (1960).

152. Eliel, E. L. and R. G. Haber: The Boiling Points of the Methylcyclo-hexanols—an Exception to the Conformational Rule. J. Organ. Chem. (USA) 23, 2041 (1958).

153. — — Conformational Analysis. VII. Reaction of Alkylcyclohexyl Bromides with Thiophenolate. The Conformational Equilibrium Constant of Bromine. J. Amer. Chem. Soc. 81, 1249 (1959).

154. Eliel, E. L. and C. A. Lukach: Conformational Analysis. II. Esterification Rates of Cyclohexanols. J. Amer. Chem. Soc. 79, 5986 (1957).

155. ELIEL, E. L. and M. N. RERICK: Reduction with Metal Hydrides. VIII. Reductions of Ketones and Epimerization of Alcohols with Lithium Aluminum Hydride—Aluminum Chloride. J. Amer. Chem. Soc. **82**, 1367 (1960).

156. ELIEL, E. L. and R. S. RO: Conformational Analysis. III. Epimerization Equilibria of Alkylcyclohexanols. J. Amer. Chem. Soc. **79**, 5992 (1957).

157. — — Conformational Analysis. IV. Bimolecular Displacement Rates of Cyclohexyl *p*-Toluenesulfonates and the Conformational Equilibrium Constant of the *p*-Toluenesulfonate Group. J. Amer. Chem. Soc. **79**, 5995 (1957).

158. ELKS, J., G. H. PHILLIPPS, D. A. H. TAYLOR and L. J. WYMAN: Studies in the Synthesis of Cortisone. Part VIII. A Wagner-Meerwein Rearrangement involving Rings C and D of the Steroid Nucleus. J. Chem. Soc. (London) **1954**, 1739.

159. ELKS, J., G. H. PHILLIPPS, T. WALKER and L. J. WYMAN: Studies in the Synthesis of Cortisone. Part XV. Improvements in the Conversion of Hecogenin into 3β : 12β-Diacetoxy-5α : 25D-spirostan-11-one and a Study of the Isomeric 11 : 12-Ketols. J. Chem. Soc. (London) **1956**, 4330.

160. EVANS, D. E. and G. H. R. SUMMERS: The Stereochemistry of the 3 : 5-*cyclo*-Cholestan-6-ylamines. J. Chem. Soc. (London) **1957**, 906.

161. FAYEZ, M. B. E., J. GRIGOR, F. S. SPRING and R. STEVENSON: Triterpenoids. Part XLIV. The Constitution of "l-α-Amyradiene". J. Chem. Soc. (London) **1955**, 3378.

162. FERRIER, R. J. and W. G. OVEREND: Newer Aspects of the Stereochemistry of Carbohydrates. Quart. Rev. Chem. Soc. (London) **13**, 265 (1959).

163. FIESER, L. F. and M. FIESER: Steroids. New York: Reinhold Publ. Corp. 1959.

164. — — preceding reference, p. 212.

165. FIESER, L. F. and S. RAJAGOPALAN: Selective Oxidation with N-Bromo-succinimide. II. Cholestane-$3\beta,5\alpha,6\beta$-triol. J. Amer. Chem. Soc. **71**, 3938 (1949).

166. FRIDRICHSONS, J. and A. McL. MATHIESON: Triterpenoids. The Crystal Structure of Lanostenyl Iodoacetate. J. Chem. Soc. (London) **1953**, 2159.

167. FROEMSDORF, D. H., C. H. COLLINS, G. S. HAMMOND and C. H. DePUY: The Direction of Elimination in the Pyrolysis of Acetates. J. Amer. Chem. Soc. **81**, 643 (1959).

168. FÜRST, A. and PL. A. PLATTNER: The Steric Course of the Reactions of Steroid Epoxides. Abstracts of Papers, 12th Intern. Congr. Pure Appl. Chem., p. 409. New York. 1951.

169. FÜRST, A. und R. SCOTONI, Jr.: Über Steroide und Sexualhormone. 191. Mitt. Die Konfiguration der 4-Oxy-cholestane. Helv. Chim. Acta **36**, 1332 (1953).

170. — — Über Steroide und Sexualhormone. 192. Mitt. Überführung von 11,12α-Oxido-Steroiden in 11-Keto-Verbindungen. Helv. Chim. Acta **36**, 1410 (1953).

171. GALLAGHER, T. F. and W. P. LONG: Partial Synthesis of Compounds Related to Adrenal Cortical Hormones. IV. An Improved Method for the Preparation of 3α,11α-Dihydroxycholanic Acid. J. Biol. Chem. **162**, 521 (1946).

172. GENT, B. B. and J. McKENNA: Stereochemical Investigations of Cyclic Bases. Part IV. Hofmann Degradation of 6α- and 6β-Cholestanyltrimethylammonium Salts. J. Chem. Soc. (London) **1959**, 137.

173. GOTO, T.: Stereochemistry of B-Norcoprostane Derivatives. J. Amer. Chem. Soc. **82**, 2005 (1960).

174. GOTO, T. and L. F. FIESER: B-Norcoprostane Derivatives. J. Amer. Chem. Soc. **81**, 2276 (1959).

175. GRABER, R. P., C. S. SNODDY, H. B. ARNOLD and N. L. WENDLER: The Oxidation of Cholesterol by Periodic Acid. J. Organ. Chem. (USA) **21**, 1517 (1956).

176. GRIGOR, J., W. LAIRD, D. MACLEAN, G. T. NEWBOLD and F. S. SPRING: Steroids. Part XIV. 7 : 8-Epoxides of 9α- and 9β-Ergostan-11-one Derivatives. J. Chem. Soc. (London) **1954**, 2333.

177. GRIMMER, G.: Konstitutionsermittlung von Hydroxysteroiden im Mikromaßstab. Liebigs Ann. Chem. **636**, 42 (1960).

178. GROB, C. A. und S. WINSTEIN: Mechanismus der Mutarotation von 5,6-Dibromocholestan. Helv. Chim. Acta **35**, 782 (1952).

179. HALLSWORTH, A. S. and H. B. HENBEST: Aspects of Stereochemistry. Part VII. Metal Reduction of Vicinal Epoxy*cyclo*hexanes. J. Chem. Soc. (London) **1957**, 4604.

180. — — Aspects of Stereochemistry. Part XVI. The Effect of a Hydroxyl Group upon the Metal-reduction of Vicinal Epoxides. J. Chem. Soc. (London) **1960**, 3571.

181. HAMMOND, G. S. and D. H. HOGLE: The Dissociation of Sterically Hindered Acids. J. Amer. Chem. Soc. **77**, 338 (1955).

182. HAMPTON, J., A. LEO and F. H. WESTHEIMER: The Mechanism of the Cleavage of Phenyl-*t*-butylcarbinol by Chromic Acid. J. Amer. Chem. Soc. **78**, 306 (1956).

183. HANNA, R., J. LEVISALLES et G. OURISSON: Stéréochimie des onocéranes. Remarques sur la dispersion rotatoire des cétones triterpéniques. Bull. soc. chim. France **1960**, 1938.

184. HANNA, R., C. SANDRIS et G. OURISSON: Étude de cetones cycliques (VI). Comparaison d'α-dicétones polycycliques. Bull. soc. chim. France **1959**, 1454.

185. HARDY, K. D. and R. J. WICKER: Chemical Reductions of Substituted Cyclohexanones. J. Amer. Chem. Soc. **80**, 640 (1958).

186. HASSEL, O.: Betydningen av De „Van der Waals'ske Atomradier" for Molekylenes Stereokjemiske Bygning. Tidsskr. Kjemi, Bergvesen Metall. **3**, 91 (1943).

187. — Stereochemistry of Cyclohexane. Research (London) **3**, 504 (1950).

188. — Stereochemistry of *cyclo*Hexane. Quart. Rev. Chem. Soc. (London) **7**, 221 (1953).

189. HASSEL, O. and B. OTTAR: The Structure of Molecules Containing Cyclohexane or Pyranose Rings. Acta Chem. Scand. **1**, 929 (1947).

190. HASSEL, O. and H. VIERVOLL: Electron Diffraction Investigations of Molecular Structures. II. Results Obtained by the Rotating Sector Method. Acta Chem. Scand. **1**, 149 (1947).

191. HAUPTMANN, H.: Some Steroid Mercaptols. J. Amer. Chem. Soc. **69**, 562 (1947).

192. HAUPTMANN, H. und F. O. BOBBIO: Über Steroidmercaptole. IV. Die Reaktion des 3,7,12-Trioxo-cholansäure-äthylesters mit substituierten Äthylendimercaptanen. Chem. Ber. **93**, 2187 (1960).

193. HAUPTMANN, H. and M. M. CAMPOS: Steroid Mercaptols. II. J. Amer. Chem. Soc. **74**, 3179 (1952).

194. HAWORTH, R. D., J. McKENNA and R. G. POWELL: The Constitution of Conessine. Part V. Synthesis of Some Basic Steroids. J. Chem. Soc. (London) **1953**, 1110.

195. HAZEBROEK, P. and L. J. OOSTERHOFF: The Isomers of Cyclohexane. Discuss. Faraday Soc. **10**, 87 (1951).

196. HENBEST, H. B. and B. J. LOVELL: Aspects of Stereochemistry. Part II. Intramolecular Electrophilic Assistance of Displacement Reactions. J. Chem. Soc. (London) **1957**, 1965.

197. HENBEST, H. B. and B. B. MILLWARD: Aspects of Stereochemistry. Part XVII. Oxetans from Monoarenesulphonyl Esters of 1,3-Diols. J. Chem. Soc. (London) **1960**, 3575.

198. HENBEST, H. B., B. NICHOLLS, W. R. JACKSON, R. A. L. WILSON, N. S. CROSSLEY, M. B. MEYERS and R. S. McELHINNEY: Directing Effects of Near and Remote Groups on the Addition of Reagents to Double Bonds. Bull. soc. chim. France **1960**, 1365.

199. HENBEST, H. B. and R. A. L. WILSON: Aspects of Stereochemistry. Part I. Stereospecificity in Formation of Epoxides from Cyclic Allylic Alcohols. J. Chem. Soc. (London) **1957**, 1958.

200. HENBEST, H. B. and T. I. WRIGLEY: Aspects of Stereochemistry. Part VI. Reactions of Some Epoxy-steroids with the Boron Trifluoride-Ether Complex. J. Chem. Soc. (London) **1957**, 4596.

201. — — Aspects of Stereochemistry. Part IX. The Formation of Fluorohydrins from the Cholesteryl 5:6-Epoxides and Boron Trifluoride-Ether Complex. J. Chem. Soc. (London) **1957**, 4765.

202. HENNION, G. F. and F. X. O'SHEA: Ethynylation of 4-*t*-Butylcyclohexanone and Kinetics of Saponification of the Ethynylcarbinol Esters. J. Amer. Chem. Soc. **80**, 614 (1958).

203. HEUSLER, K. und A. WETTSTEIN: Über Steroide. 113. Mitt. Zur Herstellung von 11-Keto-Derivaten aus im Ring C unsubstituierten Steroiden. Helv. Chim. Acta **36**, 398 (1953).

204. HEUSSER, H., K. EICHENBERGER, P. KURATH, H. R. DÄLLENBACH und O. JEGER: Über Steroide und Sexualhormone. 176. Mitt. Ein neuer Weg zur Synthese von 11-Keto-Steroiden. Helv. Chim. Acta **34**, 2106 (1951).

205. HIRSCHMANN, R., C. S. SNODDY, C. F. HISKEY and N. L. WENDLER: The Rearrangement of the Steroid C/D Rings. J. Amer. Chem. Soc. **76**, 4013 (1954).

206. HÜCKEL, W. und M. HANACK: Beiträge zur Konstellationsanalyse. I. Liebigs Ann. Chem. **616**, 18 (1958).

207. HÜCKEL, W. und Y. RIAD: Beiträge zur Konstellationsanalyse. VI. Infrarot-spektren alicyclischer Alkohole. Liebigs Ann. Chem. **637**, 33 (1960).

208. INGOLD, Sir C. K.: Structure and Mechanism in Organic Chemistry. Ithaca: Cornell Univ. Press. 1953.

209. JACKMAN, L. M.: Applications of Nuclear Magnetic Resonance Spectroscopy in Organic Chemistry. London: Pergamon Press. 1959.

210. JENSEN, F. R.: The Stereochemistry of the Cleavage of Di-*sec*-butylmercury by Mercuric Bromide. J. Amer. Chem. Soc. **82**, 2469 (1960).

211. JENSEN, F. R. and L. H. GALE: Organomercurials. V. The Conformational Preference of the Bromomercuri Group. J. Amer. Chem. Soc. **81**, 6337 (1959).

212. — — A Stereochemical Study of the Cleavage of Organomercurials by Various Brominating Agents. J. Amer. Chem. Soc. **82**, 148 (1960).

213. JENSEN, F. R., D. S. NOYCE, C. H. SEDERHOLM and A. J. BERLIN: The Energy Barrier for the Chair-Chair Interconversion of Cyclohexane. J. Amer. Chem. Soc. **82**, 1256 (1960).

214. JOHNSON, W. S.: The Relative Stability of Stereoisomeric Forms of Fused Ring Systems. Experientia **7**, 315 (1951).

215. — Energy Relationships of Fused Ring Systems. J. Amer. Chem. Soc. **75**, 1498 (1953).

216. JOHNSON, W. S., J. L. MARGRAVE, V. J. BAUER, M. A. FRISCH, L. H. DREGER and W. N. HUBBARD: The Energy Difference Between the Boat and Chair Forms of Cyclohexane. J. Amer. Chem. Soc. **82**, 1255 (1960).

217. Jones, E. R. H., G. D. Meakins and J. S. Stephenson: Studies in the Steroid Group. Part LXXI. The Preparation and Reactions of 9α-Methyl-ergostane Derivatives. J. Chem. Soc. (London) **1958**, 2156.

218. Jones, R. N. and F. Herling: The Infrared Spectra of Acetoxysteroids below 1350 cm.⁻¹ J. Amer. Chem. Soc. **78**, 1152 (1956).

219. Jones, R. N., D. A. Ramsay, F. Herling and K. Dobriner: The Infrared Spectra of α-Brominated Ketosteroids. J. Amer. Chem. Soc. **74**, 2828 (1952).

220. Jones, R. N. and G. Roberts: The Infrared Spectra of Hydroxysteroids below 1350 cm.⁻¹ J. Amer. Chem. Soc. **80**, 6121 (1958).

221. Kelly, R. B.: A Relationship Between the Conformations of Cyclohexane Derivatives and Their Physical Properties. Canad. J. Chem. **35**, 149 (1957).

222. Klinot, J. and A. Vystrčil: Conformation of Epimeric 2-Bromoallo-betulones. Chem. and Ind. **1960**, 1360.

223. Klyne, W.: The Conformations of Six-Membered Ring Systems. Progr. Stereochem. **1**, 36 (1954).

224. — Conformational Studies on *cyclo*Hexanones. Experientia **12**, 119 (1956).

225. — Optical Rotatory Dispersion in Structural and Stereochemical Studies. Bull. soc. chim. France **1960**, 1396.

226. — Optical Rotatory Dispersion and the Study of Organic Structures. Adv. Organic Chem. **1**, 239 (1960).

227. Koutchérov, V. F., V. M. Andréev et N. Y. Grigorieva: Conformations réactionelles des acides cyclohexéniques substitués. Bull. soc. chim. France **1960**, 1406.

228. Kritchevsky, T. H., D. L. Garmaise and T. F. Gallagher: Partial Synthesis of Compounds Related to Adrenal Cortical Hormones. XVI. Preparation of Cortisone and Related Compounds. J. Amer. Chem. Soc. **74**, 483 (1952).

229. Kuhn, L. P.: The Hydrogen Bond. I. Intra- and Intermolecular Hydrogen Bonds in Alcohols. J. Amer. Chem. Soc. **74**, 2492 (1952).

230. — The Hydrogen Bond. II. The Intramolecular Bond in Cyclic 1,2-Diols. J. Amer. Chem. Soc. **76**, 4323 (1954).

231. Kumler, W. D. and A. C. Huitric: Dipole Moments, Spectra and Structure of α-Halocyclohexanones, α-Halocyclopentanones and Related Compounds. J. Amer. Chem. Soc. **78**, 3369 (1956).

232. Kupchan, S. M., C. I. Ayres, M. Neeman, R. H. Hensler, T. Masamune and S. Rajagopalan: Veratrum Alkaloids. XXXVIII. The Structure and Configuration of Protoverine. J. Amer. Chem. Soc. **82**, 2242 (1960).

233. Kupchan, S. M., W. S. Johnson and S. Rajagopalan: The Configuration of Cevine. Tetrahedron **7**, 47 (1959).

234. Kupchan, S. M. and C. R. Narayanan: Veratrum Alkaloids. XXVIII. The Structure and Configuration of Germine. J. Amer. Chem. Soc. **81**, 1913 (1959).

235. Kupchan, S. M., P. Slade and R. J. Young: Intramolecular Catalysis. Facilitation of Alkaline Hydrolysis of Alicyclic 1,2-Diol Monoesters. Tetrahedron Letters **1960**, No. 24, 22.

236. Kwart, H. and P. S. Francis: Structural and Conformational Effects on the Rates of Oxidation of Secondary Alcohols by Chromic Acid. J. Amer. Chem. Soc. **81**, 2116 (1959).

237. Kwart, H. and L. B. Weisfeld: Acid Catalysis in the Isomerization of 5α,6β-Dibromocholestane. J. Amer. Chem. Soc. **78**, 635 (1956).

238. Laird, W., F. S. Spring and R. Stevenson: Pentacyclic Triterpenoid Backbone Rearrangement: Constitution of Brein. J. Amer. Chem. Soc. **82**, 4108 (1960).

239. Lardon, A., H. P. Sigg und T. Reichstein: 15-Keto- und 15-Hydroxy-ätiansäuren. Helv. Chim. Acta **42**, 1457 (1959).

240. LE FÈVRE, C. G. and R. J. W. LE FÈVRE: Molecular Polarisability. The Molar Kerr Constants and Conformations of *cis-* and *trans-*Decalins. J. Chem. Soc. (London) **1957**, 3458.

241. LEMIEUX, R. U. and P. CHU: Conformations and Relative Stabilities of Acetylated Sugars as Determined by Nuclear Magnetic Resonance Spectroscopy and Anomerization Equilibria. Abstracts of Papers, 133rd Meeting Amer. Chem. Soc., p. 31 N. San Francisco. 1958.

242. LEMIEUX, R. U., R. K. KULLNIG, H. J. BERNSTEIN and W. G. SCHNEIDER: Configurational Effects in the Proton Magnetic Resonance Spectra of Acetylated Carbohydrates. J. Amer. Chem. Soc. **79**, 1005 (1957).

243. — — — — Configurational Effects on the Proton Magnetic Resonance Spectra of Six-membered Ring Compounds. J. Amer. Chem. Soc. **80**, 6098 (1958).

244. LEVINE, S. G. and M. E. WALL: Steroidal Sapogenins. LVI. The Preparation of 12-Methyl Sapogenins. J. Amer. Chem. Soc. **82**, 3391 (1960).

245. LEVISALLES, J.: La forme bâteau dans les équilibres et les réactions des cycles hexatomiques. Bull. soc. chim. France **1960**, 551.

246. LINDE, H. und K. MEYER: Konstitution des Resibufogenins. Helv. Chim. Acta **42**, 807 (1959).

247. LINSTEAD, R. P.: Carbocyclic Compounds. Annu. Rep. Chem. Soc. (London) **32**, 305 (1935).

248. LOEWENTHAL, H. J. E.: Selective Reactions and Modification of Functional Groups in Steroid Chemistry. Tetrahedron 6, 269 (1959).

249. LWOWSKI, W.: Nachbargruppen-Effekte in der organischen Chemie. Angew. Chem. **70**, 483 (1958).

250. McALEER, W. J., M. A. KOZLOWSKI, T. H. STOUDT and J. M. CHEMERDA: Microbially Produced 7α- and 7β-Hydroxy-Δ^4-3-keto Steroids. J. Organ. Chem. (USA) **23**, 958 (1958).

251. McCOUBREY, J. C. and A. R. UBBELOHDE: The Configuration of Flexible Organic Molecules. Quart. Rev. Chem. Soc. (London) **5**, 364 (1951).

252. McKENNA, J. and A. TULLEY: Stereochemical Investigations of Cyclic Bases. Part V. Exhaustive Methylation of N-Methyl-4-aza-5α- and -5β-cholestane, and the Reaction of the Methines with Acetic Acid: the Possibility of "$\alpha'\beta$" Hofmann Elimination with Cyclic Ammonium Hydroxides. J. Chem. Soc. (London) **1960**, 945.

253. MATEOS, J. L.: Rate of Reduction of Some Steroid Ketones with Sodium Borohydride. J. Organ. Chem. (USA) **24**, 2034 (1959).

254. MATHIEU, J., M. LEGRAND et J. VALLS: Sur un effet à distance lors de la solvolyse de quelques 17β-tosyloxy stéroïdes. Bull. soc. chim. France **1960**, 549.

255. MAULI, R., H. J. RINGOLD and C. DJERASSI: Steroids. CXLV. 2-Methyl-androstane Derivatives. Demonstration of Boat Form in the Bromination of 2α-Methyl-androstan-17β-ol-3-one. J. Amer. Chem. Soc. **82**, 5494 (1960).

256. MAYOR, P. A. and G. D. MEAKINS: Steroids of Unnatural Configuration. Part IV. Oxidation of Lumisterol and Lumisteryl Acetate with Perbenzoic Acid. J. Chem. Soc. (London) **1960**, 2792.

257. — — Steroids of Unnatural Configuration. Part V. Preparation of Lumistanol A from an Oxidation Product of Lumisteryl Acetate. J. Chem. Soc. (London) **1960**, 2800.

258. MELERA, A., D. ARIGONI, A. ESCHENMOSER, O. JEGER und L. RUZICKA: Zur Kenntnis der Triterpene. 192. Mitt. Absolute Konfiguration des Kohlen-stoffatoms 20 in α-Amyrin; ein Beitrag zur Konstitution des Ringes E. Helv. Chim. Acta **39**, 441 (1956).

259. MILLS, J. A.: The Reaction of Ring-substituted *cyclo*Hexylamines with Nitrous Acid. An Interpretation based on Conformational Analysis. J. Chem. Soc. (London) **1953**, 260.

260. — The Stereochemistry of Cyclic Derivatives of Carbohydrates. Adv. Carbohydrate Chem. **10**, 1 (1955).

261. MIYAZAWA, T. and K. S. PITZER: Thermodynamic Functions for Gaseous *cis*- and *trans*-Decalins from 298 to 1000° K. J. Amer. Chem. Soc. **80**, 60 (1958).

262. MIZUSHINA, S.-I.: Structure of Molecules and Internal Rotation. New York: Academic Press Inc. 1954.

263. MOHR, E.: Die Baeyersche Spannungstheorie und die Struktur des Diamanten. J. prakt. Chem. [2] **98**, 315 (1918).

264. MORITANI, I., S. NISHIDA and M. MURAKAMI: The Effect of Conformation on Reactivity. I. Acetolysis of the *trans*-Decalyl *p*-Toluenesulphonates; 1,3-Diaxial Interactions as a Factor in the Chemical Behavior of Decalyl Derivatives. J. Amer. Chem. Soc. **81**, 3420 (1959).

265. NAVES, Y.-R.: Structure conformationelle des terpènes. Bull. soc. chim. France **1956**, 1020.

266. NEIDERHISER, D. H. and W. W. WELLS: The Structure of Methostenol and Its Distribution in Rat Tissues. Arch. Biochem. Biophys. **81**, 300 (1959).

267. NICKON, A.: A Relationship between Conformation and Infrared Absorption in 1,2-Halohydrins. J. Amer. Chem. Soc. **79**, 243 (1957).

268. NISHIDA, S.: The Effect of Conformation on Reactivity. II. Rates of Acetolysis of Isomeric Cholestanyl *p*-Toluenesulphonates. J. Amer. Chem. Soc. **82**, 4290 (1960).

269. NOYCE, D. S. and W. L. REED: Carbonyl Reactions. IX. The Rate and Mechanism of the Base-catalyzed Condensation of Benzaldehyde and Acetone. Factors Influencing the Structural Course of Condensation of Unsymmetrical Ketones. J. Amer. Chem. Soc. **81**, 624 (1959).

270. ORLOFF, H. D.: The Stereoisomerism of Cyclohexane Derivatives. Chem. Rev. **54**, 347 (1954).

271. OURISSON, G. et A. RASSAT: Réduction du camphre par les métaux dissous. Obtention sélective d'isobornéol. Tetrahedron Letters **1960**, No. 21, 16.

272. PAGE, J. E.: Studies in the Synthesis of Cortisone. Part XI. Infrared Spectra of Alkoxy- and Acetoxy-Steroids. J. Chem. Soc. (London) **1955**, 2017.

273. PAUNCZ, R. and D. GINSBURG: Conformational Analysis of Alicyclic Compounds. I. Considerations of Molecular Geometry and Energy in Medium and Large Rings. Tetrahedron **9**, 40 (1960).

274. PETROWITZ, H.-J.: Zur Kieselgelschicht — Chromatographie der stereoisomeren Menthole. Angew. Chem. **72**, 921 (1960).

275. PICKERING, R. A. and C. C. PRICE: An Estimate of the Conformational Equilibrium in Cyclohexanol from Infrared Spectra. J. Amer. Chem. Soc. **80**, 4931 (1958).

276. PITZER, K. S.: Potential Energies for Rotation About Single Bonds. Discuss. Faraday Soc. **10**, 66 (1951).

277. QUINKERT, G.: Die Stabilitätenreihenfolge der stereoisomeren Hydrindan- und Hydrindanon-Verbindungen. Experientia **13**, 381 (1957).

278. RAMIREZ, F. and S. STAFIEJ: The Nitrous Acid Deamination of 17β-Hydroxy-20-amino-C_{21} Steroids. Stereochemistry of D-Homoannulation. II. J. Amer. Chem. Soc. **78**, 644 (1956).

279. REEVES, R. E.: Chemistry of the Carbohydrates. Annu. Rev. Biochem. **27**, 15 (1958).

280. ROBERTS, I. and G. E. KIMBALL: The Halogenation of Ethylenes. J. Amer. Chem. Soc. **59,** 947 (1937).

281. ROBERTS, G. and C. W. SHOPPEE: Steroids and Walden Inversion. Part XV. The Mechanism and Stereochemical Course of Some Grignard Carboxylations and Oxygenations. J. Chem. Soc. (London) **1954,** 3418.

282. ROBINS, P. A. and J. WALKER: Eclipsed Carbonyl Oxygen-Carbon Non-Bonded Interaction in Cyclic Ketones. Chem. and Ind. **1954,** 773.

283. — — Stability Sequence of Five Stereoisomeric Perhydro-1 : 4-dioxophenanthrenes. J. Chem. Soc. (London) **1955,** 1789.

284. ROČEK, J. and J. KRUPIČKA: Oxidations with Chromium (VI) Oxide. VII. The Mechanism of Oxidation of Secondary Alcohols. Collect. Czech. Chem. Commun. **23,** 2068 (1958).

285. ROSEN, W. E., J. B. ZIEGLER, A. C. SHABICA and J. N. SHOOLERY: The Stereochemistry of Steroidal Sapogenins. III. N. m. r. Spectra. J. Amer. Chem. Soc. **81,** 1687 (1959).

286. Ross, A., P. A. S. SMITH and A. S. DREIDING: The Relative Stabilities of *cis-trans* Isomers of Bicyclic Ring Systems with Angular Methyl Groups. II. J. Organ. Chem. (USA) **20,** 905 (1955).

287. RUZICKA, L. und K. HOFMANN: Polyterpene und Polyterpenoide. C. Über Umsetzungen an den Ringen A und E der Oleanolsäure. Beiträge zur Kenntnis des Kohlenstoffgerüstes pentacyclischer Triterpene. Helv. Chim. Acta **19,** 114 (1936).

288. SACHSE, H.: Über die geometrischen Isomerien der Hexamethylenderivate. Ber. dtsch. chem. Ges. **23,** 1363 (1890).

289. SANDRIS, C. et G. OURISSON: Étude de cétones cycliques (V): Un effet conformationnel nouveau. Bull. soc. chim. France **1958,** 1524.

290. SANNIÉ, C. et H. LAPIN: Recherches sur les sapogénines à noyau stérolique. Identification de ces génines sur de petites quantités de plantes. Bull. soc. chim. France **1952,** 1080.

291. SAVARD, K.: Paper Partition Chromatography of C_{19}- and C_{21}-Ketosteroids. J. Biol. Chem. **202,** 457 (1953).

292. — Some Theoretical and Some Practical Aspects of Partition Chromatography of Ketosteroids. Recent Progr. Hormone Res. **9,** 185 (1954).

293. SCHREIBER, J. und A. ESCHENMOSER: Über die relative Geschwindigkeit der Chromosäureoxydation sekundärer, alicyclischer Alkohole. Helv. Chim. Acta **38,** 1529 (1955).

294. SHAW, J. I. and R. STEVENSON: 4-Bromo- and 4-Chloro-cholest-4-en-3-one. J. Chem. Soc. (London) **1955,** 3549.

295. SHOOLERY, J. N. and M. T. ROGERS: Nuclear Magnetic Resonance Spectra of Steroids. J. Amer. Chem. Soc. **80,** 5121 (1958).

296. SHOPPEE, C. W., R. J. W. CREMLYN, D. E. EVANS and G. H. R. SUMMERS: Steroids and Walden Inversion. Part XXXVIII. The Deamination of Epimeric Cholestan-2-, -4-, and -7-ylamines. J. Chem. Soc. (London) **1957,** 4364.

297. SHOPPEE, C. W., D. E. EVANS and G. H. R. SUMMERS: Steroids and Walden Inversion. Part XXXVI. The Mechanism of Deamination. J. Chem. Soc. (London) **1957,** 97.

298. SHOPPEE, C. W. and J. C. P. SLY: Steroids and Walden Inversion. Part XLI. The Deamination of Some A-Nor-, B-Nor-, and 17-Amino-steroids. J. Chem. Soc. (London) **1959,** 345.

299. SHRINER, R. I., R. ADAMS and C. S. MARVEL: Stereoisomerism. In: H. GILMAN, Organic Chemistry, 2nd ed., Vol. I, p. 214. New York: J. Wiley and Sons. 1943.

300. Sicher, J., M. Tichý, F. Šipoš and M. Pánková: Kinetic Evidence for a Cyclohexane Boat Intermediate during Neighbouring-group Participation. Proc. Chem. Soc. (London) 1960, 384.
301. Sigg, H. P. und Ch. Tamm: Synthese von 1-Keto-A-nor-cholestan. Beitrag zur Kenntnis der Faworski-Reaktion. Helv. Chim. Acta 43, 1402 (1960).
302. Simonsen, Sir J. and W. C. J. Ross: The Terpenes, Vol. V. Cambridge: Univ. Press. 1957.
303. Skita, A.: Über die geometrische Isomerie der Polymethylene. Ber. dtsch. chem. Ges. 53, 1792 (1920).
304. Slomp, G. and B. R. McGarvey: Nuclear Magnetic Resonance Studies on 6-Methyl Steroids. J. Amer. Chem. Soc. 81, 2200 (1959).
305. Slomp, G. and F. MacKellar: Nuclear Magnetic Resonance Studies using Pyridine Solutions. J. Amer. Chem. Soc. 82, 999 (1960).
306. Smith, L. L., M. Marx, J. J. Garbarini, T. Foell, V. E. Origoni and J. J. Goodman: 16α-Hydroxy Steroids. VII. The Isomerization of Triamcinoline. J. Amer. Chem. Soc. 82, 4616 (1960).
307. Sommer, P. F., V. P. Arya und W. Simon: Über scheinbare Dissoziationskonstanten von Carbonsäuren mit äquatorialer und axialer Lage der Carboxylgruppe. Tetrahedron Letters 1960, No. 20, 18.
308. Sondheimer, F., Y. Klibansky, Y. M. Y. Haddad, G. H. R. Summers and W. Klyne: The Boat Conformation of Ring A in the Steroids: a Further Example. Chem. and Ind. 1960, 902.
309. Stiles, M., D. Wolf and G. V. Hudson: Catalyst Selectivity in the Reactions of Unsymmetrical Ketones; Reaction of Butanone with Benzaldehyde and *p*-Nitrobenzaldehyde. J. Amer. Chem. Soc. 81, 628 (1959).
310. Stolow, R. D.: A Quantitative Relationship between Dissociation Constants and Conformational Equilibria. Cyclohexanecarboxylic Acids. J. Amer. Chem. Soc. 81, 5806 (1959).
311. Stork, G. and S. D. Darling: Stereochemistry of the Lithium-Ammonia Reduction of α,β-Unsaturated Ketones. J. Amer. Chem. Soc. 82, 1512 (1960).
312. Streitwieser, A.: Solvolytic Displacement Reactions at Saturated Carbon Atoms. Chem. Rev. 56, 571 (1956).
313. — An Interpretation of the Reaction of Aliphatic Primary Amines with Nitrous Acid. J. Organ. Chem. (USA) 22, 861 (1957).
314. Tichý, M., J. Jonáš and J. Sicher: Stereochemical Studies. XVII. Dissociation Constants of Cyclohexanecarboxylic Acids and Cyclohexylamines, and Conformational Equilibria. Collect. Czech. Chem. Comm. 24, 3434 (1959).
315. Turner, R. B.: Energy Differences in the *cis*- and *trans*-Decalins. J. Amer. Chem. Soc. 74, 2118 (1952).
316. Turner, R. B., R. Anliker, R. Helbling, J. Meier und H. Heusser: Über Steroide und Sexualhormone. 204. Mitt. Zur Stereochemie der epimeren 17a-Methyl-D-homo-testosterone. Helv. Chim. Acta 38, 411 (1955).
317. Usković, M., M. Gut, E. N. Trachtenberg, W. Klyne and R. I. Dorfman: D-Homosteroids. IV. 17β,17aβ-Dimethyl-17α,17aα-dihydroxy- and 17,17-Dimethyl-17a-keto-D-homosteroids. J. Amer. Chem. Soc. 82, 4965 (1960).
318. Villotti, R., H. J. Ringold and C. Djerassi: Optical Rotatory Dispersion Studies. XXXVII. Steroids. CXLVI. On the Mechanism and Stereochemical Course of the Bromination of 3-Keto Steroids and their Enol Acetates. J. Amer. Chem. Soc. 82, 5693 (1960).
319. Wechter, W. J.: Stereoselective *cis*-Non-Markownikoff Hydration of Steroid Double Bonds. Chem. and Ind. 1959, 294.

320. WEINSTOCK, J., R. G. PEARSON and F. G. BORDWELL: E 2 Elimination Reactions in the Cyclohexane and Cyclopentane Series. J. Amer. Chem. Soc. 76, 4748 (1954).

321. WELLS, W. W. and D. H. NEIDERHISER: Isolation and Synthesis of a New Sterol from Rat Feces. J. Amer. Chem. Soc. 79, 6569 (1957).

322. WENDLER, N. L.: D-Homosteroids: Boat Conformation of the D-Ring. Chem. and Ind. 1958, 1662.

323. — D-Homosteroids. Boat Conformation of the D-Ring. Tetrahedron 11, 213 (1960).

324. WENDLER, N. L. and D. TAUB: Group Transfer and Ring Contraction Phenomena in the Steroid Series. Chem. and Ind. 1958, 415.

325. WENDLER, N. L., D. TAUB, S. DOBRINER and D. K. FUKUSHIMA: The Structure and Synthesis of D-Homosteroids. J. Amer. Chem. Soc. 78, 5027 (1956).

326. WENDLER, N. L., D. TAUB and R. P. GRABER: Group Transfer and Ring Contraction Phenomena in the D-Homosteroid Series. Tetrahedron 7, 173 (1959).

327. WEST, R., J. J. KORST and W. S. JOHNSON: 1,3-Diaxial Hydrogen Bonding and the Intramolecular Assistance of Solvolysis. J. Organ. Chem. (USA) 25, 1976 (1960).

328. WHEELER, D. M. S. and J. W. HUFFMAN: The Stereochemistry of Reduction by Complex Metal Hydrides. Experientia 16, 516 (1960).

329. WHEELER, O. H. and V. S. GAIND: Dissociation Constants of the Cyanohydrins of Some Triterpene Ketones. Canad. J. Chem. 36, 1735 (1958).

330. WHEELER, O. H. and J. L. MATEOS: Reactivity Studies on Natural Products. V. Rates of Borohydride Reduction of Some Ring A and B Steroid Ketones. Canad. J. Chem. 36, 1049 (1958).

331. — — Stereochemistry of Reduction of Ketones by Complex Metal Hydrides. Canad. J. Chem. 36, 1431 (1958).

332. WHITHAM, G. H.: The Structure of Nyctanthic Acid. J. Chem. Soc. (London) 1960, 2016.

333. WINSTEIN, S. and collaborators: many papers under the title, The Role of Neighboring Groups in Replacement Reactions. J. Amer. Chem. Soc. 64 (1942), et seq.

334. WINSTEIN, S. and E. GRUNWALD: The Role of Neighboring Groups in Replacement Reactions. XIII. General Theory of Neighboring Groups and Reactivity. J. Amer. Chem. Soc. 70, 828 (1948).

335. WINSTEIN, S. and N. J. HOLNESS: Neighboring Carbon and Hydrogen. XIX. t-Butylcyclohexyl Derivatives. Quantitative Conformational Analysis. J. Amer. Chem. Soc. 77, 5562 (1955).

336. WINSTEIN, S. and T. G. TRAYLOR: Mechanisms of Reaction of Organomercurials. II. Electrophilic Substitution on Saturated Carbon. Acetolysis of Dialkylmercury Compounds. J. Amer. Chem. Soc. 77, 3747 (1955).

337. — — Mechanisms of Reaction of Organomercurials. III. Preparation and Substitution Reactions of Bridgehead Mercurials. J. Amer. Chem. Soc. 78, 2597 (1956).

338. WINSTEIN, S., T. G. TRAYLOR and C. S. GARNER: Mechanisms of Reaction of Organomercurials. I. Stereochemistry of Electrophilic Displacement on cis-2-Methoxycyclohexylneophylmercury by Radio-mercuric Chloride. J. Amer. Chem. Soc. 77, 3741 (1955).

339. ZIMMERMANN, H. E.: The Stereochemistry of the Ketonization Reaction of Enols. J. Organ. Chem. (USA) 20, 549 (1955), and later papers in this series.

(Received, February 20, 1961.)

Fortschritte d. Chem. org. Naturst. XIX.   16

Biogenetic-type Syntheses of Natural Products.

By E. E. VAN TAMELEN, Madison, Wisconsin.

Contents.

Hygrine, Cuscohygrine 248. — Isopelletierine 249. — Lobelanine, Arecaidine 249. — Nicotine, Anabasine 250. — Atropine, Tropine, Tropinone 251. — Pseudo-pelletierine, Meteloidine 252.

Lupinine 252. — Sparteine 253. — Cytisine, Anagyrine, Thermopsine 254.

Amyl-quinolines 255. — Vasicine 256. — Arborine 256. — Acridone alkaloids 256.

Laudanosine, Berberine, Papaverine, Glaucine 258. — Norsalsoline, Norlaudanosine, Benzyl-tetrahydroisoquinolines, Cryptopine, Corydaline 259.

Yohimbine 260. — Eleagnine 261. — Strychnine 261. — Gramine 263. — Physostigmine 263. — Rutaecarpine 264. — Cryptolepine 264. — Cinchonamine, Cinchonine 264. — Quinamine 265.

Geraniol, Geraniolenes, Iopones 266. — Limonene 266. — Farnesol derivatives, Farnesic acid 267.

Squalene 269. — Hopenone-I 270. — Friedelin 271. — Glutinone 271. — Lupeol, Germanicol 271.

α-Terpinene, Ascaridol 271. — Santonin, Verbenone 271. — Lupulone, Humulone 272. — Iridomyrmecin 272.

Fructose, Sorbose, Xylose, Arabinose, Ribose 274. — N-Acetylneuraminic acid 274.

Introduction.

During the past fifty to sixty years, tremendous advances have been made in the laboratory preparation of complex, naturally occurring organic substances. As early as the first part of the present century, pioneering work was being carried out in various laboratories, and among the early successes in this vastly challenging field may be cited the synthesis of numerous monoterpenes; alkaloids, including nicotine, dihydroquinine and the simpler opium bases; porphyrins, including the blood pigments; the common hexoses, as well as many amino acids and peptides.

Later, as the structures of steroids, nucleotides, vitamins, antibiotics, and the more complicated alkaloids and terpenes were unraveled, synthetic organic chemists found themselves faced with even more demanding trials for their ingenuity and patience. Efforts during the past twenty years have been most impressive, and have been characterized in general by the confrontation of two principal problems: first, the molecular complexity of certain natural products, in regard to not only size and number of rings, but also the presence of several to a dozen or more asymmetric centers; and second, the chemical sensitivity or perversity of the materials being sought. As a consequence of dealing with the former problem, there was developed the discipline often referred to as *stereospecific* (or *stereoselective*) *synthesis*, which has quite literally added a new dimension to the science. Among the many examples which might be provided here, the following are illustrative: non-aromatic steroids, such as cortisone, testosterone, cholesterol, and aldosterone; sesqui-, di- and triterpenes, including santonin, the resin acids, onocerin and the pentacyclic hopenone-I; alkaloids, including strychnine, yohimbine, and reserpine. Different tactics were required in the search for synthesis routes to unstable or poorly-behaved species—the development of new, subtle, and specific chemical techniques for the cases under investigation. In this connection, there come to mind such achievements as the synthesis of the penicillins; peptides, such as ACTH or gramicidin-S; and members of the nucleotide class, e. g., various coenzymes. At the present writing, many of the long-standing problems of total synthesis have been solved; and if the 1936 statement of Robinson (74), "We have traveled far since 1828, and the interest formerly attached to total synthesis has disappeared" seems in retrospect somewhat harsh and premature; certainly it is safe now to say that an approach to a molecule of moderate complexity is not likely to gain much attention unless it features some new technique or synthetic principle.

If one reviews the outstanding accomplishments in total synthesis of the last half century, he may be struck by the almost complete lack of similarity between the synthesis pathways used by the organic chemists in constructing the more intricate systems, and the methods and routes presumably employed by Nature in the genesis of such materials—a comparison which makes the methods of the laboratory worker appear somewhat artificial and contrived. Yet, while dramatic announcements of multistage syntheses appeared successively during the last several decades, a fainter appeal could be heard, emanating from a smaller group of individuals interested in natural product synthesis, but *patterned along lines considered also to represent reasonable biosynthetic routes.* As early as 1917, Robinson had devised and executed an excellent example of this school's approach, the now famous synthesis of *tropinone* (corresponding to the alkamine portion of the tropine alkaloids), proceeding in one stage

$$\begin{matrix} CHO \\ \\ \\ CHO \end{matrix} + NH_2CH_3 + \begin{matrix} CH_2COOH \\ CO \\ CH_2COOH \end{matrix} \longrightarrow NCH_3 =O + 2 H_2O + 2 CO_2$$

(XIV.) Tropinone.

from simple starting materials regarded as reasonable biogenetic precursors: succindialdehyde, methylamine and acetonedicarboxylic acid. This milestone in synthesis was particularly striking and timely in that it appeared hard on the heels of the first tropinone synthesis, an exceedingly lengthy and trying process accomplished by Willstätter. Comparison was irresistable, and vividly demonstrated at this early stage that the philosophy of synthesis exemplified by the Robinson approach had, in its beautiful simplicity, much to commend it. In the succeeding years, further examples of such syntheses appeared, but most of the cases were modest, directed toward simple natural products containing only a few rings, and at most one or two asymmetric centers.

It is the purpose here to assemble and discuss, perhaps for the first time, examples in which this philosophy of synthesis is at play; and, more important, to demonstrate, by means of the most recent developments along these lines, that the full potentialities of biogenetically patterned reactions and syntheses had heretofore not been realized. It seems likely that the most impressive gains are just beginning to be seen, and with their advent, the science of natural product synthesis will be raised to a new level.

The term *"biogenetic-type"* has been selected to describe an organic synthesis designed to follow, in at least its major aspects, biosynthetic pathways proved, or presumed, to be used in the natural construction of the end product. The suffix in the term is deliberately chosen and implies that the relationship of the laboratory synthesis to the bio-synthesis is not necessarily very close and that the in vitro route may be based on an in vivo scheme which is reasonable yet only speculative, or for which only meager evidence may be available. The important criterion is a purely chemical and practical one: does the conceived plan lead to a new or improved laboratory method? If so, then the entire venture is worthwhile. It seems hardly necessary to add that the success of a "biogenetic-type" synthesis *by itself* does not constitute evidence for the operation of a particular chemical step in nature (although in a remarkable case, the temptation to draw such a conclusion will be great); and such confirmation is not a purpose of this approach to synthesis.

As implied already, the term "biogenetic-type" is meant to refer to presumed intermediates and biosynthetic paths, and it should be stated further that little emphasis is placed on reagents and conditions. "Biogenetic-type" syntheses are thus to be distinguished from "physiological-type" syntheses, in which not only plausible bio-organic substitutes are employed, but also specific conditions of temperature, pH, dilution, etc., which supposedly compare to those obtaining in a living cell. Laboratory syntheses which proceed under these conditions are likely to correspond, at best, to in vivo syntheses which are "spontaneous", i. e., not enzyme-catalyzed. On the other hand, the striking success of certain "biogenetic-type" syntheses may depend upon utilization of reaction types which parallel enzyme-promoted processes; and in lieu of the enzyme system, the organic chemist may need to resort to reagents and conditions not available to the living system, in order that he can capitalize on the *overall* biosynthetic route. Thus, one engaged in a biogenetic-type synthesis need have no hesitation in carrying out a reaction at pH 13, or in resorting to a lithium aluminum hydride reduction or ferricyanide oxidation in order to complete the steps to the final goal.

Inspection of various biogenetic-type syntheses which have been reported reveals certain general characteristics, which are set forth below as an introduction and in order that they may serve as a guide for those unacquainted with the field:

a) The biosynthetic route which serves as a model for the laboratory synthesis may be either experimentally based or hypothetical. In the latter instance, a working knowledge of sound biochemical principles is presupposed; usually a single pathway will evolve as the most logical.

b) In a typical favorable case, at least one reaction (or series of closely connected reactions) in the model biosynthesis will be outstanding in the sense that through its agency considerable progress is made in assembling the natural product structure, i. e., an important bond, such as an amide link in a peptide, is formed; or the entire ring system of an alkaloid (if not the substance itself) is constructed. This step is the focal point in the laboratory version.

c) The intermediate required for the execution of the critical step is usually prepared by standard synthetic operations. The key intermediate may possess the exact structure proposed in the biosynthetic scheme, or it may be a simple modification—perhaps (in place of enzyme selectivity) one in which minor adjustments, e. g., blocking an otherwise reactive site, are deliberately made so as to direct the intermediate along desired channels and preclude other reaction courses.

d) In the laboratory duplication of the key biological step, any conditions or reagents as become necessary for completion of the reaction may be used (see above). Choice in these matters are dictated by the purely chemical demands of the system: in some cases, mild conditions approximating the environment in the living cell will suffice; in others, for want of enzyme influences, rather more drastic means will be employed. In the entire range of examples available, laboratory yields vary from nearly quantitative on the one hand to only a few percent on the other, in which cases practical considerations become secondary to theoretical success.

e) The key synthetic operation very likely falls near the end of the laboratory route, and, if necessary, routine means are employed to complete the synthesis.

Certain benefits may accrue by following the principles set forth. As an incidental facet, the natural product synthesis which manifestly is patterned after a reasonable biogenetic route is, to most, more esthetically pleasing and satisfying. More important, however, biogenetic-type syntheses often are neater, shorter, and more efficient than normal routes in which no attention is paid to natural processes. Indeed, it is sometimes found that the only satisfactory route to some one natural product is the biogenetic-type! The ROBINSON tropinone synthesis is an early, but still excellent case in point; and (if the writer may be permitted an example from his own work) a direct method of constructing the complex strychnine skeleton, summarized in the following scheme, is another illustration.

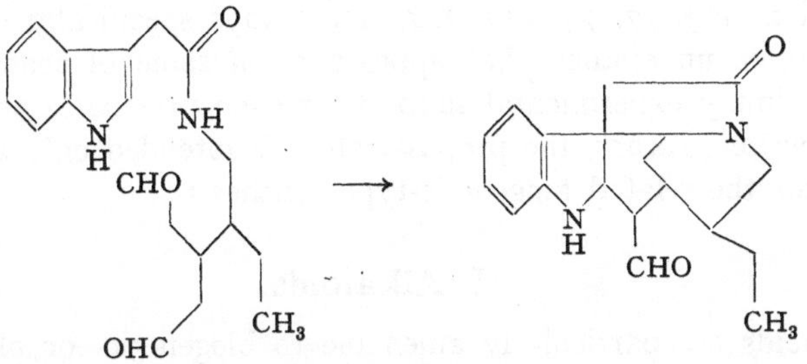

Another outcome of planning with biogenetic processes as models is the development of new reactions and synthetic techniques for organic chemistry, and moreover, ones not necessarily related only to natural product syntheses. There would seem to be little doubt that the trend of peptide synthesis during recent years has been influenced by the biochemical investigation of amino and carboxylic acid "activation". A second illustrative instance is a method for oxidation of an unactivated methyl group through the use of electrophilic oxygen,

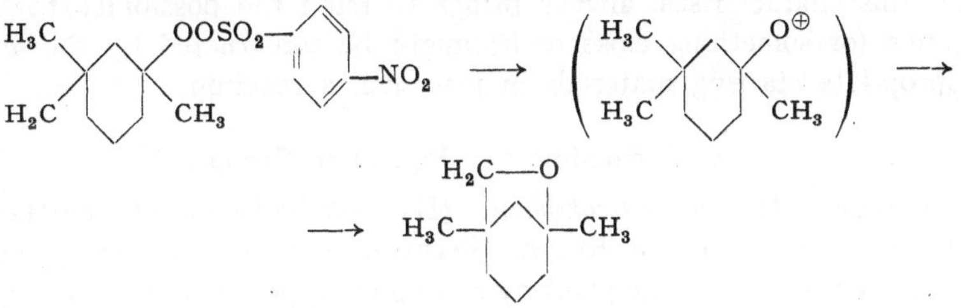

a process which parallels enzymatic hydroxylation of saturated carbon (*32*). In this review, however, emphasis will be placed on natural product synthesis, and only occasionally will other examples of the type cited be included.

Further, execution of a biogenetic-type reaction, although (as pointed out before) not substantiative per se of the biosynthetic proposal evolved as a working hypothesis, can be of value to the biochemist in that it can serve as a guide and impetus for definitive studies on said proposal.

In keeping with the survey nature of this paper and with the purpose of directing attention in a general way to the character and advantage of biogenetic-type synthesis, only the broader view of this subject will be presented; for any detailed discussion of the mechanistic and stereo-chemical aspects of the reactions described, the reader is referred to the original, and related, papers. Also, the recent literature is not wanting in biogenetic speculation, and the writer has borrowed freely from proposals of others, which are to be found in many reviews and publications (73, 77, 75, 118, 114) not always specifically cited herein. It should be understood that appearance of some biogenetic proposal does not imply experimental support, or even preference or acceptance by this writer; rather, the proposal often is intended only as a natural setting for the related biogenetic-type synthesis.

I. Alkaloids.

Alkaloids are particularly amenable to biogenetic- or physiological-type synthesis, and perhaps more success has been encountered here than in any other natural product area. Very often it is found possible to dissect the alkaloid structure into Mannich base components,

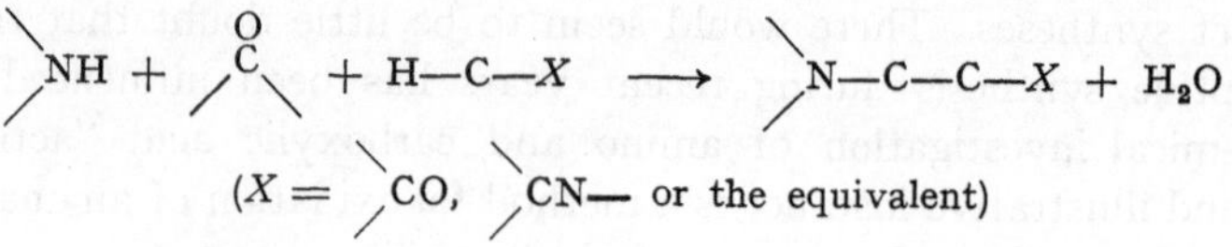

$$>NH + \overset{O}{\underset{}{C}} + H-C-X \longrightarrow \ >N-C-C-X + H_2O$$

$$(X = \ >CO, \ >CN- \text{ or the equivalent})$$

and this characteristic always brings to mind the possibility that the system (or something close to it) might be constructed by the use of appropriate starting materials in just such a reaction.

1. Pyrrolidine and Piperidine Groups.

Hygrine (I) and *cuscohygrine* (II), alkaloids found together in *Erythroxylon truxillense* Rusby (Peruvian coca-leaves), are generally considered to arise in the plant by a Mannich type interaction, a proposal which rests directly on the dissections indicated. Although a number of variants might be visualized as biological starting materials or

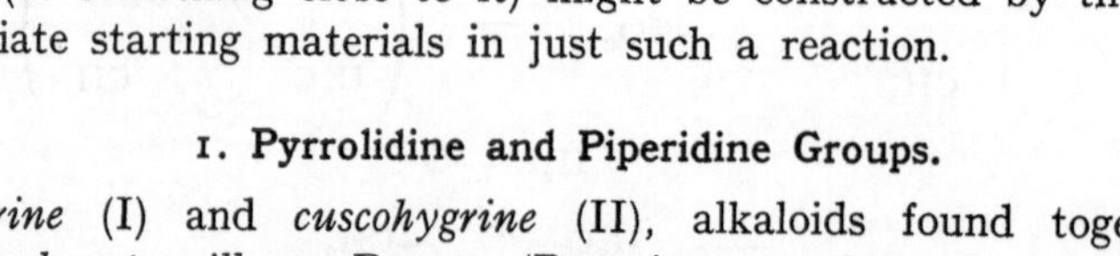

(I.) Hygrine.   (II.) Cuscohygrine.

intermediates, representative are acetone, acetoacetic acid or acetone-dicarboxylic acid for the active hydrogen component; N-methyl-γ-amino-butyraldehyde (III) or Δ¹-pyrroline (followed by N-methylation), each arising from ornithine, are exemplary intermediates for the second, basic moiety. ROBINSON, in an informal report, was the first to claim a laboratory synthesis based on these principles; condensation of (III),

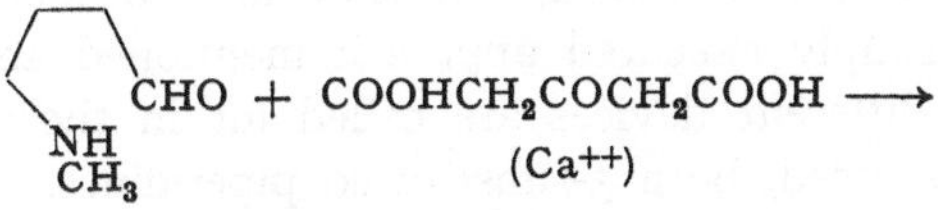

$$\text{CHO} + \text{COOHCH}_2\text{COCH}_2\text{COOH} \longrightarrow$$

$$\underset{\text{CH}_3}{\overset{\text{NH}}{|}}$$

$$(\text{Ca}^{++})$$

(III.) N-Methyl-γ-aminobutyraldehyde.

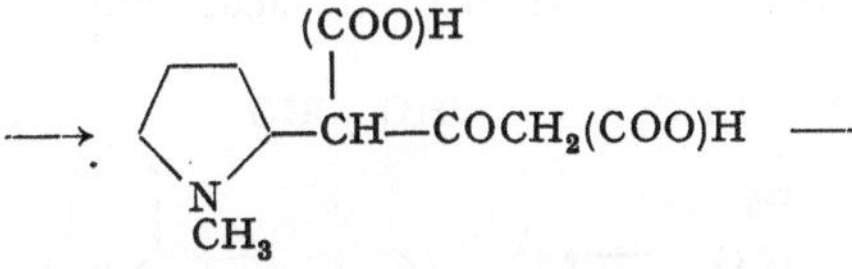

$$\longrightarrow \quad \underset{\text{CH}_3}{\overset{(\text{COO})\text{H}}{\underset{|}{\text{N}}}}\text{—CH—COCH}_2(\text{COO})\text{H} \longrightarrow \quad (\text{I.}) \text{ Hygrine.}$$

secured from the acetal, with the calcium salt of acetonedicarboxylic acid produced hygrine (74). The reaction was verified later by ANET, HUGHES and RITCHIE (4), who identified both hygrine and cuscohygrine as products; and by GALINOVSKY and coworkers (42), who generated the aminoaldehyde by selective reduction of N-methyl-pyrrolidone with lithium aluminum hydride.

A similar system is encountered in the cases of *isopelletierine* (IV) and N-methylisopelletierine (V), constituents of *Punica granatum* L. (pomegranate) which are likely to be formed in vivo from an acetone

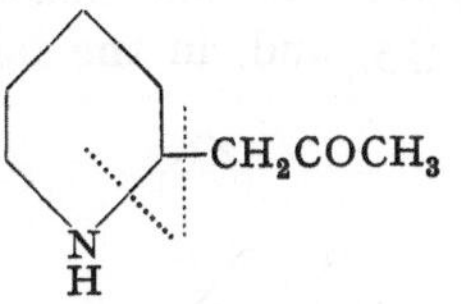

$$\underset{\overset{|}{\text{H}}}{\overset{}{\text{N}}}\text{—CH}_2\text{COCH}_3 \qquad\qquad \underset{\overset{|}{\text{CH}_3}}{\overset{}{\text{N}}}\text{—CH}_2\text{COCH}_3$$

(IV.) Isopelletierine. 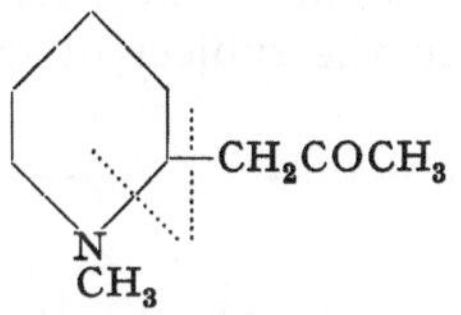   (V.) N-Methylisopelletierine.

equivalent and from lysine, via N-methyl-δ-valeraldehyde or a comparable intermediate. In the laboratory, either base can be prepared by starting with acetoacetic acid and the appropriate amino aldehyde (5, 42).

Another piperidine structure which can be subdivided nicely into Mannich base components is *lobelanine* (VI), one of the several alkaloids occurring in *Lobelia* species. In this case methylamine, glutardialdehyde and acetophenone are the hypothetical building blocks; and it is

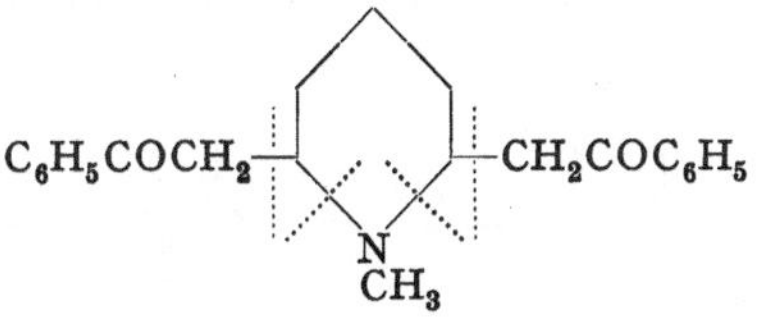

$$\text{C}_6\text{H}_5\text{COCH}_2\text{—}\underset{\overset{|}{\text{CH}_3}}{\overset{}{\text{N}}}\text{—CH}_2\text{COC}_6\text{H}_5$$

(VI.) Lobelanine.

gratifying to find that these components (except for the substitution
of benzoylacetic acid for the simple ketone) cooperate at pH 4 and 25°
to give, after some hours, the natural product (86). Other syntheses of
lobelanine have been achieved, but they are far more laborious than
the tidy and simply executed approach mentioned above.

Somewhat different devices are called for in the remaining simple
cases to be discussed, both 3-substituted piperidines. *Arecaidine* (VII),
one of the bases secured from betel nut palm *(Areca catechu)*, may be
viewed as an N-methylated, reduced nicotinic acid; but an alternative

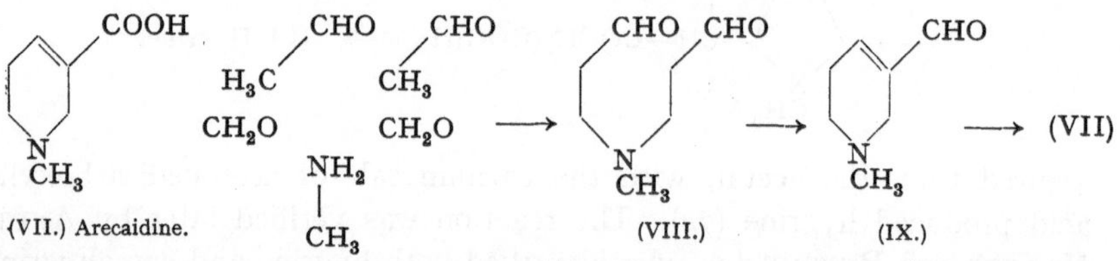

(VII.) Arecaidine.   (VIII.)   (IX.)

interpretation of its biogenesis involves interaction of formaldehyde,
acetaldehyde and methylamine—to state the case in its simplest form—to
give the amino dialdehyde (VIII), followed by cyclization to (IX) and
then oxidation. Mannich (64) has demonstrated that the simple units
listed interact at pH 3 and 70° to form directly the cyclic aldehyde (IX)
which is convertible to arecaidine.

The origin of the pyridine ring in *nicotine* (X) and *anabasine* (XI)
has been the subject of much discussion (75) and, in the former case,

(X.) Nicotine.   (XI.) Anabasine.

experimental investigation. More recently, nicotinic acid as the biological
precursor has again come into consideration, but we find that the biogenetic-type synthesis of the anabasine system is based on Δ^1-piperideine,
presumed to arise in Nature by oxidative-decarboxylation of the amino
acid lysine. Schöpf (84) has shown that this heterocycle dimerizes under
suitable conditions in a reaction involving nucleophilic attack by one

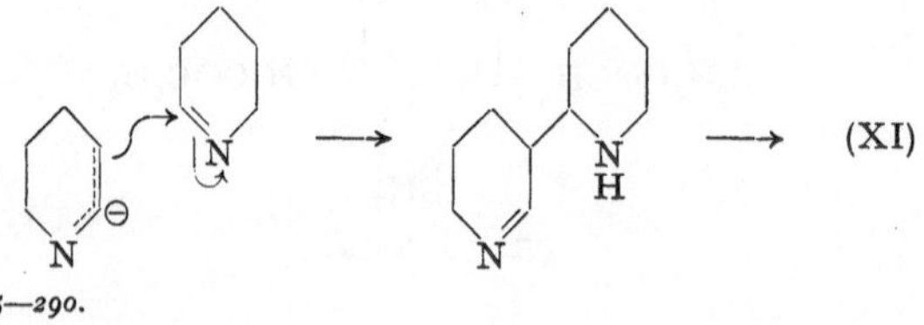

molecule on the imino group of a second. The synthesis was completed by Ag-I dehydrogenation.

Several interpretations may be placed on the inter- and intrastructural relationships evident in the tropane group, among which the mydriatic

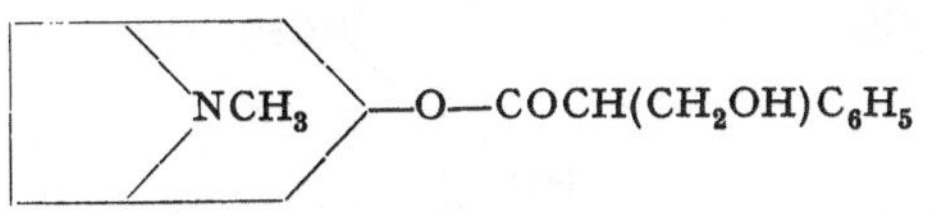

(XII.) Atropine. 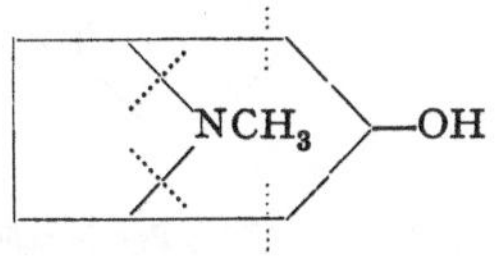   (XIII.) Tropine.

agent, *atropine* (XII), may be cited as an example. The alkamine portion, tropine (XIII), commands our attention, and one can easily discern an interesting modification of the simple Mannich base pattern. Again the methylamine and acetone requirements are clear; but in order that the bridged system can evolve, a potentially difunctional third Mannich component is required. [Ornithine can be a source of tropine, as demonstrated experimentally (62), although the actual intermediate involved during its incorporation into the tropine skeleton remains unknown.] ROBINSON (72) recognized the practical needs for reducing the implied biogenetic plan to the laboratory level and employed succindialdehyde, methylamine and the calcium salt of acetonedicarboxylic acid toward this end (p. 244).

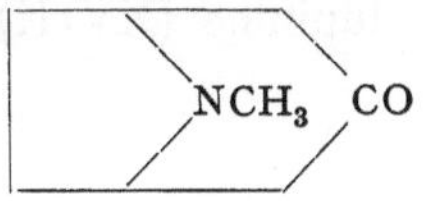

(XIV.) Tropinone.

Tropinone (XIV), the ketone corresponding to the naturally-derived alcohol, in fact evolved. In a still more remarkable experiment, ROBINSON found that when α,α'-diaminoadipic acid, citric acid and ammonia in aqueous solution were allowed to stand in the sunlight for six months, during which time dilute hydrogen peroxide was added occasionally, some nortropinone (XIVa) was produced,

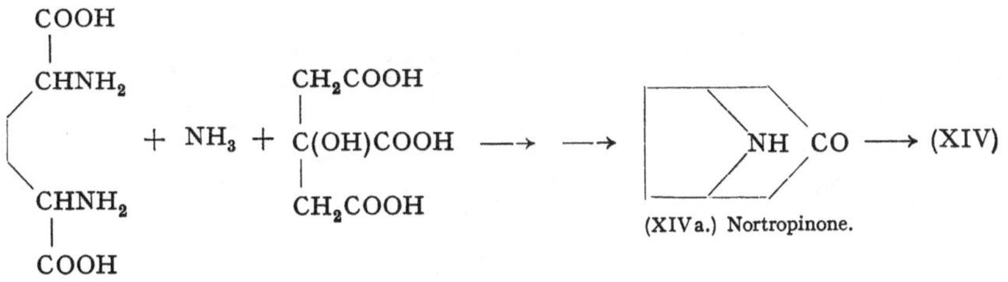

(XIVa.) Nortropinone.

as shown by the isolation of tropinone after N-methylation (74). The former sequence has been studied in more detail by SCHÖPF (86), who has shown that under the most suitable conditions the yield of tropinone is excellent. The underlying pattern of synthesis is capable of variation,

and many other bicyclic aminoketones, natural and otherwise, have
been prepared (*88, 92, 100*). For example, by utilizing glutardialdehyde,

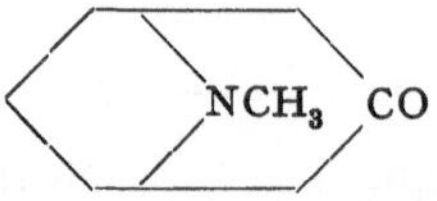

(XV.) Pseudo-pelletierine. 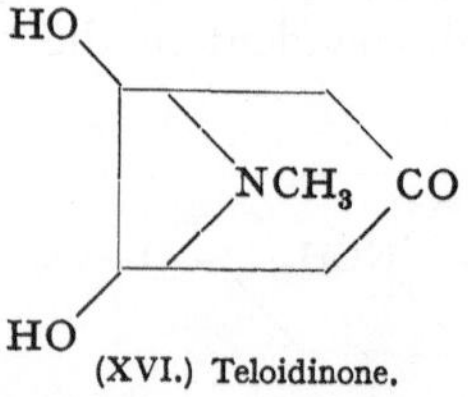   (XVI.) Teloidinone.

methylamine and acetonedicarboxylic acid, the pomegranate alkaloid
pseudo-pelletierine (XV) can be made. If *meso*-tartardialdehyde is
substituted for the simpler dialdehyde, teloidinone (XVI), reducible to
the alcoholic component of the alkaloid *meteloidine*, is produced (*84*).

2. Quinolizidine Group.

Of the Papilionaceae alkaloids, which are characterized by the
quinolizidine ring system, *lupinine* (XVII) and its *trans* isomer epilupinine (XVIII), are the simplest. These aminoalcohols are assumed

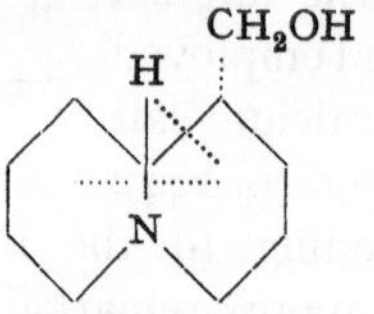

(XVII.) Lupinine. 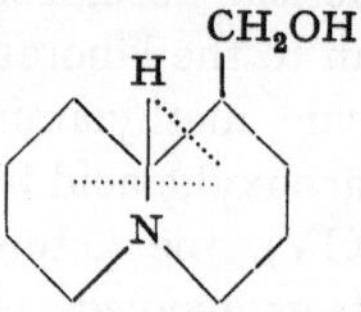   (XVIII.) Epilupinine.

to derive from lysine according to the following scheme; and this proposal
has recently found support with the experimental finding that lysine
can be a precursor of lupinine, and also sparteine. The laboratory counter-

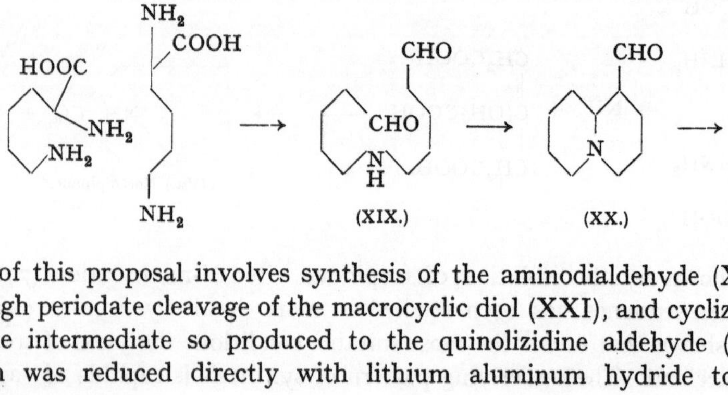

(XIX.)   (XX.)

part of this proposal involves synthesis of the aminodialdehyde (XIX),
through periodate cleavage of the macrocyclic diol (XXI), and cyclization
of the intermediate so produced to the quinolizidine aldehyde (XX),
which was reduced directly with lithium aluminum hydride to the

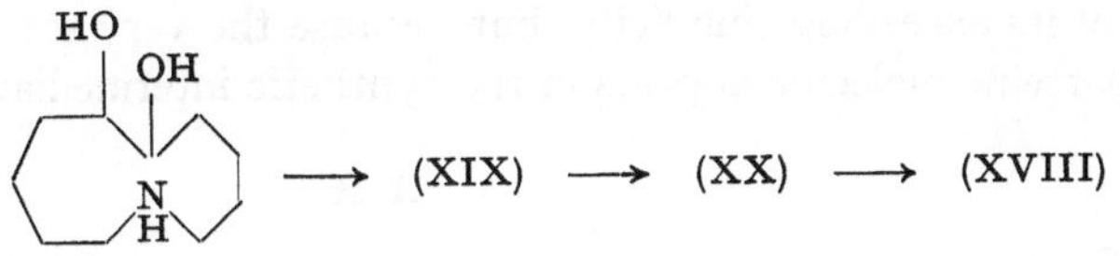

(XXI.) → (XIX) → (XX) → (XVIII)

identified product, *DL*-epilupinine (XVIII) (*108*). In a somewhat related approach, the dialdehyde (XXII), generated hydrolytically from the

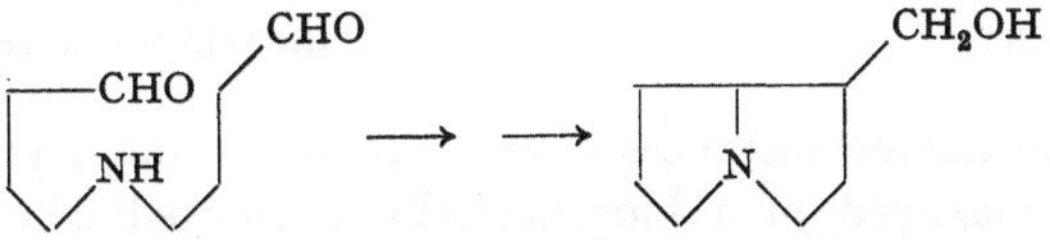

(XXII.)   (XXIII.)

diacetal, cyclized and produced, on reduction, the analogous alcohol in the pyrrolizidine series (XXIII), an alkamine belonging to the *Senecio* alkaloid class (*63*).

More uncertainty surrounds the precise methods at work in the biosynthesis of *sparteine* (XXIV), the best-known lupin alkaloid, and related tetracyclic bases; and it is entirely possible that slightly differing routes are employed by Nature in building up the variants. In regard to synthetic adaptations, a projected scheme involving δ-aminovaleraldehyde(Δ^1-piperideine), formaldehyde and acetonedicarboxylic acid, went astray and led — not to 8-ketosparteine, as claimed (*6*) — but to the spiroketal (XXV) (*83*); interestingly enough, treatment of the ketal with acetic anhydride resulted in the formation of *DL*-α-

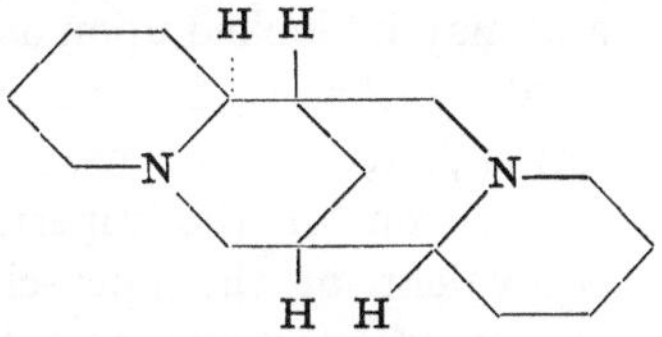

(XXIV.) Sparteine.

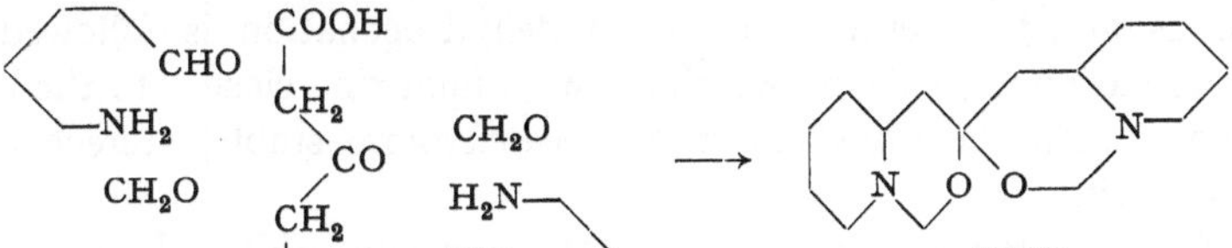

(XXV.)

isosparteine, diastereoisomeric with sparteine (*81*). In another hypothetical scheme, biosynthesis starting from lysine and the related γ-keto-α,α'-diaminopimelic acid is featured; these amino acids are appropriately decarboxylated, deaminated and condensed so that intermediate (XXVI) results, which then cyclizes Mannich-wise to 8-ketosparteine; reductive removal of oxygen completes the sequence. The laboratory duplication (*109*) of the latter, integral portion of this scheme is worthy of note, not only because of its exceeding simplicity, but because the very stereochemistry of the sparteine molecule appears in the synthetic intermediate (XXVII).

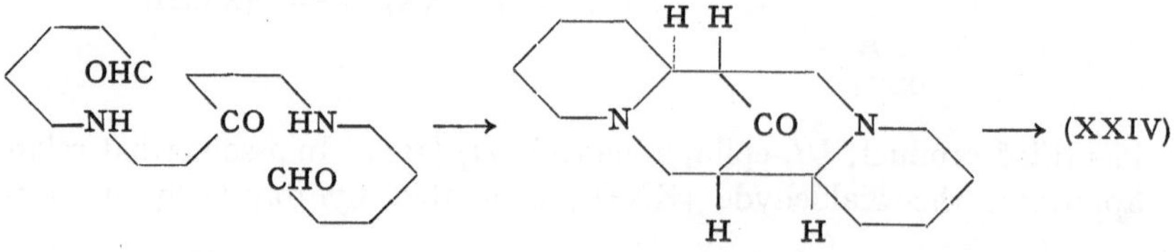

(XXVI.)   (XXVII.) 8-Ketosparteine.

This is one of the few instances where as many as four asymmetric centers have been managed in a biogenetically patterned laboratory scheme,

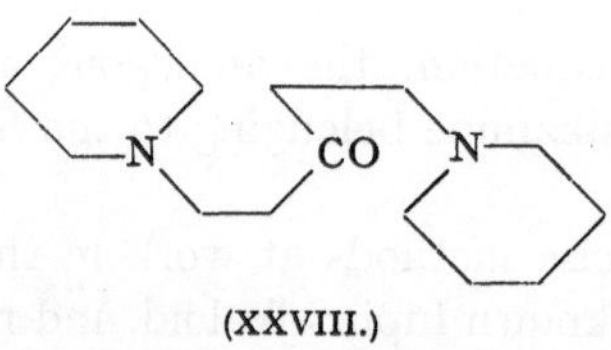

(XXVIII.) 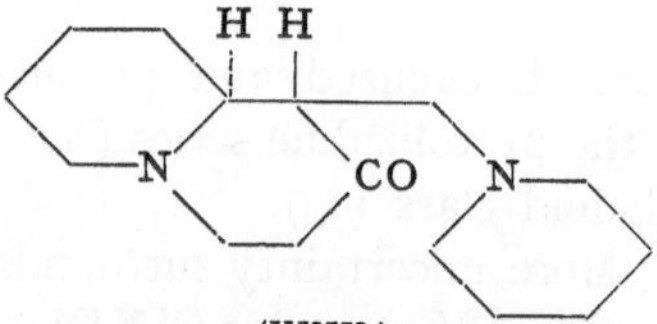   (XXIX.)

and may be looked upon as an example of a biogenetic-type stereospecific total synthesis. In the first of three steps, acetone, piperidine and formaldehyde were used to construct the Mannich base (XXVIII). Conversion of the piperidine rings to tetrahydropyridine rings—the equivalent of the open-chain amino aldehyde—was accomplished by means of mercuric acetate dehydrogenation, and the product which emerged from this treatment was the tetracyclic ketone (XXVII); Wolff-Kishner reduction then afforded *DL*-sparteine (XXIV, p. 253). It seems likely that the cyclization proceeds stepwise, first to the intermediate (XXIX), possessing the more stable *trans* relation of the two asymmetric centers; a second dehydrogenation is followed by a conformational inversion, which allows a final ring closure to the bridged system, with the *trans*(less stable)-*syn-cis*(more stable) stereochemistry of sparteine.

Different tactics are required by the partially oxidized lupin alkaloids, e. g. *cytisine* (XXX) and the epimeric *anagyrine* and *thermopsine* (XXXI), since provision must be made for the pyridone rings (*106*). However,

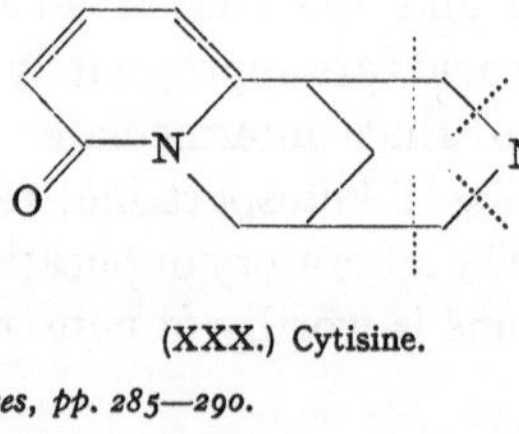

(XXX.) Cytisine. 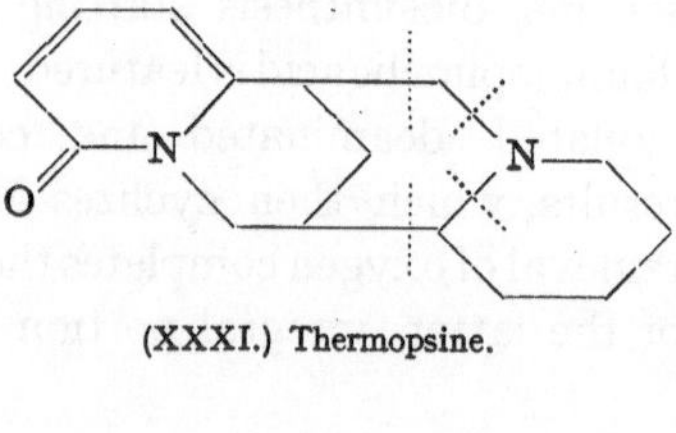   (XXXI.) Thermopsine.

the biogenetical thread is not lost, in the latter case (**XXXI**), the essentials of the desired system were assembled in one step through the interaction of Δ^1-piperideine and a pyridyl malonic acid (**XXXII**) in which the formaldehyde equivalent was already present as an exocyclic methylene group. A similar reaction, in which formaldehyde and benzyl-amine replace piperideine, afforded the precursor (**XXXIV**) of cytisine.

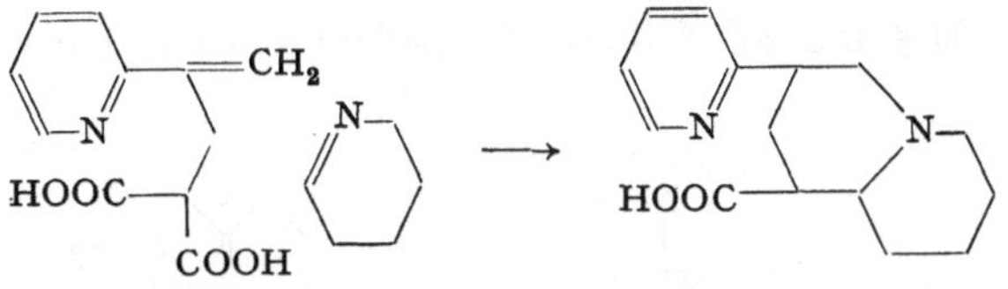

(XXXII.)　　　　　　　　　　　(XXXIII.)

In both cases (**XXXIII** and **XXXIV**), completion of the total syntheses involved a more or less standard sequence of operations, including formation of the pyridinium salt (**XXXV**), and ferricyanide oxidation of the latter to the corresponding pyridone.

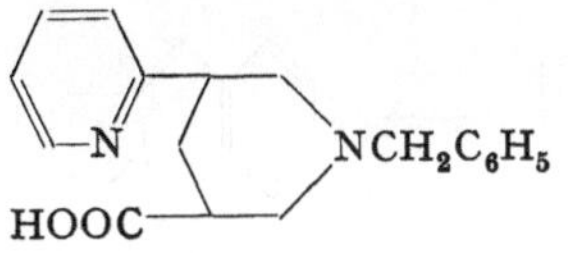

(XXXIV.) 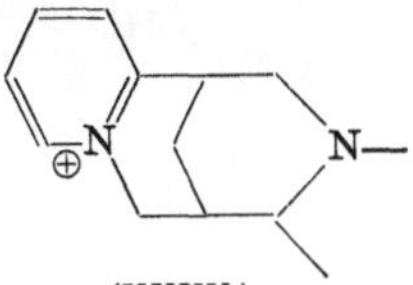　　　　　　　　　　　(XXXV.)

3. Anthranilic Acid Group.

This alkaloid type was reviewed in the present Series by Price (*70a*).

A recurring unit in the alkaloid class is the anthranilic acid nucleus, thought to be associated with tryptophan in a degraded state. Again, the biologically active species has not yet been identified, but *o*-amino-benzaldehyde has been a favorite representative in laboratory syntheses.

Angostura provides a family of simple quinoline derivatives, of which 2-*n*-amylquinoline (**XXXVI**) is exemplary. Schöpf (*85*) has found that the condensation of *o*-aminobenzaldehyde with methyl-*n*-amyl ketone

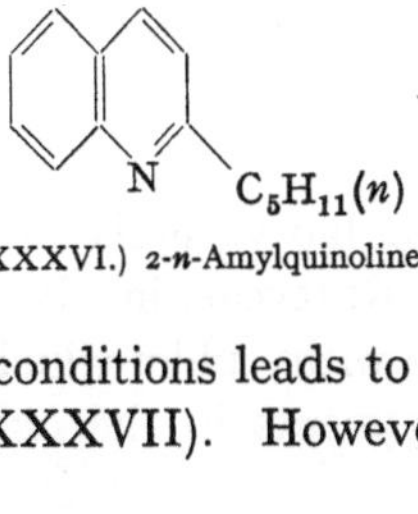

(XXXVI.) 2-*n*-Amylquinoline. 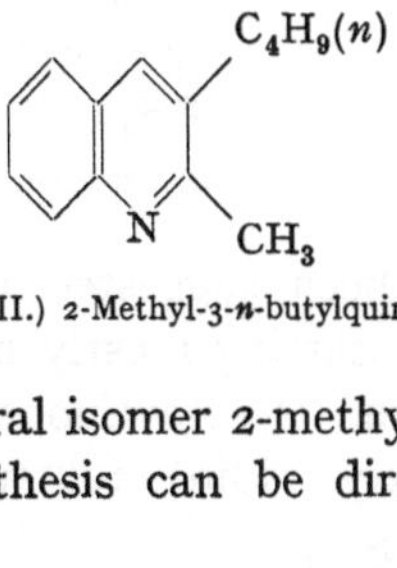　　　(XXXVII.) 2-Methyl-3-*n*-butylquinoline.

under mild conditions leads to the unnatural isomer 2-methyl-3-*n*-butyl-quinoline (**XXXVII**). However, the synthesis can be directed along proper lines by substituting a β-ketoacid for the simple ketone—thus caproylacetic acid led readily to the desired product (XXXVI).

Vasicine (XXXVIII) is considered to arise in Nature by interaction of an anthranilic acid equivalent and an aliphatic unit such as α-hydroxy-γ-aminobutyraldehyde, possibly derived from ornithine, β-hydroxy-ornithine or proline. The closest laboratory parallel (*87*) is condensation of o-aminobenzaldehyde with γ-aminobutyraldehyde (as the diethyl acetal) at pH 5 and 30°, followed by palladium-catalyzed rearrangement

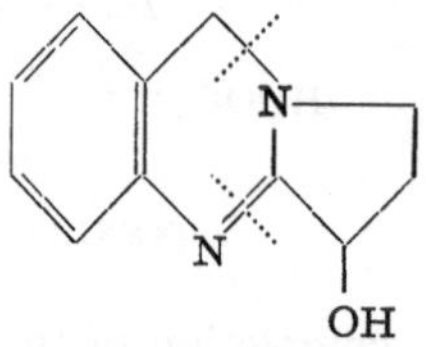

(XXXVIII.) Vasicine. 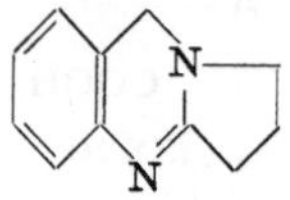   (XL.) Deoxyvasicine.

of the intermediate (XXXIX) to deoxyvasicine (XL), known also from the natural source.

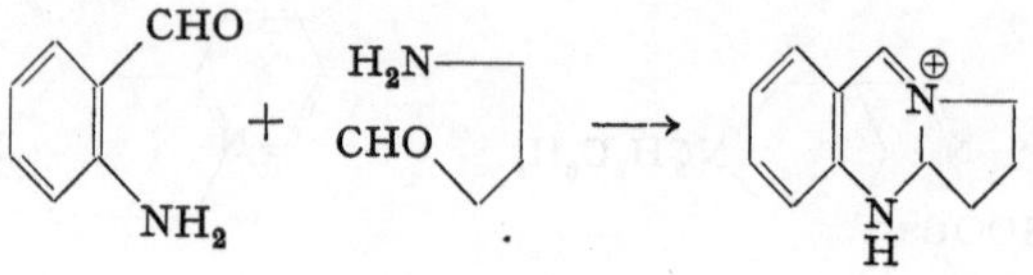

(XXXIX.)

Appreciation of the biogenetic-type approach to synthesis is evident in the method applied to *arborine* (XLI), in which the elements of anthranilic acid, ammonia and phenylacetic acid (from phenylalanine) are detected. The amide (XLII), prepared by phenylacetylation of N-methylanthranilamide, gave rise on heating to the natural product (*25*).

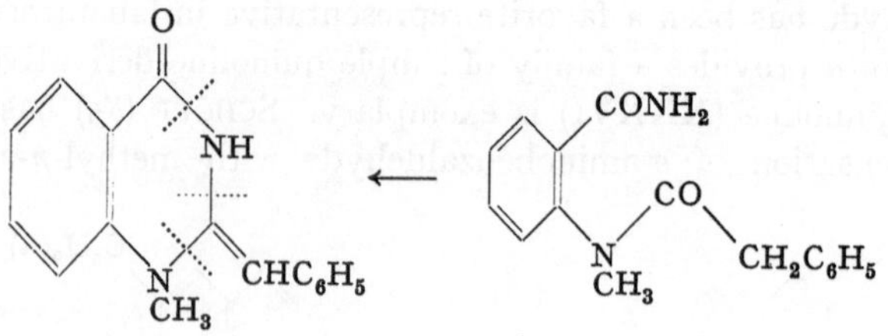

(XLI.) Arborine.   (XLII.)

It has been suggested (*57*) that the biosynthesis of the *acridone alkaloids*, occurring mostly in Australian Rutaceae, involves in principle condensation of o-aminobenzaldehyde with phloroglucinol to give 2,4-dihydroxyacridine, followed by adjustments in the oxidation state, including conversion to the acridone. The key reaction mentioned has actually been known for some time (*38*), and has been shown to proceed

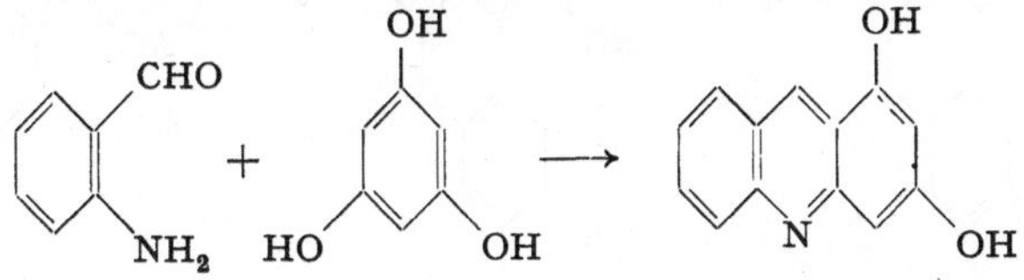

2,4-Dihydroxyacridine.

in high yield at pH 8 and room temperature. By means of a variation in the basic plan, a synthesis of a naturally occurring acridone was

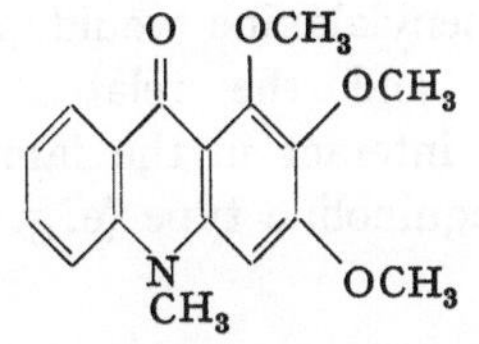

(XLIII.) 2,3,4-Trimethoxy-10-methylacridone. 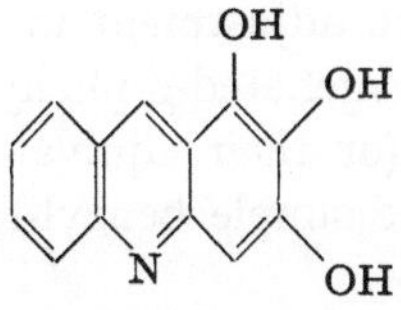   (XLIV.) 2,3,4-Trihydroxyacridine.

achieved (*57*), viz. that of 2,3,4-trimethoxy-10-methylacridone (XLIII), which has been detected in the leaves of *Evodia alata*. The acridine (XLIV) was secured through the use of 1,2,3,5-tetrahydroxybenzene; and after O- and N-methylation, ferricyanide oxidation was employed to produce (XLIII).

4. Isoquinoline Group.

Reproduced below are some of the structural clues which allowed earlier investigators such as ROBINSON, WINTERSTEIN and TRIER to formulate reasonable biosynthetic routes for most of the isoquinoline

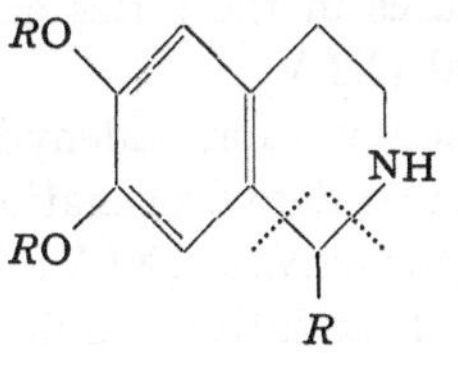

(XLVI.)
R = H, R' = CH$_3$. Norsalsoline.

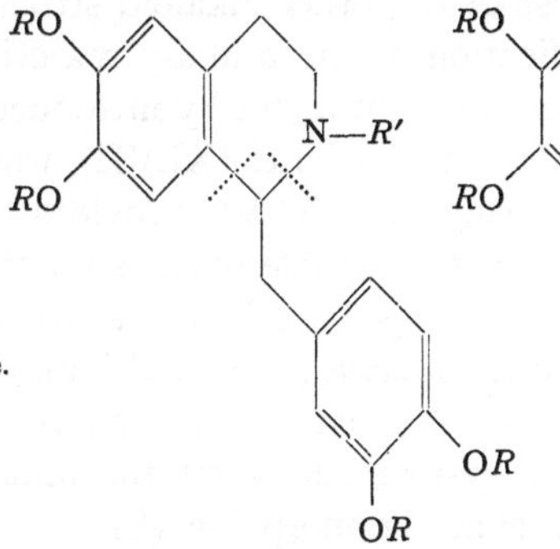

(XLV.)
$R = R' =$ CH$_3$. Laudanosine.
$R =$ CH$_3$, $R' =$ H. Norlaudanosine.
$R =$ H $R' =$ CH$_3$. N-Methyl-laudanosoline.

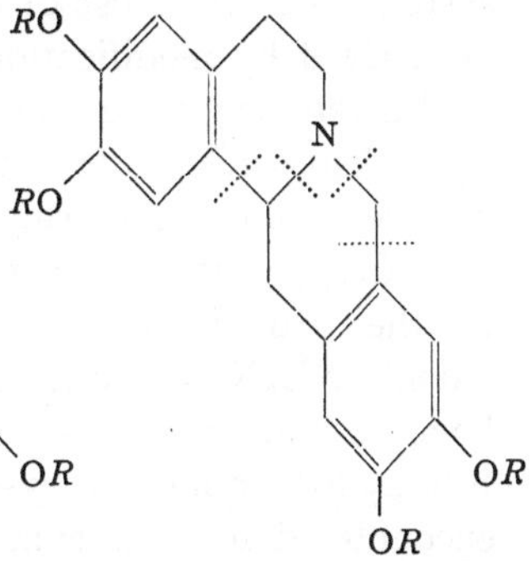

(XLVII.) Berberine alkaloids.
$R =$ CH$_3$. Norcoralydine.

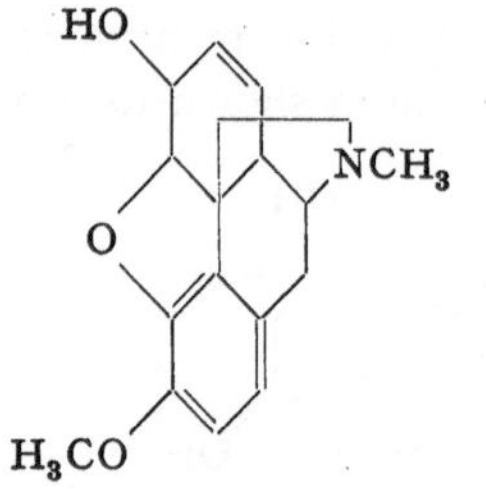

(L.) Morphine. 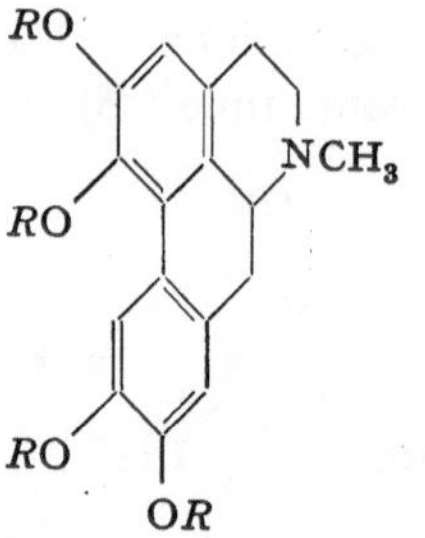   (XLIX.) Aporphine alkaloids.   (XLVIII.) $R = CH_3$. Papaverine.
                          $R = CH_3$. Glaucine.

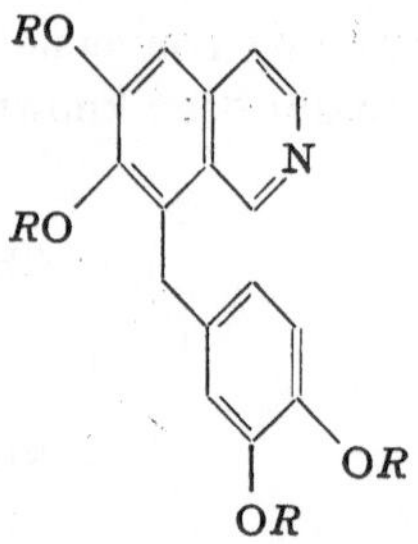

alkaloids, well-known as constituents of *opium*. It was considered that appropriate adjustment in dihydroxyphenylalanine would provide both the *ar*-oxygenated-β-phenylethylamine and the related phenylacet-aldehyde (or their equivalents), which interact in the familiar fashion to give the simple benzyl-tetrahydroisoquinoline type (e. g. *laudanosine*,

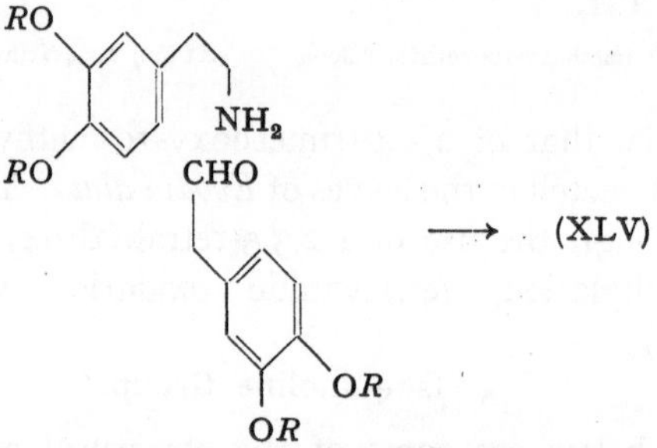

XLV, $R = R' = CH_3$). Union of the phenylethylamine with a building stone derived from a purely aliphatic amino acid affords the simpler system (XLVI). Most of the other alkaloid structures in the series can be derived by modification of the primary material (XLV).

Further ring closure brought about by introduction of a formaldehyde equivalent leads to the *berberine* series (XLVII), whereas dehydrogenation of the heterocyclic ring in (XLV) provides *papaverine* (XLVIII, $R = CH_3$). Another reaction course open is internal oxidative coupling of the two benzenoid rings, which takes us to the aporphine al-kaloids (XLIX), of which *glaucine* ($R = CH_3$) is an example; alternatively, bond formation at a substituted position on the aromatic half of the tetrahydroisoquinoline system allows for the unusual bridged structure encountered in sinomenine or morphine (L).

A certain amount of success has been achieved in attempts to simulate the implied phytochemical operations. Thus, there was no difficulty in converting 3,4-dihydroxyphenylethylamine by treatment with acetaldehyde under mild conditions (*82*) to *norsalsoline* (XLVI, $R = H$, $R' = CH_3$), or in producing *norlaudanosine* (XLV, $R = CH_3$, $R' = H$) by condensing 3,4-dimethoxyphenylethylamine with 3,4-dimethoxy-phenylacetaldehyde (*47*, *95*). Aromatization to (XLVIII) is trivial; and the laboratory conversion of norlaudanosine to norcoralydine (XLVII, $R = CH_3$) by means of formaldehyde has been accomplished (*70*, *96*).

Synthesis of the aporphine or morphine type through laboratory oxidation of a laudanosine derivative has met with difficulty, however. For example, dehydrogenation of N-methyl-laudanosoline (XLV, $R = H$, $R' = CH_3$) afforded not (XLIX, $R = H$), but the product of nitrogen-carbon coupling (LI) (*76*, *90*). An unexpected dividend resulted

(LI.)

from this speculative venture when it later developed that the system so produced occurs naturally (*40*), doubtless arising by an equivalent biochemical process.

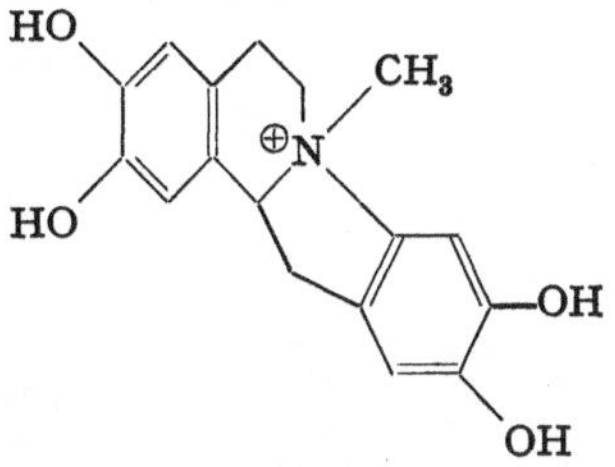

Likewise, the in vitro oxidative ring closure of a suitably substituted *benzyl-tetrahydroisoquinoline* to a member of the morphine-sinomenine class remains a tantalizing possibility (*75*). The closest analogy which might be cited here is the *acid-catalyzed* ring closure of the benzyl-octa-hydroisoquinoline (LII) to *DL*-tetrahydro-deoxycodeine (LIIa) (*44*).

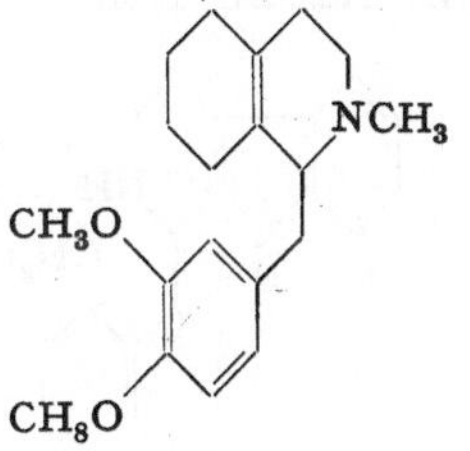

(LII.)

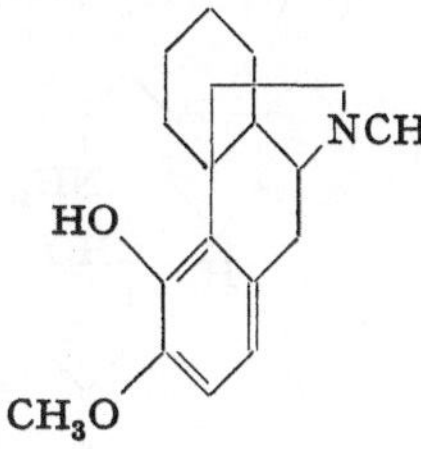

(LIIa.) Tetrahydro-deoxycodeine.

Finally, the nuances which lend interest to the berberine series find their parallel, in some instances, in the laboratory. *Cryptopine* (LIII)

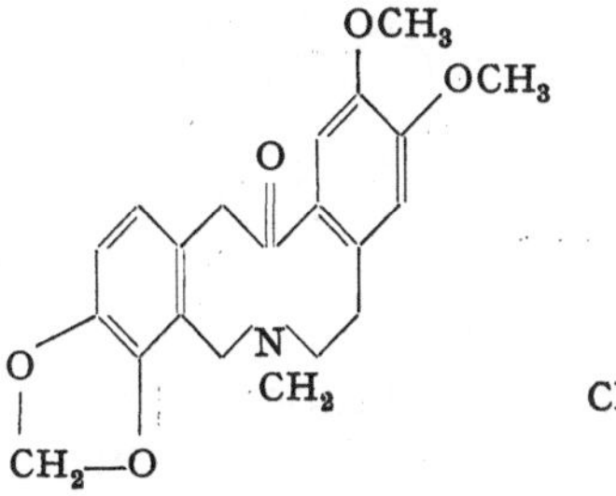

(LIII.) Cryptopine.

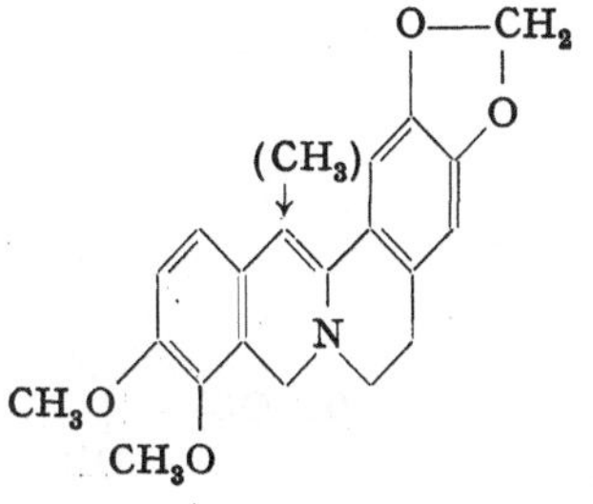

(LIV.) Dihydroberberine.

17*

must be formed in Nature by N-methylation and oxidative cleavage
to the ten-membered ring, a sequence which has its parallel in the test
tube. Also, dihydroberberine (LIV), being an enamine, has been methylated
at the indicated position (48, 111); a similar process must be operative
in the natural production of *corydaline*, a *Corydalis* alkaloid having
the structure (LV).

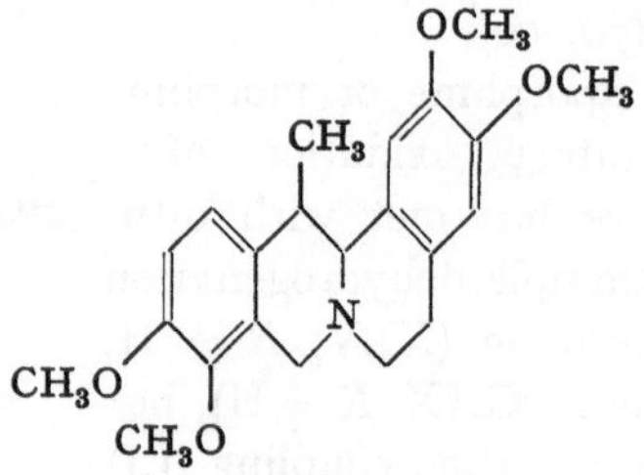

(LV.) Corydaline.

5. Indole Group.

Multiform, and reminiscent in many respects of the isoquinoline
alkaloids, the indole group of plant products reveal architectural features
which have influenced the synthetic organic chemist. Since space does
not permit description of all the members in this family, only those
cases which have inspired biogenetic-type syntheses will be presented.

Although allowance must be now made for the possibility that the
yohimbine group (LVIII) is derived in Nature from a non-aromatic

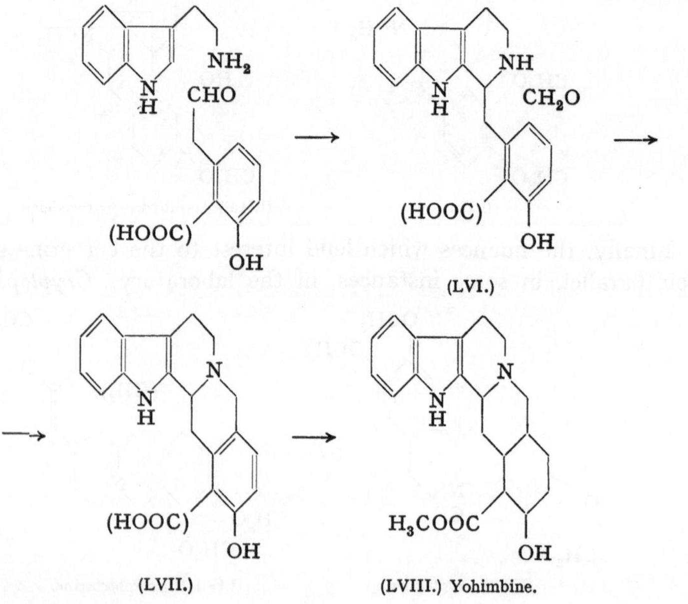

(LVI.)

(LVII.)   (LVIII.) Yohimbine.

ring-*E* precursor (*114*), it has been generally assumed that tryptamine (from tryptophan) and a phenylacetaldehyde (from the corresponding phenylalanine), or a related species such as the pyruvic acid, are the raw materials, utilized as shown below. The essentials of this scheme have been realized in the laboratory by HAHN (*45*). Condensation of tryptamine with *m*-hydroxyphenylacetaldehyde (or preferably, *m*-hydroxyphenylpyruvic acid) yielded under "physiological conditions" the tetrahydro-β-carboline (cf. LVI), which on treatment with formaldehyde afforded the pentacyclic yohimbine skeleton (LVII, without the carboxyl group). Alternatively—much as in the tetrahydroisoquinoline series—tryptamine and acetaldehyde can be combined at pH 5–6 to give the simple indole alkaloid *eleagnine* (*46*).

One of the most complex cases in the indole series which has been analyzed is *strychnine* (LXII). The first suggestion (*118*) featured origin again from tryptamine and a hydroxylated phenylalanine, but with

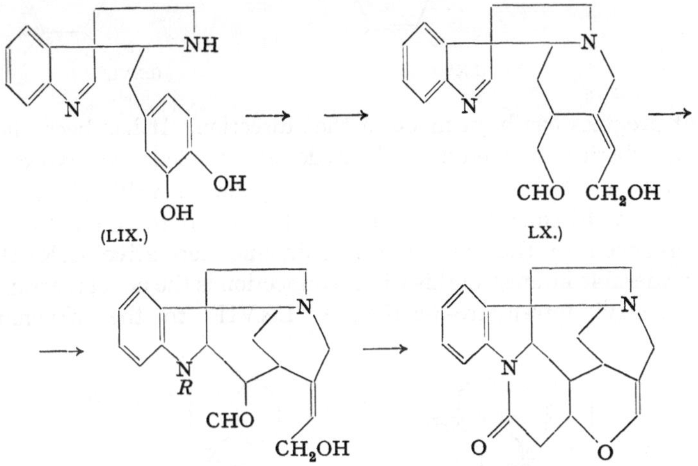

(LIX.)   LX.)

(LXI.)   (LXII.) Strychnine.

the difference that the β- rather than the α-position of the indole nucleus is involved. Formation of an intermediate such as (LIX) is accompanied by introduction of the methylene bridge and fission of the oxygenated ring (LX), followed by cyclization to (LXI). An acetic acid equivalent is needed for formation of the piperidone unit and the seven-membered ether cycle; and this process has been carried out in the laboratory, starting with the Wieland-Gumlich aldehyde, a naturally occurring substance (LXI, R = H) (*7*). Alternatives to the above scheme include (a) replacement of dihydroxyphenylalanine by a non-aromatic precursor, shikimic or prephenic acid (*114*), and (b) initial β-cyclization involving an oxindole rather than a true indole ring.

There has been much informal interest evidenced in accomplishing a biogenetic-type laboratory synthesis of the strychnine system, and

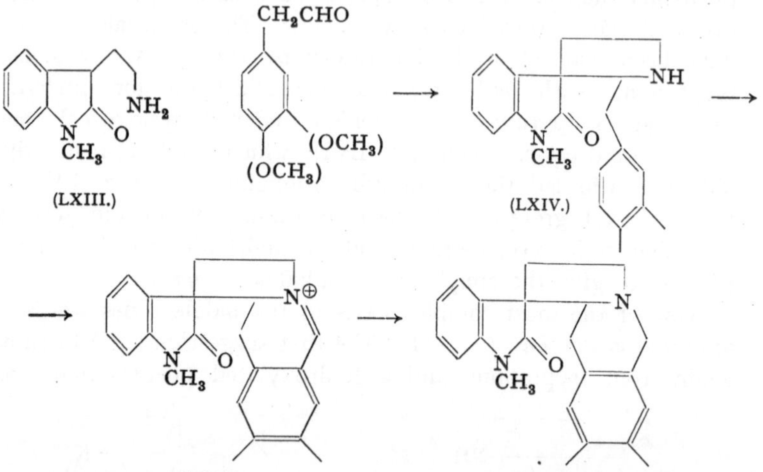

(LXIII.)   (LXIV.)

(LXV.)   (LXVI.)

recent progress has been made in that direction. It has been shown (8) that a substituted phenylacetaldehyde will condense as expected with the oxindole (LXIII) to yield the spiro compound (LXIV). N-Formylation followed by Bischler-Napieralski ring closure led to (LXV), which was then reduced to the tetrahydroisoquinoline derivative (LXVI). Also of considerable interest in this whole connection is the novel rearrangement of a benzyl tetrahydro-β-carboline (LXVII) to the strychnine-like

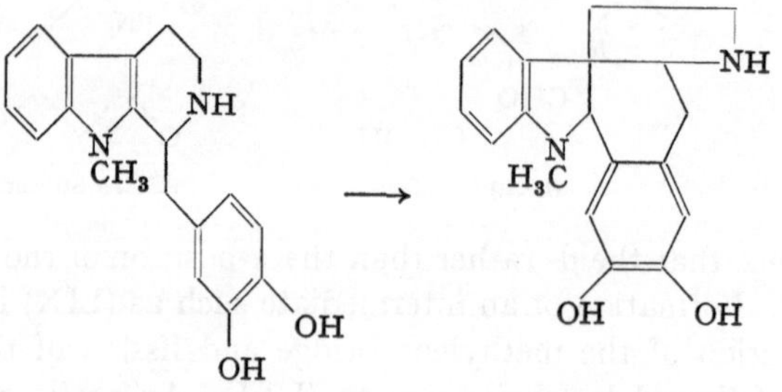

(LXVII.)   (LXVIII.)

product (LXVIII), brought about by the action of hot hydrochloric acid (50). Perhaps one of the most interesting performances a molecule has ever staged is the acid-catalyzed ring closure sequence of the simple indole derivative (LXIX) to (LXX), a close relative of the Wieland-Gumlich aldehyde, which duplicates all of the essential molecular and stereochemical features of strychnine as well as the curare bases (107).

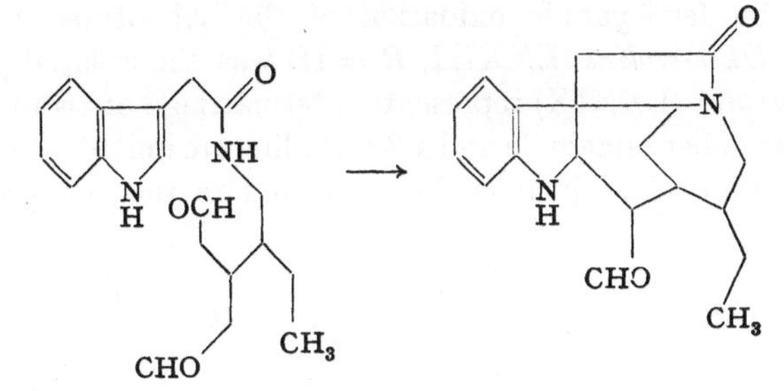

(LXIX.) (LXX.)

The variety of structures encountered in the indole group is not exhausted by the yohimbine family, but includes cases of definitely diverse ancestry. One of the simpler examples, *gramine* (LXXI), is, by

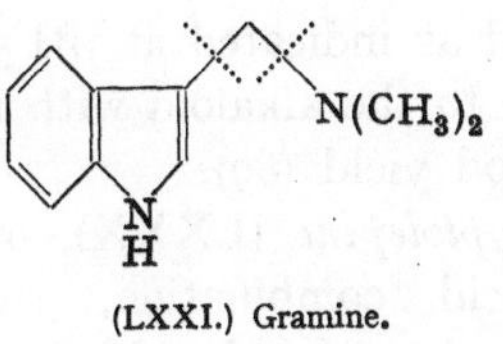

(LXXI.) Gramine. 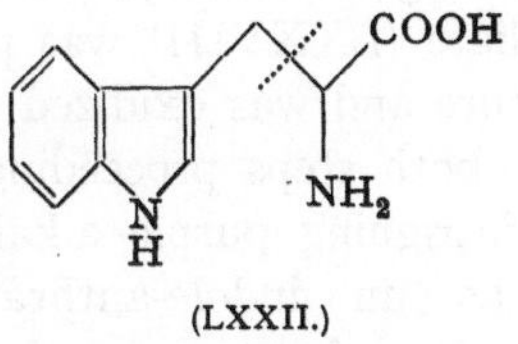 (LXXII.)

inspection and actual synthesis, a Mannich base; however, despite the implication, the alkaloid has been shown to arise in barley plants from tryptophan, by means of the cleavage (LXXII) (*13*).

The *physostigmine* (LXXIII, R = CONHCH$_3$) biogenesis has been considered to include N-methylation and β-C-methylation of a 5-oxygenated

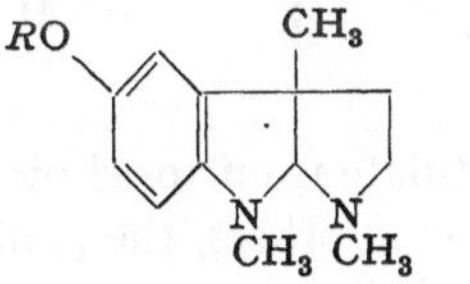

(LXXIII.) Physostigmine (R = CONHCH$_3$). (LXXIV.)
Eseroline (R = H).

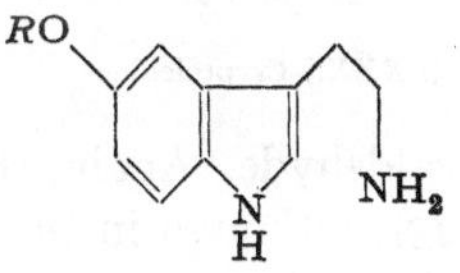

tryptamine (LXXIV), providing the indolenine (LXXV), which should cyclize spontaneously to the alkaloidal skeleton (LXXIII). In the laboratory, an intermediate akin, if not identical to (LXXV) can be

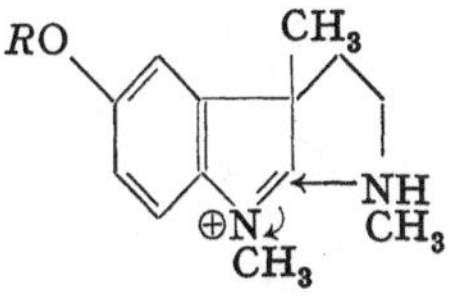

(LXXV.) 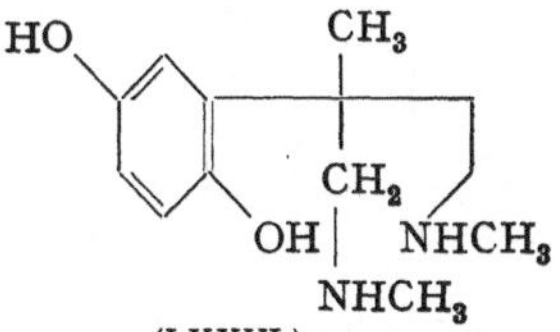 (LXXVI.)

generated by ferricyanide oxidation of the substituted hydroquinone (LXXVI); *DL-eseroline* (LXXIII, *R* = H) was the isolated product (53).

Rutaecarpine (LXXIX) represents intermarriage of two great families in that *o*-aminobenzaldehyde and a β-carboline are united, as in (LXXVII), to give (LXXVIII), which is then oxidized to the end product. In a

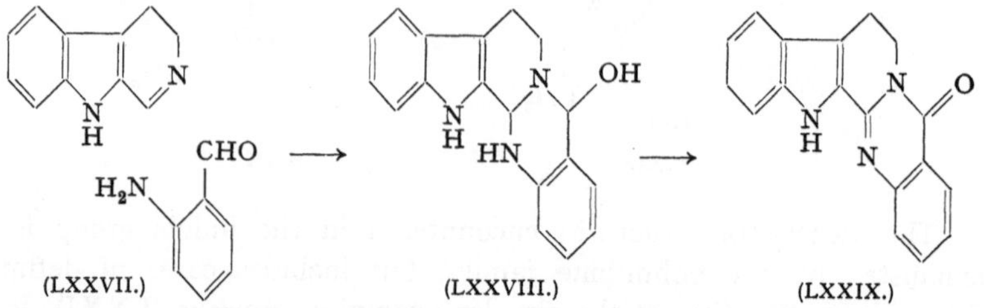

(LXXVII.)   (LXXVIII.)   (LXXIX.)

very pleasing laboratory synthesis, this route was followed exactly: intermediate (LXXVIII) was produced as indicated at pH 5 and room temperature and was oxidized directly to the alkaloid with ferricyanide at pH 7, both steps proceeding in good yield (89).

The intriguing purple alkaloid, *cryptolepine* (LXXX), may also be viewed as an indole-anthranilic acid combination, the simplest interpretation of its genesis being reaction of indoxyl with N-methyl-

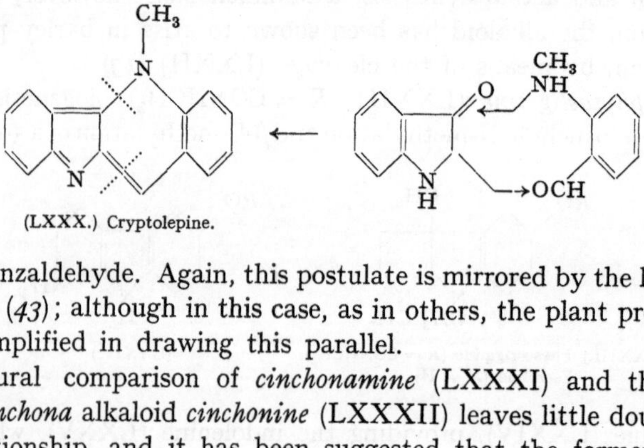

(LXXX.) Cryptolepine.

o-aminobenzaldehyde. Again, this postulate is mirrored by the laboratory chemistry (43); although in this case, as in others, the plant process may be oversimplified in drawing this parallel.

Structural comparison of *cinchonamine* (LXXXI) and the better-known *Cinchona* alkaloid *cinchonine* (LXXXII) leaves little doubt about their relationship, and it has been suggested that the former alkaloid

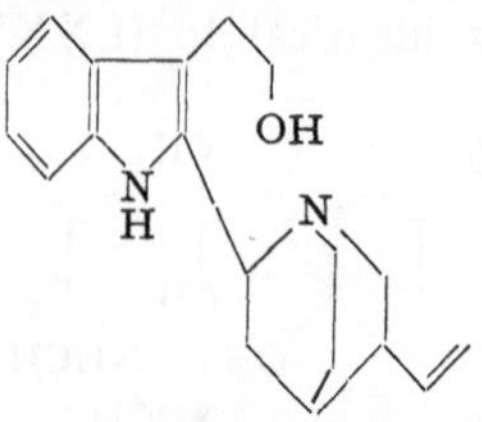

(LXXXI.) Cinchonamine. 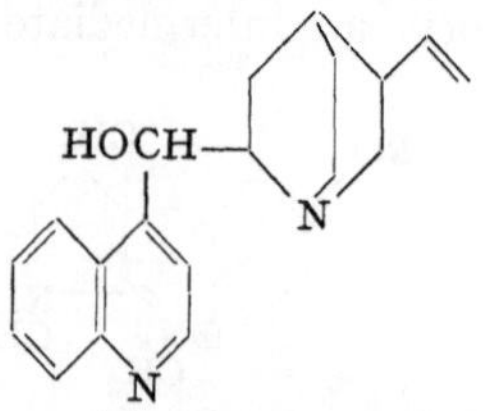   (LXXXII.) Cinchonine.

is the natural precursor of the latter. The change may be detailed in the following fashion, with no specific implication as to order of steps:

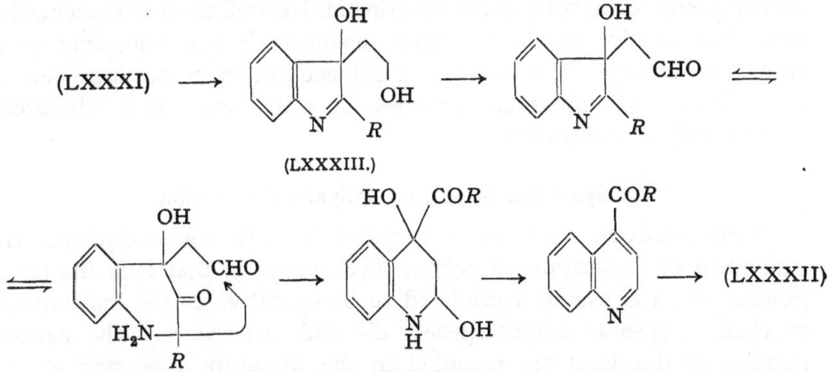

(LXXXI) →

(LXXXIII.)

⇌ → → (LXXXII)

Support for the initial oxidation in the proposed route is embodied in quinamine, an alkaloid corresponding in oxidation state to (LXXXIII), but existing in the cyclic form (LXXXIV). *Quinamine* can be secured in the laboratory by peracetic acid oxidation of cinchonamine (*116*).

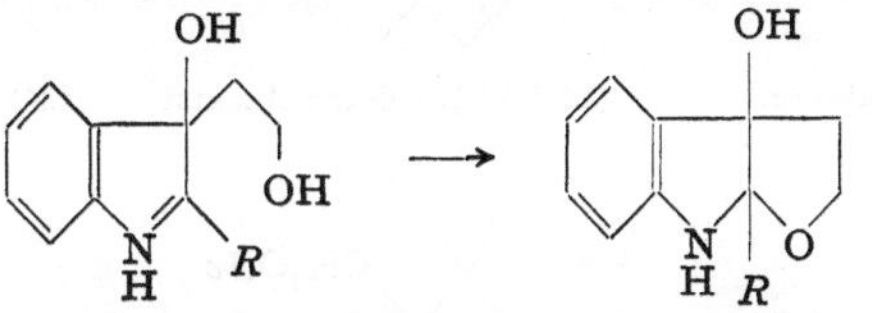

(LXXXIV.) Quinamine.

In an attempt to develop a laboratory parallel for the overall conversion (LXXXI → LXXXII), a new rearrangement was uncovered: the direct oxidative conversion of 2-methyltryptophan to 4-acetyl-quinoline, proceeding most probably through the indoleacetaldehyde or imine. The change involves a minimum of seven to eight separate reactions, corresponding in principle to those laid out above, and can be brought about through the use of aqueous alkaline hypochlorite (*110*).

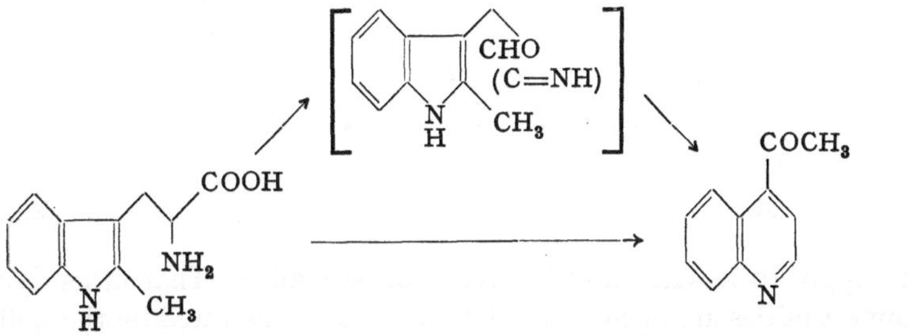

(LXXXIVa.) 2-Methyltryptophan.

(LXXXIVb.) 4-Acetylquinoline.

II. Terpenes.

Biosynthesis of terpenes and steroids has received increasing attention during recent years, both on the experimental as well as on the speculative side. The relationship of the many members is now becoming quite clear (77); however, this fascinating subject cannot be outlined here in its entirety, and only those schemes for which there is a laboratory parallel will be considered.

1. Syntheses Involving Polyene Cyclization.

Biogenetic-type synthesis of terpenes is, with few exceptions, the equivalent of acid-catalyzed polyene cyclization, in that it is the latter process which is usually considered to be operative in the biosynthesis of cyclic terpenes, sesquiterpenes, di- and triterpenes. The earliest findings of this kind are recorded in the literature appearing at the turn of the century (94), and there may be mentioned as examples the

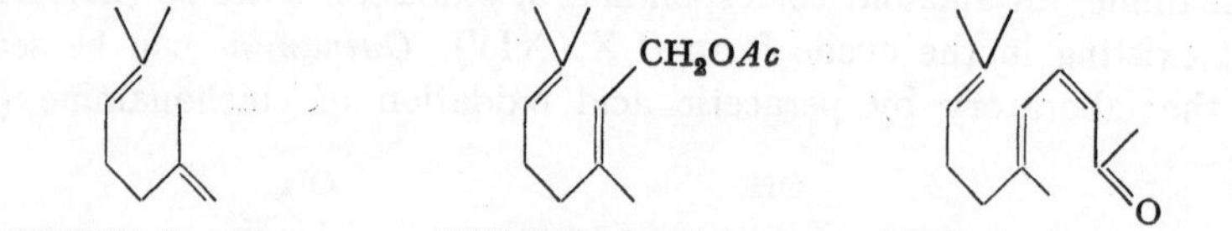

(LXXXV.) Geraniolene.   (LXXXVII.) Geraniol acetate.   (LXXXIX.) Pseudoionone.

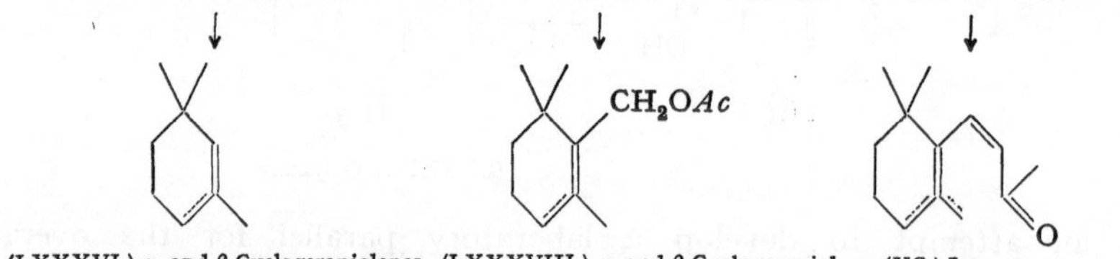

(LXXXVI.) α- and β-Cyclogeraniolenes. (LXXXVIII.) α- and β-Cyclogeraniol   (XC.) Ionones.
acetates.

cyclization of: *geraniolene* (LXXXV) to α- and β-cyclogeraniolenes (LXXXVI) (*105*); *geraniol* acetate (LXXXVII) to α- and β-cyclogeraniol acetates (LXXXVIII); and *pseudoionone* (LXXXIX) to the *ionones* (XC) (*104*), which are of importance in perfumery. All these cyclizations are based on the mechanistic course (XCI → XCII), which is supposed

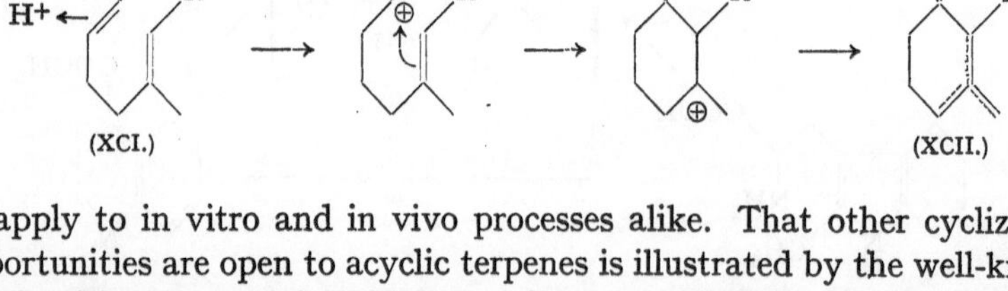

(XCI.)   (XCII.)

to apply to in vitro and in vivo processes alike. That other cyclization opportunities are open to acyclic terpenes is illustrated by the well-known relationship of geraniol (XCIII) and *limonene* (XCIV). Finally, mention should be made of an early attempt to capitalize on the biogenetic implications of the original isoprene rule, the acid-induced polymerization

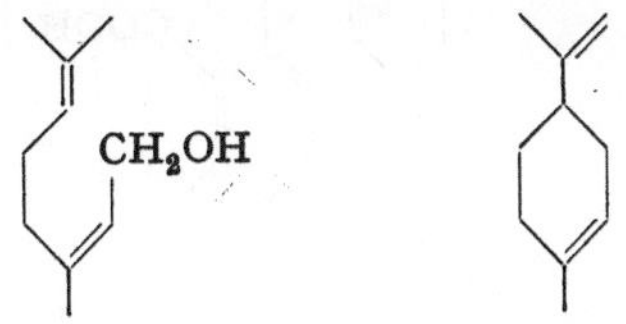

(XCIII.) Geraniol.   (XCIV.) Limonene.

of *isoprene* itself. Through the agency of acetic acid-sulfuric acid, this hydrocarbon was converted to a mixture of products, which included geraniol, cyclogeraniol, linalool, α-terpineol, 1:4- and 1:8-cineol as well as C_{15}-hydrocarbons (*113*).

A great number of acid-catalyzed polyene cyclizations have been patterned on a biogenetic postulate involving *farnesol* derivatives

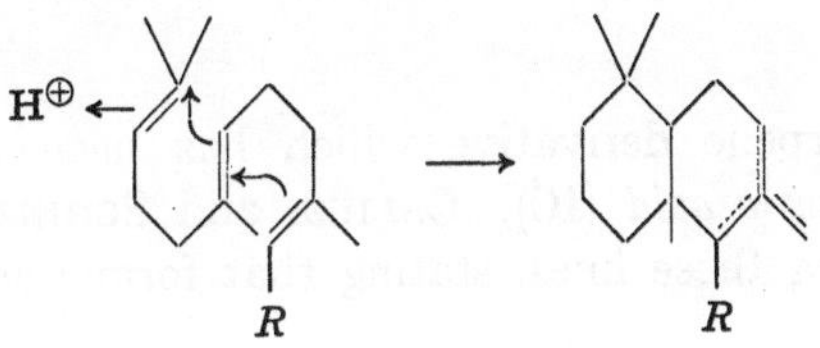

(XCV.)   (XCVI.)

(XCV → XCVI). However, one of the first experiments in sesquiterpene cyclization, designed and executed by RUZICKA and CAPATO (*78*), and

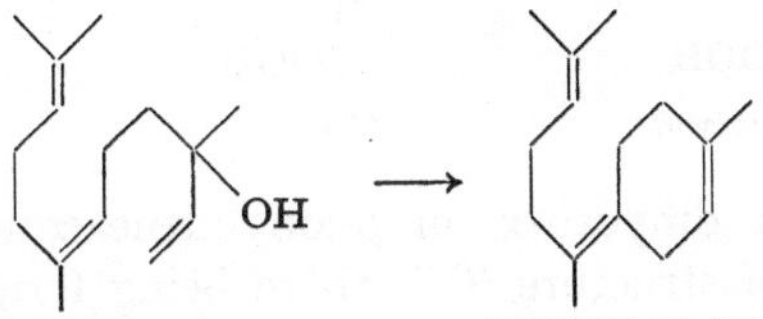

(XCVII.) Nerolidol.   (XCVIII.) Bisabolene.

published in 1925, was based on the scheme nerolidol (XCVII) → bisabolene (XCVIII), the latter being produced along with other materials on treatment of racemic (XVII) with acid. Reproduced below are a share of the farnesol-type materials studied in the laboratory, along with the structures proposed for the products isolated.

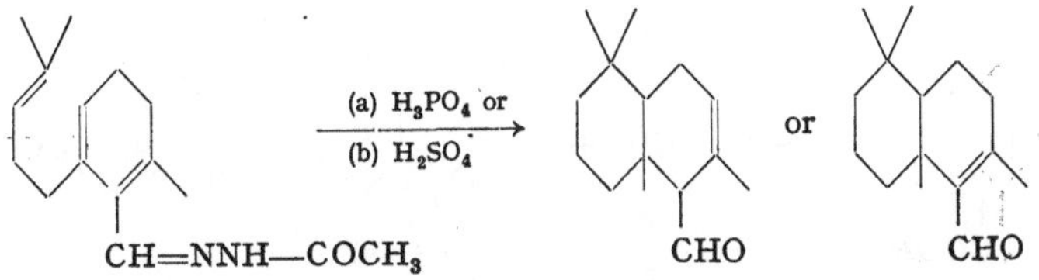

CH=NNH—COCH₃

Farnesal semicarbazone (*101*).

α-Bicyclofarnesal.   β-Bicyclofarnesal.

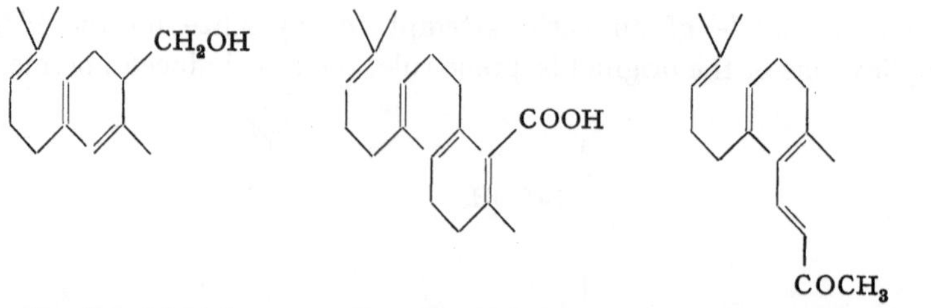

Sesquilavandulol (28).

ω-Geranylgeranic acid (24).

Farnesylidene acetone (119).

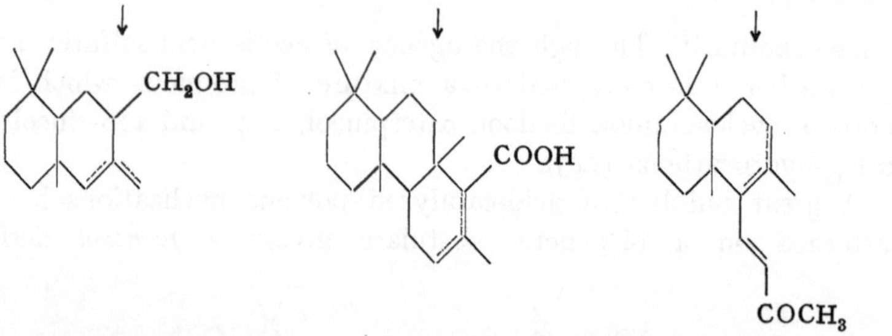

The sesquiterpene derivative which has been the object of most attention is *farnesic acid* (IC). CALIEZI and SCHINZ (21) first reported experiments along these lines, stating that formic acid converted either

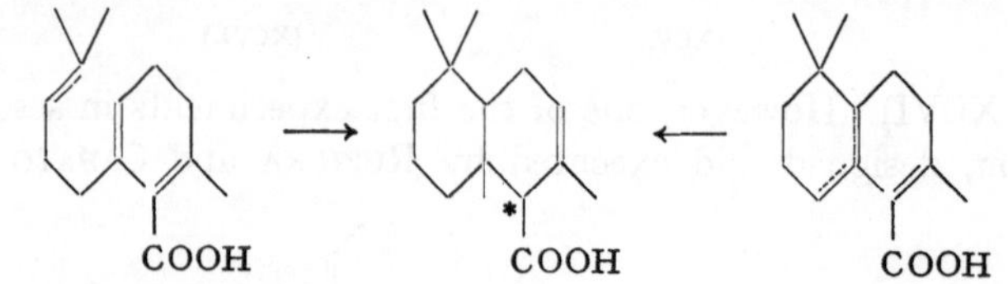

(IC.) Farnesic acid.   (CI.)   (C.)

the above acid or dihydro-α- or β-ionylideneacetic acid (C) (22) to a bicyclic product of structure (CI), there being formed two isomers, the proportion of which depended on the starting material and the reaction conditions. The hydroxyester (CII) was reported to give a mixture of acids (CI), (CIII) and (CIV) (23). These investigators also treated the consequences of the geometry of the starting unsaturated acid: the *trans* form was reported to give rise to the α-bicycloacid, whereas the *cis* isomer was converted to the *allo* acid (CIV). The same sesquiterpenoid

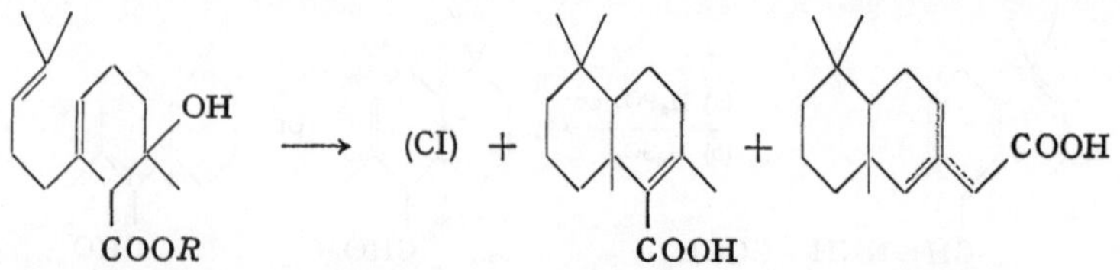

(CII.)   (CIII.)   (CIV.)

acid (IC) was studied in detail by STORK and BURGSTAHLER (*102*), who, using boron trifluoride as the catalyst, obtained the bicyclic acids corresponding to structure (CI), and showed them to be C* epimers. Furthermore, the monocyclic acids (*α,β-cis* and *trans*) (CV) actually could be isolated by carrying out the boron trifluoride reaction under mild conditions, and could be cyclized separately to one or the other bicyclic epimer (XVII). Subsequently, ESCHENMOSER and his collaborators (*99*) described their incisive explorations of the trienoic

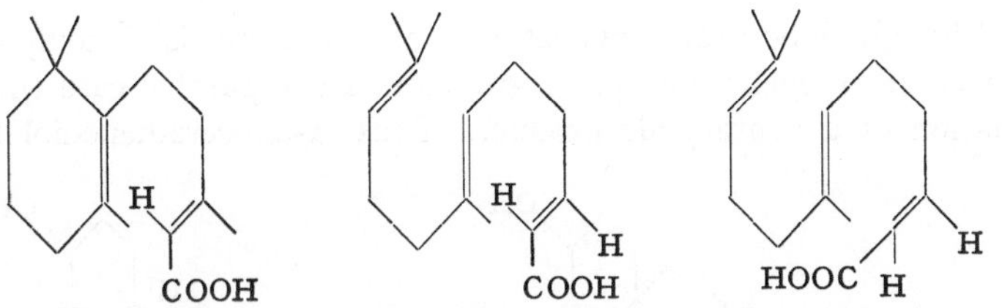

(CV.)   (CVIa.)   (CVIb.)

acids (CVIa) and (CVIb). In this case as well as in that of farnesic acid (*98, 117*) it is the *trans* decalin system which is produced; and the cyclization process as a whole is found to be amenable to stereochemical and conformational treatment (*39, 98, 99*).

Despite the lack of stereospecificity and the low yield (2–3%), one of the best examples in this class is the direct conversion of farnesyl-

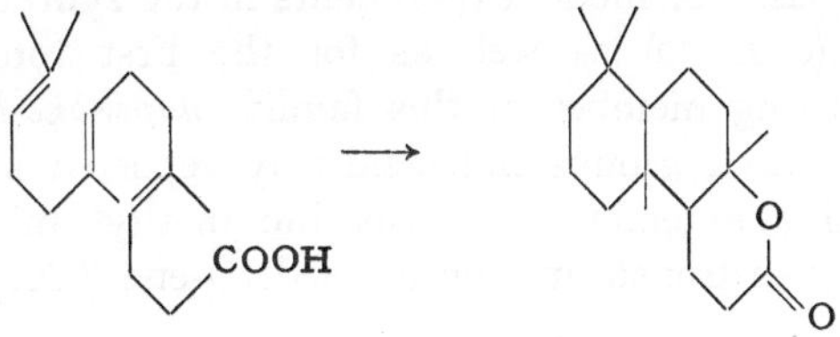

(CVII.) Farnesyl-acetic acid.   (CVIII.) Ambreinolide.

acetic acid (CVII) to *DL-ambreinolide* (CVIII) (*36, 102*), a key degradation product of ambrein. The cyclization to (CVIII) is probably, but not necessarily, concerted.

2. Conversions and Partial Syntheses.

Cyclization of *squalene* (CIX) has emerged as a key process in the biosynthesis of both terpenes and steroids, and it is therefore of interest to note that the non-enzymatic process has also received attention. HEILBRON and coworkers (*54*) determined that with formic acid, the acyclic polyene was brought to the tetracyclo stage; and more recently, the structure of tetracyclo-squalene was shown to be (CX) (*77*).

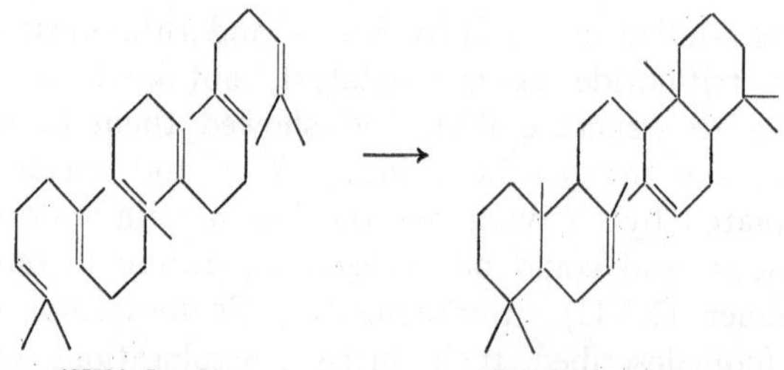

(CIX.) Squalene.   (CX.) Tetracyclo-squalene.

Although laboratory cyclization of squalene is incomplete, the presence of exocyclic methylene groups in a comparable case will permit formation of a pentacyclic product. Thus, α-onoceradienediol (CXI) is

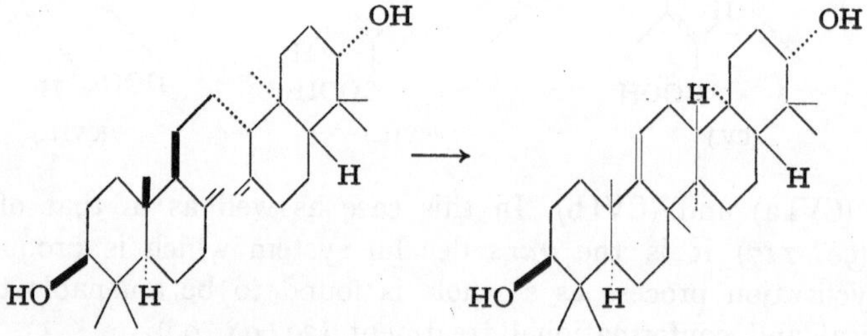

(CXI.) α-Onoceradienediol.   (CXII.)

convertible to the pentacyclic diol (CXII) (11). This cyclization has served as the basis for model experiments in the synthesis of pentacyclic triterpenoids (e. g. 30) as well as for the first total synthesis of a naturally-occurring member of this family, *hopenone-I* (103).

Several research groups independently reported some years ago a most impressive extended, stereospecific methyl migration sequence, the acid-catalyzed transformation of friedel-3-ene (CXIII) (alternatively,

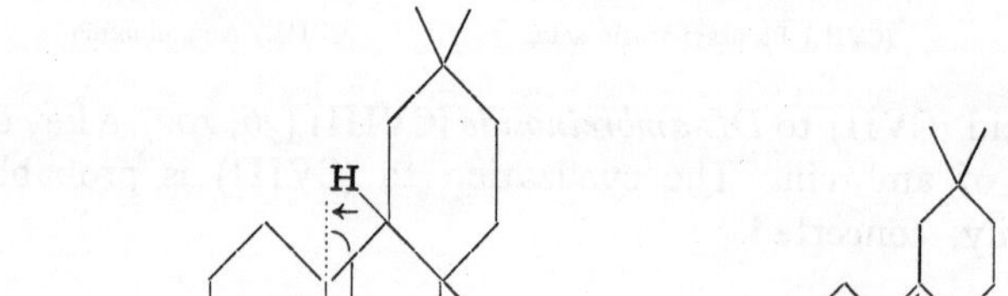

(CXIII.) Friedel-3-ene.   (CXIV.) Olean-13(18)-ene.

-2-ene or -3β-ol) to olean-13(18)-ene (CXIV) (*17*, *31*, *37*). This finding prompted the proposal that *friedelin* arises in Nature from β-amyrin

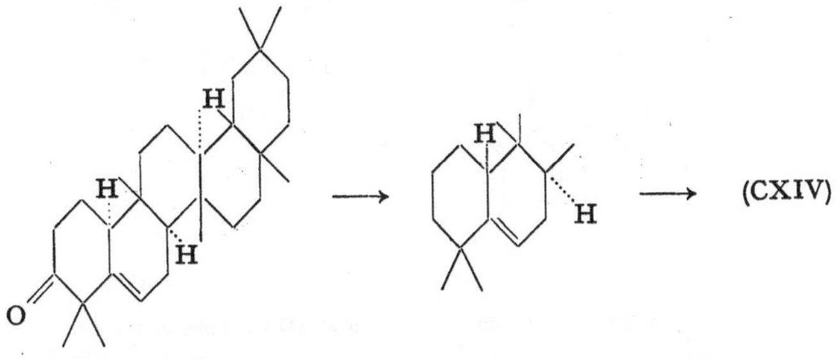

CXV.) Glutinone.

by means of a like rearrangement but operating in the reverse direction. A similar, although somewhat simpler, example is found in the case of *glutinone* (CXV), from alder bark. Wolff-Kishner reduction of this ketone afforded a hydrocarbon, which on acid treatment gave rise to the oleanene (CXIV) (*97*). A final example of a triterpene interconversion which may correspond to an in vivo process is the formation of germanicol from lupeol, proceeding as depicted (*1*):

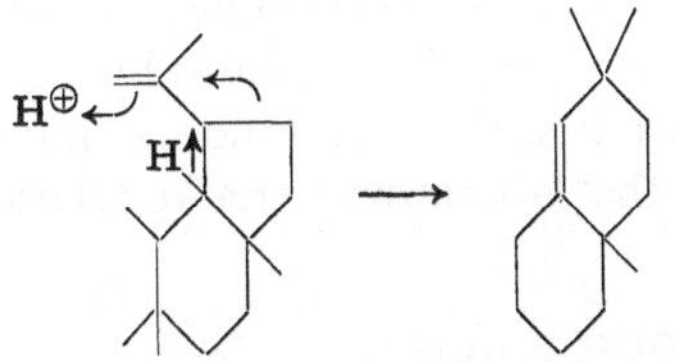

Lupeol.   Germanicol.

3. Miscellaneous.

An unusual example of an artificial process which, by all odds, corresponds to the natural phenomenon, is the oxidation of *α-terpinene*

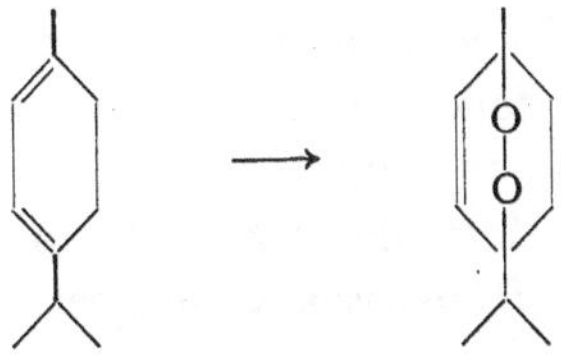

(CXVI.) α-Terpinene.   (CXVIa.) Ascaridol.

(CXVI) to the naturally-occurring peroxide ascaridol (CXVI), carried out with molecular oxygen in the presence of chlorophyll and light (*80*). Whether the irradiation-induced rearrangement of *santonin* (CXVII) to

iso-photosantonin (CXVIII) (*10*), or *verbenone* (CXIX) to chrysanthenone
(CXX) (*58*), corresponds in any way to biochemical events is problematical;

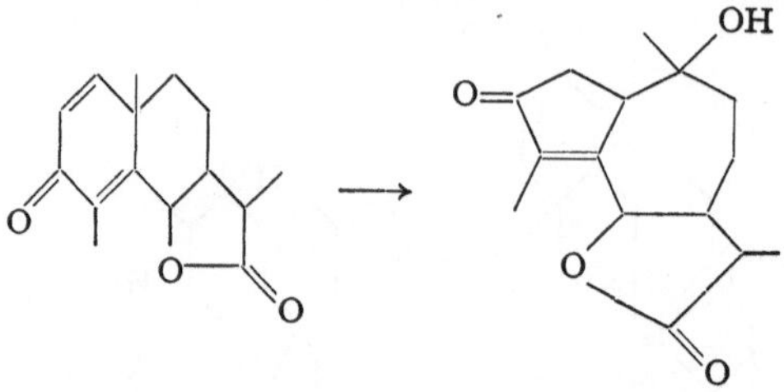

(CXVII.) Santonin.   (CXVIII.) Iso-photosantonin.

(CXIX.) Verbenone.   (CXX.) Chrysanthenone.

however, the appearance of unusual structures which also correspond
to naturally-occurring systems remains a tempting coincidence.

The structures of the hop constituents *lupulone* (CXXI) and
humulone (CXXII) suggest the biochemical union of an aromatic moiety
with isoprene units, very probably supplied by a pentenyl or γ,γ-dimethyl-
allyl pyrophosphate alkylation. A crude imitation of the process can
be accomplished in the laboratory: trialkenylation of 2,4,6-trihydroxy-

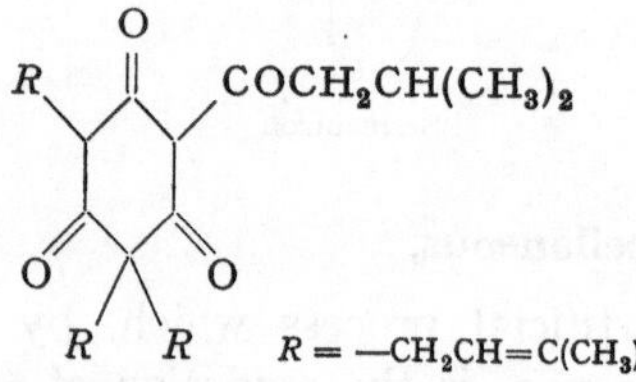

(CXXI.) Lupulone. $R = -CH_2CH=C(CH_3)_2$ (CXXII.) Humulone.

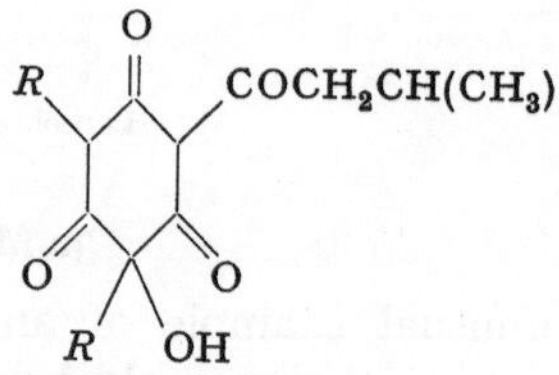

isovalerophenone with γ,γ-dimethylallyl bromide afforded lupulone,
whereas dialkenylation followed by air oxidation in the presence of
lead diacetate, gave rise to humulone (*71*).

Examination of the formula (CXXIII) for *iridomyrmecin*, isolated
from ants of the genus *Iridomyrmex*, suggested origin from citronellal,

(CXXIII.) Iridomyrmecin.   (CXXVII.)

by way of conversion to the cyclic dialdehyde (CXXVI) followed by disproportionation to the hydroxy acid (CXXVII). This scenario was developed into a laboratory performance in which citronellal was converted

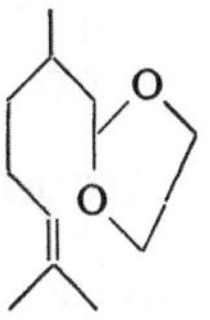

(CXXIV.) 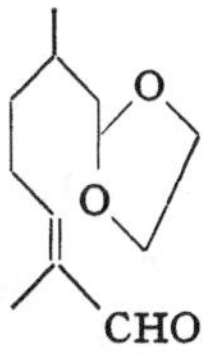   (CXXV.) 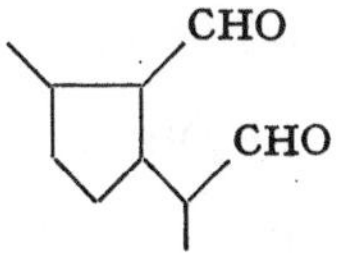   (CXXVI.)

first to the ethylene acetal (CXXIV), which was then oxidized selectively with selenium dioxide to the unsaturated aldehyde (CXXV). Through treatment with aqueous acetic acid, the latter intermediate was converted to a mixture of the acylic dialdehyde and the dialdehyde (CXXVI) resulting from internal addition. On being warmed with aqueous base, the latter suffered oxidation-reduction, being converted to the δ-hydroxy acid (CXXVII). Lactonization gave iso-iridomyrmecin, the epimer (*) of the natural material (26).

Consideration of a plausible mechanism (CXXVIII) for the coupling of isoprene units in terpene biosynthesis led to the concept of an interesting alkylation variant, of which the reaction (CXXIX) → (CXXX) was provided as an example (61).

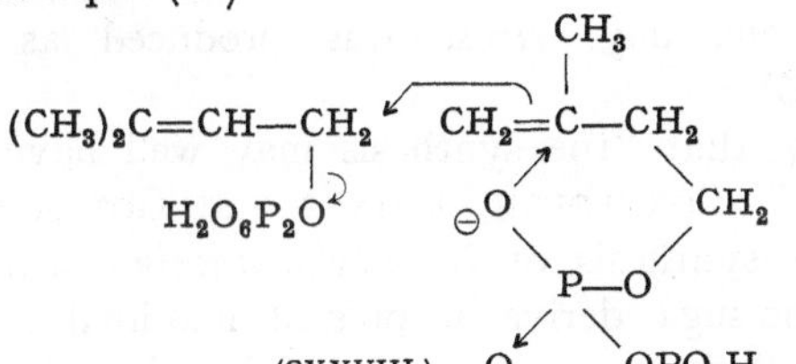

(CXXVIII.)

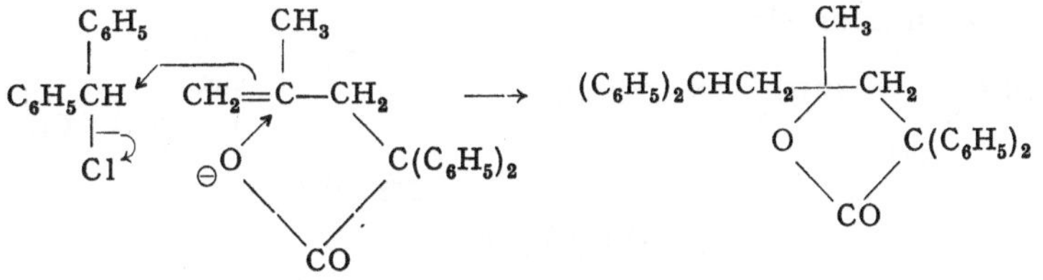

(CXXIX.)   (CXXX.)

III. Some Other Natural Products.

1. Monosaccharides.

E. FISCHER early anticipated the elucidation of one aspect of carbo-hydrate metabolism when he discovered that DL-glyceraldehyde, on treatment with dilute aqueous alkali, was converted to *DL-fructose* (CXXXI) and *DL-sorbose* (CXXXII), presumably by the process depicted:

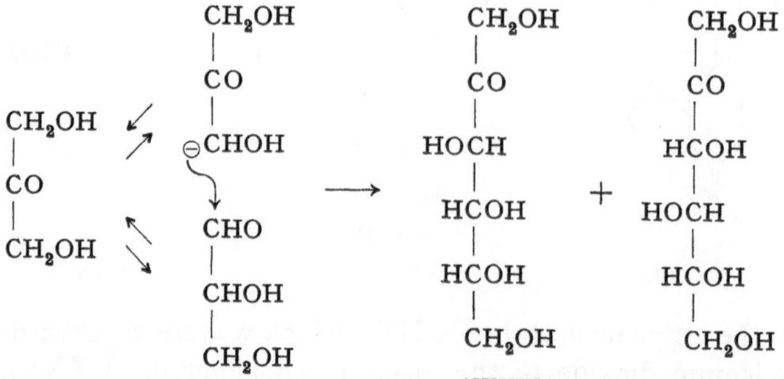

$$
\begin{array}{ccccc}
& & CH_2OH & CH_2OH & CH_2OH \\
& & | & | & | \\
& & CO & CO & CO \\
CH_2OH & & | & | & | \\
| & & \ominus CHOH & HOCH & HCOH \\
CO & & | & | & | \\
| & & CHO & HCOH & HOCH \\
CH_2OH & & | & | & | \\
& & CHOH & HCOH & HCOH \\
& & | & | & | \\
& & CH_2OH & CH_2OH & CH_2OH \\
& & & (CXXXI.) & (CXXXII.)
\end{array}
$$

Some years later H. O. L. Fischer (*41*) further demonstrated that the reaction proceeded with retention of optical activity, i. e., *D*-glyceraldehyde gave rise to optically active ketohexoses. Still more recently, this simple and direct method of synthesis was extended to the pentoses, it being shown that members of this class are formed from glycolic aldehyde and dihydroxyacetone or glyceraldehyde. Either pair of substances, on treatment for ten days with dilute sodium hydroxide, was converted to a mixture of C_5-sugars, from which *xylose* and *arabinose* were isolated. If lime water was used instead, and the reaction time shortened to one day, *ribose* was produced as the predominant product (*55, 56*).

Commenting that "the synthesis may well have followed the biosynthetic path", Cornforth, Daines and Gottschalk (*33*) recently announced the synthesis of *N-acetylneuraminic acid* (CXXXIII), the important amino sugar derivative present in animal mucoprotein, starting from N-acetyl-*D*-glucosamine and oxalylacetic acid (at pH 10–11 for two to three days).

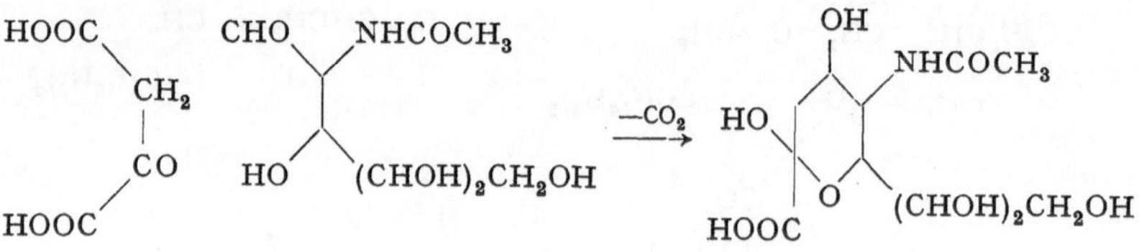

$$
\begin{array}{l}
HOOC \\
\quad\searrow \\
\qquad CH_2 \\
\qquad | \\
\qquad CO \\
\quad\nearrow \\
HOOC
\end{array}
\qquad
\begin{array}{l}
CHO \quad NHCOCH_3 \\
\\
HO \quad (CHOH)_2CH_2OH
\end{array}
\xrightarrow{-CO_2}
\begin{array}{l}
\qquad OH \\
\qquad | \\
\qquad\qquad NHCOCH_3 \\
HO \\
\qquad\quad O \\
HOOC \quad (CHOH)_2CH_2OH
\end{array}
$$

(CXXXIII.) N-Acetylneuraminic acid.

2. Porphyrins.

The brutal but successful methods developed by H. Fischer for the study and synthesis of porphyrins underlined the high, aromatic stability of this group of biologically essential molecules, and implicitly suggested that the natural propensity for their formation was great. From this, it might be concluded that Nature's method for constructing the porphyrin is no more complicated than necessary, and that therefore purely synthetic routes could closely duplicate her simple, direct means. This retrospective observation is fortified by such examples as the virtually spontaneous intermolecular condensation of the simple pyrrole (CXXXIV) to give a *porphyrinogen*, which could be air-oxidized without isolation to *etioporphyrin* (93).

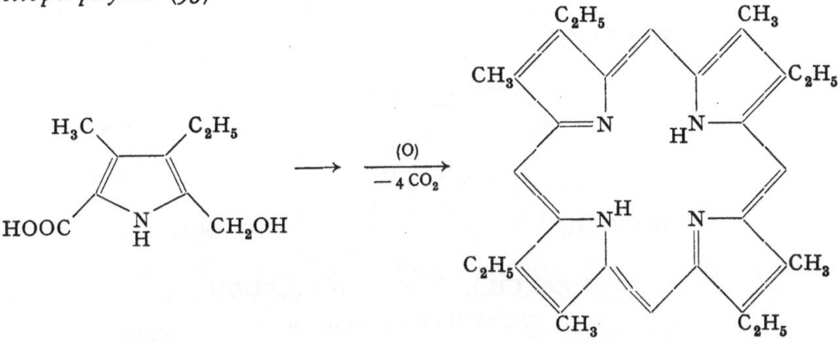

(CXXXIV.)   (CXXXV.) Etioporphyrin.

In recent years, extensive tracer and other experiments have shown that porphyrin biosynthesis starts from glycine and succinate, which condense to give δ-aminolevulinic acid (CXXXVI); two molecules of the latter interact to provide the substituted pyrrole, *porphobilinogen* (CXXXVII). Through manifold displacement of the "benzylic" amino group from the α-aminomethylene substituent by the nucleophilic pyrrole ring of a second molecule of porphobilinogen, the *uroporphyrinogen* (CXXXVIII) is ultimately built up; although it should be noted here that the positioning of one pyrrole unit (*) is abnormal, indicating that one bridging methylene group has suffered overall migration from its normal position, adjacent to the acetic acid substituent, to the other α-position (encircled) on the pyrrole ring. Dehydrogenation leads to *uroporphyrin-III* (CXXXIX), which is regarded as the biological parent of heme, chlorophyll and other natural products.

It has now transpired—in a sequence which represents a powerful vindication for the method of biogenetic-type synthesis—that, in two simple laboratory operations, δ-aminolevulinic acid (CXXXVI) can be converted by way of porphobilinogen (CXXXVII) to uroporphyrin-III, during which process the unusual ring transposition described must have been duplicated, in addition to condensation to, and dehydrogenation of, the porphyrinogen. The preparation of pyrrole from aminolevulinic acid was carried out as a separate step (91); porphobilinogen, on brief treatment with dilute aqueous hydrochloric acid at 100°, followed by air

18*

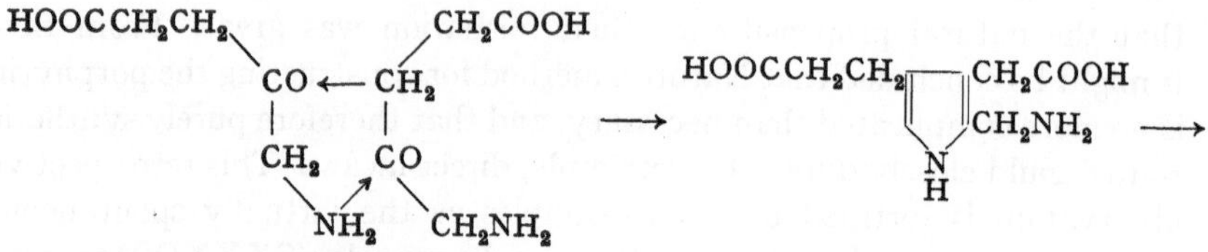

HOOCCH$_2$CH$_2$ CH$_2$COOH

CO ← CH$_2$

CH$_2$ CO

NH$_2$ CH$_2$NH$_2$

(CXXXVI.) δ-Aminolevulinic acid.

HOOCCH$_2$CH$_2$ ⌐ CH$_2$COOH

CH$_2$NH$_2$

(CXXXVII.) Porphobilinogen.

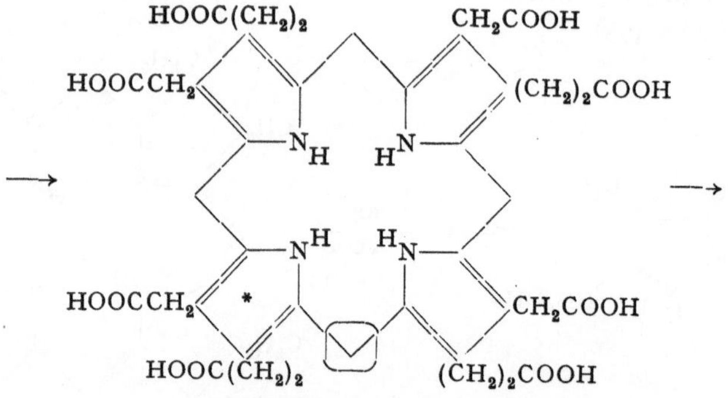

HOOC(CH$_2$)$_2$ CH$_2$COOH

HOOCCH$_2$ (CH$_2$)$_2$COOH

HOOCCH$_2$ *   CH$_2$COOH

HOOC(CH$_2$)$_2$ (CH$_2$)$_2$COOH

(CXXXVIII.) Uroporphyrinogen.

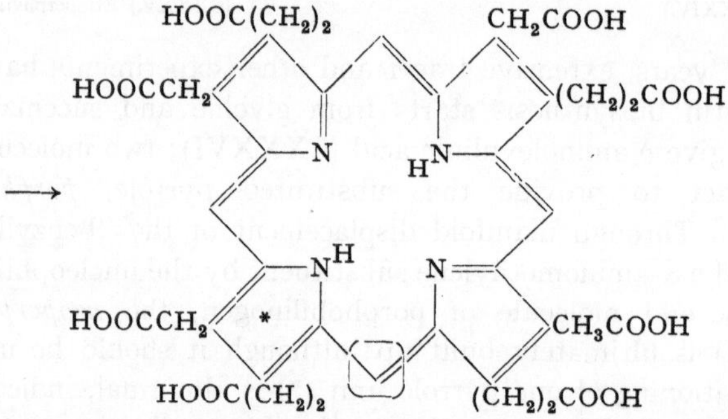

HOOC(CH$_2$)$_2$ CH$_2$COOH

HOOCCH$_2$ (CH$_2$)$_2$COOH

HOOCCH$_2$ *   CH$_3$COOH

HOOC(CH$_2$)$_2$ (CH$_2$)$_2$COOH

(CXXXIX.) Uroporphyrin-III.

oxidation, was transformed in good yield into a porphyrin mixture, which consisted mostly of uroporphyrin-III (29). Mechanistic aspects of the porphyrin formation have been studied and discussed (66).

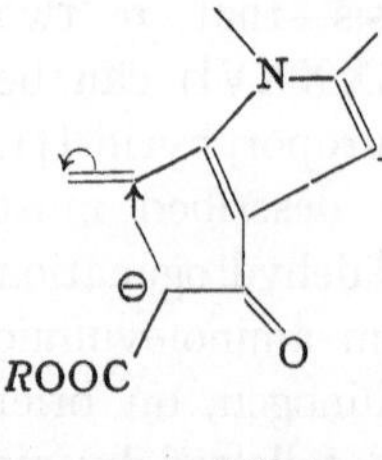

CH$_3$

ROOC

(CXL.)

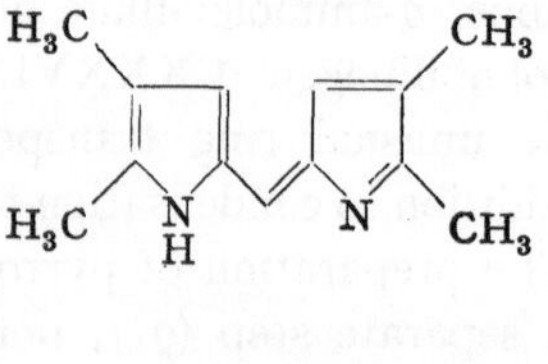

H$_3$C   CH$_3$

H$_3$C   N   N   CH$_3$
      H

(CXLI.)

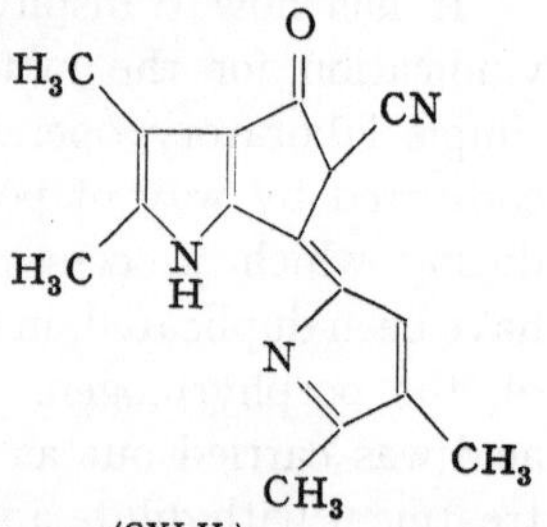

H$_3$C   O

CN

H$_3$C   N
     H

CH$_3$

CH$_3$

(CXLII.)

As a model (CXL) for a proposal relating to biogenesis of the five-membered β-keto ester ring in *chlorophyll*, the addition of cyanoacetic ester to the dipyrromethene (CXLI) was carried out, with the resultant formation of the tricyclic ketone (CXLII) (*60*).

3. Vitamins.

Several diverse substances shown to be accessory dietary factors have been obtained through syntheses which closely resemble reasonable biosynthetic routes. In addition to the well-known artificial conversion of *ergosterol* to *vitamin D$_2$* (CXLIV) by irradiation, attention is directed to two examples.

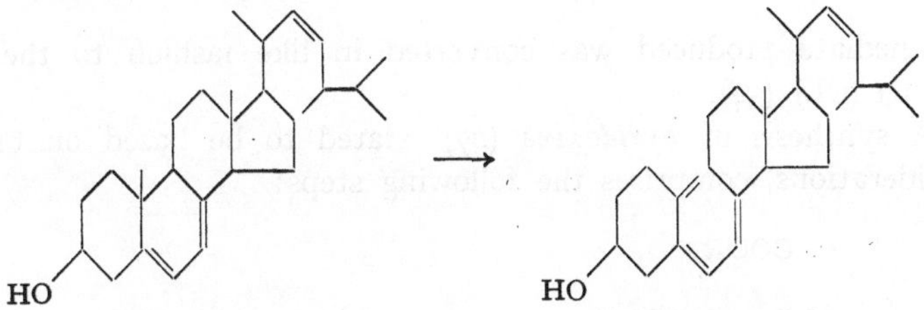

(CXLIII.) Ergosterol.   (CXLIV.) Vitamin D$_2$.

In a sequence which served as a model for vitamin B$_2$ *(riboflavin)* synthesis, the dimer (CXLV) obtained by treatment of diacetyl with

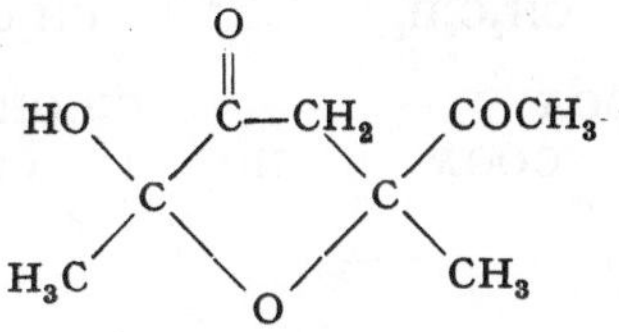

(CXLV.) 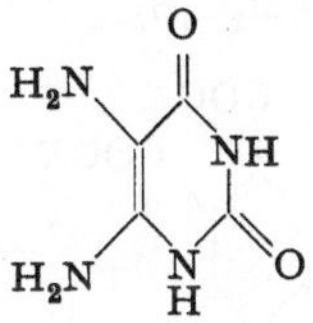   (CXLVI.) 4,5-Diaminouracil.

alkali was allowed to react with 4,5-diaminouracil (XLV). The intermediate produced thereby (possibly CXLVII), on treatment with more alkali gave rise to the tricyclic *lumichrome* (CXLVIII) (*12*). As a direct modification of this route, riboflavin itself was obtained by substituting

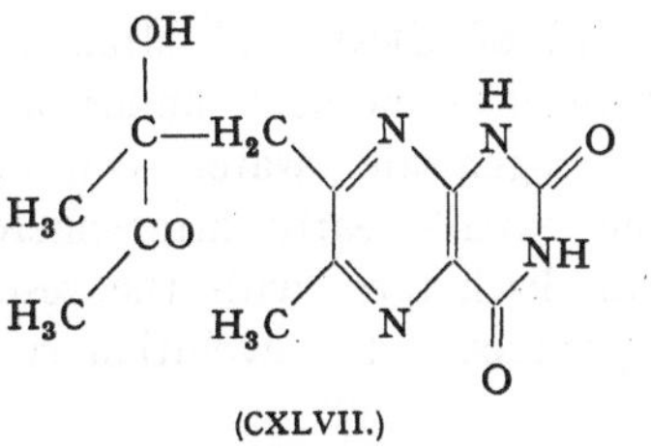

(CXLVII.) 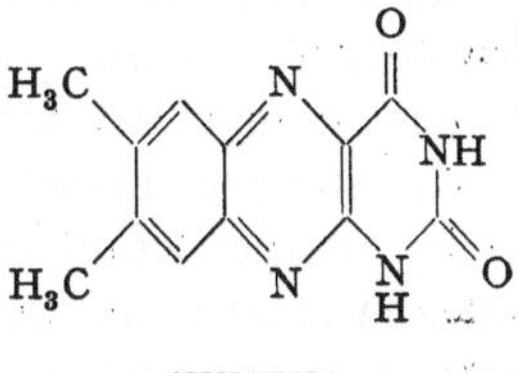   (CXLVIII.)

5-amino-4-D-ribityl-lumazine (CIL) for the simpler uracil (CXLVI), and a diacetyl trimer (of unknown structure) for (CXLV). The comparable

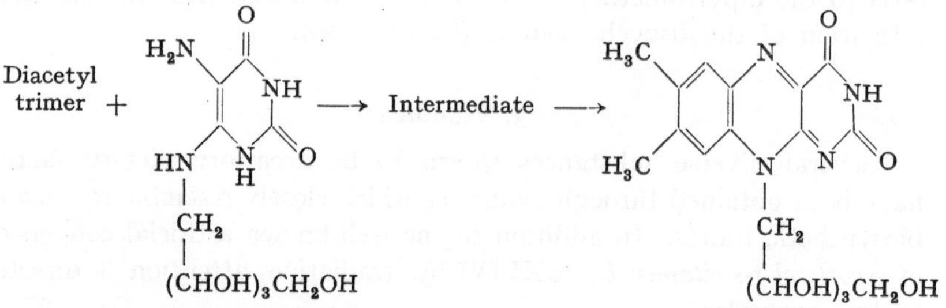

Diacetyl trimer +

(CIL.) 5-Amino-4-D-ribityl-lumazine.

(CL.) Riboflavin.

intermediate produced was converted in like fashion to the natural product (CL) (34).

A synthesis of *pyridoxine* (27), stated to be based on biogenetic considerations, comprises the following steps:

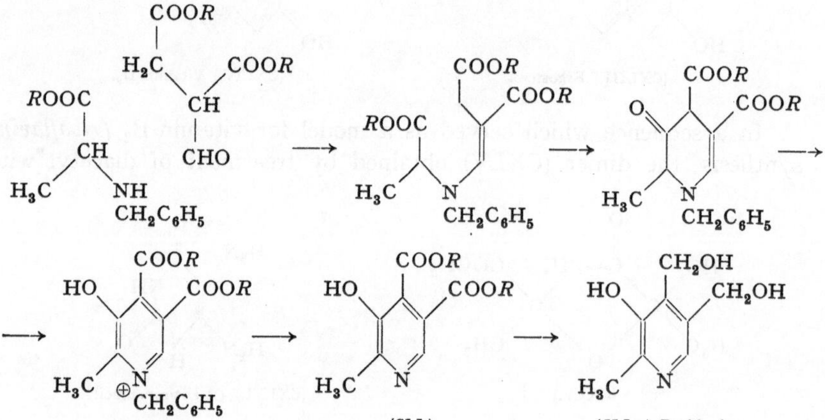

(CLI.)

(CLIa.) Pyridoxine.

The key intermediate was the substituted pyridine (CLI) or a related compound, which was convertible to pyridoxine (CLI a) through standard means.

4. Amino Acids and Peptides.

Mention may be made first of all of Urey and Miller's novel experiments (69) on the possible origin of amino acids under primordial conditions. Ammonia, methane, hydrogen and water—considered to be likely starting materials in the earth's early atmosphere—were subjected to the action of electrical discharges, with the result that alanine and other amino acids were produced. The evolutionary aspects of the results are evident.

That it is possible to utilize vitamin B_6 (pyridoxal), along with metal salts, to reproduce biochemical processes in the laboratory was demonstrated by SNELL (*68*), the pertinent case being the reversible catalytic cleavage of β-hydroxyamino acids.

$$R\text{—CHOHCH(NH}_2)\text{COOH} \rightleftarrows R\text{CHO} + \text{CH}_2(\text{NH}_2)\text{COOH}$$

Under suitable conditions the equilibrium can be affected so as to favor the β-hydroxy acid, and the synthesis of serine and other such substances was demonstrated. Somewhat later, the pyridoxal was shown to be unnecessary for the laboratory process, in that it is sufficient to warm glycine copper complex with the aldehyde in weakly basic solution for several hours. Through the modified method, *serine, threonine* (and *allo-threonine*) and *β-hydroxyleucine* were obtained (*59, 79*).

In the cellular manufacture of *thyroxine* (CLIII) it has been demonstrated that two molecules of 3,5-diiodotyrosine (CLII) are joined by oxygen-carbon bond formation, with extrusion of a C_3-side-chain

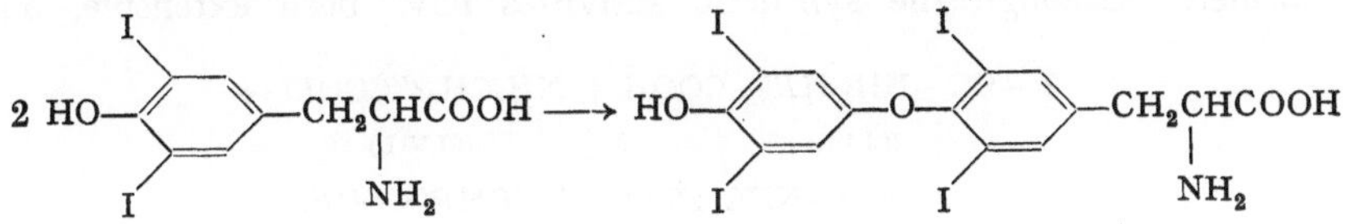

(CLII.) 3,5-Diiodotyrosine.　　　(CLIII.) Thyroxine.

from one of the starting amino acids. Interestingly enough, this whole process can be simulated by allowing the diiodotyrosine to stand in weakly alkaline solution at 37°, under which conditions thyroxine is gradually formed (*112*). It is assumed that the change is initiated by aerial oxidative coupling of the phenolic moieties, leading to (CLIV), from which the amino acid unit blocking the terminal aromatic ring is subsequently lost. Mechanistic and other aspects of the overall reaction have been studied (*20, 65*).

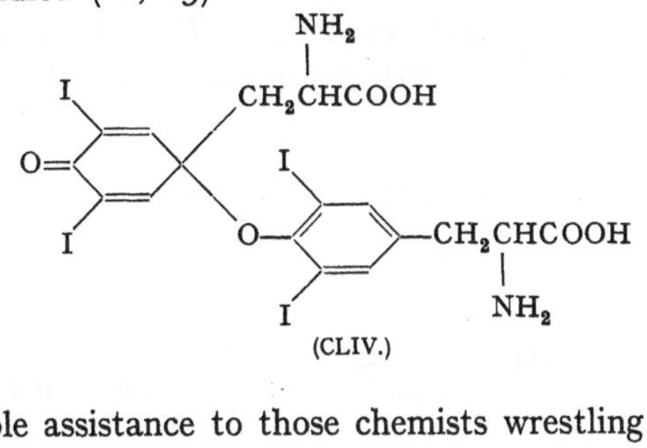

(CLIV.)

Considerable assistance to those chemists wrestling with the general problem of improving *peptide* formation from amino acids or derivatives thereof, has been lent by biological scientists whose work has dealt with "activated" carboxylic acids of one kind or another. The whole area of investigation is represented by the general expression starting from (CLV), where (PR) is a protecting group, and (ACT) symbolizes

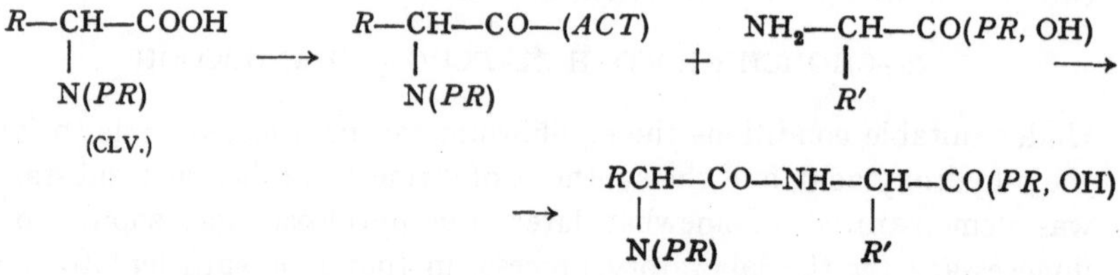

$$
\begin{array}{ccc}
\text{R—CH—COOH} & \text{R—CH—CO—(ACT)} & \text{NH}_2\text{—CH—CO(PR, OH)} \\
\mid & \longrightarrow \quad \mid & + \quad \mid \quad \longrightarrow \\
\text{N(PR)} & \text{N(PR)} & \text{R'} \\
\text{(CLV.)} & &
\end{array}
$$

$$
\longrightarrow \quad
\begin{array}{c}
\text{RCH—CO—NH—CH—CO(PR, OH)} \\
\mid \qquad\qquad \mid \\
\text{N(PR)} \qquad\quad \text{R'}
\end{array}
$$

a substituent which "activates" the carbonyl group of the amino acid and allows facile amide bond formation by attack of the free amino group of a second amino acid derivative. The use of mixed carboxyl-phosphoric (phosphorous) acid anhydrides as active intermediates in peptide synthesis brings to mind recent trends in nucleic acid chemistry, and developments in the latter field surely have stimulated work in the former. Although the synthetic activities have been extensive, one

$$
\text{R—CO—NHCH}R'\text{—COOH} + \text{NH}_2\text{CH}R''\text{COOEt} \rightarrow
$$

(CLVI.)   (CLVII.)

$$
\rightarrow \text{R—CO—NHCH}R'\text{CO—NHCH}R''\text{COOEt}
$$

example will need to suffice here: amide formation from N-acylamino acid (CLVI) and amino acid or ester (CLVII), brought about by the action of tetramethyl pyrophosphite (2).

In the hydrolysis of an ester, a *histidine* residue seemingly plays a role by forming an N-acyl complex, easily cleaved to the starting histidine

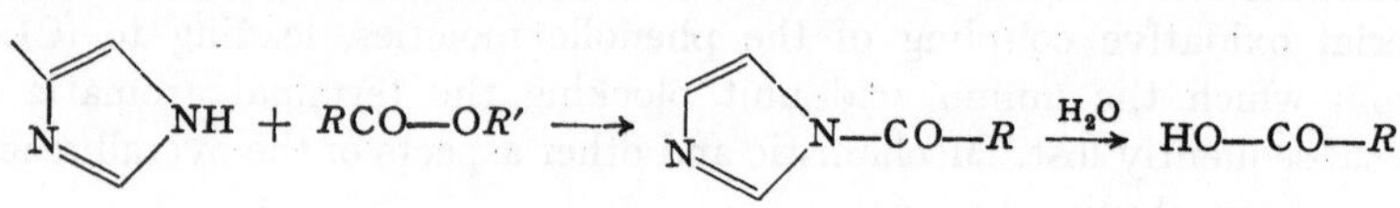

system and the product carboxylic acid. This effect of the imidazole ring was applied in a new method for peptide construction (3):

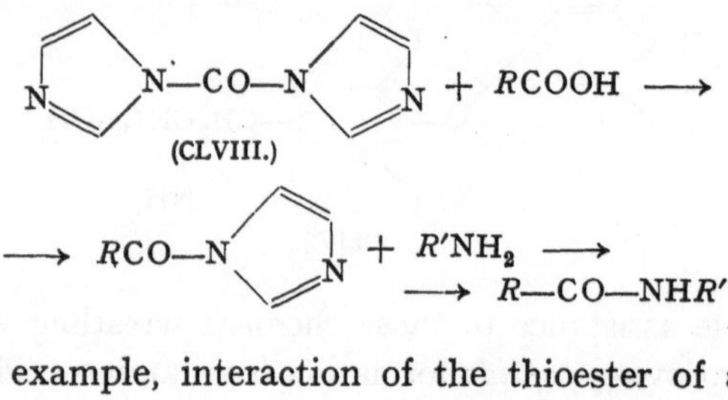

As a final example, interaction of the thioester of an N-protected amino acid with a second, unmodified amino acid (CLIX) may be

$$C_6H_5\text{—}CH_2OCO\text{—}NHCH(CH_3)\text{—}CO\text{—}SC_6H_5 + NH_2CH(CH_3)COOH \rightarrow$$
(CLIX.)

$$\rightarrow C_6H_5CH_2OCO\text{—}NHCH(CH_3)\text{—}CO\text{—}NHCH(CH_3)COOH$$

given (*115*), which method depends on the lability of the thioester group, as does the in vivo formation of higher fatty acids from acetic acid, via acetyl Co-A (or malonyl Co-A).

One of the most unusual devices for constructing a peptide chain, referred to as "Aminoacyl-Einlagerung" (*14*), features the rearrangement

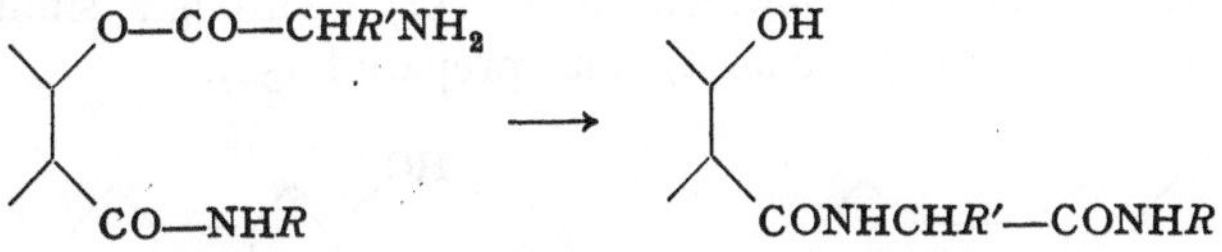

(CLX.)   (CLXI.)

of a simple β-hydroxy acid derivative (CLX) (or a corresponding o-salicylic acid system) to the peptide (CLXI). The suggestion is made that serine, threonine or cysteine units may be interpolated, in a similar way, into already intact protein chains during the biosynthetic process.

5. Phenol Oxidation Products.

Various investigators have demonstrated, by means of well-conceived, quite often elegant, synthetic schemes, that natural processes considered to involve oxidation of phenols as key steps are readily adaptable to laboratory execution. In certain cases, oxidation occurs initially to a quinone which then goes on to further transformation products; in other instances, the starting phenol is subjected to a single electron loss to give a phenoxy radical, which couples with a second actual or potential radical in going to the end product.

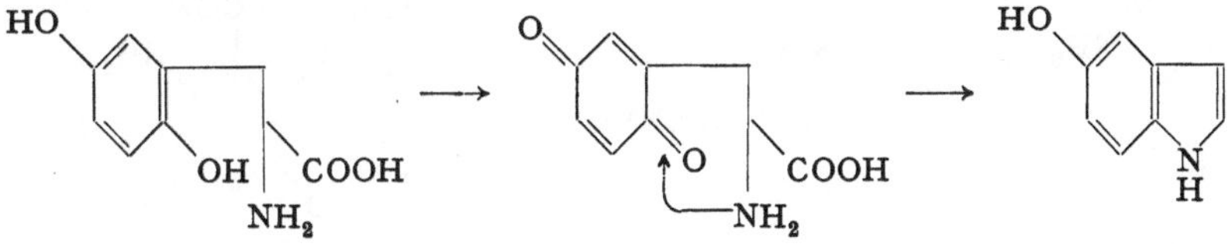

(CLXII.) 2,5-Dihydroxyphenylalanine.   (CLXIIa.) 5-Hydroxyindole.

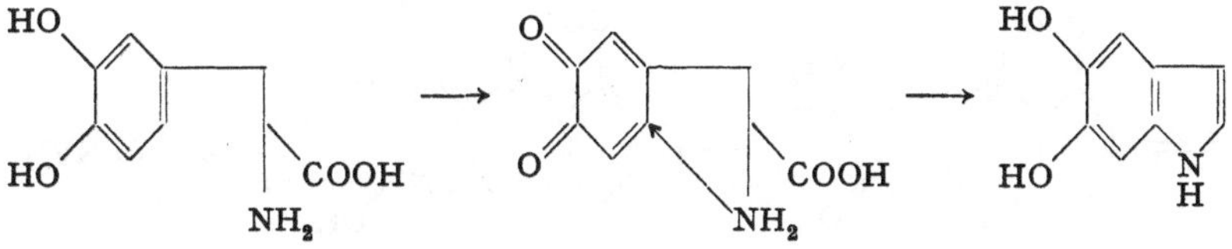

(CLXIII.) 3,4-Dihydroxyphenylalanine.   (CLXIIIa.) 5,6-Dihydroxyindole.

Basing his experiments on the bio-formation of *hydroxyindoles* from *ar*-hydroxylated phenylalanines, such as tyrosine or 3,4-dihydroxyalanine, Harley-Mason (*49*) has carried out a number of ferricyanide oxidations of such amino acids. 2,5-Dihydroxyphenylalanine (CLXII), assumed to be formed in vivo from tyrosine, was converted to 5-hydroxyindole (CLXIIa). Similarly, 3,4-dihydroxyphenylalanine gave rise to 5,6-dihydroxyindole (CLXIIIa), of interest because of its relation to melanin (*18*). As an application to natural product synthesis, the diamine (CLXIV) was transformed by the ferricyanide reagent to *bufotenin* (CLXV), a pressor amine present in the common toad (*51*). Through similar means, 5-hydroxytryptamine *(serotonin)* was prepared (*52*).

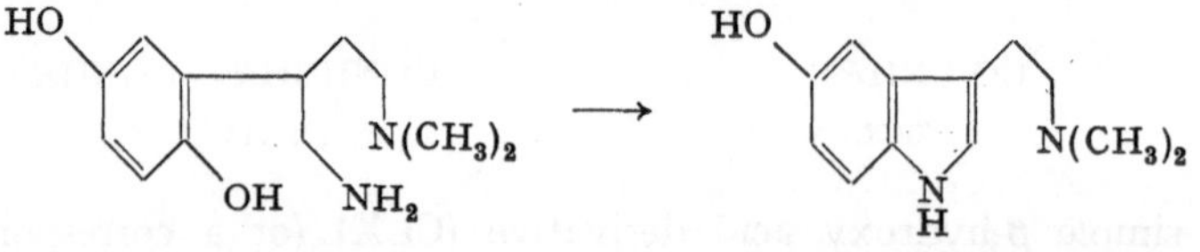

(CLXIV.)   (CLXV.) Bufotenin.

The same basic principles are operative in the artificial oxidative conversion of anthranilic acid types to naturally occurring systems.

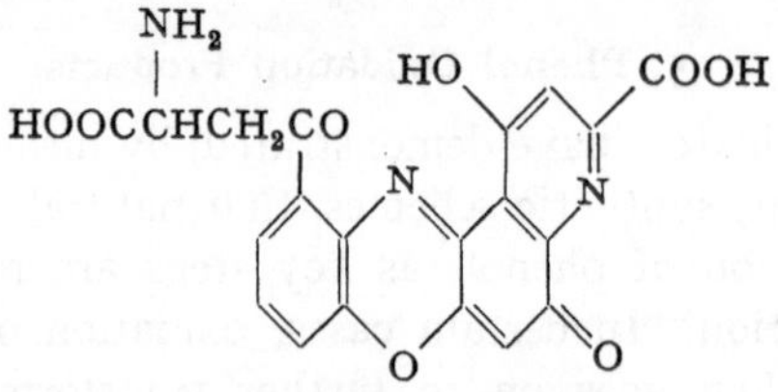

(CLXVI.) Xanthommatin.

Xanthommatin (CLXVI) belongs to the class of *ommochromes*, insect pigments considered to be end-products of tryptophan metabolism. In

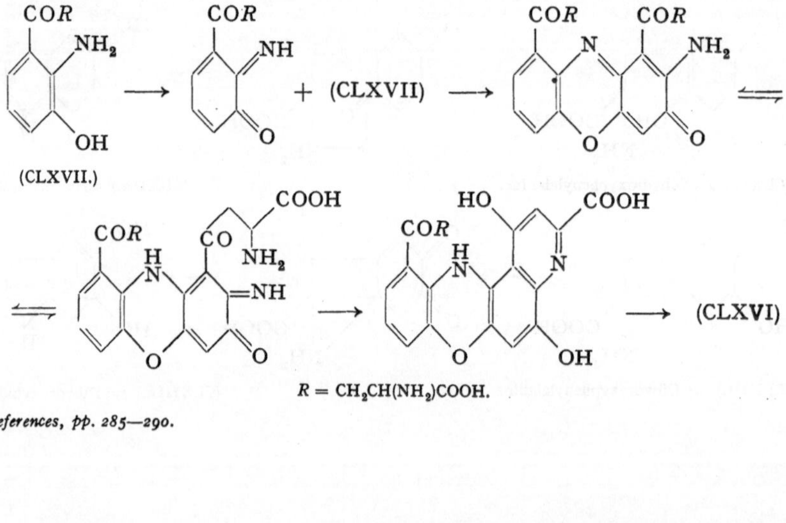

(CLXVII.)

$R = CH_2CH(NH_2)COOH.$

an astonishing sequence of in vitro chemical changes which begins with hydroxy-kynurenine (CLXVII), an established intermediate in tryptophan breakdown, structure (CLXVI) is built up through the agency of potassium ferricyanide as the sole chemical reagent (*19*). Similarly, actinocinyl-bis-glycine dimethyl ester (CLXIX), which represents the

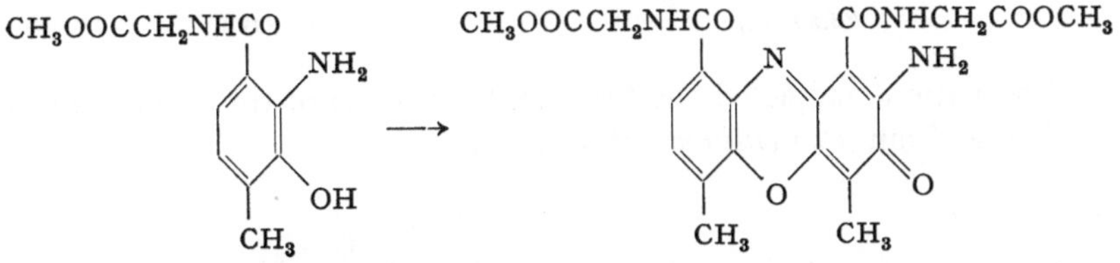

CH₃OOCCH₂NHCO

(CLXVIII.) 2-Amino-3-hydroxy-4-methylbenzoyl-glycine. (CLXIX.) Actinocinyl-bis-glycine dimethyl ester.

chromophore of the *actinomycin* antibiotics, can be formed through air oxidation of the methyl ester of 2-amino-3-hydroxy-4-methylbenzoyl-glycine (CLXVIII) (*16*).

BARTON'S (*9*) excellent total synthesis of *usnic acid* (CLXXII), involving oxidative coupling of phenol radicals, probably proceeds in

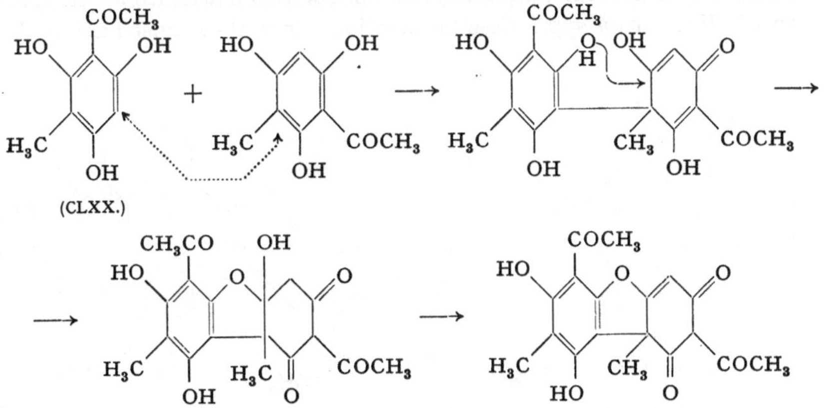

(CLXX.)

(CLXXI.)

(CLXXII.) Usnic acid.

much the same fashion as does the biosynthesis of this lichen product. Bond formation by way of radical coupling (CLXX) leads to the blocked aromatic dimeric methylphloroacetophenone, which cyclizes further to (CLXXI), an isolable intermediate. Formation of racemic usnic acid was achieved by dehydrating this intermediate.

In the case of *griseofulvin* (CLXXV), the ferricyanide oxidation was carried out an a suitably substituted benzophenone (CLXXIII), which gave rise in good yield to the cyclohexadienone (CLXXIV). Catalytic

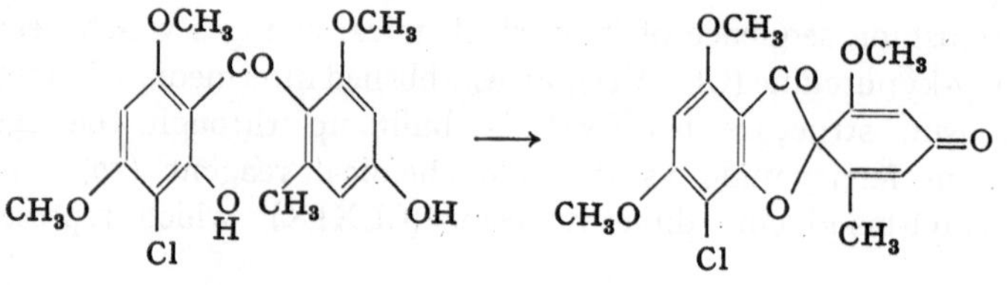

(CLXXIII.)   CLXXIV.)

hydrogenation completed the first total synthesis of this important oral antibiotic from *Penicillium patulum* (*35*).

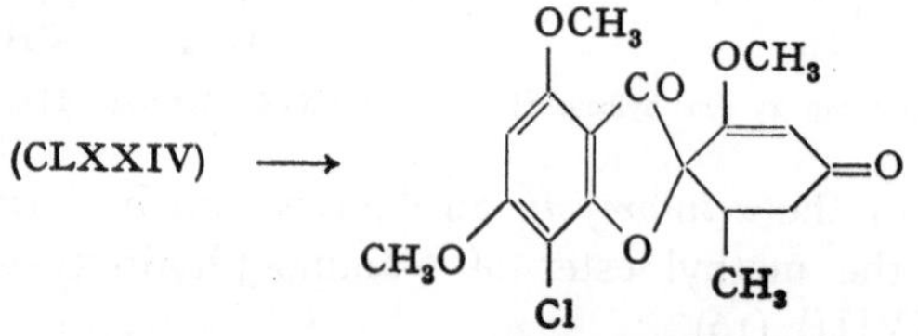

(CLXXIV) $\longrightarrow$

(CLXXV.) Griseofulvin.

Two dimerizations observed in connection with the study of *hypericin* (CLXXVIII), the photodynamic coloring matter from "St. John's wort" (*Hypericum* spp.), deserve mention, since they must have a close

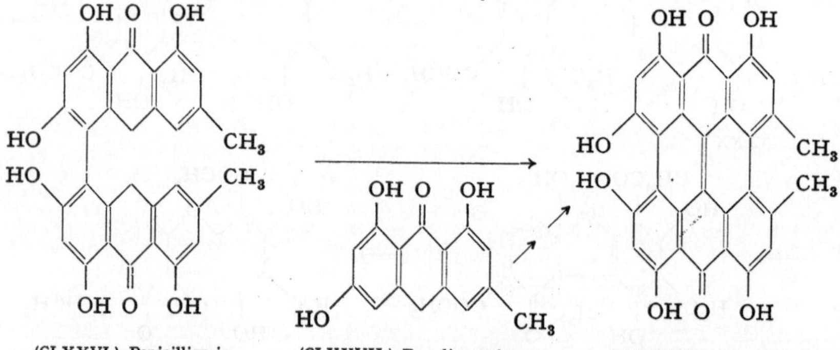

(CLXXVI.) Penicilliopsin.   (CLXXVII.) Emodin anthrone.   (CLXXVIII.) Hypericin.

relationship with phytochemical processes. The naturally-occurring penicilliopsin can be transformed by irradiation and air oxidation to hypericin; and, in a total synthesis, emodin anthrone (CLXXVII) was air oxidized to protohypericin, which on irradiation again gave the same octacyclic end product (*15*).

As in the previous examples, oxidative dimerization via radicals permits virtual duplication of self-evident biosynthetic pathways.

See also Brockmann's survey article in this Series (*14a*).

References.

1. AMES, T. R., G. S. DAVY, T. G. HALSALL and E. R. H. JONES: The Chemistry of the Triterpenes. Part XII. The Action of Formic Acid on Lupeol. J. Chem. Soc. (London) **1952**, 2868.

2. ANDERSON, G. W., J. BLODINGER and A. D. WELCHER: Tetraethyl Pyrophosphite as a Reagent for Peptide Syntheses. J. Amer. Chem. Soc. **74**, 5309 (1952).

3. ANDERSON, G. W. and R. PAUL: N,N'-Carbonyldiimidazole, a New Reagent for Peptide Synthesis. J. Amer. Chem. Soc. **80**, 4423 (1958).

4. ANET, E., G. K. HUGHES and E. RITCHIE: Syntheses of Hygrine and Cuscohygrine. Nature (London) **163**, 289 (1949).

5. — — — A Synthesis of *iso*-Pelletierine and Methyl *iso*-Pelletierine. Nature (London) **164**, 501 (1949).

6. — — — A Synthesis of Sparteine. Nature (London) **165**, 35 (1950).

7. ANET, F. A. L. and R. ROBINSON: Conversion of the Wieland-Gumlich Aldehyde into Strychnine. Chem. and Ind. **1953**, 245.

8. BAN, Y. and T. OISHI: Synthesis of 3-Spiro-oxindole Derivatives. Chem. and Ind. **1960**, 349.

9. BARTON, D. H. R., A. M. DEFLORIN and O. E. EDWARDS: The Synthesis of Usnic Acid. Chem. and Ind. **1955**, 1039.

10. BARTON, D. H. R., P. DE MAYO and M. SHAFIQ: Photochemical Transformations. Part I. Some Preliminary Investigations. J. Chem. Soc. (London) **1957**, 929.

11. BARTON, D. H. R. and K. H. OVERTON: Triterpenoids. Part XX. The Constitution and Stereochemistry of a Novel Tetracyclic Triterpenoid. J. Chem. Soc. (London) **1955**, 2639.

12. BIRCH, A. J. and C. J. MOYE: Studies in Relation to Biosynthesis. Part X. A Synthesis of Lumichrome from Non-benzenoid Precursors. J. Chem. Soc. (London) **1957**, 412.

13. BOWDEN, K. and L. MARION: The Biogenesis of Alkaloids. IV. The Formation of Gramine from Tryptophan in Barley. Canad. J. Chem. **29**, 1037 (1951).

14. BRENNER, M., J. P. ZIMMERMANN, J. WEHRMÜLLER, P. QUITT, A. HARTMANN, W. SCHNEIDER und U. BEGLINGER: Aminoacyl-Einlagerung. 1. Mitt. Definition, Übersicht und Beziehung zur Peptidsynthese. Helv. Chim. Acta **40**, 1497 (1957).

14a. BROCKMANN, H.: Photodynamisch wirksame Pflanzenfarbstoffe. Fortschr. Chem. organ. Naturstoffe **14**, 141 (1957).

15. BROCKMANN, H. und H. EGGERS: Synthese des Proto-hypericins und Hypericins aus Emodin-anthron-(9). Chem. Ber. **91**, 547 (1958).

16. BROCKMANN, H. und H. MUXFELDT: Konstitution und Synthese des Actinomycin-Chromophors. Chem. Ber. **91**, 1242 (1958).

17. BROWNLIE, G., F. S. SPRING, R. STEVENSON and W. S. STRACHAN: Friedelin and Cerin. Chem. and Ind. **1955**, 1156.

18. BU'LOCK, J. D. and J. HARLEY-MASON: Melanin and its Precursors. Part III. New Syntheses of 5:6-Dihydroxyindole and its Derivatives. J. Chem. Soc. (London) **1951**, 2248.

19. BUTENANDT, A.: Über Ommochrome, eine Klasse natürlicher Phenoxazon-Farbstoffe. Angew. Chem. **69**, 16 (1957).

20. CAHNMANN, H. J. and T. MATSUURA: Model Reactions for the Biosynthesis of Thyroxine. II. The Fate of the Aliphatic Side Chain in the Conversion of 3,5-Diiodophloretic Acid to 3,5,3',5'-Tetraiodothyropropionic Acid. J. Amer. Chem. Soc. **82**, 2050 (1960).

21. Caliezi, A. und H. Schinz: Zur Kenntnis der Sesquiterpene und Azulene. 90. Mitt. Die Cyclisation der Farnesylsäure. Helv. Chim. Acta **32**, 2556 (1949).

22. — — Zur Kenntnis der Sesquiterpene und Azulene. 92. Mitt. Die Cyclisation der Dihydro-β- und der Dihydro-α-jonyliden-essigsäure zur α-Bicyclofarnesylsäure. Helv. Chim. Acta **33**, 1129 (1950).

23. — — Zur Kenntnis der Sesquiterpene und Azulene. 102. Mitt. Über α-, β- und Allo-bicyclofarnesylsäure. Helv. Chim. Acta **35**, 1637 (1952).

24. — — Zur Kenntnis der Diterpene. 63. Mitt. Über zwei Cyclisationen in der Diterpenreihe. Helv. Chim. Acta **35**, 1649 (1952).

25. Chakravarti, D., R. N. Chakravarti and S. C. Chakravarti: Alkaloids of *Glycosmis arborea*. Part I. Isolation of Arborine and Arborinine: The Structure of Arborine. J. Chem. Soc. (London) **1953**, 3337.

26. Clark, K. J., G. I. Fray, R. H. Jaeger and R. Robinson: Synthesis of *D*- and *L-iso*Iridomyrmecin and Related Compounds. Tetrahedron **6**, 217 (1959).

27. Cohen, A., J. W. Haworth and E. G. Hughes: Synthetical Experiments in the B Group of Vitamins. Part IV. A Synthesis of Pyridoxine. J. Chem. Soc. (London) **1952**, 4374.

28. Colombi, L. und H. Schinz: Zur Kenntnis der Sesquiterpene und Azulene. 101. Mitt. Synthese und Cyclisation von ,,Sesquilavandulol". Helv. Chim. Acta **35**, 1066 (1952).

29. Cookson, G. H. and C. Rimington: Porphobilinogen: Chemical Constitution. Nature (London) **171**, 875 (1953).

30. Corey, E. J. and R. R. Sauers: Total Synthesis of Pentacyclosqualene. J. Amer. Chem. Soc. **79**, 3925 (1957).

31. Corey, E. J. and J. J. Ursprung: Proof of the Constitution of Friedelin by Multi-group Rearrangement of Friedelan-3β-ol to Olean-13(18)-ene. J. Amer. Chem. Soc. **77**, 3668 (1955).

32. Corey, E. J. and R. W. White: Cationic Displacement of Hydrogen by Oxygen at a Saturated Carbon Atom. J. Amer. Chem. Soc. **80**, 6686 (1958).

33. Cornforth, J. W., M. E. Daines and A. Gottschalk: Synthesis of N-Acetylneuraminic Acid (Lactaminic Acid, O-Sialic Acid). Proc. Chem. Soc. (London) **1957**, 25.

34. Cresswell, R. M. and H. C. S. Wood: Chemical Studies of the Biosynthesis of Riboflavin. Proc. Chem. Soc. (London) **1959**, 387.

35. Day, A. C., J. Nabney and A. I. Scott: The Total Synthesis of Griseofulvin. Proc. Chem. Soc. (London) **1960**, 284.

36. Dietrich, P. et E. Lederer: Synthèse totale de l'ambréinolide racémique et de quelques-uns de ses dérivés. Helv. Chim. Acta **35**, 1148 (1952).

37. Dutler, H., O. Jeger und L. Ruzicka: Zur Kenntnis der Triterpene. 184. Mitt. Zur Konstitution und Konfiguration von Friedelin und Cerin; ein Beitrag zur Biogenese pentacyclischer Triterpene. Helv. Chim. Acta **38**, 1268 (1955).

38. Eliasberg, J. und P. Friedländer: Über einige Condensationen des o-Amidobenzaldehyds. Ber. dtsch. chem. Ges. **25**, 1752 (1892).

39. Eschenmoser, A., L. Ruzicka, O. Jeger und D. Arigoni: Zur Kenntnis der Triterpene. 190. Mitt. Eine stereochemische Interpretation der biogenetischen Isoprenregel bei den Triterpenen. Helv. Chim. Acta **38**, 1890 (1955).

40. Ewing, J., G. K. Hughes, E. Ritchie and W. C. Taylor: An Alkaloid Related to Dehydrolaudanosoline. Nature (London) **169**, 618 (1952).

41. FISCHER, H. O. L. und E. BAER: Synthese von d-Fructose und d-Sorbose aus d-Glycerinaldehyd, bzw. aus d-Glycerinaldehyd und Dioxy-aceton; über Aceton-glycerinaldehyd, III. Helv. Chim. Acta 19, 519 (1936).

42. GALINOVSKY, F., A. WAGNER und R. WEISER: Die Umsetzung von N-Methyl-α-pyrrolidon und N-Methyl-α-piperidon mit LiAlH₄ zu den ω-Methylamino-aldehyden. Synthesen von Hygrin, Cuskhygrin und Methylisopelletierin. Monatsh. Chem. 82, 551 (1951).

43. GELLÉRT, E., RAYMOND-HAMET und E. SCHLITTLER: Die Konstitution des Alkaloids Cryptolepin. Helv. Chim. Acta 34, 642 (1951).

44. GREWE, R., A. MONDON und E. NOLTE: Die Totalsynthese des Tetrahydro-deoxycodeins. Liebigs Ann. Chem. 564, 161 (1949).

45. HAHN, G. und A. HANSEL: Synthese von 5.6.3.14-Tetrahydro-yobyrinen. Ber. dtsch. chem. Ges. 71, 2192 (1938).

46. HAHN, G. und H. LUDEWIG: Synthese von Tetrahydro-harman-Derivaten unter physiologischen Bedingungen, I. (vorl.) Mitt. Ber. dtsch. chem. Ges. 67, 2031 (1934).

47. HAHN, G. und O. SCHALES: Über β-(Oxy-phenyl)-äthylamine und ihre Um-wandlungen, III. Mitt.: Synthese von Benzylisochinolinen unter physiologischen Bedingungen. Ber. dtsch. chem. Ges. 68, 24 (1935).

48. HAMILTON, E. E. P. and R. ROBINSON: An Extension of the Theory of Addition to Conjugated Unsaturated Systems. Part I. Note on the Constitution of the Salts of 1-Benzylidene-2-methyl-1:2:3:4-tetrahydroisoquinoline. J. Chem. Soc. (London) 109, 1029 (1916).

49. HARLEY-MASON, J.: Oxidation of 2:5-Dihydroxyphenylalanine and its Biogenetic Significance. Chem. and Ind. 1952, 173.

50. — Private communication (in press).

51. HARLEY-MASON, J. and A. H. JACKSON: A New Synthesis of Bufotenine. Chem. and Ind. 1952, 954.

52. — — Hydroxytryptamines. Part I. Bufotenine, 6-Hydroxybufotenine and Serotonin. J. Chem. Soc. (London) 1954, 1165.

53. — — Hydroxytryptamines. Part II. A New Synthesis of Physostigmine. J. Chem. Soc. (London) 1954, 3651.

54. HEILBRON, I. M., E. D. KAMM and W. M. OWENS: The Unsaponifiable Matter from the Oils of Elasmobranch Fish. Part I. A Contribution to the Study of the Constitution of Squalene (Spinacene). J. Chem. Soc. (London) 1926, 1630.

55. HOUGH, L. and J. K. N. JONES: The Synthesis of Sugars from Simpler Substances. Part I. The in vitro Synthesis of the Pentoses. J. Chem. Soc. (London) 1951, 1122.

56. — — The Synthesis of Sugars from Simpler Substances. Part II. The Synthesis of DL-Ribose in vitro from D-Glyceraldehyde and Glycollic Aldehyde. J. Chem. Soc. (London) 1951, 3191.

57. HUGHES, G. K. and E. RITCHIE: Experiments on the Synthesis of the Acridone Alkaloids. Austral. J. Sci. Research A 4, 423 (1951).

58. HURST, J. J. and G. H. WHITHAM: Synthesis of Chrysanthenone by the Photo-isomerisation of Verbenone. Proc. Chem. Soc. (London) 1959, 160.

59. IKUTANI, Y., T. OKUDA and S. AKABORI: β-Hydroxyleucine. I. Synthesis by Means of Copper Complex and Separation of the Diastereomeric Racemates. Bull. Chem. Soc. Japan 33, 582 (1960).

60. JAIN, A. C. and G. W. KENNER: Pyrroles and Related Compounds. Part II. Michael Addition to Pyrromethenes. J. Chem. Soc. (London) 1959, 185.

61. JOHNSON, W. S. and R. A. BELL: A Biogenetically Patterned Concept for the Laboratory Synthesis of Cyclic Compounds. Tetrahedron Letters **1960**, No. 12, 27.

62. LEETE, E., L. MARION and I. D. SPENSER: Biogenesis of Hyoscyamine. Nature (London) **174**, 650 (1954).

63. LEONARD, N. J. and S. W. BLUM: Laboratory Realization of the Robinson-Schöpf Scheme of Alkaloid Synthesis. The Pyrrolizidine Alkaloids. J. Amer. Chem. Soc. **82**, 503 (1960).

64. MANNICH, C.: Eine Synthese des Arecaidinaldehyds und des Arecolins. Ber. dtsch. chem. Ges. **75**, 1480 (1942).

65. MATSUURA, T. and H. J. CAHNMANN: Model Reactions for the Biosynthesis of Thyroxine. III. The Synthesis of Hindered Quinol Ethers and their Conversion to Hindered Analogs of Thyroxine. J. Amer. Chem. Soc. **82**, 2055 (1960).

66. MAUZERALL, D.: The Condensation of Porphobilinogen to Uroporphyrinogen. J. Amer. Chem. Soc. **82**, 2605 (1960).

67. MENZIES, R. C. and R. ROBINSON: A Synthesis of ψ-Pelletierine. J. Chem. Soc. (London) **125**, 2163 (1924).

68. METZLER, D. E., J. B. LONGENECKER and E. E. SNELL: The Reversible Catalytic Cleavage of Hydroxyamino Acids by Pyridoxal and Metal Salts. J. Amer. Chem. Soc. **76**, 639 (1954).

69. MILLER, S. L.: Production of Some Organic Compounds under Possible Primitive Earth Conditions. J. Amer. Chem. Soc. **77**, 2351 (1955).

70. PICTET, A. und T. Q. CHOU: Über die Einwirkung von Methylal auf Tetrahydro-papaverin. Ber. dtsch. chem. Ges. **49**, 370 (1916).

70a. PRICE, J. R.: Alkaloids Related to Anthranilic Acid. Fortschr. Chem. organ. Naturstoffe **13**, 302 (1956).

71. RIEDL, W.: Konstitution und Synthese der Hopfenbitterstoffe *d,l*-Humulon und Lupulon sowie einiger Analoga (V. Mitt.). Chem. Ber. **85**, 692 (1952).

72. ROBINSON, R.: A Synthesis of Tropinone. J. Chem. Soc. (London) **111**, 762 (1917).

73. — A Theory of the Mechanism of the Phytochemical Synthesis of Certain Alkaloids. J. Chem. Soc. (London) **111**, 876 (1917).

74. — Synthesis in Biochemistry (Fifth Pedler Lecture). J. Chem. Soc. (London) **1936**, 1079.

75. — The Structural Relations of Natural Products. Oxford: Clarendon Press. 1955.

76. ROBINSON, R. and S. SUGASAWA: Preliminary Synthetic Experiments in the Morphine Group. Part IV. A Dehydro-derivative of Laudanosoline Hydrochloride and its Constitution. J. Chem. Soc. (London) **1932**, 789.

77. RUZICKA, L.: The Isoprene Rule and the Biogenesis of Terpenic Compounds. Experientia **9**, 357 (1953).

78. RUZICKA, L. und E. CAPATO: Höhere Terpenverbindungen. XXIV. Ringbildungen bei Sesquiterpenen. Totalsynthese des Bisabolens und eines Hexahydro-cadalins. Helv. Chim. Acta **8**, 259 (1925).

79. SATO, M., K. OKAWA and S. AKABORI: A New Synthesis of Threonine. Bull. Chem. Soc. Japan **30**, 937 (1957).

80. SCHENCK, G. O.: Autoxydation von Furan und anderen Dienen. (Die Synthese des Ascaridols.) Angew. Chem. **57**, 101 (1944).

81. SCHÖPF, CL.: Die Überführung von Δ^1-Piperidein in rac. Lupinin und rac. α-Isospartein sowie in Nebenalkaloide der *Lobelia inflata*. Angew. Chem. **69**, 69 (1957).

82. Schöpf, Cl. und H. Bayerle: Zur Frage der Biogenese der Isochinolin-alkaloide. Die Synthese des 1-Methyl-6,7-dioxy-1,2,3,4-tetrahydro-isochinolins unter physiologischen Bedingungen. Liebigs Ann. Chem. **513,** 190 (1934).

83. Schöpf, Cl., G. Benz, Fr. Braun, H. Hinkel und R. Rokohl: Die Konden-sation von Δ^1-Piperidein mit Acetondicarbonsäure und Formaldehyd. Angew. Chem. **65,** 161 (1953).

84. Schöpf, Cl., A. Komzak, Fr. Braun und E. Jacobi: Über die Polymeren des Δ^1-Piperideins. Liebigs Ann. Chem. **559,** 1 (1948).

85. Schöpf, Cl. und G. Lehmann: Über die Alkaloide der Angosturarinde: Die Synthese des Chinaldins und α-n-Amylchinolins unter physiologischen Be-dingungen. Liebigs Ann. Chem. **497,** 7 (1932).

86. — — Die Synthese des Tropinons, Pseudopelletierins, Lobelanins und ver-wandter Alkaloide unter physiologischen Bedingungen. Liebigs Ann. Chem. **518,** 1 (1935).

87. Schöpf, Cl. und F. Oechler: Zur Frage der Biogenese des Vasicins (Peganins). Die Synthese des Deoxyvasicins unter physiologischen Bedingungen. Liebigs Ann. Chem. **523,** 1 (1936).

88. Schöpf, Cl. und A. Schmetterling: Versuche zur Synthese des Scopinons. Angew. Chem. **64,** 591 (1952).

89. Schöpf, Cl. und H. Steuer: Zur Frage der Biogenese des Rutaecarpins und Evodiamins. Die Synthese des Rutaecarpins unter zellmöglichen Bedingungen. Liebigs Ann. Chem. **558,** 124 (1947).

90. Schöpf, Cl. und K. Thierfelder: Die Dehydrierung des Laudanosolins und des Laudanosolin-3′,4′-dimethyläthers. Liebigs Ann. Chem. **497,** 22 (1932).

91. Scott, J. J.: Synthesis of Crystallizable Porphobilinogen. Biochemic. J. **62,** 6 P (1956).

92. Sheehan, J. C. and B. M. Bloom: The Synthesis of Teloidinone and 6-Hydroxy-tropinone. J. Amer. Chem. Soc. **74,** 3825 (1952).

93. Siedel, W. und F. Winkler: Oxydation von Pyrrolderivaten mit Bleitetra-acetat. Neuartige Porphyrinsynthesen. Liebigs Ann. Chem. **554,** 162 (1943).

94. Simonsen, J. L. and L. N. Owen: The Terpenes, Vol. I, p. 107. London: Cambridge Univ. Press. 1947.

95. Späth, E. und F. Berger: Über eine für die Phytochemie bemerkenswerte Synthese des d,l-Tetrahydropapaverins. Ber. dtsch. chem. Ges. **63,** 2098 (1930).

96. Späth, E. und E. Kruta: Über die Synthese von berberinartigen Basen aus Verbindungen vom Typus des Tetrahydropapaverins. Monatsh. Chem. **50,** 341 (1928).

97. Spring, F. S., J. M. Beaton, R. Stevenson and J. L. Stewart: The Consti-tution of Alnusenone (Glutinone), a New Type of Pentacyclic Triterpenoid. Chem. and Ind. **1956,** 1054.

98. Stadler, P. A., A. Eschenmoser, H. Schinz und G. Stork: Untersuchungen über den sterischen Verlauf säure-katalysierter Cyclisationen bei terpenoiden Polyenverbindungen. 3. Mitt. Zur Stereochemie der Bicyclofarnesylsäuren. Helv. Chim. Acta **40,** 2191 (1957).

99. Stadler, P. A., A. Nechvatal, A. J. Frey und A. Eschenmoser: Unter-suchungen über den sterischen Verlauf säure-katalysierter Cyclisationen bei terpenoiden Polyenverbindungen. 1. Mitt. Cyclisation der 7,11-Dimethyl-2(trans),6(trans),10-dodecatrien- und der 7,11-Dimethyl-2(cis),6(trans),10-do-decatrien-säure. Helv. Chim. Acta **40,** 1373 (1957).

100. Stoll, A., B. Becker und E. Jucker: Synthese des Äpfelsäure-dialdehyds und des 3,6-Dioxy-tropans. 1. Mitt. über Alkaloidsynthesen. Helv. Chim. Acta **35**, 1263 (1952).

101. Stoll, M. et A. Commarmont: Odeur et constitution. II. α- et β-bicyclo-farnésal et β-bicyclofarnésol. Helv. Chim. Acta **32**, 1836 (1949).

102. Stork, G. and A. W. Burgstahler: The Stereochemistry of Polyene Cycli-zation. J. Amer. Chem. Soc. **77**, 5068 (1955).

103. Stork, G., J. E. Davies and A. Meisels: The Total Synthesis of a Naturally Occurring Pentacyclic Triterpene System. J. Amer. Chem. Soc. **81**, 5516 (1959).

104. Tiemann, F. und P. Krüger: Über Veilchenaroma. Ber. dtsch. chem. Ges. **26**, 2675 (1893).

105. Tiemann, F. und F. W. Semmler: Über Verbindungen der Citral- (Geranial-) reihe. Ber. dtsch. chem. Ges. **26**, 2708 (1893).

106. Van Tamelen, E. E. and J. S. Baran: Total Synthesis ofO xygenated Lupin Alkaloids. J. Amer. Chem. Soc. **80**, 4659 (1958).

107. Van Tamelen, E. E., L. J. Dolby and R. G. Lawton: A Biogenetically-Patterned Laboratory Synthesis in the Strychnine-Curare Alkaloid Series. Tetrahedron Letters **1960**, No. 19, 30.

108. Van Tamelen, E. E. and R. L. Foltz: Synthesis of the Lupinine System Patterned after the Biogenetic Scheme of Schöpf and Robinson. J. Amer. Chem. Soc. **82**, 502 (1960).

109. — — The Biogenetic-type Synthesis of *dl*-Sparteine. J. Amer. Chem. Soc. **82**, 2400 (1960).

110. Van Tamelen, E. E. and V. Haarstad: Unpublished observations.

111. von Bruchhausen, F.: Über die Synthese von *r*-Corydalin. Arch. Pharmaz. **261**, 31 (1923).

112. von Mutzenbecher, P.: Über die Bildung von Thyroxin aus Dijodtyrosin. Z. physiol. Chem. (Hoppe-Seyler) **261**, 253 (1939).

113. Wagner-Jauregg, T.: Synthesen von Terpenen aus Isopren Liebigs Ann. Chem. **496**, 52 (1932).

114. Wenkert, E.: Alkaloid Biosynthesis. Experientia **15**, 165 (1959).

115. Wieland, Th., E. Schäfer und E. Bokelmann: Über Peptidsynthesen. V. Über eine bequeme Darstellungsweise von Acylthiophenolen und ihre Verwendung zu Amid- und Peptid-Synthesen. Liebigs Ann. Chem. **573**, 99 (1951).

116. Witkop, B.: The Structure of Quinamine. J. Amer. Chem. Soc. **72**, 2311 (1950).

117. Wolff, R. E. et E. Lederer: Nouvelle synthèse totale de l'ambréinolide racémique. Bull. soc. chim. France **1955**, 1466.

118. Woodward, R. B.: Neuere Entwicklungen in der Chemie der Naturstoffe. Angew. Chem. **68**, 13 (1956).

119. Zobrist, F. und H. Schinz: Zur Kenntnis der Sesquiterpene. 86. Mitt. Die Cyclisation von Farnesyliden-aceton. Helv. Chim. Acta **32**, 1192 (1949).

(Received, January 13, 1961.)

Der Kohlenhydratstoffwechsel im Roggen und Weizen.

Von **H. H. Schlubach**, München.

Mit 6 Abbildungen.

Inhaltsübersicht.

I. Die löslichen Kohlenhydrate im Roggen und ihre Konstitution.

Die Erforschung des Kohlenhydratstoffwechsels im Roggen beginnt mit einer von MÜNTZ (*27*) im Jahre 1878 mitgeteilten Beobachtung. Aus milchreifen Roggenkörnern gewann er eine optisch inaktive Verbindung, die Fehlingsche Lösung nicht reduzierte und nach Säurehydrolyse eine Drehung von — 52° zeigte. Er verfolgte ihre Bildung im Laufe der Vegetationsperiode und stellte fest, daß sie in dem Maße abnimmt, in dem die Stärkebildung im reifenden Korn zunimmt. Er schloß daraus, daß sich die Stärke auf Kosten der Verbindung bildet. Aber auch im reifen Roggenkorn verschwand die Verbindung nicht vollständig. Je nach der Roggensorte blieb vielmehr ein Gehalt von 2,0—5,2% an ihr zurück. In einem weißen Handelsmehl wurden 2,3% gefunden. Dextrine konnten nicht beobachtet werden. Die Verbindung wurde für identisch mit der Synanthrose gehalten, die acht Jahre zuvor von POPP (*29*) aus der Knolle von *Dahlia variabilis* gewonnen war. Im Weizen, dem Hafer, der Gerste und dem Mais konnte sie nicht festgestellt werden. MÜNTZ hat deshalb vorgeschlagen, den betrügerischen Zusatz von Roggenmehl zu Weizenmehl an dem Gehalt an Synanthrose zu erkennen. Er spricht seine Verwunderung darüber aus, daß sich zwei so nahe verwandte Arten wie Roggen und Weizen für die gleiche Funktion so verschiedener Zuckerverbindungen bedienen.

Erst zwölf Jahre später (1890) hat TANRET (*50*) es unternommen, ohne auf die Mitteilung von MÜNTZ Bezug zu nehmen, die Verbindung aus Roggenmehl weiter zu reinigen und näher zu charakterisieren. Nach Abtrennung über die Barytverbindung drehte sie bei $[\alpha]_D = -36°$ (Wasser, c = 5), reduzierte Fehlingsche Lösung nicht und wurde weder durch Bierhefe noch durch Diastase vergoren. Das Molekulargewicht wurde mit 652 gemessen. Bei der ebenso leicht wie bei der Saccharose verlaufenden Säurehydrolyse sank die Drehung auf $-76°$. Aus dem Hydrolysat wurde nach Entfernung der Fructose durch Kalkwasser ein Rückstand von der Drehung $+5°$ erhalten. Auf Grund dieses Befundes hat TANRET angenommen, daß die Verbindung ein Tetrasaccharid sei, das zu drei Viertel aus Fructose und zu einem Sechstel aus einem anderen Kohlenhydrat bestände; der Rest sei durch die Kalkbehandlung zerstört. Wegen der Linksdrehung der Verbindung wurde sie „Levosin" genannt. Ihr Gehalt betrug im jungen Roggenkorn im Juni etwa 0,3%, bei der vollständigen Reife bis 0,7%. Da aber gleichzeitig der Wassergehalt von 65% auf 16% sank, bewegte sich der Gehalt, bezogen auf die Trockensubstanz, nahezu konstant um 0,8%. Im Gegensatz zu MÜNTZ stellte aber TANRET fest, daß auch im Weizen und in der Gerste die gleiche Verbindung gebildet wird, nicht aber im Hafer und im Mais.

Wiederum unabhängig von MÜNTZ und TANRET haben im Jahre 1894 und 1895 SCHULZE und FRANKFURT (*47, 48*) aus den Achsen von Roggenpflanzen über die Strontiumverbindung neben Saccharose ein in Alkohol schwerlösliches Kohlenhydrat isoliert, das sie anfangs „β-Lävulin", später „Secalose" benannt haben. Sie haben die Verbindung als eine weiße, hygroskopische Masse beschrieben, die $[\alpha]_D = -28,6°$ bis $-28,9°$ drehte, Fehlingsche Lösung nicht reduzierte und nach Säurehydrolyse nur Fructose ergab, die $-81°$ drehte.

Eingehender hat sich darauf JENSEN-HANSEN (*19*) mit den im Roggen in verschiedenen Stadien der Entwicklung gebildeten löslichen Kohlenhydraten beschäftigt. Durch Auszug von unreifem Roggen mit 90%igem Alkohol erhielt er neben Glucose, Fructose und Saccharose eine Verbindung, die er als wahrscheinlich identisch mit der von SCHULZE und FRANKFURT beschriebenen, angesehen hat. Aus einem Auszug mit 70%igem Alkohol erhielt er nach Barytfällung neben anderen Kohlenhydraten eine „Apeponin" genannte Verbindung. Er hat sie als eine weiße, amorphe, nicht reduzierende Verbindung beschrieben, welche bei $[\alpha]_D = -41,3°$ drehte. Sie wurde als von der Secalose wahrscheinlich verschieden aber mit dem Lävosin von TANRET identisch angesehen.

Vom Jahre 1922 an haben COLIN und BELVAL (*12, 13*), später auch DE CUGNAC (*15*) den Kohlenhydratstoffwechsel in den verschiedenen Organen und den verschiedenen Phasen des Wachstums hauptsächlich in der Weizenpflanze, daneben aber auch im Roggen, eingehend verfolgt.

Literaturverzeichnis: SS. 312—315.

In den Blättern wurde nur Saccharose und deren Inversionsprodukte angetroffen, in keinem Falle Stärke. Die reduzierenden Zucker machten hierbei ein Viertel bis die Hälfte des Gesamtzuckers aus. In den Blattscheiden reicherte sich vom Mai bis Juni ein linksdrehender Zucker an. Da in den Blatträndern stets mehr Saccharose gefunden wurde als in den Blattscheiden, wurde geschlossen, daß die Saccharose das erste greifbare Produkt der Chlorophyll-Assimilation sei. Die Hauptansammlung des „Levosins" erfolgt in den Achsen und zwar stärker in den unteren als in den oberen Abschnitten. Nach dem Erscheinen der Körner enthielten diese bis zu 8% an Levosin, entsprechend 30—40% der Trockensubstanz. Im Laufe des Juli sinkt der Gehalt an Levosin in dem Maße, in dem derjenige an Stärke im reifenden Korn ansteigt. Die Achsen werden daher als Reservestoffbehälter angesehen, in denen transitorisch Kohlenhydrate in einer Form gespeichert werden, welche eine höhere Konzentration unter Herabsetzung des osmotischen Druckes ermöglicht. Nach dem Erscheinen der Körner dient es in ihnen zur Ausbildung der Stärke. Ähnlich wie das Inulin von den Synanthrinen, wird auch das Levosin von anderen ähnlichen Polysacchariden begleitet.

Im Jahre 1928 hat dann TILLMANS (51) eine Verbindung beschrieben, die er durch Extraktion von Roggenmehl mit 70%igem Alkohol erhalten hatte. Sie drehte $[\alpha]_D = -43{,}9°$ und nach Säurehydrolyse $-92{,}7°$. Da die Verbindung von Hypojodit nicht angegriffen wurde, nahm TILLMANS an, daß sie nur aus Fructose bestünde. Auf Grund der Elementaranalyse und von Molekulargewichtsbestimmungen (488) wurde sie als ein Trifructosan angesprochen. Da TILLMANS weiter annahm, daß sie im Weizenmehl nicht oder nur in Spuren enthalten sei, hielt er, wie dies schon MÜNTZ getan hatte, in dem Nachweis des „Trifructosans" ein Verfahren für gegeben, um Verfälschungen von Weizenmehl durch Roggenmehl nachweisen zu können.

Im gleichen Jahre haben CHRZASZCZ und MICHALSKI (11) über die löslichen Kohlenhydrate im Roggenmehl berichtet. Später hat sich KRUISHEER (23) eingehend mit der von TILLMANS angegebenen Methode beschäftigt. Im Roggenmehl fand er einen Gehalt von 1,5—2% „Trifructosan"; aber, abweichend von TILLMANS, auch im Weizenmehl einen solchen von 0,1—0,3%, im Mittel 0,22%. Beim Lagern wurde ein Ansteigen des Gehaltes beobachtet.

1934 wurde von SCHLUBACH und KOENIG (39) erstmalig der Versuch unternommen, die Konstitution der Polysaccharide des Roggens näher zu bestimmen. Durch Extraktion von Roggenmehl mit 70%igem Alkohol wurde ein Rohprodukt erhalten, das nach Enteiweißung durch Fällung der wässerigen Lösungen mit Alkohol so lange fraktioniert wurde, bis sich die Drehung nicht mehr änderte. Sie betrug dann, übereinstimmend mit TANRET, $[\alpha]_D^{20} = -36{,}6°$. Nach mehrfachem Abdampfen mit

294 H. H. Schlubach:

Wasser und Trocknung im Hochvakuum konnte sie auf — 40,0° erniedrigt werden. Außer diesem direkten wurde noch der Weg einer indirekten Reinigung über die Acetylverbindung eingeschlagen. Nach mehrfacher Fällung ihrer benzolischen Lösungen mit Petroläther drehte die Acetylverbindung $[\alpha]_D^{20} = -7,2°$ (Chloroform, c = 1,0). Das nach der Entacetylierung erhaltene Polysaccharid drehte, ebenso wie die direkt gereinigte Verbindung, bei — 36,6°. Die an dem Polysaccharid wie an seiner Acetylverbindung gemessenen Molekulargewichte ergaben übereinstimmend Werte, die auf einen Polymerisationsgrad von 9—10 Hexoseeinheiten schließen ließen. Da bei der Säurehydrolyse nur ein Aldosewert von 0,59% gefunden wurde, wurde geschlossen, daß die Verbindung nur aus Fructoseresten aufgebaut sei. In der Annahme, daß es sich um eine allgemein in den Gramineen vorkommende Verbindung handle, wurde sie „Graminin" benannt.

Aus der bedeutenden Differenz zwischen den Halbumsatzzeiten der Hydrolyse nach der Drehung und dem Reduktionswert wurde auf einen komplizierten Bau des Graminins geschlossen. Diese Folgerung wurde durch das Ergebnis der Bausteinanalyse nach der Methylierungsmethode bestätigt. Es wurde ein Gemisch von Tetra-, Tri- und Dimethyl-hexosen im Verhältnis von 1 : 1 : 1 oder wahrscheinlicher 2 : 1 : 2 gefunden. Die Tetramethylverbindung erwies sich als die 1,3,4,6-Tetramethyl-fructose, die Trimethylverbindung als die 3,4,6-Trimethyl-fructose. Demnach war die Bindung im Graminin wie im Inulin zwischen den einzelnen Fructoseresten an der zweiten und ersten Hydroxylgruppe erfolgt. Da das Graminin nicht reduzierte, wurde ein großer, stark verzweigter Ring angenommen, für den ein Formelbild gegeben wurde.

Im gleichen Jahre hat Kretowitsch (21) gefunden, daß in den Blättern der Roggenpflanze Fructose die Glucose überwiegt und hat ein Verhältnis von 1,79 : 1,07 festgestellt.

1939 haben Tschertok und Schapiro (52) den Gehalt von ukrainischem Roggen an „Trifructosan" untersucht; sie fanden einen solchen von 1,63—1,77%. Da sie gleichzeitig in Weizenmehl einen Gehalt von 0,55—0,66% feststellten, folgerten sie, daß erst ein Gehalt von mehr als 0,66% im Weizenmehl auf einen Zusatz von Roggenmehl schließen läßt, die Methode von Tillmans also dieser Begrenzung bedarf.

Um festzustellen, ob die von Schulze und Frankfurt aus den Achsen von Roggenpflanzen isolierte Secalose mit dem Graminin der Ähren identisch sei, haben Schlubach und Bandmann (31) von den Ähren befreite Roggenhalme 14 Tage nach der Blüte in der üblichen Weise mit Alkohol extrahiert. Die direkt oder über die Acetylverbindung gereinigte Verbindung drehte $[\alpha]_D^{20} = -37,6°$ (Wasser, c = 1) und erwies sich dadurch sowie durch die Drehung der Acetylverbindung von $[\alpha]_D^{20} = +3°$ (Chloroform, c = 1) als vom Graminin verschieden. In

Anlehnung an den von SCHULZE und FRANKFURT gewählten Namen Secalose wurde sie Secalin genannt.

Noch deutlicher offenbarte sich der Unterschied gegenüber dem Graminin bei der Bausteinanalyse nach der Methylierungsmethode. Denn es wurde bei ihr außer der gleichen 1,3,4,6-Tetramethyl-fructose die bei 75° schmelzende 1,3,4-Trimethyl-fructose gefunden. Auch die Dimethyl-fructose war verschieden. Im Gegensatz zum Graminin, das, wie erwähnt, wie das Inulin gebaut ist und zum Inulintyp (S. 294) gehört, ist also im Secalin die Bindung der Hexosereste über die zweite und sechste Hydroxylgruppe erfolgt; es gehört dem Phleintyp an. Da es Fehlingsche Lösung nicht reduziert und bei der Säurehydrolyse nur Fructose erhalten wurde und die Molekulargewichtsbestimmungen auf eine Molekelgröße von 4—5 Fructoseeinheiten schließen ließen, wurde die Konstitution eines Tetrafructose-anhydrids angenommen und hierfür ein Formelbild vorgeschlagen.

Um die Frage zu klären, ob das aus Roggenähren erhaltene Graminin mit dem aus Weizenähren isolierten Sitosin (43) identisch ist, haben SCHLUBACH und MÜLLER (44) das Graminin noch einmal sorgfältig gereinigt. Die Drehung der Acetylverbindung sank hierbei auf — 10,2°, das Polysaccharid drehte — 39,9°. Der Aldosewert war auf Null gesunken. Neue Molekulargewichtsbestimmungen an der Acetylverbindung ergaben Werte, die 15 Fructoseeinheiten entsprachen. Das Verhältnis der bei der Hydrolyse des Methyläthers erhaltenen Tetra-, Tri- und Dimethyl-fructosen wurde mit 1 : 1 : 1 festgelegt. Obgleich danach Graminin und Sitosin nach ihrem Bau identisch sein könnten, wurde doch hinsichtlich der Teilchengrößen ein Unterschied in dem Sinne festgestellt, daß sie beim Sitosin niedriger, bei etwa 12 Fructoseeinheiten lag. Dem entsprach auch, daß das Sitosin durch Hefeinvertin erheblich rascher gespalten wurde als das Graminin. Denn nach den an den Polyfructosanen der Gräser gewonnenen Erfahrungen verhalten sich die Halbwertszeiten der Hydrolyse durch Invertin umgekehrt proportional den Teilchengrößen (30).

1957 haben SCHLUBACH und HABERLAND (33) das Graminin erneut einer sorgfältigen Reinigung unterzogen. Für die Acetylverbindung wurde $[\alpha]_D^{20} = — 5,8°$ (Chloroform, c = 1) gefunden, beim Graminin selber wurde eine Tiefstdrehung von — 41,9° erreicht. Das Verhältnis der Methyläther-Spaltstücke wurde mit 1 : 1 : 1 festgelegt. Abweichungen von den früheren Werten wurden bei der Dimethyl-fructose und im Aldosewert gefunden. Für diesen wurde, und zwar übereinstimmend mit Hypojodit und Perjodat, ein Wert von 3,1% gemessen. Auf Grund dieser Befunde, nach dem Verhalten bei der Papierchromatographie und den Molekulargewichtsbestimmungen, wurde auf eine Molekelgröße von 30 Hexoseeinheiten geschlossen und unter Verwertung der neuen analytischen Daten die folgende Formel vorgeschlagen:

$$Gl\ 1 \underbrace{\begin{bmatrix} 2\ Fru\ 1 \underline{\quad\quad} 2\ Fru\ 1 \\ 6 \\ | \\ 2\ Fru \end{bmatrix}}_{9} 2\ Fru\ 1 \underline{\quad} 2\ Fru \\ 6 \underline{\quad\quad} 1$$

Da sich so für das Graminin ein stark verzweigter Bau ergeben hatte, bestand wenig Aussicht, die Art dieser Verzweigung bei den höchst-

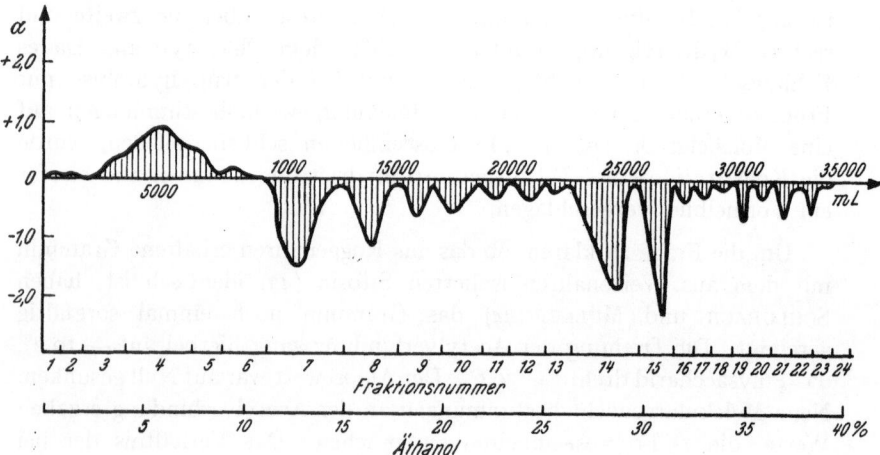

Abb. 1. Zerlegung von Graminin-Komponenten durch Desorption mit Äthanol in kontinuierlich ansteigender Konzentration; die Fraktionen wurden nach den Nullpunkten der Drehungen geschnitten. [Aus: Liebigs Ann. Chem. 614, 126 (1958).]

molekularen Komponenten des Gemisches klären zu können. SCHLUBACH und KOEHN (38) haben deshalb versucht, die niederen Glieder, bei denen die Verzweigung beginnt, zu isolieren und ihre Konstitution zu bestimmen. Zu diesem Zwecke war es nötig, das so sehr komplexe Gemisch in seine sämtlichen Komponenten zu zerlegen. Dies ist, durch Einsatz der von WHISTLER und DURSO (54) angegebenen, in einer etwas modifizierten Form angewandten Methode einer selektiven Desorption von einer Kohle-Celite-Säule, gelungen. Das Gemisch konnte so in 24 Fraktionen zerlegt werden (Abb. 1). Diese Fraktionen fielen dabei sämtlich in so ausreichenden Mengen an, daß ihre Konstitutionen nach der Methylierungs-methode bestimmt werden konnten (Abb. 2). Bei den niederen Gliedern wurde, da sie sich zunächst papierchromatographisch noch nicht als einheitlich erwiesen, eine Nachtrennung an einer Cellulosesäule durch-geführt. Es konnten so zwei Trisaccharide, ein Tetra- und ein Penta-saccharid isoliert werden, die sich nun papierchromatographisch als einheitlich erwiesen (Abb. 3). Die beiden Trisaccharide wurden als die von BARKER, BOURNE und CARRINGTON (6) entdeckte 1-Kestose oder Isokestose (I), und die von ALBON und Mitarb. (2) gefundene 6-Kestose oder Kestose (II) identifiziert. Von besonderem Interesse ist aber die Auffindung des Tetrasaccharids gewesen, für das die Konstitution (III) einer Bis-

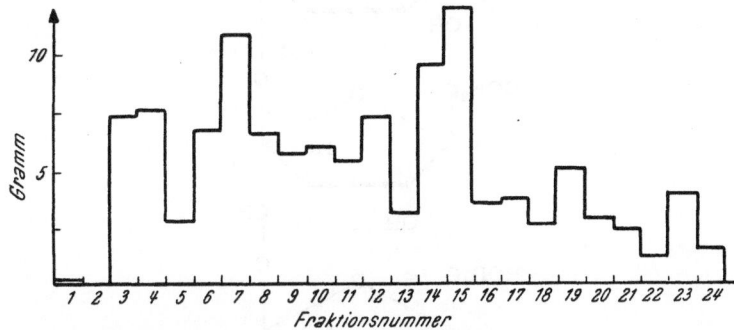

Abb. 2. Mengen der 24 Fraktionen von Graminin-Komponenten (in Grammen). [Aus: Liebigs Ann. Chem. 614, 126 (1958).]

(2—1 Fru, 2—6 Fru)-*D*-fructofuranosyl-saccharose festgestellt wurde — beginnt doch mit ihr die Verzweigung. Wegen der Bedeutung, die dieser Verbindung für das Verständnis der Art der Verzweigung zukommt,

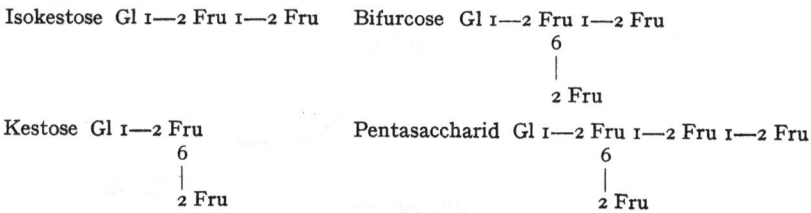

Abb. 3. Di-, Tri-, Tetra- und Pentasaccharide aus Graminin. [Aus: Liebigs Ann. Chem. 614, 126 (1958).}

die auch von ASPINALL und DAS GUPTA (5) hervorgehoben wurde, hat sie den Namen „Bifurcose" (von Bifurcation = Gabelung) erhalten.

Die Konstitution des Pentasaccharids wurde ebenfalls als diejenige eines verzweigten Oligosaccharids bestimmt, bei dem eines der beiden Fructose-endglieder der Bifurcose eine Verlängerung um einen Fructose-rest erfahren hat. Schematisch lassen sich also die vier Oligosaccharide folgendermaßen wiedergeben (vgl. nächste Seite):

Isokestose Gl 1—2 Fru 1—2 Fru   Bifurcose Gl 1—2 Fru 1—2 Fru
                                              6
                                              |
                                            2 Fru

Kestose Gl 1—2 Fru   Pentasaccharid Gl 1—2 Fru 1—2 Fru 1—2 Fru
        6   6
        | |
      2 Fru   2 Fru

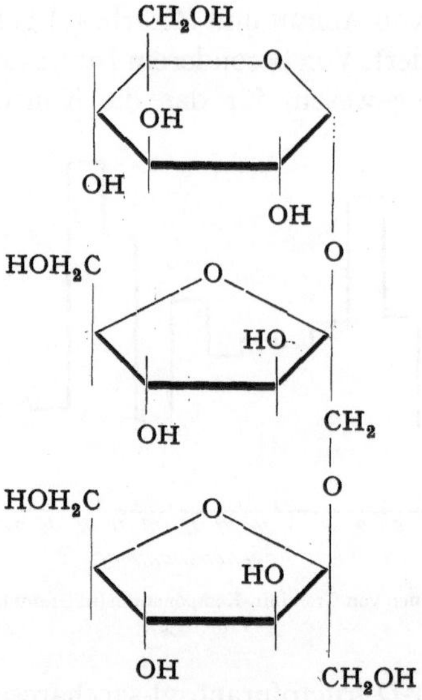

(I.) Isokestose.

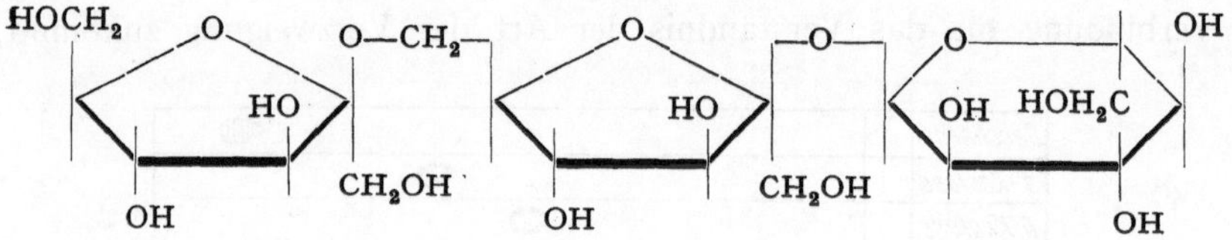

(II.) Kestose.

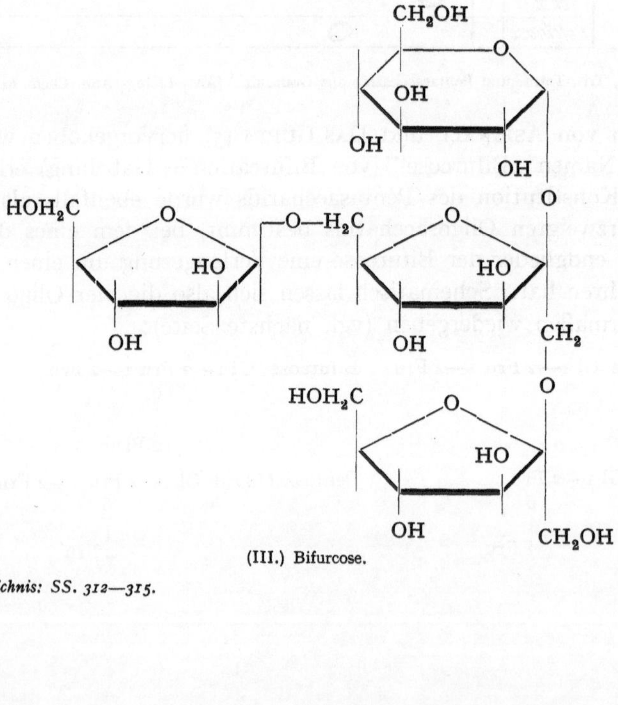

(III.) Bifurcose.

Literaturverzeichnis: SS. 312—315.

Nachdem so die ersten Stufen der Verzweigung in den Oligosacchariden geklärt waren, ergab sich die Frage, wie sich der Aufbau zu den Polysacchariden fortsetzt. Sie konnte beantwortet werden, indem die spezifischen Drehungen der höheren Fraktionen gegen die P-Werte (Polymerisationsgrad) aufgetragen wurden, wie sie sich aus den R_S-Werten ($R_S = R_f$: Saccharose) ergeben. Es zeigte sich, daß sie sich in zwei Geraden anordnen lassen, die mit der Bifurcose als dem gemeinsamen

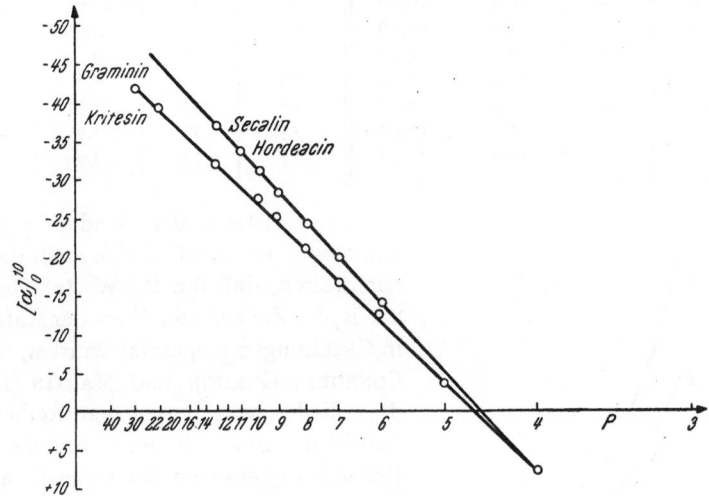

Abb. 4. Beziehungen zwischen den Drehungen und Polymerisationsstufen der beiden Reihen von Polysacchariden in den Roggenhalmen. (Es sind auch die Werte der höchsten aus der Gerste isolierten Glieder, Kritesin und Hordeacin, eingefügt.) [Aus: Liebigs Ann. Chem. 614, 126 (1958).]

Ausgangspunkt beginnen und sich bis zum Graminin einerseits, dem Secalin andererseits als den höchsten Gliedern der beiden Reihen fortführen lassen *(Abb. 4)*.

Da das Graminin dem Inulintyp, das Secalin dem Phleintyp angehört, folgert daraus für die Bildung der beiden Reihen, daß sie durch Fortsetzung der Transfructosidierung auf der Bifurcose entweder an der ersten oder der sechsten Hydroxylgruppe entstanden sind, wobei in bestimmten Abständen, vermutlich der jeweils übernächsten Fructose, eine weitere Verzweigung erfolgt. Daß derartige Verzweigungen in gewissen Abständen in einer Kette eintreten können, ist ja aus der Chemie der Stärke her wohlbekannt. Aber auch für Polyfructosane ist dies von HESTRIN und GOLDBLUM (*18*) für ein von ihnen erhaltenes bakterielles Lävan angenommen worden.

Da die Glieder der beiden Reihen mit einem P von 6—9 immer paarweise zusammengehören, sollten sie, wenn sie durch Transfructosidierung auf Saccharose entstanden sind, auch paarweise den gleichen Aldosewert aufweisen. Die Richtigkeit dieses Schlusses konnte, wie *Tabelle 1* zeigt, bestätigt werden.

Tabelle 1. Polymerisationsgrade und Aldosewerte beim Roggen.

Fraktion	P	Berechnet	Gefunden	Fraktion	P	Berechnet	Gefunden
8	6	16,7	16,62	16	10	10,0	10,1
9	6	16,7	16,60	17	10	10,0	10,0
10	7	14,3	14,59	18	11	9,09	9,36
11	7	14,3	14,29	19	11	9,09	9,04
12	8	12,5	12,90	20	12	8,3	8,79
13	8	12,5	12,51	22	13	7,7	7,98
14	9	11,1	11,02	23	13	7,7	7,63
15	9	11,1	11,01	24	13	7,7	7,75

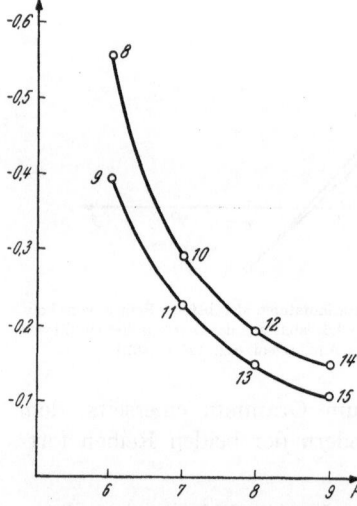

Abb. 5. $\log 1 \, (R_R - 1)$ in Abhängigkeit vom Polymerisationsgrad P für die Fraktionen 8—15. [Aus: Liebigs Ann. Chem. **614**, 126 (1958).]

Die Existenz der beiden polymer-homologen Reihen ließ sich auch dadurch nachweisen, daß die R_R-Werte (Quotient aus R_f der Zucker : R_F-Wert der Raffinose) in Gleichungen eingesetzt wurden, die von Consden, Gordon und Martin (*14*) für die Glieder von homologen Reihen von Peptiden aufgestellt sind. Für die Fraktionen 8—15 ergeben sich so zwei parallele Kurven *(Abb. 5)*.

Eine weitere Bestätigung der beiden polymer-homologen Reihen wurde von Elias und Schlubach (*16*) durch Messung der Molekulargewichte in der Ultrazentrifuge, also einer physikalischen Methode, nach Archibald (*4*) erbracht. Wenn auch die gefundenen Werte durchweg höher lagen als die nach den chemischen Methoden erhaltenen, so bewegten sie sich doch in der gleichen Größenordnung und unterscheiden sich untereinander durch die gleichen Abstufungen. Die Methode erlaubte auch, zu erkennen, daß die Fraktionen, insbesondere bei den höheren Gliedern, nicht polymer-einheitlich waren. Es wurden ferner Andeutungen dafür gefunden, daß außer dieser Uneinheitlichkeit auch eine solche im chemischen Bau vorhanden war. Dies kann z. B. sich dadurch ergeben, daß ein Bruchteil der Polysaccharide nicht auf Saccharose, sondern auf einer Difructose aufgebaut ist, wie dies ja bei den Gräsern nachgewiesen ist (*30*). Eine Trennung dieser glucosefreien Polyfructosane von den glucosehaltigen der gleichen Polymerisationsstufe ist bei den höheren

Gliedern nicht möglich. Daß mit den beiden so erkannten Aufbauwegen der Polyfructosane im Roggen noch nicht alle tatsächlichen Möglichkeiten erschöpft sind, geht aus dem Befund von SCHLUBACH und HABERLAND (*34*) hervor, daß in den Roggenähren neben dem Graminin, das nunmehr zweckmäßig als Graminin A zu bezeichnen wäre, noch ein Graminin B vorkommt, das dem Phleintyp angehört. Nach seinen analytischen Daten hat es sich bisher in keine der beiden Reihen einordnen lassen. Möglicherweise sind in ihm, wie aus der vorgeschlagenen Formel hervorgeht, die Verzweigungen in größeren Abständen erfolgt als im Secalin.

Mit der Untersuchung von SCHLUBACH und KOEHN (*37*) ist es, 80 Jahre nach der Entdeckung der Polyfructosane im Roggen durch MÜNTZ, gelungen, die Gesamtheit der in ihm gebildeten löslichen Kohlenhydrate zu erfassen und durch Feststellung ihrer Konstitution und Zuordnung in zwei polymer-homologe Reihen, die beiden Hauptwege ihrer Bildung zu erkennen. Die so komplexe Zusammensetzung, die im Laufe der Vegetationsperiode einem ständigen Wandel unterworfen und obendrein von den jeweiligen Wachstumsbedingungen abhängig ist, lassen es rückblickend verständlich erscheinen, warum die große Zahl von Forschern, die sich in der langen Zeit mit der Frage beschäftigt haben, zu so verschiedenen Ergebnissen gelangt sind. Und erst die Einführung der durch die selektive Desorption gegebenen Trennungsmethode in Verbindung mit den vertieften Erkenntnissen über den Aufbau polymerer Verbindungen hat es ermöglicht, das Problem einer Lösung entgegenzuführen.

II. Die löslichen Kohlenhydrate im Weizen und ihre Konstitution.

Im Gegensatz zu MÜNTZ, der im Weizen außer Saccharose keine Polysaccharide entdecken konnte, hat TANRET (*50*) 1890 erwähnt, daß er im Weizen das gleiche Levosin angetroffen habe, das er aus Roggen isoliert hatte. Den Gehalt an ihm fand er Anfang Juli ungefähr ebenso hoch wie Ende Juni beim Roggen. JENSEN-HANSEN (*19*) bestätigte, daß in jungen Weizenkörnern das gleiche „Apeponin" enthalten sei wie im Roggen. Aber wiederum abweichend von diesen Feststellungen hat GIRARD (*17*) in Auszügen aus Weizenmehl vergeblich nach den von MÜNTZ (*27*) beschriebenen „Dextrinen" gesucht und in ihnen nur Saccharose nachgewiesen. Anstatt dessen erwähnt er, eine „Galactin" genannte Verbindung gefunden zu haben, die von FLEURENT (*16a*) ebenfalls beschrieben wurde.

Eingehend haben, vom Jahre 1922 an, COLIN und BELVAL (*12, 13*) die Bildung des „Levosins" in den verschiedenen Organen der Weizenpflanze und im Laufe der Vegetationsperiode verfolgt. In den Blättern wurde es zu keiner Zeit angetroffen, aber von Mitte Mai bis Mitte Juni ansteigend von 0,52 bis 2,71% in den Blattscheiden. In den Achsen stieg der Gehalt, und zwar von oben nach unten fortschreitend, von Ende Mai bis Anfang Juli, an. Am höchsten wurde er in den Weizenkörnern, unmittelbar nach der Befruchtung, gefunden. Anfang Juli erreichte er 7%. Da die frischen Körner 80% Wasser enthielten, entspricht dies einem Gehalt von 35% der Trockensubstanz. In dem Maße, in dem im Laufe des Juli das Levosin schwindet, steigt der Gehalt an Stärke in den Körnern an. In einer Reihe von Weizensorten verschiedenen Ursprunges wurden die Gehalte nur wenig voneinander abweichend gefunden. Da in den Blättern nur Saccharose nachzuweisen war, wurde sie als das erste greifbare Produkt der Chlorophyll-Assimilation angesehen. Aus ihr erfolgt die Bildung der Fructosane, um eine Ansammlung zu höheren Konzentrationen ohne Erhöhung des osmotischen Druckes zu ermöglichen. Die Achsen dienen hierbei vorübergehend als Speicher für die Kohlenhydratreserven, welche den Aufbau der Stärke im reifenden Korn ermöglichen.

In einer sorgfältigen Untersuchung wurden diese Ergebnisse durch Barnell (7, 8) im Jahre 1938 bestätigt. Sie wichen nur darin ab, daß auch in den Blättern ein geringer, während der ganzen Vegetationsperiode nur wenig wechselnder Gehalt an Fructosanen festgestellt wurde.

In den Blattscheiden wurde, nach einem kurzen vorübergehenden Anstieg Mitte März, der höchste Gehalt mit 0,69% der Trockensubstanz am 8. Juni gemessen, um die gleiche Zeit, in welcher er auch in den Ähren mit 3,02% seinen Höhepunkt erreichte. Während des Mai und Juni wiesen die Achsen in ihren unteren Abschnitten einen höheren Gehalt auf als in den oberen. In den Ähren fiel er im Laufe des Juli mit zunehmender Reife rasch ab. In diesen Abwandlungen hat Barnell eine Bestätigung der von Belval (8a) geäußerten Ansicht gefunden, daß die fructosebildenden Polysaccharide eine wichtige Rolle als transitorische Reservekohlenhydrate besonders in den jungen Ähren spielen.

Lissitzyn (25) stellte in assimilierenden Weizenblättern einen Überschuß von Fructose über Glucose fest. Er hat hierin eine Bestätigung der Ansicht von Kirsch erblickt, daß die Fructofuranose das erste greifbare Assimilationsprodukt sei.

Die bereits erwähnte Annahme von Tillmans (51), daß im Weizenmehl Fructosane nicht oder nur in Spuren enthalten seien, ist durch die Untersuchung von Kruisheer (23) und später von Tschertok und Schapiro (52) dahingehend berichtigt worden, daß ebenso wie im reifen Roggen auch im Weizen ein gewisser Rest an Fructosanen verbleibt, nur ist sein Anteil im Weizen geringer.

Auf Grund der Beobachtung, daß erst nach vorangegangener Säurehydrolyse durch tryptische Verdauung Zucker in Freiheit gesetzt wurden, hat Krotkov (22) die Bildung eines Kohlenhydrat-Protein-Komplexes angenommen.

Literaturverzeichnis: SS. 312—315.

1940 hat ARCHBOLD (3) eine umfassende, sehr sorgfältige Zusammenfassung der Fructosane in den Monokotyledonen gebracht, in der auch die Getreidepolysaccharide eingehend behandelt werden. Mit den Fermentwirkungen in den reifenden Weizenkörnern haben sich im gleichen Jahre KNJAGINITSCHEW, MUTUL und PALILOWA (20) sowie KURSSANOW und BRJUSCHKOWA (24) beschäftigt.

Den ersten Versuch, die Konstitution der im Weizen enthaltenen löslichen Polysaccharide näher aufzuklären, haben 1948 SCHLUBACH und HUCHTING (36a) unternommen. Aus Weizenhalmen haben sie sowohl durch direkte Fraktionierung der wässerigen Lösungen mit Alkohol als auch durch Fraktionierung der Acetylverbindung sehr verlustreich ein Polysaccharid herausgearbeitet, das eine Drehung von $[\alpha]_D = -30,0°$ (Wasser, c = 1) aufwies, während die zugehörige Acetylverbindung $[\alpha]_D = +8,5°$ (Chloroform, c = 1) zeigte. Die Verbindung, welche „Pyrosin" benannt wurde, reduzierte nicht und ergab bei der Säurehydrolyse nur Fructose. Aus Molekulargewichtsbestimmungen sowie aus der Bausteinanalyse nach der Methylierungsmethode ergab sich ein Mindestgewicht von 5 Fructoseeinheiten. Da unter den Produkten der Hydrolyse des Methyläthers außer der 1,3,4,6-Tetramethyl-fructose die 1,3,4-Trimethyl-fructose gefunden wurde, gehört das Pyrosin dem Phleintyp an. Für seine Konstitution wurde ein verzweigter, vielgliedriger Ring angenommen.

1952 wurden von SCHLUBACH und MÜLLER (43) Weizenähren im Stadium der beginnenden Milchreife untersucht. Durch Kombination einer Fraktionierung des Polysaccharids sowie der Acetylverbindung mittels Gegenstromverteilung wurde, ebenfalls sehr verlustreich, ein „Sitosin" benanntes Polysaccharid herausgearbeitet, das ein $[\alpha]_D^{20} = -41,2°$ (Wasser, c = 1) und seine Acetylverbindung $[\alpha]_D^{20} = -10,1°$ (Chloroform, c = 1) zeigte. Nach den Molekulargewichtsbestimmungen wurde eine Teilchengröße von mindestens 9, wahrscheinlich 12 Hexoseeinheiten angenommen. Der Aldosegehalt betrug 3%. Als Komponenten der Hydrolyse des Methyläthers wurden die 1,3,4,6-Tetramethyl-fructose, die 3,4,6-Trimethyl-fructose und eine Dimethyl-fructose im Verhältnis von 1:1:1 gefunden. Im Gegensatz zum Pyrosin der Halme gehört das Sitosin der Ähren daher dem Inulintyp an. Da in allen drei Hydrolysenprodukten Aldosewerte, und zwar im Verhältnis von 2:3:2, gefunden wurden, wurde angenommen, daß das Polyfructosan von einem nicht abzutrennenden Glucosan begleitet war. Weiter unten wird gezeigt, daß neuere Befunde eine andere Deutung ermöglichen.

Eine papierchromatographische Zerlegung der Gesamtheit eines alkalischen Extraktes aus Weizenmehl und auch aus Weizen haben WHITE und SECOR (55) durchgeführt. Sie gelangten zu dem Schluß, daß im Weizen zwei homologe Reihen von Oligosacchariden gebildet werden, von denen die Tri- und Tetrasaccharide mit den durch Invertin-

wirkung auf Saccharose erhaltenen identisch seien. Es wurde auch ein
Tetrasaccharid beschrieben, das als eine Fructosyl-Raffinose angesprochen
wurde.

Von Bishop (9) wurde aus Weizenstroh-xylan durch enzymatischen Abbau
mit *Mycotherium verrucei* ein Trisaccharid erhalten; von Carter, McCluer und
Slifer (10) wurden mit Benzolextrakten aus Weizenmehl zwei Galactosylglycerine
isoliert.

Eingehender haben Montgomery und Smith (26) ein Glucofructosan
untersucht, das sie aus Weizenmehl durch Alkoholextraktion und
Fraktionierung des Baryumhydroxydkomplexes sowie der Acetylverbindung isoliert haben. Aus 12 Fraktionen der Acetylverbindung haben
sie die mengenmäßig überwiegende mit einer Drehung von $[\alpha]_D = + 3°$
herausgegriffen. Nach Entacetylierung erhielten sie ein Polysaccharid
von $[\alpha]_D = — 21°$ (Wasser), das sie anschließend analytisch untersuchten.
Der durch Methylierung der Acetylverbindung erhaltene Methyläther
erwies sich bei einer Fraktionierung nicht als einheitlich. Die Haupt-
fraktion von $[\alpha]_D^{28} = — 36°$ (Aceton, c = 1,3) wurde in der üblichen
Weise hydrolysiert und die Spaltprodukte chromatographisch getrennt.
Es wurden die 2,3,4,6-Tetra-O-methyl-D-glucose, die 1,3,4,6-Tetra-O-
methyl-D-fructose, die 1,3,4-Tri-O-methyl-D-fructose und die 3,4-Di-O-
methyl-D-fructose im Verhältnis 1 : 3 : 2 : 2 oder 1 : 3 : 3 : 2 erhalten.
Auf Grund dieser Daten sowie der Perjodatoxydation, bei der 1 Mol
Ameisensäure auf 10 Hexoseeinheiten frei wurden, wurde die folgende
Formel vorgeschlagen:

```
Gl 1—2 Fru 6—2 Fru 6—2 Fru 6—2 Fru 6—2 Fru
           I                    I
           |                    |
         2 Fru                2 Fru
```

Die von Schlubach und Müller bei der Hydrolyse des Methyläthers
ihres „Sitosins" erhaltene 3,4,6-Tri-O-methyl-D-fructose konnte nicht
gefunden, eine Deutung für diese Diskrepanz nicht gegeben werden.
Es wurde nur auf den Befund von zwei homologen Reihen von Glucose-
fructosanen durch White und Secor (55) hingewiesen.

Eine auch präparative Zerlegung der Gesamtheit des komplexen
Gemisches von löslichen Kohlenhydraten im Weizen ist nach dem beim
Roggen durchgeführten Vorbild von Schlubach und Lederer vor-
genommen worden (40), und zwar in zwei verschiedenen Stadien des
Wachstums, 14 Tage vor und 14 Tage nach der Blüte. Bei der Trennung
an einer Kohle-Celite-Säule wurde bei dem vor der Blüte gewonnenen
Material 70% an Monosacchariden und Saccharose sowie 24% an Oligo-
und Polysacchariden erhalten; der Verlust betrug 6%. Die Oligo- und
Polysaccharide wurden nach dem etwas modifizierten Verfahren von
Whistler und Durso (54) in 41 Hauptfraktionen zerlegt, die Oligo-

saccharide noch weiter an einer Cellulosesäule. Von den hierbei erhaltenen
11 Fraktionen erwies sich eine papierchromatographisch als einheitlich.
Mit ihrer Drehung von $[\alpha]_D^{20} = +8,3°$ (Wasser, c = 1), einem Glucose-
wert von 25,1% und einem R_S-Wert von 0,21 wurde sie als die von
SCHLUBACH und KOEHN (*38*) entdeckte Bifurcose diagnostiziert. Aus den
R_S-Werten einer anderen Fraktion wurde auf die Trisaccharide Kestose
(II, S. 298) und, als eine neue Beobachtung, die Neokestose (IV) geschlossen. Diese Neokestose (IV) wurde von GROSS, BLANCHARD
und BELL (*17 a*) entdeckt und ihre Konstitution wurde als die einer
O-β-D-Fructofuranosyl-(2—6)-α-D-glucopyranosyl-(1—2)-β-D-fructofura-
nose bestimmt, also eines Trisacchrids, bei dem im Unterschied zur
Kestose und Isokestose die Transfructosidierung nicht auf der Fructose,
sondern der Glucosehälfte der Saccharose erfolgt ist.

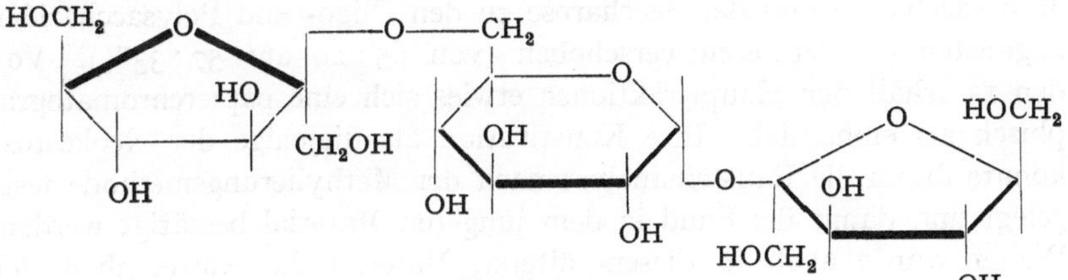

(IV.) Neokestose.

Durch eine weitere Fraktionierung des Gemisches an höheren Gliedern
wurden zwei Komponenten herausgearbeitet. Die leichter lösliche
Fraktion K mit einer Drehung von $[\alpha]_D^{20} = -16°$ (Wasser, c = 1),
der Acetylverbindung von $[\alpha]_D^{20} = +3°$ (Chloroform, c = 1) und einem
Glucosewert von 11%, entsprechend einem P von 9, ergab bei der Bausteinanalyse die Tetra-, Tri- und Dimethyl-hexosen im Verhältnis 2 : 1 : 1.
Die Tetramethyl-hexosen erwiesen sich als ein Gemisch der 2,3,4,6-
Tetra-O-methyl-D-glucose mit der 1,3,4,6-Tetra-O-methyl-D-fructose im
Verhältnis 1 : 3, die Trimethyl-hexose als die 1,3,4-O-Methyl-D-fructose
und die Dimethyl-hexose als die 3,4-O-Dimethyl-fructose. Auf Grund
dieser Daten wurde die Verbindung als ein verzweigtes Octasaccharid
angesprochen und die folgende Formel vorgeschlagen:

$$Gl\ 1—2\ Fru\ 6—2\ Fru\ 6—2\ Fru\ 6—2\ Fru\ 6—2\ Fru$$
$$\quad\ 1 \qquad\qquad\qquad\quad 1$$
$$\quad\ | \qquad\qquad\qquad\quad |$$
$$\quad 2\ Fru \qquad\qquad\quad 2\ Fru$$

Die schwerer lösliche Fraktion L von der Drehung $[\alpha]_D^{20} = -29,4°$
(Wasser, c = 1), dem Glucosewert 7,3% und dem Aldosewert 7,8%
ergab bei der Spaltung ihres Methyläthers die drei Fraktionen im Verhältnis von 5 : 4 : 3. Ihm entsprach am besten eine Mischung eines

Gluco-fructosans mit einem glucosefreien Fructosan im Verhältnis 5 : 1 von folgenden Konstitutionen:

$^5/_6$ Gl 1—2 Fru 6—2 Fru 6—2 Fru 6—2 Fru 6—2 Fru 6—2 Fru 6—2 Fru 6—2 Fru
　　　　 1　　　　　　　　　 1　　　　　　　　　 1
　　　　 |　　　　　　　　　 |　　　　　　　　　 |
　　　　 2 Fru　　　　　　　 2 Fru　　　　　　　 2 Fru

$^1/_6$ Fru 2—2 Fru 6—2 Fru 6—2 Fru 6—2 Fru 6—2 Fru 6—2 Fru 6—2 Fru
　　　　 1　　　　　　　　　 1　　　　　　　　　 1
　　　　 |　　　　　　　　　 |　　　　　　　　　 |
　　　　 2 Fru　　　　　　　 2 Fru　　　　　　　 2 Fru

Es handelt sich also um zwei Dodekasaccharide, die ebenso wie das Octasaccharid der Phleinreihe angehören.

Das 14 Tage nach der Blüte, also 4 Wochen später gewonnene Material wurde in analoger Weise in seine Komponenten zerlegt. Wie bei dem fortgeschrittenen Wachstum zu erwarten, hatte sich das Verhältnis der Monosaccharide und der Saccharose zu den Oligo- und Polysacchariden zugunsten der letzteren verschoben (von 65 : 29 auf 57 : 33%). Von den 13 erhaltenen Hauptfraktionen erwies sich eine papierchromatographisch als einheitlich. Ihre Konstitution als diejenige der Neokestose konnte durch die Bausteinanalyse nach der Methylierungsmethode festgelegt und damit ihr Fund in dem jüngeren Material bestätigt werden. Weiter wurde auch in diesem älteren Material die Anwesenheit der Bifurcose wahrscheinlich gemacht und ein Pentasaccharid isoliert, für das die folgende Konstitution bestimmt wurde:

　　　　　 Gl 1—2 Fru 6—2 Fru 6—2 Fru
　　　　　　　　　　　　　　 1
　　　　　　　　　　　　　　 |
　　　　　　　　　　　　　 2 Fru

Es ist das Isomere in der Phleinreihe des von Schlubach und Koehn (38) im Roggen gefundenen Pentasaccharids der Inulinreihe. Für das von Montgomery und Smith (26) beschriebene und als ein Octasaccharid angesprochene Polysaccharid wurde die Konstitution eines Deka- oder Undekasaccharids wahrscheinlich gemacht.

Im älteren Material wurden also die folgenden Verbindungen aufgefunden und in ihren Konstitutionen bestimmt: Das Trisaccharid Neokestose, das verzweigte Tetrasaccharid Bifurcose, ein Penta-, Octa- und Dodekasaccharid. Sie gehören sämtlich dem Phleintyp an und haben die folgende Bauweise gemeinsam:

　　　　　 Gl 1—⌈—2 Fru 6——2 Fru ⌉
　　　　　　　　 |　　 1　　　　 |
　　　　　　　　 |　　 |　　　　 |
　　　　　　　　 ⌊　 2 Fru　　 ⌋$_n$

wobei n = 1 der Bifurcose und n = 4 dem Dodekasaccharid entspricht. Dieser analogen Bauart entsprechend ergaben ihre Drehungen, gegenüber den 1/P-Werten aufgetragen, eine Gerade (Abb. 6).

Im Gegensatz zum Roggen wurde in den Weizenhalmen nur eine Reihe von Oligo- und Polysacchariden, und zwar vom Phleintyp gefunden.

Da aber von SCHLUBACH und MÜLLER (*43*) in den Weizen-ähren ein Polyfructosan vom Inulintyp nachgewiesen wur-de, ist es wahrscheinlich, daß in ihnen auch eine polymer-homologe Reihe von Oligo- und Polysacchariden vom Inulintyp enthalten ist. Mit dieser Annahme stimmt über-ein, daß von WHITE und SECOR (*55*) im Weizenmehl zwei polymerhomologe Rei-hen beobachtet wurden. Der Kohlenhydratstoffwechsel in den Weizenhalmen entspricht insofern demjenigen in den

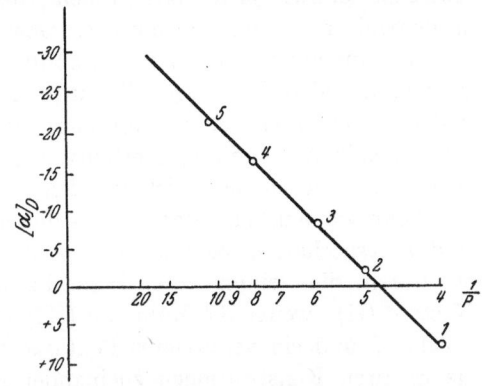

Abb. 6. $[\alpha]_D$ als Funktion von $1/P$: 1. Bifurcose. 2. Penta-saccharid. 3. Hexasaccharid. 4. Octasaccharid. 5. Decasaccha-rid, nach MONTGOMERY und SMITH (*26*), korr. [Aus: Liebigs Ann. Chem. 635, 154 (1960).]

Gräsern, als in ihnen nur Oligo- und Polysaccharide vom Phleintyp gebildet werden; er unterscheidet sich aber dadurch von den letzteren, daß in ihnen verzweigte Polysaccharide auftreten.

III. Die Biogenese der Polyfructosane im Roggen und Weizen.

Auf Grund der im Roggen und Weizen gefundenen Oligo- und Poly-saccharide und dem Nachweis der bei ihrem Aufbau beteiligten Enzyme läßt sich jetzt ein Bild ihrer Biogenese entwerfen.

Der Ausgangspunkt für die Umwandlungen ist die Saccharose. Ihre überwiegende Anwesenheit in den assimilierenden Blättern der Roggen- und Weizenpflanzen ist von allen Forschern übereinstimmend beobachtet. Obgleich eine ganze Reihe kurzlebiger Zwischenstufen zwischen dem Eintritt des Kohlendioxyds in die Spaltöffnungen der Blätter und dem Erscheinen der Saccharose in ihnen nachgewiesen wurde, bedeutet diese doch diejenige Verbindung, bei der diese raschen Umwandlungen zunächst haltmachen. Sie vermag sich daher in größerer Konzentration an-zusammeln. Aber es bleibt weiter schwer verständlich, warum es gerade ein so kompliziert gebautes und labiles Disaccharid wie die Saccharose ist, die so von der Pflanze bevorzugt wird. Wo doch bei anderen physiolo-gischen Vorgängen gebildete Disaccharide, wie die Lactose, weit be-ständiger sind und die neuerdings im Stoffwechsel der Insekten nach-gewiesene Trehalose (*56*) das gegen Hydrolyse stabilste Disaccharid ist, das wir kennen. Eine dahingehende allgemeine Deutung für diese

Sonderstellung der Saccharose, daß ihr nur eine transitorische Funktion zukommt und sie deshalb so labil ist, findet ihre Widerlegung in der Tatsache, daß sie ja in einer großen Anzahl von Pflanzen, insbesondere in denjenigen, die ihrer technischen Gewinnung dienen, das Endprodukt bildet. Andererseits kann ihre Labilität aber für die Fälle, in denen sie wie in anderen wichtigen Kulturpflanzen (Gräser und Getreidearten) alsbald nach ihrer Bildung im Laufe der weiteren Vegetationsperiode weitgehende Abwandlungen erfährt, als Erklärung für die Leichtigkeit dienen, mit denen diese sich vollziehen.

Wenn wir nun zum Verständnis dieser Vorgänge zunächst die Oligo- und Polysaccharide ins Auge fassen, die im Roggen gefunden wurden, so waren dies einmal die Trisaccharide Isokestose (I, S. 298) und Kestose (II); weiter die Bifurcose (III) als das erste verzweigte Tetrasaccharid und ein verzweigtes Pentasaccharid. Das Verhältnis, in dem sie in ihren Konstitutionen zueinander stehen, ist schematisch bereits auf S. 297 gegeben. Bei den beiden Trisacchariden sind demnach die Fructosereste entweder an die erste oder die sechste Hydroxylgruppe der Fructosehälfte der Saccharose angelagert, und bei der Bifurcose zugleich an die beiden primären Hydroxylgruppen der Fructose. Im Pentasaccharid endlich ist die weitere Angliederung an die erste Hydroxylgruppe des Fructoserestes erfolgt, der seinerseits mit der ersten Hydroxylgruppe an die Saccharose gebunden war. Und diese Verlängerung der Fructoseketten hat nun bei den höheren Oligo- und Polysacchariden ihre Fortsetzung in zwei verschiedenen Richtungen gefunden: Entweder wie bei dem Pentasaccharid durch weitere Anlagerungen an den ersten Hydroxylgruppen oder aber an den sechsten Hydroxylgruppen, bis beide Reihen im Graminin A einerseits, im Secalin andererseits ihren Abschluß finden. Entsprechend dem Bau der Bifurcose setzen sich die Verzweigungen in bestimmten, aber noch nicht sicher festgelegten Abständen fort.

Diese beiden parallelen Aufbauwege lassen sich nach Schlubach und Koehn (38) am einfachsten durch die Annahme deuten, daß zwei verschiedene Enzyme vorhanden sind, von denen das eine die Anlagerung an die erste, das andere eine solche an die sechste Hydroxylgruppe bewirkt.

Der Nachweis der so geforderten zwei verschiedenen Enzyme ist erstmalig von Schlubach, Hohn und Repenning (35) geführt worden. Sie konnten zeigen, daß von den beiden in *Aspergillus niger* enthaltenen Enzymen das eine, die Inulase, welches die 2—1-Bindungen im Inulin zu hydrolysieren vermag, durch Streptomycin gehemmt wird, während das andere, die Phleinase, das die 2—6-Bindungen in den Phleinen spaltet, durch Streptomycin in seiner Wirksamkeit nicht beeinflußt wird. Inzwischen wurde, wie aus den nachstehenden Angaben hervorgeht, gefunden (*32*), daß der gleiche Unterschied auch bei den Enzymen der Hefe festgestellt werden kann:

Ansätze: Je 0,32408 g Inulin von — 33,8° bzw. Loliin$_{23}$ von — 36° wurden in 2 ml Wasser gelöst, 4 ml Acetatpuffer vom pH 4,7 zugegeben, 0,8 ml Invertin Merck hinzugefügt und mit Wasser auf 20 ml aufgefüllt. Die Ansätze wurden bei 30° gehalten, die Drehungen im 1-dm-Rohr verfolgt.

Minuten:	0	60	120	180	1080	2880	Differenz
Inulin + Invertin........	— 0,40°	— 0,42°	— 0,43°	— 0,43°	— 0,45°	— 0,44°	— 0,04°
Loliin + Invertin........	— 0,40°	— 0,43°	— 0,46°	— 0,53°	— 0,70°	— 0,73°	— 0,33°

Ebenso wie bei den Enzymen aus *Aspergillus niger* ließ sich die Inulase aus Hefe durch Zusatz von 1,6 ml einer 8%igen Streptomycinlösung hemmen:

Minuten:	0	120	1080	2880	3060	Differenz
Inulin + Invertin + Streptomycin........	— 1,69°	— 1,68°	— 1,69°	— 1,69°	— 1,69°	0,0°
Loliin + Invertin + Streptomycin........	— 1,68°	— 1,75°	— 1,98°	— 2,00°	— 2,01°	— 0,33°

Zum Vergleich der Wirkungen der beiden Enzyme aus Hefe und aus *Aspergillus niger* wird die letztere, ohne Streptomycinzusatz, wiedergegeben:

Minuten:	30	120	210	300	Differenz
Inulin..............	— 0,47°	— 1,37°	— 1,41°	— 1,46°	— 0,99°
Loliin..............	— 0,54°	— 0,97°	— 0,97°	— 0,98°	— 0,42°

Wie ersichtlich, ist der Wirkungsgrad der beiden Enzyme aus *Aspergillus niger* und aus Hefe ein recht verschiedener. Während das Enzym aus *Aspergillus* Inulin ungefähr doppelt so rasch hydrolysiert wie das Loliin, spaltet das Enzym aus Hefe Loliin um das Mehrfache rascher als das Inulin. Die Konzentrationen der beiden Enzyme sind daher in den beiden Organismen sehr verschieden: im *Aspergillus* überwiegt die Inulase, in der Hefe herrscht die Phleinase vor.

Demnach sollte die Inulase durch Transfructosidierung auf Saccharose Oligosaccharide vom Inulintyp, also mit 2—1-Bindungen, die Phleinase solche vom Phleintyp, also mit 2—6-Bindungen, bilden. Das trifft in der Tat zu. Denn von BARKER, BOURNE und CARRINGTON (*6*) wurde durch Einwirkung eines Enzyms aus *Aspergillus niger* auf konzentrierte Saccharoselösungen ein Trisaccharid vom Bau eines O-α-D-Gluco-pyranosyl-(1—2)-O-β-D-fructofuranosyl-(1—2)-β-D-fructofuranosids, die sogenannte Isokestose (III), also mit 2—1-Bindungen, erhalten. ALBON, BELL, BLANCHARD, GROSS und RUNDELL (*2*) entdeckten bei der Einwirkung von Hefenenzym auf Saccharoselösungen ein Trisaccharid von der Konstitution eines O-α-D-Glucopyranosyl-(1—2)-O-β-fructofuranosyl-(6—2)-D-fructofuranosids, die sogenannte Kestose (II, S. 298), also mit 6—2-Bindungen.

Die Fortsetzung dieser transfructosidierenden Wirkung von Hefenenzym zu höheren Polyfructosanen vom Phleintyp ist von Schlubach und Lübbers (41) gezeigt worden, indem durch Einwirkung von Invertin auf konzentrierte Saccharoselösungen ein Loliin vom P = 25, dieses in zwei Stufen zu Polyfructosanen vom P = 34 und weiter P = 41, aufgebaut werden konnte.

Wenn man, wie Wallenfels (53, 53a) dies an der β-Galaktosidase nachgewiesen hat, annimmt, daß es das gleiche Enzym ist, welches den Aufbau durch Transglycosidierung wie den Abbau durch Hydrolyse bewirkt, wobei die letztere nur als der Sonderfall einer Transglycosidierung auf Wasser anzusehen ist, sollte Hefeninvertin auch den Abbau der Polyfructosane herbeiführen. Schlubach und Holzer (36) haben nun schon früher gezeigt, daß der Abbau in der Tat in der Weise erfolgt, daß von der Fructosekette eines Phleins ein Fructoseglied nach dem anderen abgebaut wird, also die Umkehrung des Aufbaues stattfindet. Da danach die Geschwindigkeit der Hydrolyse umgekehrt proportional der Kettenlänge erfolgt, ist die Halbumsatzzeit der Invertinspaltung als ein Maßstab für den Polymerisationsgrad herangezogen worden.

Die Annahme der beiden Enzymarten macht auch die Bildung der verzweigten Polyfructosane verständlich, wie sie für die Getreidearten charakteristisch sind. Augenscheinlich sind in ihnen beide Enzymarten nebeneinander enthalten, wie aus der Auffindung der Kestose neben der Isokestose in den Roggenhalmen hervorgeht. Die überwiegende Bildung des Graminins andererseits in den Roggenähren läßt darauf schließen, daß in ihnen die Konzentration der Inulase eine höhere ist als in den Halmen. Ebenso ist in den Weizenhalmen eine verhältnismäßig geringe Konzentration der Inulase anzunehmen, da ja in ihnen ebenso wie in den Gräsern nur Polyfructosane vom Phleintyp angetroffen wurden.

Einer besonderen Klärung bedarf noch die Frage, ob die die Saccharose und nahe verwandte Oligosaccharide wie die Raffinose spaltende, eigentliche Saccharase mit einer der beiden Transfructosidasen identisch oder aber von ihnen verschieden ist. Hinsichtlich der Inulase scheint das letztere der Fall zu sein. Denn Adams, Richtmyer und Hudson (1) haben schon 1943 auf Grund der Verschiedenheit der pH-Aktivitätskurven des Inulins gegenüber derjenigen der Saccharose bestimmt ausgesprochen, daß die Hefeninulase ein spezifisches Enzym sei, nicht identisch mit der β-Fructofuranosidase, welche die Saccharose hydrolysiert. Und Pigman (28) hat gezeigt, daß der Quotient zwischen der Saccharasewirkung und der Inulase zwischen 5 : 1 bei *Aspergillus niger* und 4000 : 1

bei Hefe schwanken kann. Weiter konnten SCHLUBACH und NEURATH (45) den Quotienten aus der Fructosanase- gegenüber der Saccharase-wirkung durch Metallsalzvergiftung erheblich verschieben. HESTRIN und GOLD- BLUM (18) haben über eine Lävanpolyase berichtet, welche Inulin langsam, Saccharose aber gar nicht hydrolysiert. Bei der Hydrolyse von Lävan machte diese bei der Bildung eines Lävan-tri- und -tetra-saccharids halt. Diese letzteren konnten dann aber durch Hefenenzym vollständig zu Fructose hydrolysiert werden.

Offen bleibt noch die Frage, ob die Saccharase mit der Phleinase identisch ist. Wenn es sich erweisen sollte, daß dies ebensowenig wie mit der Inulase der Fall ist, würden sich daraus neue Gesichtspunkte für die Beurteilung des Mißerfolges, das Saccharose spaltende Enzym in reiner Form zu isolieren, gewinnen lassen. Denn bei all den zahlreichen Versuchen zur Reindarstellung dieses Enzyms, von denen SUMNER und O'KANE (49) mit Recht hervorgehoben haben, daß sie weit häufiger und intensiver unternommen sind als bei irgendeinem anderen Enzym, ist als Kriterium des erreichten Reinheitsgrades neben allgemeinen analytischen Prüfungen nur die Saccharose als Testsubstanz verwandt worden. Da nicht gegen Inulin oder Loliin getestet wurde, ist es sehr wohl möglich, daß Inulase oder Phleinase mitgeschleppt wurde, da diese vermutlich in ihrem Bau von der Saccharase nicht sehr verschieden sind und sie daher begleiten. Es würde damit auch verständlich werden, warum die Bemühungen WILLSTÄTTERs und seiner Mitarbeiter trotz der umfangreichen Reihen ihrer Untersuchungen nicht zum Ziel geführt haben.

Mit der so dargelegten Biogenese der Oligo- und Polysaccharide im Roggen und Weizen ist der Hauptweg ihres Werdens und Vergehens gegeben. Aber es gibt noch zwei Varianten, die zwar im allgemeinen nur in geringem Ausmaße begangen wurden, aber doch Berücksichtigung verdienen.

Einmal wird, ausgehend von der Saccharose, durch Transfructosidierung auf Fructose eine Difructose gebildet. Für diese Verbindung konnten Anzeichen gefunden werden. Es gelang aber bisher nicht, sie zu isolieren, da sie, wie sich schätzen läßt, noch weit leichter der Hydrolyse unterliegt als die Saccharose. Wohl aber konnte aus *Lolium perenne* ein höheres Homologe, ein nur aus vier Fructose- resten bestehendes Tetrasaccharid (42), und aus *Avena flavescens* das entsprechende Pentasaccharid (37) isoliert werden. Auf einer derartigen Difructose, die sich von der Saccharose in ihrem Bau nur dadurch unterscheidet, daß in ihr an Stelle der Glucosehälfte eine zweite Fructofuranose getreten ist, werden in ganz analoger Weise durch Transfructosidierung Oligo- und Polysaccharide aufgebaut, die nun glucosefrei sind. Sie verraten ihre Gegenwart dadurch, daß der Glucosegehalt in den Gemischen, in denen sie enthalten sind, niedriger ist, als er sich auf Grund der Berechnung nach dem Polymerisationsgrad ergibt. So konnte von SCHLUBACH und TRAUTSCHOLD (46) unter Zuhilfenahme der Ergebnisse der Perjodatoxydation bei dem Dactylin III aus *Dactylis glomerata* bewiesen werden, daß dieses zu 76%

aus einem glucosehaltigen und zu 24% aus einem glucosefreien Polyfructosan besteht. Wenn dieser Weg des Aufbaues im allgemeinen nur in untergeordnetem Maße beschritten wird, ist dies auf die große Labilität der Difructose zurückzuführen. Da andernfalls zu hohe Schätzungen des aus dem Glucosegehalt errechneten Polymerisationsgrades erfolgen können, bedarf diese Möglichkeit dennoch der Berücksichtigung.

Eine zweite mögliche Nebenreaktion hat sich aus der Auffindung der Neokestose unter den Oligosacchariden des Weizens ergeben. In den auf diesem Trisaccharid anstatt auf der Isokestose oder der Kestose aufgebauten Polysacchariden steht die Glucose nicht am Ende der Kette, sondern mehr in der Mitte. Infolgedessen erscheint unter den Spaltprodukten der Methyläther neben einer Trimethyl-fructose auch eine Trimethyl-glucose. Dies ist in der Tat mehrfach beobachtet worden (40). Wenn endlich an der so in der Kette stehenden Glucose eine Verzweigung erfolgt, sollte auch in der Dimethyl-fraktion eine Dimethyl-glucose enthalten sein. Dies ist in erheblichem Maße bei den Polysacchariden des Hafers beobachtet worden (32).

Literaturverzeichnis.

1. ADAMS, M., N. K. RICHTMYER and C. S. HUDSON: Some Enzymes Present in Highly Purified Invertase Preparations; a Contribution to the Study of Fructofuranosidases, Galactosidases, Glucosidases and Mannosidases. J. Amer. Chem. Soc. 65, 1369 (1943) (especially p. 1371).
2. ALBON, N., D. J. BELL, P. H. BLANCHARD, D. GROSS and J. T. RUNDELL: Kestose: a Trisaccharide formed from Sucrose by Yeast Invertase. J. Chem. Soc. (London) 1953, 24.
3. ARCHBOLD, H. K.: Fructosans in the Monocotyledons. A Review. New Phytologist 39, 185 (1940).
4. ARCHIBALD, W. J.: A Demonstration of Some New Methods of Determining Molecular Weights from the Data of the Ultracentrifuge. J. Physic. Coll. Chem. 51, 1204 (1947).
5. ASPINALL, G. O. and P. C. DAS GUPTA: The Structure of the Fructosan from *Agave vera cruz* MILL. J. Chem. Soc. (London) 1959, 718.
6. BARKER, S. A., E. J. BOURNE and T. R. CARRINGTON: Studies of *Aspergillus niger*. III. The Structure of a Trisaccharide Synthesised from Sucrose. J. Chem. Soc. (London) 1954, 2125.
7. BARNELL, H. R.: Seasonal Changes in the Carbohydrates of the Wheat Plant. New Phytologist 35, 229 (1936).
8. — Distribution of Carbohydrates between Component Parts of the Wheat Plant at Various Times during the Season. New Phytologist 37, 85 (1938).
8a. BELVAL, H.: Sur la genèse de l'Amidon dans les Céréales. Rev. gén. Bot. 36, 308, 337, 395 (1924).
9. BISHOP, C. T.: Structure of a Trisaccharide from Wheat Straw Xylan. J. Amer. Chem. Soc. 78, 2840 (1956).
10. CARTER, H. E., R. H. McCLUER and E. D. SLIFER: Lipids of Wheat Flour. I. Characterization of Galactosylglycerol Components. J. Amer. Chem. Soc. 78, 3735 (1956).
11. CHRZASZCZ, T. und W. MICHALSKI: Die löslichen Kohlenhydrate des Roggenmehles und Feststellung der Dichte des Teiges. Przemysl Chemiczny 12, 389 (1928) [Chem. Zbl. 1929, I, 2364].
12. COLIN, H. et H. BELVAL: La genèse des hydrates de carbone dans le blé. Présence de lévulosanes dans la tige. C. R. hebd. Séances Acad. Sci. 175, 1441 (1922).
13. — — Les lévulosanes dans les céréales. C. R. hebd. Séances Acad. Sci. 177, 973 (1923).

14. CONSDEN, R., A. H. GORDON and A. J. P. MARTIN: Qualitative Analysis of Proteins: a Partition Chromatographic Method Using Paper. Biochemic. J. **38,** 224 (1944).

15. CUGNAC, A. DE: Recherches sur les glucides des Graminées. Ann. sci. nat. bot. **13,** 1 (1931).

16. ELIAS, H. G. und H. H. SCHLUBACH: Untersuchungen über Polyfructosane. LVII: Molekulargewichtsbestimmungen an Polyfructosanen. Liebigs Ann. Chem. **627,** 126 (1959).

16a. FLEURENT, E.: Sur la composition des blés durs et sur la constitution physique de leur gluten. C. R. hebd. Séances Acad. Sci. **133,** 944 (1901).

17. GIRARD, A.: Recherches sur la composition des blés et sur leur analyse. C. R. hebd. Séances Acad. Sci. **124,** 876 (1897).

17a. GROSS, D., P. H. BLANCHARD and D. J. BELL: *neo*Kestose: a Trisaccharide Formed from Sucrose by Yeast Invertase. J. Chem. Soc. (London) **1954,** 1727.

18. HESTRIN, S. and J. GOLDBLUM: Laevanpolyase. Nature (London) **172,** 1046 (1953).

19. JENSEN-HANSEN, H.: Studien über die im Roggen, dem Weizen und der Gerste in den verschiedenen Entwicklungsstadien vorkommenden Kohlenhydrate. Carlsberg Lab. Medd. **4,** 145 (1896).

20. KNJAGINITSCHEW, M. I., I. F. MUTUL und I. K. PALILOWA: Aktivität der Carbohydrasen im reifenden Korn verschiedener Weizenarten. Biokhimiya (USSR) **5,** 288 (1940).

21. KRETOWITSCH, W. L.: Über die Zucker assimilierenden Blätter. Z. physiol. Chem. (Hoppe-Seyler) **231,** 265 (1935).

22. KROTKOV, G.: Carbohydrates of Wheat Leaves. Plant Physiol. **14,** 559 (1939).

23. KRUISHEER, C. I.: Zur Kenntnis und Bestimmung der Polyfructosen. II. Die Bestimmung des Trifructoseanhydrids zur Ermittlung des Roggenmehlgehaltes von Mahlprodukten. Rec. trav. chim. Pays-Bas **50,** 153 (1931).

24. KURSSANOW, A. und K. BRJUSCHKOWA: Die Wirkung der Fermente in reifenden Weizenkörnern. Biokhimiya (USSR) **5,** 681 (1940) [Chem. Zbl. **1942,** I, 1270].

25. LISSITZYN, D. I.: Die Zucker assimilierenden Blätter. II. Über das Verhältnis von Glucose zu Fructose in den assimilierenden Blättern. Biokhimiya (USSR) **2,** 908 (1937) [Chem. Zbl. **1938,** II, 871].

26. MONTGOMERY, R. and F. SMITH: The Carbohydrates of Gramineae. IX. The Constitution of a Glucofructan of the Endosperm of Wheat *(Triticum vulgare)*. J. Amer. Chem. Soc. **79,** 446 (1957).

27. MÜNTZ, A.: Sur la maturation de la graine du seigle. C. R. hebd. Séances Acad. Sci. **87,** 679 (1878).

28. PIGMAN, W. W.: Occurrence of Sucrose and Inulin-hydrolyzing Enzymes in Commercial Enzyme Preparations. J. Res. Nat. Bur. Stand. **30,** 159 (1943).

29. POPP, O.: Über die Synanthrose, ein neues Kohlehydrat der Synanthereen. Liebigs Ann. Chem. **156,** 180 (1870).

30. SCHLUBACH, H. H.: Der Kohlenhydratstoffwechsel der Gräser. Fortschr. Chem. organ. Naturstoffe **15,** 1 (1958).

31. SCHLUBACH, H. H. und CHR. BANDMANN: Untersuchungen über Fructose-anhydride. XXII: Über das Secalin. Liebigs Ann. Chem. **540,** 285 (1939).

32. SCHLUBACH, H. H. und M. GREHN: (im Druck).

33. SCHLUBACH, H. H. und E. HABERLAND: Untersuchungen über Polyfructosane. XLIX: Über das Graminin III. Liebigs Ann. Chem. **604,** 22 (1957).

34. — — Untersuchungen über Polyfructosane. LIV: Über das Graminin B. Liebigs Ann. Chem. **614,** 123 (1958).

35. Schlubach, H. H., G. Hohn und K. Repenning: Untersuchungen über Poly-fructosane. LVI: Zwei Polyfructosanasen aus *Aspergillus niger*. Liebigs Ann. Chem. 627, 123 (1959).

36. Schlubach, H. H. und K. Holzer: Untersuchungen über Polyfructosane. XXXIII: Über das Phlein II. Die Invertinspaltung der Polyfructosane. Liebigs Ann. Chem. 578, 213 (1952).

36a. Schlubach, H. H. und I. Huchting: Untersuchungen über Fructoseanhydride. XXV: Über das Pyrosin. Liebigs Ann. Chem. 561, 173 (1948).

37. Schlubach, H. H. und H. O. A. Koehn: Untersuchungen über Polyfructosane. LI: Über die niedermolekularen Kohlenhydrate in *Avena flavescens*. Liebigs Ann. Chem. 606, 130 (1957).

38. — — Untersuchungen über Polyfructosane. LV: Die Bildung der verzweigten Polyfructosane in den Roggenhalmen. Liebigs Ann. Chem. 614, 126 (1958).

39. Schlubach, H. H. und K. Koenig: Untersuchungen über Fructoseanhydride. XV: Die Konstitution des Graminins I. Liebigs Ann. Chem. 514, 182 (1934).

40. Schlubach, H. H. und F. Lederer: Untersuchungen über Polyfructosane. XLIX: Der Kohlenhydratstoffwechsel im Weizen. Liebigs Ann. Chem. 635, 154 (1960).

41. Schlubach, H. H. und H. Lübbers: Untersuchungen über Polyfructosane. XLII: Über die Phleine und ihren enzymatischen Aufbau. Liebigs Ann. Chem. 594, 54 (1955).

42. Schlubach, H. H., H. Lübbers und H. Borowski: Untersuchungen über Polyfructosane. XLIV: Die niedermolekularen Kohlenhydrate in *Lolium perenne*. Liebigs Ann. Chem. 595, 229 (1955).

43. Schlubach, H. H. und H. Müller: Untersuchungen über Polyfructosane. XXIX: Über das Sitosin. Liebigs Ann. Chem. 578, 194 (1952).

44. — — Untersuchungen über Polyfructosane. XXX: Über das Graminin II. Liebigs Ann. Chem. 578, 198 (1952).

45. Schlubach, H. H. und G. Neurath: Untersuchungen über Polyfructosane. LII: Die Polyfructosanase-Aktivität des Hefeinvertins. Liebigs Ann. Chem. 606, 134 (1957).

46. Schlubach, H. H. und E. W. Trautschold: Untersuchungen über Poly-fructosane. L: Die Polyfructosane in *Dactylis glomerata*. Liebigs Ann. Chem. 606, 124 (1957).

47. Schulze, E. und S. Frankfurt: Über krystallisiertes Lävulin. — Über β-Lävulin. Ber. dtsch. chem. Ges. 27, 65, 3525 (1894).

48. — — Über die Verbreitung des Rohrzuckers in den Pflanzen, über seine physiologische Rolle und über lösliche Kohlenhydrate, die ihn begleiten. Z. physiol. Chem. (Hoppe-Seyler) 20, 511 (1895) (und zwar S. 527).

49. Sumner, J. B. and D. J. O'Kane: The Chemical Nature of Yeast Saccharase. Enzymologia 12, 251 (1946—1948).

50. Tanret, Ch.: Sur la lévosine, nouveau principe immédiat de quelques céréales. C. R. hebd. Séances Acad. Sci. 112, 293 (1891); Bull. soc. chim. France [3] 5, 724 (1891).

51. Tillmans, J., H. Holl und L. Jariwala: Ein neues Kohlenhydrat in Roggen-mehl und ein darauf aufgebautes Verfahren zum Nachweis von Roggenmehl in Weizenmehl und anderen Mehlarten. Z. Unters. Lebensmittel 56, 26 (1928).

52. Tschertok, W. R. und J. A. Schapiro: Über den Gehalt von ukrainischen Weizen und Roggen an Trifructosan und die Bestimmung von Roggenmehl-zusatz in Weizenmehl. Problems Nutrit. (USSR) 8, 74 (1939) [Chem. Zbl. 1939, II, 1795].

53. WALLENFELS, K.: Untersuchung über milchzucker-spaltende Enzyme. Angew. Chem. **71**, 526 (1959).

53a. WALLENFELS, K. und M. L. ZARNITZ: Kristallisation der β-Galaktosidase aus *Escherichia coli*. Angew. Chem. **69**, 482 (1957).

54. WHISTLER, R. L. and D. F. DURSO: Chromatographic Separation of Sugars on Charcoal. J. Amer. Chem. Soc. **72**, 677 (1950).

55. WHITE, L. M. and G. E. SECOR: Occurrence of Two Similar Homologous Series of Oligosaccharides in Wheat Flour and Wheat. Arch. Biochem. Biophys. **43**, 60 (1953).

56. WYATT, G. R. and G. F. KALF: The Chemistry of Insect Hemolymph. II. Trehalose and Other Carbohydrates. J. Gen. Physiol. **40**, 833 (1957).

(Eingelaufen am 28. September 1960.)

Les phosphatases des végétaux supérieurs: répartition et action.

Par Jean Émile Courtois et Andréa Lino, Paris.

Avec 5 figures.

Sommaire.

Introduction.

Définition. Les phosphatases appartiennent au groupe des enzymes hydrolysantes ou hydrolases. On classe d'une façon générale dans les phosphatases les enzymes susceptibles de mettre en liberté par voie hydrolytique de l'acide orthophosphorique à partir de ses combinaisons minérales ou organiques. Une telle définition n'englobe pas toutes les phosphatases car l'on range aussi dans ce groupe les phosphodiestérases qui hydrolysent une molécule de diester orthophosphorique (I) en une molécule de monoester orthophosphorique (II) et un composé hydroxylé selon la réaction (1):

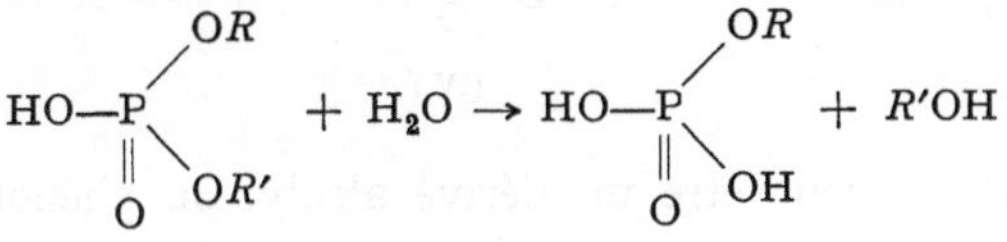

$$HO-\underset{\underset{(I.)}{O}}{\overset{OR}{P}}{\diagdown}_{OR'} + H_2O \rightarrow HO-\underset{\underset{(II.)}{O}}{\overset{OR}{P}}{\diagdown}_{OH} + R'OH \qquad (1)$$

La notion de phosphatases enzymes hydrolysantes est demeurée pratiquement sans ambiguité jusque vers 1950. Depuis cette époque il a été montré que les phosphatases pouvaient transférer le radical orthophosphoryl sur d'autres accepteurs que l'hydroxyle de l'eau.

L'on désigne sous le nom générique de *transphosphatases* les enzymes transférant un reste orthophosphoryl d'un substrat donateur sur un substrat accepteur selon la réaction générale (2):

$$R-PO_3H_2 + R'-H \rightleftarrows R-H + R'-PO_3H_2. \qquad (2)$$

Le nombre des transphosphatases décrites n'a cessé de s'accroître au fur et à mesure que progressaient nos connaissances sur le métabolisme intermédiaire. Pour toutes ces nouvelles transphosphatases R' est exceptionnellement OH; R' est un groupement organique et parfois minéral, ce qui fait que $R'H$ est un accepteur le plus souvent spécifique du groupement PO_3H_2.

Comme l'indique leur nom, toutes les transphosphatases sont par définition des phosphatases. Toutefois l'usage s'est établi de dénommer phosphatases sans indications complémentaires les transphosphatases pour lesquels l'accepteur habituel est l'eau dans la réaction (2) qui se formulera ainsi selon (3):

$$R-PO_3H_2 + HO-H \rightleftarrows R-H + PO_4H_3. \qquad (3)$$

Le terme accepteur habituel doit être entendu dans le sens qu'en présence d'une forte proportion d'eau la réaction d'hydrolyse prédominera très largement sinon presque exclusivement.

La plus grande partie des travaux effectués sur les phosphatases végétales ont d'ailleurs été réalisés en déterminant analytiquement l'hydrolyse des substrats.

Afin de ne pas compliquer inutilement la terminologie utilisée dans cette revue nous désignerons par action phosphatasique une réaction d'hydrolyse. La réversibilité d'action des phosphatases et leur action transférante seront l'objet d'un paragraphe spécial. La liaison rompue par les phosphatases sera du type X—O—P ou du type N—P.

Dans les composés du type X—O—P l'on trouve les corps de formule générale (III) et (IV).

$$
\begin{array}{ccc}
X\!-\!O\!-\!P\!\!\begin{array}{l}{}^{OH}\\{}_{\displaystyle O}{}^{OH}\end{array} &
\begin{array}{l}X\!-\!O\\X'\!-\!O\end{array}\!\!P\!\!\begin{array}{l}-OH\\O\end{array} &
\begin{array}{l}R\\R'\end{array}\!\!N\!-\!P\!\!\begin{array}{l}{}^{OH}\\{}_{\displaystyle O}{}^{OH}\end{array}\\
\text{(III.)} & \text{(IV.)} & \text{(V.)}
\end{array}
$$

Dans (III), X peut être un dérivé alcolyl ou phénolyl, ce sont les monoesters orthophosphoriques. X peut aussi être un autre radical orthophosphorique, et (III) est alors l'acide pyrophosphorique. Le second phosphore de X pouvant lui-même être relié soit à d'autres radicaux orthophosphoriques : acides polyphosphoriques, soit à d'autres corps organiques.

Les corps des types (III) et (IV) sont les principaux substrats qui ont été utilisés pour l'étude des phosphatases végétales. Toutefois on a l'habitude de ranger aussi dans le groupe des phosphatases les phosphamidases hydrolysant la liaison P—N des phosphamides du type (V).

Historique. Un certain nombre d'hydrolases furent décelées au début des recherches sur l'enzymologie entre 1833 et 1850: amylase, émulsine, pepsine, etc. Il fallut attendre beaucoup plus longtemps pour la découverte de la phosphatase. Les phosphatases sont largement distribuées dans les organes animaux, mais c'est cependant avec un matériel d'origine végétale qu'en 1906 Suzuki, Yoshimura et Takaishi (*195*) décelèrent pour la première fois une activité phosphatasique. Ces auteurs réussirent à mettre en évidence une enzyme présente dans les céréales qui libérait de l'acide orthophosphorique à partir de l'acide phytique (ester hexa-mono-orthophosphorique du mésoinositol). Comme nous le verrons par la suite, l'acide phytique n'est pas hydrolysé par la plupart des préparations phosphatasiques des végétaux supérieurs; la phytase est fort peu répandue et ce fut cependant la première phosphatase identifiée. Deux ans plus tard, McCollum et Hart (*134*) décelèrent une activité phosphatasique dans le foie et le sang de bœuf. Depuis lors, plus de deux mille publications ont permis de préciser que les phosphatases ont été caractérisées dans tous les organismes vivants examinés, ceci aussi bien pour le règne animal que pour le règne végétal.

La plupart de ces études ont été réalisées avec des animaux supérieurs, étant donné le rôle joué par les phosphatases dans la calcification de l'os, le travail musculaire, le métabolisme des glucides, etc. Enfin, la détermination de l'activité phosphatasique du sérum est une épreuve classique des laboratoires de Chimie Clinique; le taux des phosphatases sériques varie au cours d'affections nombreuses et variées.

Les publications sur les phosphatases des Végétaux supérieurs sont comparativement moins nombreuses, bien que le métabolisme du phosphore s'y montre aussi important que chez les animaux. Les principaux organes des Végétaux supérieurs: feuilles, tubercules, racines, et surtout les graines, possèdent une forte activité phosphatasique. Cette teneur en phosphatase d'un tissu végétal subit des variations au cours du développement.

Classification générale des phosphatases. Le terme général de phosphatase s'applique à toute une classe d'enzymes différentes ayant des spécificités diverses et des caractéristiques parfois très éloignées les unes des autres.

Il est possible de subdiviser tout d'abord les phosphatases en deux principaux groupes: (A) celles scindant une liaison entre un atome de phosphore et un radical organique et (B) celles scindant une liaison entre deux atomes de phosphore et qui sont désignées sous la dénomination générale de polyphosphatases. Ces deux groupes peuvent ensuite être divisés selon le schéma général suivant:

(A)	Phosphomonoestérases Phosphodiestérases Phosphamidases Acylphosphatases Osylphosphatases	(B)	Pyrophosphatases Triphosphatases Métaphosphatases ATPases Apyrases Nucléotide-pyrophosphatases

Aux phosphatases l'on peut aussi rattacher les enzymes agissant sur les triesters-orthophosphoriques. Le sérum des Mammifères contient par exemple une enzyme qui hydrolyse l'ester diéthyl-*p*-nitrophényl orthophosphorique en *p*-nitrophénol et acide diéthylphosphorique. A notre connaissance une enzyme de ce type n'a pas été signalée chez les végétaux.

Divers auteurs ayant étudié l'action de préparations végétales riches en monoestérases et même en diestérases sur les triesters (triphénylphosphate — tricrésylphosphate) n'ont pas observé d'hydrolyse de ces corps.

Etant donné le but limité de notre exposé, nous nous bornerons à l'étude des seuls types de phosphatases qui sont le plus largement répandus chez les Végétaux supérieurs.

La structure des phosphatases demeure encore inconnue; les préparations les plus purifiées d'origine animale n'ont pas permis d'y déceler de constituant absolument caractéristique. Toutefois, LORA-TAMAYO (*129*) a présenté des arguments indiquant que dans la phosphatase rénale l'acide uridylique est essentiel pour l'activité enzymatique et doit constituer une partie de l'enzyme.

I. Les phosphomonoestérases.

1. Classification des principaux groupes de phosphomonoestérases.

Bases de la classification. Les phosphomonoestérases sont de beaucoup les plus répandues et les mieux connues parmi les diverses classes de phosphatases. Il en est résulté que le terme phosphatase est presque toujours synonyme de phosphomonoestérase. Nous signalerons dès maintenant que chez les Végétaux les autres classes de phosphatases ont le plus souvent des propriétés générales communes avec les monoestérases: activité en fonction du pH, activateurs et inhibiteurs similaires, etc. Il est par exemple très difficile de différencier phosphomonoestérases et pyrophosphatases associées.

Parmi les phosphomonoestérases on peut envisager deux catégories:

(a) Celles qui ne présentent pas de spécificité préférentielle pour un seul substrat ou un groupe étroitement délimité de substrats. Avec ces enzymes il ne se manifeste pas d'écarts très accentués entre les vitesses d'hydrolyse de concentrations équimoléculaires d'esters de monoalcools, polyalcools, mono- et disaccharides, composés phénoliques. Ces enzymes qui sont le plus largement distribuées, et par ce fait les plus fréquemment étudiées, sont les phosphomonoestérases à large spécificité.

(b) D'autres phosphomonoestérases manifestent par contre une assez étroite spécificité de substrat. Ces enzymes sont en général dénommées d'après la nature du substrat qu'elles hydrolysent assez préférentiellement; c'est ainsi que l'on distingue des phytases, hexosediphosphatases, glucose-6-phosphatases, etc.

La distinction entre ces deux catégories est souvent délicate; nous la suivrons cependant dans le seul but de simplifier la présentation. Nous examinerons donc en premier lieu les phosphomonoestérases à large spécificité, nous décrirons ensuite celles à action plus spécifique.

Dans ces deux catégories il peut exister dans la même plante et parfois les mêmes organes des „phosphatases isodynames". ROCHE a proposé de désigner ainsi des enzymes ayant sensiblement le même champ d'action mais se différenciant par leur pH optimum et l'action des effecteurs.

Les phosphomonoestérases fournissent l'exemple le plus complet d'enzymes isodynames. Ces enzymes hydrolysent une gamme étendue de monoesters d'alcools ou phénols. On relève de très nettes différences entre leurs courbes d'activité en fonction du pH, ce qui a fourni un premier élément pour subdiviser ces enzymes isodynames. D'autres éléments, tels que l'action de certains effecteurs caractéristiques, ont permis de compléter la classification. Ces caractères sont à la base de la classification actuelle en cinq principaux groupes. Dès 1936, FOLLEY et KAY (90) ont proposé de différencier plusieurs types de phosphatases; cette classification fut reprise par ROCHE et COURTOIS (184) en 1944 et adoptée dans ses grandes lignes par la suite.

Cette classification n'avait pour but que d'apporter un certain ordre dans un domaine complexe, et ce but elle l'a rempli. Il est certes indiscutable que dans le même type se trouvent réunis des enzymes à caractères parfois un peu divergents. C'est ainsi, par exemple, que les enzymes du type II des animaux ont quelques propriétés communes avec les phosphatases II végétales, mais diffèrent par un certain nombre de caractères. Par contre, les diverses phosphatases II végétales présentent une assez grande homogénéité entre elles. Il est de même plus aisé de classer ensemble les phosphatases du type III des végétaux supérieurs que d'envisager dans un même groupe les phosphatases du type III des végétaux supérieurs, des moisissures et des animaux.

La classification en les types principaux I—V reste cependant celle qui correspond le mieux à nos connaissances sur les phosphatases, c'est pourquoi nous l'adopterons dans cette revue.

Les principaux types de phosphomonoestérases.

Phosphomonoestérases type I. Ces enzymes, couramment dénommées „phosphatases alcalines", ont un pH optimum voisin de 8,0—9,0, qui d'ailleurs peut varier dans certaines limites en fonction de l'origine de l'enzyme, de son degré de purification, de la présence d'effecteurs, de la nature ou de la concentration du substrat utilisé. Les phosphatases I manifestent leur activité à partir de pH 6,5—7,0 jusqu'à pH 10,0—11,0; elles se différencient ainsi nettement des enzymes type II—V qui, réagissant en milieu acide plus ou moins élevé, sont désignées sous l'appellation générale de „phosphatases acides".

Les phosphatases alcalines sont très répandues chez les Animaux et il a été possible de les déceler dans presque tous les organes des Mammifères. Des enzymes actives en milieu alcalin ont été également caractérisées chez les levures et certaines bactéries, tandis que les Végétaux supérieurs en semblent complètement dépourvus.

On pouvait se demander si dans les graines une phosphatase de ce type n'était pas masquée éventuellement par l'acide phytique, qui représente la plus grande partie du phosphore acido-soluble des graines et qui pourrait dissimuler l'enzyme en bloquant les cations métalliques activateurs des phosphatases alcalines. Mais des recherches, effectuées par Mme RICAUD-MANOUVRIER (*179*) sur les graines d'Amandier, ont démontré qu'après avoir éliminé l'acide phytique il n'y a aucune possibilité de déceler une phosphatase du type I, même si on ajoute du Mg^{++}, activateur le plus constant des phosphatases alcalines.

Phosphomonoestérases type II. Les phosphatases de ce type ont un pH optimum vers 5,0—5,8 et manifestent leur activité de pH 4,0 à 7,0. Elles n'ont pas besoin d'être activées par les ions métalliques et sont très fortement inhibées par les fluorures et les molybdates.

Chez les Animaux les phosphatases du type II sont presque aussi répandues que les phosphatases alcalines, mais elles se montrent souvent moins actives. Le foie, la rate, le cerveau et la plupart des autres organes manifestent une activité phosphatasique à pH 5 à côté de celle, plus élevée, existant à un pH alcalin. Seuls les os et la muqueuse intestinale, si riche en phosphatase alcaline, sont pratiquement dépourvus d'enzyme II ; par contre, la prostate de l'Homme et des Singes anthropomorphes ne manifestent qu'une activité minime à pH alcalin mais sont très riches en phosphatase II.

Des phosphatases de ce type II se trouvent aussi chez certaines bactéries, chez les Champignons et surtout chez les Végétaux supérieurs, où, sauf quelques exemples d'enzyme III, il n'y a pas d'autres types des phosphatases. Il se trouve ainsi que la phosphatase du type II est par excellence la phosphatase des Végétaux. Elle a été décelée dans tous les tissus des Végétaux chlorophylliens où on l'a recherchée : graines, tubercules, racines, tiges, feuilles, fleurs, pollen, fruits, etc.

A notre connaissance, il n'existe qu'une exception : le surnageant obtenu par décantation du latex de *Rhus succedanea* (arbre à laque) qui ne paraît renfermer aucune activité phosphatasique, tandis que le même liquide est riche en autres enzymes et, en particulier, en laccase. Ces essais, réalisés par BOSSARD (*32*), mériteraient d'être vérifiés avec du latex fraîchement récolté, car il pourrait bien s'agir d'une inactivation de l'enzyme semblable à celle observée par PHILPOTT (*165*) dans le latex de Caoutchouc gardé à basse température pendant deux semaines.

Phosphomonoestérases type III. Il s'agit d'enzymes à pH optimum vers 3,5—4,0, nettement moins répandues que les précédentes. Des phosphatases du type III ont été caractérisées dans certains organes animaux, mais la source principale de ces enzymes demeure les Champignons et surtout la takadiastase de l'*Aspergillus oryzae*.

NGUYEN VAN THOAI (*154*) a présenté le premier en 1942 des arguments tendant à indiquer la présence de phosphatases du type III dans les graines d'*Amygdalus* et de Soja. Par la suite, COURTOIS et KHORSAND (*58, 68, 121*) ont découvert, dans les feuilles de certaines plantes, des phosphatases de type III inhibées par les fluorures et les molybdates, mais, ce qui est exceptionnel pour des enzymes de ce type, nettement activées par les cations bivalents, activateurs réguliers des phosphatases alcalines.

Phosphomonoestérases type IV à pH optimum vers 6,5. Décelées dans les levures, les hématies et certaines bactéries; les Végétaux supérieurs en semblent dépourvus.

Phosphomonoestérases type V à pH optimum d'action en milieu très acide, vers 2,5. Une phosphatase de ce type a été décelée par Courtois et Manet (69) chez le Colibacille.

2. Répartition dans les organes végétaux.

Nous avons signalé précédemment que tous les tissus des Phanérogames renferment des enzymes susceptibles d'hydrolyser les monoesters orthophosphoriques. Les travaux sur les enzymes des graines sont les plus approfondis et nous les examinerons en premier lieu, puis envisagerons ensuite l'étude des autres organes.

a) Graines.

Variations d'activité selon l'origine botanique. On a décelé une activité phosphatasique dans une longue liste de graines (9, 12, 29, 32, 41—45, 47, 49, 66, 71, 73, 115, 119, 150, 151, 158, 188, 203, 204, 211). Jusqu'ici, à notre connaissance, toutes les graines étudiées contiennent des phosphatases du type II avec pH optimum vers 5,0—5,8. Les extraits aqueux de la plupart des graines hydrolysent plus rapidement l'acide β-glycérophosphorique (VI) que son isomère α (VII).

$$CH_2OH—CH(O—PO_3H_2)—CH_2OH \qquad CH_2OH—CHOH—CH_2—O—PO_3H_2$$
$$\text{(VI.)} \qquad\qquad\qquad\qquad \text{(VII.)}$$

Cependant quelques graines hydrolysent à des vitesses pratiquement égales les deux isomères. Il est exceptionnel que la forme α soit hydrolysée le plus rapidement, ceci s'observe d'ailleurs aussi avec les phosphatases alcalines I d'origine animale.

Il existe d'assez nettes différences entre les intensités d'action des diverses graines. A titre d'exemple, nous avons reproduit dans le *tableau 1* (p. 324) les activités phosphatasiques de quelques graines.

La richesse en enzymes varie beaucoup selon l'espèce botanique; c'est ainsi que les graines oléagineuses se montrent plus actives que les graines albumineuses, qui sont elles-mêmes plus riches que les graines amylacées (41). Parfois on remarque des activités enzymatiques différentes suivant les variétés d'une même espèce, comme dans le cas du Blé (71) ou du Mil (74).

Tableau 1. Activités phosphatasiques sur le β-glycérophosphate à pH 5,2 d'extraits d'un même poids de graines [Courtois et Pérez (72)].

Source enzymatique	mg de P libérés	Source enzymatique	mg de P libérés
Phoenix dactylifera	0,61	*Citrus nobilis*	5,50
Hordeum vulgare	1,75	*Citrus vulgaris*	1,70
Bromus pratensis	0,65	*Citrus aurantium*	4,80
Zea mays	4,80	*Pistacia atlantica*	2,50
Ricinus communis	12,40	*Vicia faba*	6,10
Raphanus sativus	3,40	*Lens esculenta*	5,40
Sinapis alba	13,40	*Cicer arietinum*	4,50

Purification et champ d'action. Les phosphatases des graines coexistent dans ces organes avec de nombreuses autres protéines dont certaines sont douées d'activité enzymatique.

Jusqu'ici les techniques de purification des phosphatases de graines n'ont pas permis d'obtenir des préparations dont l'activité sur un substrat par rapport à la teneur en azote protéique soit très élevée. Les préparations les plus actives obtenues à partir des graines ont à ce point de vue une activité inférieure à celles d'autres préparations phosphatasiques d'origine animale: phosphatase II de la prostate humaine ou phosphatases I alcalines du rein et de l'intestin.

La purification des phosphatases de graines a cependant permis de mettre en évidence deux faits principaux: (a) Il n'existe pas de co-enzyme facilement dissociable; (b) les fractions obtenues ont un champ d'action assez étendu, ce qui tend à indiquer qu'une même enzyme est susceptible d'hydrolyser une gamme assez étendue de monoesters. Cette indication se trouve corroborée par le fait que dans certains cas l'on obtient à partir de la même graine deux fractions phosphatasiques distinctes. Ces deux fractions à propriétés différentes ont un champ d'action en grande partie commun; elles agissent l'une et l'autre sur une série assez étendue de monoesters orthophosphoriques.

Trois phosphatases de graines ont été plus spécialement étudiées dans notre laboratoire: celles d'Amandier, Moutarde-Blanche et son de Blé.

Phosphatase d'Amandier. Barré (*18, 22*) a différencié dans les graines d'Amandier trois principaux groupes de fractions protéiques qu'il dénomma amando-albumine, amando-globuline et émulsine.

Lorsque l'on acidifie par l'acide acétique un extrait aqueux de graines d'amandes délipidées il se précipite une combinaison d'acide phytique et d'amando-albumine; la presque totalité de l'activité phosphatasique demeure dans le liquide surnageant.

La fraction émulsine contient plusieurs protéines douées d'activités enzymatiques variées, principalement des glucidases et des phosphatases. Elle est insolubilisée pour des molarités salines supérieures à 2,18. Elle peut être scindée d'après ses courbes de relargage salines en au moins deux fractions. La première précipite pour des molarités salines comprises entre 2,18 et 2,50 et la seconde entre 2,50 et 2,87. L'activité phosphatasique semble être solidaire de la première fraction et l'activité β-glucosidasique de la seconde.

La présence des protéines associées des Amandes complique la purification. Lorsque l'on soumet un extrait aqueux de poudre d'Amandes délipidée à un relargage fractionné par le sulfate d'ammonium, la phosphatase est entraînée en grande partie par le précipité d'amando-albumine.

Tableau 2. Séparation de la phosphatase et de la β-glucosidase des Amandes par relargage fractionné au sulfate d'ammonium [BARRÉ et WORMSER (22)].

Zone de relargage exprimée en pourcentage de saturation en $SO_4(NH_4)_2$	Activité phosphatasique exprimée en mg de PO_4H_3 libérés du β-glycérophosphate par mg de poudre sèche en 24 h à 37° et pH 5,2	Activité β-glucosidasique, exprimée en % d'hydrolyse du salicoside par 4 mg de poudre sèche en 3 h à 37° et pH 5,3
0 → 40	1,368	0
38 → 45	0,485	0
40 → 48	1,170	0
50 → 55	3,76	57,8
55 → 60	1,62	100,0
60 → 70	0,395	100,0

Le *tableau 2* montre que la phosphatase est inégalement partagée entre les différents précipités. Au contraire, la β-glucosidase ne précipite qu'à partir d'une concentration de sulfate d'ammonium supérieure à 48% de saturation. Il est ainsi possible d'obtenir des préparations de phosphatases dépourvues de β-glucosidase. La phosphatase est précipitée en ajoutant du sulfate d'ammonium jusqu'à 48% de saturation, elle est mélangée aux protéines globulaires de la graine. Le précipité est repris par de l'eau distillée et acidifié à pH 5,0, correspondant au point isoélectrique des protéines globulaires; ces protéines sont insolubilisées et la phosphatase demeure dans le filtrat d'où elle est précipitée par l'acétone.

Tableau 3. Esters phosphoriques hydrolysables par la phosphatase des graines d'amandier [COURTOIS (55)].

Méthylphosphate	β-Glycérophosphate
n-Propylphosphate	Fructose-1,6-diphosphate
Isopropylphosphate	Aminoéthanolphosphate
n-Butylphosphate	Diméthylaminoéthanolphosphate
Allylphosphate	Diéthylaminoéthanolphosphate
Glycolphosphate	Morpholinoéthanolphosphate
Aldéhyde glycolique phosphate	Cholinephosphate
Acide phosphoglycolique	Inositotriphosphate
α-Glycérophosphate	Inositotétraphosphate

Nous indiquons dans la *figure 1* (p. 326) la courbe d'activité en fonction du pH de la phosphatase d'Amandes et dans le *tableau 3* une liste d'esters phosphoriques d'alcools hydrolysés par cette enzyme.

La phosphatase d'Amande s'est montrée active sur tous les esters d'alcools essayés à l'exclusion des esters penta- et hexaphosphoriques du mésoinositol. Elle est donc dépourvue d'activité phytasique.

La phosphatase d'Amande présente de plus une activité sur le pyro-phosphate (*85, 86*); cette pyrophosphatase n'est pas séparable de la phosphomonoestérase par les techniques habituelles de fractionnement

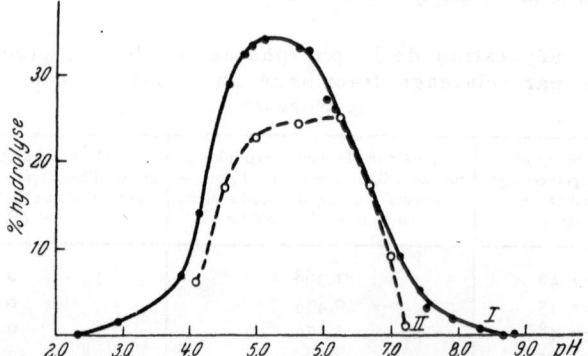

Fig. 1. Hydrolyse du glycérophosphate β par une préparation d'amandes purifiée en fonction du pH: ————, sans addition d'effecteur; — — —, avec addition d'acide phytique M/50 en acide phosphorique estérifié dans le milieu. [Ricaud-Manouvrier (*179*).]

des enzymes. Cependant la pyrophosphatase est plus sensible que la phosphatase à l'inactivation en milieu alcalin; il a été ainsi possible d'obtenir des échantillons pratiquement dépourvus d'activité pyro-phosphatasique (*86*). En effet il existe un net parallélisme entre les deux enzymes, qui tend à indiquer une similarité, sinon une identité, de leurs supports protéiques.

Les tentatives pour se débarrasser de la pyrophosphatase, puis les essais effectués dans le but de vérifier si les deux glycérophosphates isomères α et β étaient dédoublés par deux enzymes distinctes, ont conduit à une purification assez poussée de la phosphatase d'Amande, ainsi qu'à l'étude très détaillée de ses caractéristiques, en particulier l'action des effecteurs.

Phosphatase de Moutarde-Blanche. Une autre série de recherches a été effectuée dans notre laboratoire sur les graines de Moutarde-Blanche (*Sinapis alba*, Crucifères). Dès 1926, Fleury et Sutu (*89*) avaient observé que ces graines hydrolysaient les α- et β-glycérophosphates. Dès 1934, Courtois (*41, 44, 50, 51*) étudia les propriétés de cette enzyme; son pH optimum vers 5,6 permet de la classer dans les phosphomono-estérases du type II.

Anagnostopoulos (*6*) a réussi à purifier notablement la phosphatase de Moutarde-Blanche. Une préparation purifiée selon ces données est décrite dans les publications de SERVANT (*8, 192*). Extraction de la poudre de graine délipidées par une solution de chlorure de sodium à 9⁰/₀₀, élimination de la fraction protéique insolubilisée à 33% de saturation en sulfate d'ammonium; précipitation de la phosphatase entre 33 et 80% de saturation en $SO_4(NH_4)_2$. Le précipité est repris par l'eau et dialysé; l'enzyme est alors insolubilisée par addition de trois volumes d'acétone. Elle est à nouveau purifiée par reprise à l'eau, élimination de la fraction précipitant avec 20% d'acétone dans le milieu; puis finalement précipitée pour une concentration de 40% d'acétone. Toutes ces opérations étant réalisées entre 0° et — 10°.

La phosphatase de Moutarde-Blanche hydrolyse toujours le β-glycérophosphate plus rapidement que l'α, ceci quels que soient le pH, la température, la durée de réaction. Cette phosphatase hydrolyse de nombreux monoesters phosphoriques d'alcools (*tableau 4*). Elle est de plus apte à hydrolyser l'acide phytique.

Tableau 4. Esters phosphoriques d'alcools hydrolysables par la phosphatase des graines de Moutarde-Blanche. (Les chiffres correspondent aux mg de P libérés sous forme d'acide orthophosphorique par la même dose d'enzyme réagissant sur des quantités équimoléculaires des différents substrats.)
[COURTOIS (*51*).]

Substrats	Enzyme brute	Enzyme purifiée
Glycolphosphate	3,6	2,15
Phosphoglycolate	6,5	3,9
Isopropylphosphate	1,0	1,1
β-Glycérophosphate	5,7	5,6
Saccharosephosphate	3,2	1,7
Fructosediphosphate	5,0	4,7
Inositotriphosphate	7,8	6,5
Inositotétraphosphate	6,2	4,0
Inositopentaphosphate	3,9	1,0
Inositohexaphosphate	3,6	traces

Le problème s'est posé de rechercher si l'acide phytique était ou non hydrolysé par une enzyme autre que la phosphomonoestérase. Les extraits de Moutarde-Blanche ont été fractionnés par relargage avec les sels alcalins, par addition de solvants organiques miscibles à l'eau, par des essais d'adsorption puis élution sur différents adsorbants. L'alternance et la succession de ces opérations ont permis de séparer des fractions actives à la fois sur le glycérophosphate et l'acide phytique et d'autres fractions actives sur le glycérophosphate mais incapables d'hydrolyser l'acide phytique (*51*).

Il ressort du tableau 4 que les préparations de phosphomonoestérase de Moutarde-Blanche pratiquement débarrassées de phytase sont capables d'hydrolyser les inosito-tri- et tétra-phosphates presque aussi vite que le glycérophosphate, tandis que le dédoublement de l'inosito-pentaphosphate subit un net ralentissement. On peut en déduire que la capacité de dédoubler les inosito-penta- et hexa-phosphates est une particularité de la phytase et nous verrons par la suite que l'activité de cette enzyme n'est pas limitée à ces deux esters, mais elle s'étend aussi aux autres esters dédoublés par la phosphomonoestérase banale.

La phosphatase de Moutarde-Blanche est aussi capable d'hydrolyser des monoesters phosphoriques de phénols comme le phénylphosphate et son dérivé p-nitré (8). Cette propriété s'observe aussi avec la presque totalité des autres phosphomonoestérases végétales. Il semble bien que comme pour les phosphatases alcalines I c'est la même phosphatase qui hydrolyse les monoesters d'alcools et de phénols, elle rompt la liaison PO_3H_2—O—R en PO_4H_3 et R—OH qui peut être un alcool ou un phénol. En règle générale les esters de phénols sont hydrolysés plus rapidement que ceux des alcools; ceci s'observe avec les diverses classes de phosphatases.

Cette propriété présente par ailleurs un intérêt analytique; les esters phénoliques comme ceux du phénol, du p-nitrophénol, de l'α-bromonaphtol, sont utilisés couramment pour caractériser et doser les phosphatases. Ils sont rapidement hydrolysés et le composé phénolique libéré se prête assez bien à un dosage colorimétrique précis.

Les hétérosides de phénols, les esters de phénols avec des acides aromatiques sont fort nombreux dans le règne végétal. Ils sont hydrolysables respectivement par des glucidases et estérases largement répandues dans les végétaux. Les esters phosphoriques d'alcools sont assez nombreux chez les végétaux: glycérophosphates et acides phosphatidiques, inositohexaphosphates, esters phosphoriques d'aldoses et cétoses. Il n'est donc pas surprenant de trouver dans les végétaux des enzymes les hydrolysant. Par contre, les esters phosphoriques de phénols sont rarissimes dans les végétaux. Ce n'est que tout récemment que l'on a identifié la psilocybine, extraite d'un Champignon du Mexique à l'ester phosphorique d'une fonction phénol fixée sur un cycle indolique (O-phosphoryl-4-hydroxy-ω-N-N-diméthyl-tryptamine). Il est certes permis d'envisager que l'action de phosphatases spécifiques des esters de phénols a pour résultat de les hydrolyser dès qu'ils se forment dans le végétal. Il est cependant plus probable que les phosphomonoestérases hydrolysent ces esters parce qu'ils possèdent comme les esters d'alcools le type de liaison qui convient au groupe actif de ces enzymes.

Son de Blé. Comme pour la Moutarde-Blanche il a été possible de séparer deux fractions de pH optimum vers 5,4—5,6, l'une active sur la plupart des monoesters d'alcools et phénols mais incapable d'hydrolyser l'acide phytique, l'autre hydrolysant les mêmes monoesters et en plus l'acide phytique (*10, 21, 49, 64, 87*).

Courtois (*52*) a appliqué aux phosphatases du son de Blé les principales techniques de fractionnement des enzymes: précipitation fractionnée par les sels et les solvants, adsorptions et élutions.

L'activité des préparations obtenues a été comparée vis-à-vis de dix composés phosphorés: glycolmonophosphate, n-propylphosphate, saccharosemonophosphate, fructose-1,6-diphosphate, β-glycérophosphate, acide phosphoglycolique, benzylphosphate, inositohexaphosphate, phénylphosphate et pyrophosphate.

Tout se passe comme si le son de Blé renfermait au moins deux fractions phosphatasiques: (a) Une fraction phosphomonoestérasique inactive sur l'inositohexaphosphate, mais dédoublant les neuf autres substrats. (b) Une fraction phytophosphatasique, sans action sur les esters des alcools *n*-propylique et benzylique et probablement le glycolphosphate. Cette phytophosphatase dédouble par contre les sept autres substrats.

Il est toujours difficile de séparer des enzymes de spécificités très voisines. Les techniques de fractionnement des enzymes sont en réalité des techniques de séparation des protéines; il est possible de concevoir que des enzymes de spécificités rapprochées se comportent comme des protéines à propriétés physico-chimiques voisines sinon presque identiques. Il n'est donc pas exclu que des techniques plus perfectionnées n'arrivent à dissocier des fractions phosphomonoestérasiques à spécificités plus limitées et qui sont habituellement associées dans les graines.

Influence de la germination sur l'activité phosphatasique. Le réveil de l'activité métabolique des graines en dormance par la germination s'accompagne d'une façon générale d'une augmentation de l'activité des enzymes préexistantes, et même de l'apparition de certaines activités enzymatiques qui ne sont pas décelables dans les graines avant leur germination.

L'augmentation de l'activité amylasique des graines d'Orge au cours de la germination est connue depuis les temps les plus reculés. Ceci fut étendu aux phosphatases en 1914 par H. et B. v. EULER (*84*); ils observèrent que la germination de l'Orge provoquait une augmentation de l'activité phosphatasique qui atteignait son maximum vers le quatrième jour de germination. Ensuite NĚMEC (*152*) détermina l'augmentation d'activité phosphatasique en fonction du temps de germination de quatre sortes de graines. Il observa que si l'Orge et le Maïs se comportaient de la même façon, avec une baisse de leur activité (dans les premières 24 heures) suivie d'une augmentation, les deux autres par contre, bien qu'augmentant d'activité tout de suite sans aucune baisse préalable, ne présentaient pas la même allure. En effet, si l'activité des graines de Soja augmentait rapidement jusqu'au quatrième jour pour ralentir ensuite, donnant ainsi une courbe convexe, les graines de Pois avaient un comportement tout-à-fait différent, présentant une courbe concave.

Un résultat comparable à celui des graines d'Orge et de Maïs a été obtenu par COURTOIS et PÉREZ (*72*) avec des graines de Radis. Ils ont noté que l'activité de ces graines envers le glycérophosphate subit une baisse dans les deux premiers jours pour augmenter ensuite jusqu'à arriver à un plafond vers le cinquième jour.

L'activité phosphomonoestérasique au cours de la germination fût étudiée dans un grand nombre de graines (*10, 24, 37, 39, 72, 82, 84, 93, 95, 99, 112, 130, 137, 152*), et c'est surtout l'Orge que plusieurs chercheurs ont examinée. Il a été ainsi vu et confirmé que l'activité phosphatasique atteint son maximum vers le quatrième-cinquième jour de germination (*82, 84, 137*), avec une augmentation de 8 à 11 fois par rapport aux graines non germées (*130*).

Il reste encore aujourd'hui à savoir si l'augmentation de l'activité est due à la formation ,,ex novo" de protéines enzymatiques ou bien si l'enzyme préexiste déjà dans la graine sous forme de précurseur ou lié à d'autres complexes insolubles et si cette forme ,,liée" se transforme en forme ,,libre" pendant la germination, comme il a été envisagé par SARMA et GIRI (*188*).

b) Feuilles.

A notre connaissance des phosphatases ont été décelées dans toutes les feuilles essayées à ce point de vue (*15, 30, 31, 58, 63, 68, 75, 79—81, 84, 104—106, 108, 111, 112, 135, 144, 153, 157, 161, 166, 172—175, 180—182, 196, 198, 199, 208*).

Il existe d'assez nettes différences d'activité entre les espèces. De plus, il y a de très nets écarts entre l'extractibilité par l'eau des phosphatases des feuilles d'espèces différentes. Tandis que pour certaines feuilles la presque totalité de l'enzyme peut être extraite par l'eau distillée ou par des solutions tampons, pour d'autres, au contraire, la majeure partie de l'activité phosphatasique reste liée à la partie insoluble des feuilles que ces dernières soient à l'état frais ou à l'état sec. Dans le cas des feuilles de Belladone, par exemple, COURTOIS, ANAGNOSTOPOULOS et KHORSAND (*58*) ont trouvé qu'une forte proportion de l'activité phosphatasique est extractible par l'eau distillée et peut être purifiée par relargage au sulfate d'ammonium. Dans les feuilles de Marronnier, par contre, la fraction extractible par l'eau est limitée à moins de 20% de l'activité phosphatasique totale. COURTOIS et TRAN VAN AI (*75*) ont tenté vainement de solubiliser l'enzyme en mettant en œuvre les principales techniques habituellement utilisées pour extraire les enzymes solidement liées aux constituants insolubles des cellules.

Cette phosphatase ne paraît pas être incluse dans les mitochondries, car leur destruction par une série de congélations suivies de dégels, ou encore le traitement par le *n*-butanol, n'ont pu favoriser la solubilisation de l'enzyme. Il est aussi peu probable que la phosphatase soit sous une forme soluble dans les chloroplastes, car après leur destruction et l'extraction de la chlorophylle par le benzène, il n'y a pas eu d'accroissement de la solubilité de la phosphatase. Il en est de même d'un traitement par l'acétone qui élimine les tanins susceptibles de former des combinaisons insolubles avec les protéines. L'emploi des ultra-sons, de détergents ou de solutions neutres d'électrolytes a été également négatif.

Les seuls résultats partiellement positifs ont été obtenus avec un tampon boraté alcalin, qui permet de doubler la quantité d'enzyme extractible en milieu aqueux. Des essais complémentaires ont montré que cette action du borate pouvait être rapprochée de l'action solubilisante qu'il possède sur les polysaccharides peu solubles et non ramifiés. On pourrait ainsi envisager que la phosphatase des feuilles de Marronnier se trouve liée à des polysaccharides. Ce n'est qu'une simple hypothèse de travail, qui n'est encore appuyée par aucune preuve directe, mais qu'il serait intéressant de vérifier pour ses relations éventuelles avec le rôle physiologique peu connu des phosphatases des feuilles.

Les feuilles à cycle de vie annuel montrent des variations sensibles dans leur activité enzymatique (*183*). IGNATIEFF (*111*) a fait germer

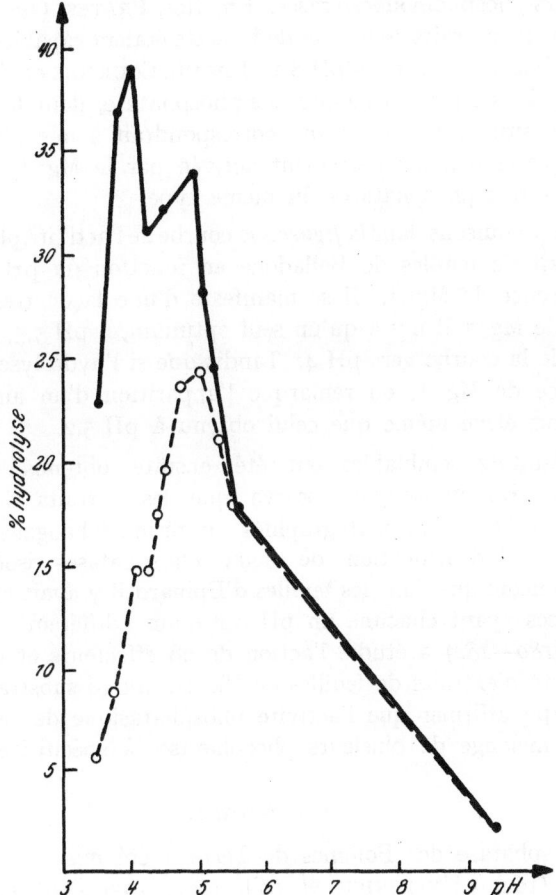

Fig. 2. Activité de la phosphatase de feuilles de Belladone en fonction du pH: — — —, en l'absence de sel de magnésium; ————, en présence de SO_4Mg $1 \cdot 10^{-2}$ M- dans le milieu [COURTOIS, ANAGNOSTOPOULOS et KHORSAND (*58*).]

une variété de Haricot rouge et il a dosé la phosphatase des feuilles en fonction de l'âge de la plante. L'activité augmente parallèlement au développement de la plante pour atteindre son maximum au 35e jour après l'ensemencement, puis elle diminue graduellement.

Les variations sont bien moins accentuées dans les plantes à cycle évolutif plus lent. COURTOIS et TRAN VAN AI (*75, 199*) ont suivi l'activité phosphatasique de divers échantillons de feuilles de Marronnier d'Inde au cours des différentes saisons; les feuilles récoltées soit au printemps, soit en été, soit en automne (au moment de leur chute), présentent à peu de chose près la même activité sur le glycérophosphate.

La question se posait de savoir si dans les feuilles il y avait une seule ou plusieurs phosphomonoestérases. En effet, PRATESI (172) avait signalé en 1937 que les extraits de feuilles de Violette étaient capables d'hydrolyser le glycérophosphate, même à pH 8,1. Ensuite COURTOIS et KHORSAND (68) ont décelé dans certaines feuilles une phosphatase, dont le pH optimum d'action se situe vers 4,0, ce qui correspondrait à une phosphatase du type III; cette enzyme nettement activée par le Mg^{++}, se différencie ainsi des autres phosphatases du même type.

Nous reproduisons dans la *figure 2* la courbe de l'activité phosphatasique d'un extrait de feuilles de Belladone en fonction du pH, en présence et en l'absence de Mg^{++}. Il se manifeste d'une façon très nette qu'en l'absence de Mg^{++} il n'y a qu'un seul optimum, à pH 5,5, et une petite inflexion de la courbe vers pH 4. Tandis que si l'hydrolyse est effectuée en présence de Mg^{++}, on remarque l'apparition d'un autre pic, vers pH 4,0, plus élevé même que celui obtenu à pH 5,2.

Des résultats semblables ont été ensuite obtenus par d'autres chercheurs. BOROUGHS (30) observa que les extraits de feuilles de Betterave et Tabac chromatographiés sur résine échangeuse de cations, permettaient l'identification de deux phosphatases isodynames. Il remarqua encore que dans les feuilles d'Epinard il y avait au moins trois phosphatases ayant chacune un pH optimum différent. Récemment ROBERTS (180—182) a étudié l'action de 28 effecteurs et de la chaleur sur l'activité d'extraits de feuilles de Blé envers 16 substrats différents. Il a ainsi pu affirmer que l'activité phosphatasique de ces feuilles est due à un mélange de plusieurs phosphatases à spécificité de substrat restreinte.

c) Tubercules.

La phosphatase des Pommes de Terre a été mise en évidence par PFANKUCH (164). HELFERICH et coll. (101, 102) l'ont purifiée d'une façon assez poussée par précipitations successives avec l'acétone, le tanin, le sulfate d'ammonium, l'alcool. La phosphatase de Pommes de Terre a été fréquemment utilisée comme réactif pour déphosphoryler enzymatiquement des corps. C'est probablement la phosphatase végétale utilisée le plus fréquemment comme réactif. La matière première est facile à se procurer dans le commerce, l'enzyme est aisée à préparer sous une forme assez active; par ailleurs le champ d'action de l'enzyme est assez vaste. NAGANNA et coll. (145) se sont intéressés accessoirement à la phosphatase de ces tubercules, l'objet principal de leurs travaux étant surtout la recherche de la pyrophosphatase alcaline. Ils ont trouvé qu'il y a une seule phosphomonoestérase du type II (leur courbe de pH optimum présente un aspect typique en cloche), elle est légèrement inhibée par le Mg^{++}. La recherche d'une phosphomonoestérase alcaline, même avec des substrats autres que le β-glycérophosphate, a été négative.

Notons ici en passant que la pyrophosphatase alcaline est séparable de la pyrophosphatase acide et de la phosphatase acide, tandis qu'une séparation de ces deux dernières n'a pas été possible jusqu'à maintenant. On peut rapprocher de ces résultats ceux de Ito et coll. (*113, 114*), qui n'ont pu obtenir une activité étroitement monoestérasique dans les préparations très purifiées de phosphatase de Patates douces. Leurs extraits enzymatiques attaquent le phénylphosphate, le β-glycéro-phosphate, le pyrophosphate inorganique, l'adénosine-triphosphate, les polymétaphosphates, le diphénylphosphate. Le fractionnement par l'éthanol ou le sulfate d'ammonium, l'électrophorèse ou la chromatographie sur colonne d'amidon ne changent pas les rapports entre les vitesses d'hydrolyse de ces différents substrats. Ces chercheurs en ont déduit qu'il n'existait probablement qu'un seul type d'enzyme à spécificité assez étendue pour une gamme variée de dérivés phosphorés.

La phosphatase de Pommes de Terre a été récemment purifiée de façon très poussée par Andreu, Fernández Alvarez et Lora-Tamayo (*11a*).

Les étapes de la purification sont des précipitations par l'acétone, puis l'éthanol, adsorption d'impuretés sur charbon, fractionnements chromatographiques sur phosphate de calcium et diéthylaminoéthylcellulose, puis électrophorèse sur colonne. La préparation obtenue est l'une des phosphatases végétales d'activité spécifique la plus élevée, sinon la plus active.

L'action de divers effecteurs indique que l'activité paraît dépendre de groupes carboxyles à faible acidité, du groupe imidazole de l'histidine et d'un groupe non identifié qui est peut être un ester phosphorique.

d) Racines, fleurs et fruits.

Les recherches effectuées au sujet des racines (*16, 28, 100, 142, 143, 176, 185, 200*), des fleurs (*97, 110, 149, 179*), des fruits (*12, 77, 207, 209*) sont beaucoup moins importantes que celles que nous venons de décrire au sujet des graines.

Nous nous bornerons ici à souligner que Boman et Westlund (*28*), par l'emploi d'une résine échangeuse d'ions, ont réussi à séparer à partir du Raifort deux phosphatases tout-à-fait semblables comme activité enzymatique, mais qui sont éluées par des concentrations différentes du tampon utilisé pour l'élution; elles ont probablement une structure moléculaire différente, tout en présentant les mêmes groupements actifs.

Au sujet des fruits nous signalons le remarquable travail de Axelrod (*12*) sur le jus d'Orange, qui nous a permis d'approfondir nos connaissances sur les différentes activités phosphatasiques de ce fruit. Mais nous reviendrons sur ce sujet pour l'étudier d'une façon plus détaillée dans la section sur les transferts (p. 354).

3. Phosphomonoestérases à action spécifique caractéristique.

C'est une règle générale que des enzymes d'origines différentes et à propriétés générales assez voisines se distinguent par leur comportement vis-à-vis d'un groupe de substrats de la même famille chimique. Soit une série d'hydrolases réalisant la scission hydrolytique de la liaison $X—Y$ de corps du type général $R—X—Y$ selon

$$R—X—Y + H_2O = R—X—H + Y—OH.$$

Pour des concentrations équimoléculaires de $R—X—Y$ la vitesse d'hydrolyse varie avec la structure de R. L'usage s'est établi de classer dans un même goupe général les enzymes scindant le même type de liaison, c'est la règle que nous avons suivie pour les phosphomonoestérases où $Y = PO_3H_2$.

Mais des préparations d'origines différentes se distinguent très souvent par leur aptitude à hydrolyser plus ou moins rapidement la liaison $X—Y$ selon la nature de R. C'est ainsi, par exemple, que certaines glucidases agissent plus rapidement sur la liaison osidique lorsque R est un composé holocyclique que lorsque R est un composé acyclique ou hétérocyclique. Inversement d'autres glucidases agissent plus rapidement lorsque R est acyclique ou hétérocyclique que lorsque R est holocyclique.
· Les écarts entre les vitesses d'hydrolyse de deux substrats différents par des enzymes distinctes sont parfois très accentués. L'étude des glucidases a. montré que deux substrats hydrolysés à des vitesses sensiblement égales par une enzyme étaient hydrolysés à des vitesses, dont le rapport pouvait devenir de 1 à 100 par une autre enzyme d'origine différente. Des écarts aussi accentués s'observent rarement avec les phosphomonoestérases des végétaux.

La spécificité de substrat est en général moins accentuée que pour les phosphatases alcalines d'origine animale. Certaines phosphatases alcalines agissent très spécifiquement sur un substrat ou un nombre limité de substrats et ne possèdent qu'une action infime, parfois même presque pratiquement nulle, sur une gamme variée d'autres monoesters orthophosphoriques.

Les phosphomonoestérases des végétaux ne présentent qu'assez exceptionnellement une spécificité limitée à quelques monoesters. Les tableaux 3 et 4 (pp. 325, 327) indiquent que les phosphatases des graines d'Amandier et Moutarde-Blanche sont capables d'hydrolyser toutes deux une gamme variée de monoesters.

D'une façon générale si l'on compare les spécificités des préparations des diverses graines c'est avec les esters phosphoriques de l'inositol que l'on observe les différences les plus accentuées. Ceci a conduit à examiner plus spécialement les enzymes dénommés phytases.

a) Les phytases.

L'acide phytique, ou *acide inositohexaphosphorique* $(CHOPO_3H_2)_6$, est le principal constituant phosphoré des graines, dont il représente, selon les espèces, de 50 à 90% du phosphore organique total.

L'acide phytique est susceptible d'être hydrolysé en une molécule de mésoinositol, $(CHOH)_6$, et six d'acide phosphorique par une phosphomonoestérase particulière, dénommée phytase. Le mécanisme d'action de cette enzyme a été étudié tout d'abord par S. et T. POSTERNAK (*171*); ils ont trouvé que le clivage des six liaisons ester de l'inositohexaphosphate par la phytase s'effectuait progressivement, de façon à scinder un seul radical phosphorique à la fois. COURTOIS et JOSEPH (*67*) ont établi que la phytase attaque en réalité les liaisons ester disposées en méta par rapport à la première fonction alcoolique libérée. Ces recherches ont été récemment confirmées par DESJOBERT et PETEK (*78*) qui, à l'aide de la séparation chromatographique sur papier, ont pu déceler la formation progressive par hydrolyse enzymatique des cinq différents inositophosphates avant d'arriver à la libération finale d'une molécule d'inositol pour six molécules d'acide orthophosphorique.

Répartition. La phytase, déjà fort irrégulièrement distribuée chez les Végétaux, apparaît être une enzyme assez exceptionnelle chez les Animaux. Elle fait défaut dans les sécrétions digestives de la plupart des animaux supérieurs, seul le petit intestin du Rat paraît contenir régulièrement de la phytase. La phytase a été décelée par PILEGGI (*168*) dans le rein, le cerveau, la rate, le foie et le plasma du Rat, mais c'est la muqueuse intestinale qui est le tissu le plus riche en phytase. La phytase est pratiquement absente du plasma d'autres Mammifères mais des phytases assez actives ont été signalées dans les plasmas d'Oie, Pigeon et Perche.

Il apparaît cependant utile de bien préciser la désignation de phytase. Cette désignation doit être réservée à une enzyme hydrolysant l'acide phytique à une vitesse du même ordre de grandeur que pour d'autres monoesters phosphoriques étudiés à la même concentration moléculaire en acide phosphorique estérifié. Une phytase d'activité moyenne hydrolysera sensiblement à la même vitesse une solution M/x d'acide glycérophosphorique et une solution M/6 x d'acide inositohexaphosphorique; une phytase active hydrolysera d'ailleurs plus rapidement le glycérophosphate que le phytate.

En 1951, COURTOIS (*53*) a signalé qu'il était assez curieux de noter qu'à l'exception de la phytase intestinale du Rat, toutes les autres préparations de phytases (Végétaux supérieurs, Bactéries, Moisissures) agissent en milieu plus ou moins fortement acide. Il est dès lors permis de se demander si l'incapacité des phosphatases alcalines à dédoubler l'acide phytique ne peut être reliée d'une part à l'intervention de cations divalents au cours de l'action des phosphatases alcalines et d'autre part à l'aptitude de l'acide phytique à bloquer ou précipiter ces métaux. Il serait séduisant d'envisager que l'inaptitude des phosphatases alcalines à hydrolyser aisément l'acide phytique ne résulterait pas d'une incapacité du centre actif d'hydrolyser les liaisons monoester de l'acide phytique; elle proviendrait plus d'une paralysie de la phosphatase par l'acide phytique s'emparant des cations métalliques nécessaires à l'activité de l'enzyme. De nombreuses préparations de phosphatases alcalines finissent en effet après une longue durée d'incubation par hydrolyser d'une façon très minime l'acide phytique (*169*).

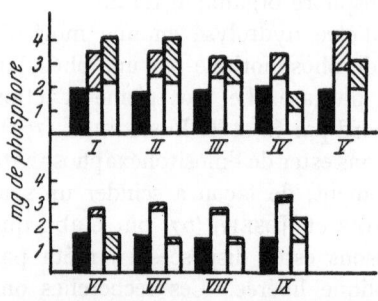

Fig. 3. Teneur en inositophosphates et activités glycérophosphatasique et phytasique de diverses variétés de Blé tunisien. Les rectangles noirs représentent le nombre de mg de P contenus sous forme d'inositohexaphosphates dans 1 g de poudre de graines stabilisée, délipidée et desséchée. Les rectangles centraux correspondent au nombre de mg de P libérés sous forme d'acide orthophosphorique à partir du glycérophosphate. Partie inférieure non hachurée: quantité libérée durant les premières 24 heures. Partie supérieure hachurée: quantité libérée entre 24 heures et 48 heures. Les rectangles les plus à droite correspondent à l'activité phytasique déterminée d'une façon similaire avec l'inositohexaphosphate et représentée d'une façon semblable. I variété „de Kasserine", II variété „de Chili", III variété „Mahmoudi + Kokini", IV variété „Mahmoudi", V variété „Sindjouk + Mahmoudi", VI variété „Guelma", VII variété „Baroota", VIII variété „Ariana", IX variété „Mahon". Variétés I à V blés durs; variétés VI à IX blés tendres. [Courtois et Pérez (*71*).]

Pour les préparations enzymatiques de végétaux supérieurs nous ne considérerons donc comme douées d'activité phytasique que celles qui ont une activité manifeste (*1—3, 9—12, 17, 24—26, 38, 40, 48—52, 64, 66, 71—74, 76, 83, 87, 94, 103, 109, 130—133, 160, 187, 190, 195, 210, 211*). Ce sont presque toujours les graines qui ont été étudiées.

Cependant la phytase n'arrive pas à avoir la même diffusion que nous avons observée pour la phosphomonoestérase dite „banale". Ce sont les graines qui représentent la source la plus importante de phytase, mais l'on observe de nets écarts entre les espèces rentrant dans le cadre d'une même famille botanique. L'exemple le plus typique en est fourni par les Graminées: l'Avoine, le Maïs et certains Mils sont presque dépourvus d'activité phytasique, tandis que l'Orge, ou le Dactyle, ont une activité moyenne. Le Blé et le Seigle renferment les deux phytases végétales les plus actives actuellement connues. L'activité phytasique est susceptible de varier notablement dans le cadre d'une même espèce; par exemple les Blés durs en sont plus riches que les variétés tendres

(*71, 160*). Il ne paraît pas non plus exister de relation nette entre l'activité phytasique des graines et leur teneur en acide phytique ou leur activité glycérophosphatasique ou pyrophosphatasique (*71, 74, 85*) *(figure 3)*.

Plusieurs chercheurs ont trouvé qu'au cours de la germination il y avait une diminution progressive de l'acide phytique jusqu'à disparition complète. Par contre, on pouvait remarquer une augmentation de l'activité phytasique; il arrivait même que des graines qui n'étaient pas capables d'hydrolyser l'acide phytique avant leur germination, montraient ensuite une remarquable activité envers ce substrat. Par exemple les graines d'Avoine en repos n'ont pratiquement aucune activité sur l'acide phytique (*9, 71, 103*); mais au cours de leur germination il y aura alors apparition d'une activité phytasique (*3*).

Parallèlement le taux d'acide phytique de la graine diminue progressivement au cours de la germination et finit assez souvent par disparaître. Le même phénomène s'observe avec beaucoup d'autres Graminées (*71, 72, 74, 83, 160*) et Légumineuses (*17, 24, 71*) qu'il y ait ou non une activité phytasique dans la graine avant la germination. En effet, même dans les graines à forte activité phytasique, comme le Blé ou l'Orge, la germination s'accompagne d'une augmentation de 6 à 11 fois cette activité.

Les autres organes des Végétaux sont en général très pauvres en phytase. Les feuilles n'ont qu'une activité minime, parfois même nulle (*48, 121, 161*); il n'a été signalé jusqu'à maintenant qu'une seule exception avec les feuilles d'Epinard qui, selon WILDMAN et BONNER (*208*) hydrolysent le phytate plus rapidement que le glycérophosphate.

La phytase n'a été qu'assez rarement recherchée dans les autres organes, elle a été décelée dans le suc de Pomme de Terre (*164*), dans les fruits du Cocotier (*187*) et dans le suc d'Orange (*12*); il ne s'agit le plus souvent que d'observations marginales sur des activités phytasiques d'ailleurs très faibles.

Purification. Dans le chapitre consacré aux phosphomonoestérases (p. 320) nous avons déjà signalé que certaines préparations hydrolysaient aussi bien l'acide phytique que d'autres monoesters. Il existe aussi des préparations actives sur une gamme étendue de monoesters d'alcools mais ne possédant qu'une action minime ou nulle sur les esters penta- et hexaphosphoriques de l'inositol; c'est par exemple le cas de la préparation purifiée de Moutarde du tableau 4 (p. 327).

De 1944 à 1956 COURTOIS et collaborateurs (*21, 48—51, 66, 71—74, 87, 161—163*) ont recherché si dans les graines contenant des enzymes hydrolysant l'acide phytique il serait possible de séparer une enzyme spécifique de ce corps.

Diverses techniques de séparation des enzymes furent appliquées à des graines d'activité phytasique plus ou moins accentuée: Blé, Seigle, Moutarde-Blanche et Radis. Il fut procédé à de multiples essais de précipitation fractionnée par le sulfate d'ammonium, l'acétone, l'éthanol, le méthylal, ainsi qu'à des opérations d'adsorption et élution dans diverses conditions avec diverses alumines, le kaolin, etc. comme agents adsorbants.

Avec les quatre espèces de graines il fut possible de dissocier deux types d'activités phosphatasiques.

Tout se passe comme si les graines actives vis-à-vis de l'acide phytique renfermaient associées deux phosphatases distinctes.

(1) Une *phosphomonoestérase banale* susceptible d'hydrolyser une grande variété de monoesters phosphoriques. Cette enzyme est sans action sur les esters penta- et hexaphosphoriques de l'inositol, elle dédouble assez difficilement l'inositotétraphosphate et plus aisément l'inositotriphosphate. L'inositomonophosphate est hydrolysé aussi facilement que les autres monoesters ne possédant pas plus d'une fonction alcoolique estérifiée par molécule. C'est cette phosphomonoestérase banale du type II de la classification des phosphatases qui se rencontre sous une forme plus ou moins active dans les diverses graines.

(2) Une *phytase* susceptible d'hydrolyser non seulement les mêmes esters que la phosphomonoestérase banale mais en plus les esters penta- et hexaphosphoriques de l'inositol. Cette phytase a sensiblement le même pH optimum que la phosphomonoestérase.

L'emploi de techniques de purification fort variées n'a jamais permis de séparer une fraction active sur l'acide phytique et sans action sur le glycérophosphate. Par contre, il a été séparé de toutes les graines hydrolysant l'acide phytique des fractions susceptibles d'hydrolyser le glycérophosphate ou d'autres monoesters mais dépourvues d'action sur l'acide phytique.

Spécificité. Quelques faits caractéristiques conduisirent à rapprocher les résultats obtenus avec les enzymes avec ceux observés par Barré et Courtois dans leur étude des combinaisons des dérivés phosphorés et des protéines. Ils ont observé (*18*, *19*, *59*, *60*) qu'en milieu acide les dérivés phosphorés se combinaient aux protéines. Entre pH 6,0 et 2,0 la courbe de solubilité des protéines en fonction du pH est modifiée, la zone de précipitation de la protéine est élargie et le précipité formé contient de la protéine associée au dérivé phosphoré.

Il se manifeste de nets écarts entre les comportements des dérivés phosphorés étudiés comparativement à des concentrations équimoléculaires.

(a) Avec les monoesters substrats habituels des phosphomonoestérases: β-glycérophosphate, phénylphosphate, p-nitrophénylphosphate, etc. la zone de précipitation des protéines n'est que faiblement élargie. La teneur en phosphore du précipité est constante tout le long de la zone de précipitation. Elle est relativement faible et représente en moyenne 1,3 à 1,75% de phosphore. (b) Avec les corps renfermant plusieurs radicaux phosphorés dans la même molécule la précipitation est plus accentuée.

C'est ce qui s'observe avec les acides tripolyphosphorique, trimétaphosphorique, tétramétaphosphorique, méthyl-2-naphtohydroquinonediphosphorique, inositotriphosphorique et inositohexaphosphorique.

En leur présence la courbe de solubilité des protéines en fonction du pH du milieu est fortement modifiée. Les protéines sont insolubilisées

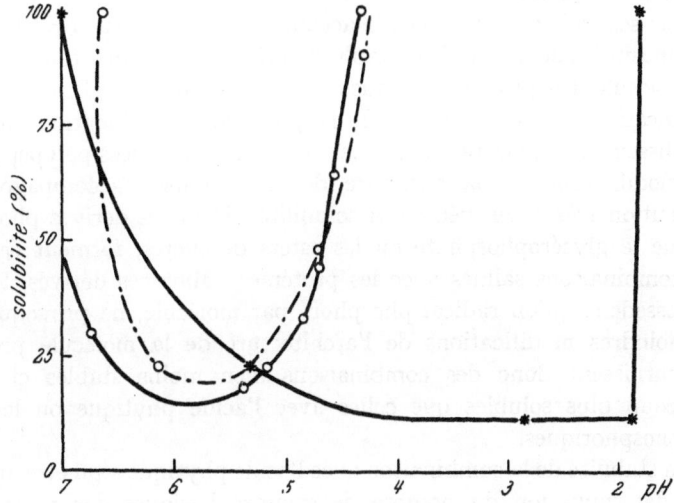

Fig. 4. Courbe de solubilité de l'insuline en présence des acides β-glycérophosphorique et phytique, en fonction du pH du milieu. En ordonnées : solubilité de l'insuline exprimée en pourcentage de protéine demeurant en solution. o—·—·—o, protéine seule; o—o—o, acide β-glycérophosphorique; *——*——*, acide inositohexaphosphorique.

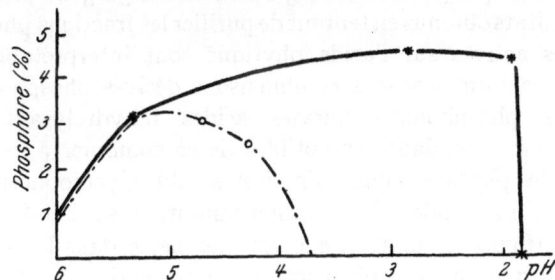

Fig. 5. Influence des acides β-glycérophosphorique et phytique sur la teneur en phosphore du précipité avec l'insuline. En ordonnées : phosphore, % du précipité protéine-ester phosphorique; o—·—o, acide β-glycérophosphorique; *————*, acide inositohexaphosphorique.

en-dessous de leur point isoélectrique jusqu'à des pH très acides, voisins de 2,0. La teneur en phosphore du précipité protéique est élevée, elle s'accroît de pH 4,5 jusqu'à des pH voisins de 2,0 pour atteindre des teneurs de l'ordre de 4,0 à 4,5%.

Ces résultats furent obtenus avec diverses protéines végétales ou animales. Les *figures 4* et *5* ont été établies avec l'insuline.

Parmi les multiples dérivés minéraux et organiques essayés, c'est l'acide phytique qui forme avec les protéines les combinaisons les plus insolubles et les plus stables. Une étude poussée de ces combinaisons (*18—20, 22, 59, 60*) a montré que l'acide phytique se combine aux groupements basiques libres de la protéine, en particulier ceux de l'arginine, la lysine et l'histidine.

Par ses radicaux phosphorés accumulés sur le noyau cyclohexanique de l'inositol l'acide inositohexaphosphorique rassemble tout autour de sa molécule les groupes basiques de la protéine. Ceci provoque une modification de l'architecture de la protéine, la salification entraînant des plissements et des rapprochements dans les séquences polypeptidiques. La modification de la structure de la protéine s'accompagne d'une diminution très accentuée de sa solubilité. D'autres dérivés phosphorés, comme le glycérophosphate ou les esters de sucres, forment également des combinaisons salines avec les protéines. Mais ces dérivés, lorsqu'ils ne possèdent qu'un radical phosphoré par molécule, ne provoquent que de moindres modifications de l'architecture de la molécule protéique; ils fournissent donc des combinaisons bien moins stables et surtout beaucoup plus solubles que celles avec l'acide phytique ou les acides polyphosphoriques.

La stabilité de la combinaison avec l'acide phytique a pu être appliquée à la détermination du nombre de groupes basiques des protéines; la teneur en phosphore phytique de la combinaison insoluble acide phytique-protéine est directement proportionnelle à la teneur en groupes basiques: un groupe basique par molécule-g s'unissant à 31 g de phosphore.

Les résultats obtenus en tentant de purifier les fractions phosphatasiques des graines actives sur l'acide phytique sont interprétables en tenant compte de la formation des combinaisons dérivés phosphorés-protéines.

(1) Les phosphomonoestérases acides n'hydrolysant pas l'acide phytique sont cependant susceptibles de se combiner à ce corps. C'est ainsi que le phytate inhibe l'hydrolyse du glycérophosphate par la phosphatase d'Amande. Cette action inhibitrice se manifeste en-dessous de pH 6,0 (figure 1, p. 326), elle est très nette dans la zone où l'acide phytique forme les combinaisons les plus stables avec la protéine. L'inhibition augmente avec la concentration en acide phytique. C'est une inhibition compétitive; en effet, pour une dose constante d'acide phytique l'inhibition est inversement proportionnelle au rapport des concentrations moléculaires glycérophosphate/phytate (*tableau 5*).

Avec la phosphatase des hampes florales d'*Helleborus foetidus* l'inhibition par l'acide phytique est de même plus accentuée à pH 4,0 qu'à pH 5,2 (*179*).

Ces résultats indiquent que les acides glycérophosphorique et phytique se fixent sur des zones identiques de l'enzyme. Il est logique de considérer

Tableau 5. Inhibition de la phosphomonoestérase purifiée des Amandes par une dose constante d'acide phytique (M/100 en acide phosphorique estérifié) en présence de concentrations variables en acide glycérophosphorique [Ricaud-Manouvrier (*179*)].

Concentration moléculaire en acide β-glycérophosphorique dans le milieu de pH 4,0		$4 \cdot 10^{-3}$	$1 \cdot 10^{-2}$	$2 \cdot 10^{-2}$	$4 \cdot 10^{-2}$	$6 \cdot 10^{-2}$
mg de P libérés sous forme de PO_4H_3	en l'absence d'acide phytique..........	1,54	2,86	3,63	3,96	3,96
	en présence d'acide phytique..........	0,22	0,44	1,98	2,09	2,31
Rapport d'activité en présence d'acide phytique/activité en l'absence d'acide phytique		0,14	0,15	0,54	0,52	0,58

que parmi ces zones figurent les sites actifs de l'enzyme. L'on pourrait envisager que par suite de la taille de sa molécule l'acide phytique est incapable de s'insérer dans les cavités du centre actif où pénétreraient les substrats des phosphomonoestérases banales. Cette interprétation prête à discussion car les phosphomonoestérases hydrolysent des substrats dont les molécules sont plus volumineuses que celle de l'acide phytique.

Il apparaît beaucoup plus vraisemblable que, par suite de la présence de plusieurs fonctions acides, l'acide phytique contracte des liaisons salines qui modifient la structure de la protéine enzymatique. Ces modifications de l'architecture protéique empêchent le centre actif de jouer son rôle. Il en résulte que d'une part l'acide phytique n'est pas hydrolysé par le centre actif de l'enzyme, et que d'autre part en présence d'acide phytique les modifications provoquées empêchent le centre actif d'hydrolyser les monoesters. Cette inhibition étant d'autant plus marquée que le rapport moléculaire acide phytique/monoesters est plus élevé.

(2) Cette conception trouve un appui dans le comportement des phosphatases vis-à-vis des divers esters phosphoriques de l'inositol. Les préparations à faible activité phytasique hydrolysent assez lentement les acides inosito-penta- ou hexaphosphoriques; la vitesse d'hydrolyse est un peu plus élevée pour l'acide inositotétraphosphorique et augmente notablement pour l'acide inositotriphosphorique [Courtois et Joseph (*66*), Courtois (*51*)]. Les acides inositomonophosphoriques sont hydrolysés rapidement, à des vitesses du même ordre de grandeur que pour les autres monoesters comme par exemple les acides glycérophosphoriques.

Avec la série de ces mêmes esters de l'inositol les préparations phosphatasiques inactives sur les dérivés hexa et penta, ont une faible action sur les esters tétra et tri et hydrolysent fort bien les inosito-monophosphates.

L'inaptitude à hydrolyser les esters penta- et hexaphosphoriques de l'inositol ne paraît pas résulter du fait que c'est le groupe actif qui est incapable de rompre des liaisons entre l'acide orthophosphorique et les fonctions alcool secondaire du mésoinositol. L'inaptitude doit plutôt être envisagée comme résultant de l'accumulation des restes d'acide phosphorique sur une molécule ne possédant que six atomes de carbone.

L'acide inositomonophosphorique ne donne pas de combinaisons stables et peu solubles avec les protéines tandis que les acides inosito-tri- et hexa-phosphoriques fournissent des combinaisons très stables et très insolubles. L'augmentation du nombre de liaisons ester phosphorique autour du cycle de l'inositol accroit les possibilités de formation de liaisons salines avec les groupes basiques de la protéine. La formation de ces liaisons entrave le fonctionnement du groupe actif de l'enzyme. Les phytases seraient ainsi des phosphomonoestérases où la salification des groupes basiques par l'acide phytique ne nuit pas au fonctionnement du centre actif.

(3) Lors du fractionnement de la phosphomonoestérase banale et de la phytase toutes les graines étudiées (Blé, Seigle, Moutarde-Blanche, Radis) ont conduit à des résultats similaires. Les fractions dépourvues de phytase et contenant la phosphomonoestérase banale ont toujours été insolubilisées avant les fractions phytasiques. Ceci fut observé aussi bien lors des relargages fractionnés par les sels neutres (sulfates d'ammonium et magnésium) que lors de la précipitation par addition progressive d'un solvant organique (méthanol, éthanol, acétone, méthylal, dioxane).

Ces résultats indiquent que la phosphomonoestérase banale a une structure moléculaire différente de celle de la phytase. La monoestérase possède soit une structure plus compacte, soit un poids moléculaire plus élevé, soit même les deux caractéristiques, ce qui lui confère un caractère hydrophobe plus accentué. Ces différences de propriétés sont comparables aux résultats obtenus avec l'acide phytique.

L'acide phytique insolubilise les protéines sans précipiter leurs produits de protéolyse, propriété mise à profit par Courtois et Villiers-Huiban (57 a) pour doser la pepsine dans le suc gastrique ou les préparations pharmaceutiques. L'acide phytique n'insolubilise pas la gélatine et les mucoprotéines sériques peu acides. Cependant il insolubilise des protéines de faible poids moléculaire lorsqu'elles sont particulièrement riches en groupes basiques; c'est le cas des protamines qui forment des combinaisons très insolubles. Mais lorsque l'acide phytique se trouve en présence de protéines de poids moléculaire différents et renfermant des pourcentages voisins d'amino-acides basiques il forme des combinaisons plus insolubles avec les protéines de poids moléculaire élevé qu'avec celles de plus faible poids moléculaire.

Bibliographie: pp. 363—373.

Il est dès lors permis d'envisager que les phosphomonoestérases banales ont des groupes actifs qui pourraient éventuellement scinder les liaisons ester de l'acide phytique; mais cet acide se combine à la protéine enzymatique en provoquant des distorsions qui placent le groupe actif dans une position incompatible à son action.

Avec les phytases, de taille moléculaire plus réduite ou moins compacte, les modifications provoquées par l'acide phytique sont moins accentuées et les sites actifs de l'enzyme continuent de pouvoir remplir leur rôle. Selon cette conception l'activité phytasique serait en relation avec la taille et la compacité de la molécule. Il n'y aurait pas de phytase au sens strict d'enzyme spécifique d'un type de liaison; toutes les phospho-monoestérases seraient théoriquement aptes à hydrolyser l'acide phytique lorsque ce dernier n'intervient pas comme un inhibiteur de sa propre hydrolyse en modifiant la structure de la protéine par salification des groupes basiques.

Une observation de Courtois (51) est interprétable dans ces conditions: En présence de thiocyanate d'ammonium M/25 dans le milieu la vitesse d'hydrolyse du glycérophosphate n'est pas modifiée. Dans le cas des inositophosphates il se manifeste une inhibition qui s'accentue avec le nombre de radicaux phosphorés; le pourcentage d'inhibition est de 9 pour l'inositotriphosphate, 14 pour le tétra, 20 pour le penta et 70 pour l'hexa. Le thiocyanate a provoqué une agrégation des molécules protéiques, de ce fait l'entrave apportée au groupe actif pour intervenir est d'autant plus manifeste que les substrats renferment un plus grand nombre de liaisons ester.

(4) Cette notion d'acide phytique inhibiteur de sa propre hydrolyse a été confirmée par Perlès (163). Cet auteur réussit à démasquer une activité phytasique dans des préparations enzymatiques réputées comme à peu près inactives sur l'acide phytique. En opposant des doses constantes

Tableau 6. Hydrolyse phytasique et glycérophosphatasique en fonction de la concentration initiale en substrat. (Durée de l'hydrolyse 24 heures, à 37°, et pH 5,15: Volume réactionnel 50 ml. Enzymes: 1 ml de solution d'enzyme purifiée à 5°/₀₀. Les chiffres représentent les pourcentages d'hydrolyse du substrat. IHP = inositohexaphosphate. GP = glycérophosphate.) [Perlès (163).]

Substrat	IHP	GP	IHP	GP	IHP	GP	IHP	GP	IHP	GP	IHP	GP
Concentration moléculaire en substrat dans le milieu (en acide phosphorique estérifié)	M/5000		M/2500		M/1000		M/500		M/250		M/100	
Origine de l'enzyme — Sorgho...	77,5	77,5	79,7	79	39,6	70,5	0	50,7	0	25,4	0	11,2
Avoine...	77,5	67,6	0	75,2	0	75,5	0	55	0	25,2	0	8,2
Dactyle ..	77,5	77,5	70,5	86	31,6	94,2	0	86	0	55,5	0	19

d'enzyme à des doses croissantes de substrat (M/5000 à M/100 en acide phosphorique estérifié) il observa une hydrolyse phytasique pour les faibles concentrations en acide phytique. Cette hydrolyse ne se manifeste plus au-dessus d'une certaine concentration en acide phytique *(tableau 6)*.

Dans les mêmes conditions l'hydrolyse du glycérophosphate est notable quelle que soit sa concentration. Tout se passe comme si la phosphatase formait une combinaison stable et inactive avec l'acide phytique dès que la concentration de ce dernier dépasse un certain taux.

Avec les préparations de Sorgho et Dactyle la zone d'hydrolyse est comprise entre les concentrations M/5000 et M/1000 en acide phosphorique estérifié (soit M/30000 à M/60000 en acide phytique). Dans le cas de l'Avoine l'hydrolyse phytasique ne se manifeste qu'à de plus faibles concentrations en substrat, elle est alors du même ordre de grandeur qu'avec les deux autres enzymes. Pour une concentration en substrat supérieure à M/5000 il n'y a plus aucun dédoublement de l'acide phytique.

De cet ensemble de données il paraît résulter que l'activité phytasique ne représenterait qu'un aspect particulier de l'activité phosphomono-estérasique. L'activité phytasique, au sens strict d'une action hydrolysante nette sur l'acide phytique, ne se manifesterait que pour les phosphomono-estérases dont la structure permet à l'acide phytique d'atteindre les sites actifs sans entraver leur fonctionnement.

Il semble donc permis de substituer à la notion de phytase spécifique celle de phosphomonoestérase dont la structure permet l'hydrolyse de l'acide phytique. L'acide phytique n'exige pas un groupe actif spécifique de ses liaisons esters mais exige une structure de la protéine enzymatique dont il ne risque pas de perturber l'architecture au point d'être l'inhibiteur de sa propre hydrolyse.

b) Les osylphosphatases.

Les oses reliés à l'acide orthophosphorique par leur groupe réducteur forment un groupe particulier de dérivés phosphorés. Dans ces osyl-phosphates l'hydroxyle du groupement réducteur est combiné à l'acide orthophosphorique, les dérivés d'aldoses ayant la structure générale (VIII).

$$PO_3H_2—O—C—(CHOH)_n—CH—R \quad (R = H \text{ ou } CH_2OH)$$

$$H \; \lfloor—O—\rfloor$$

(VIII.)

Au point de vue richesse de la liaison phosphorée, les osylphosphates occupent une position intermédiaire entre les dérivés à liaison riche (phosphamides, acides polyphosphoriques, anhydrides mixtes entre l'acide phosphorique et un acide organique) et les esters phosphoriques d'alcools à liaison pauvre.

La quantité d'énergie incluse dans la liaison de l'acide glucosyl-1-phosphorique (ester de Cori) est d'environ 4800 calories.

Bibliographie: pp. 363—373.

La liaison entre l'ose et l'acide phosphorique est stable en milieu alcalin. En milieu acide elle est beaucoup plus labile que celle des monoesters phosphoriques d'alcools.

Tandis que la liaison monoester orthophosphorique d'alcool ou de phénol peut être rapprochée de la liaison éther-oxyde [COURTOIS (54)], la liaison entre l'acide orthophosphorique et le carbone réducteur d'un ose présente les caractères d'un acétal. Par leurs propriétés chimiques et certaines de leurs réactions biochimiques, les osylphosphates s'apparentent aux hétérosides. Les essais d'hydrolyse enzymatique des osylphosphates ont été réalisés pour la plupart avec comme substrat l'acide glucose-1-phosphorique. De nombreuses préparations phosphatasiques impures en libèrent l'acide phosphorique.

Le problème s'est posé de savoir s'il existait une osylphosphatase spécifique ou si la libération de l'acide phosphorique résultait d'une suite de deux réactions enzymatiques: (a) isomérisation du glucose-1-phosphate en glucose-6-phosphate par la phosphoglucomutase, avec le glucose-1,6-diphosphate comme stade intermédiaire; (b) hydrolyse par les phosphomonoestérases du glucose-6-phosphate, qui est en général un bon substrat de ces enzymes.

Diverses préparations phosphatasiques végétales libèrent l'acide phosphorique du glucose-1-phosphate: graines de Blé, Moutarde-Blanche, Amande [FLEURY, COURTOIS, ANAGNOSTOPOULOS et DESJOBERT (88)], feuilles de Tomate [SPENCER (194)], de Blé [ROBERTS (180)]. Par contre, le suc d'Orange, qui hydrolyse de nombreux monoesters phosphoriques, demeure sans action sur le glucose-1-phosphate même après contact prolongé (12).

Ces résultats ne permettent pas de conclure s'il existe dans les préparations actives une osylphosphatase spécifique, ou si la phosphomonoestérase est intervenue sur le glucose-1-phosphate, ou même encore si une phosphoglucomutase associée l'avait isomérisé en glucose-6-phosphate.

L'existence d'osylphosphatases spécifiques a été démontrée dans le règne animal. Le ver à soie renferme une hexose-1-phosphatase spécifique. A partir du foie du Rat il est possible d'obtenir des préparations qui ont conservé une action sur le glycérophosphate mais n'hydrolysent pas le glucose-1-phosphate.

Récemment TURNER et TURNER (202) ont étudié l'action des graines de Pois sur divers esters phosphoriques d'oses. Une préparation purifiée par relargage au sulfate d'ammonium, dialyse, adsorption, continue d'agir sur toute une série d'esters. Les vitesses relatives d'hydrolyse par l'enzyme purifié sont (en prenant comme unité celle du glucose-1-phosphate): 1 pour le galactose-1-phosphate, 6 pour le glucose-6-phosphate, 8 pour le fructose-6-phosphate, 15 pour le fructose-1,6-diphosphate et

38 pour le p-nitrophénylphosphate. Les courbes d'activité en fonction du pH sont analogues avec les glucoses 1-ou 6-phosphate; les effecteurs agissent d'une façon similaire sur l'hydrolyse de ces divers substrats.

Le problème de l'existence d'une osylphosphatase spécifique différente de la phosphomonoestérase demeure à résoudre chez les Végétaux. Nous verrons que les pyrophosphatases végétales sont extrêmement difficiles et parfois même impossibles à séparer des phosphomonoestérases associées. Par contre, il est beaucoup plus facile de séparer pyrophosphatases et phosphomonoestérases de préparations d'origine animale.

Il semble que dans les végétaux l'osylphosphatase et la pyrophosphatase aient des propriétés physico-chimiques presque identiques à celles des phosphomonoestérases associées. Cette identité de propriétés est telle qu'il n'est pas exclu d'envisager qu'une même enzyme possède les trois activités phosphatasiques.

4. Action des effecteurs.

a) Généralités.

Dans la grande majorité des mémoires se rapportant aux phosphatases végétales figure la description d'essais en présence d'effecteurs chimiques. En présence de nombreux composés chimiques les diverses phosphomonoestérases se comportent d'une façon similaire. L'inhibition par les orthophosphates, fluorures, molybdates est observable très régulièrement. Pour d'autres corps les résultats varient selon l'origine des préparations phosphatasiques.

D'une façon assez générale l'on connaît toute une série de corps inhibiteurs énergiques et réguliers des phosphomonoestérases végétales. Par contre, il n'a pas été décrit d'activateurs intervenant de façon constante sur toutes les phosphatases végétales des types II ou III. Les données de la littérature ne sont pas toujours concordantes et ces divergences résultent de deux principales causes.

(1) Les essais réalisés à des pH différents ne sont pas comparables entre eux; selon le pH ce sera soit l'enzyme du type II soit celui du type III qui interviendra. Si entre pH 3,0 et 4,0 l'action est rapportable au comportement d'une enzyme du type III, entre pH 5,0 et 6,0 c'est le type II qui intervient, mais entre pH 4,0 et 5,0 les deux types réagissent simultanément et il est difficile de dissocier leurs participations respectives.

Dans toutes ces études il ne faut jamais perdre de vue que les substrats des phosphatases ont un caractère polaire très accentué. Alors qu'entre pH 3,0 et 4,0 le substrat est principalement sous la forme $R\!-\!O\!-\!PO_3H_2$ associé à $R\!-\!O\!-\!PO_3HM$, on se trouve en présence d'un mélange $R\!-\!O\!-\!PO_3HM$ et $R\!-\!O\!-\!PO_3M_2$ au-dessus de pH 4,0 à 4,5 selon les substrats.

Bibliographie: pp. 363—373.

Comme par ailleurs l'enzyme possède son point isoélectrique à l'intérieur de la zone d'activité des phosphatases II, le comportement d'un effecteur de ces enzymes variera avec le pH. L'inhibition de la phosphatase d'Amande par l'acide phytique de la figure 1 (p. 326) en a fourni un exemple.

L'influence des cations bivalents est de même fonction du pH [BARRÉ, COURTOIS et WORMSER (20—23)]. Avec les sels de calcium, magnésium et zinc le métal stabilise les liaisons intramoléculaires du type salin, ou hydrogène, de la protéine par formation d'un complexe (carboxyle — métal — groupe polaire). Il provoque une diminution des points de fixation pour le substrat sur l'enzyme en empêchant ou retardant l'ionisation des groupes cationiques protéiques inclus dans ces liaisons intramoléculaires. Ceci a pour conséquence de diminuer l'affinité K_M* de l'enzyme pour le substrat sans modifier la vitesse maximum d'hydrolyse par les centres actifs.

Les carboxyles libres des protéines capables de contracter des liaisons sont essentiellement les deuxièmes carboxyles des acides aspartique et glutamique de pK voisin de 4,0. La stabilisation par les métaux des liaisons intramoléculaires va donc se manifester quand le pH descend de 4,5 à 3,5. Par ailleurs, le métal stabilise la liaison enzyme-substrat par formation d'un complexe avec le radical phosphorique du substrat d'une part et un groupe basique d'autre part. Ceci provoque une diminution de la constante de Michaelis par recul d'ionisation, ce qui se traduit par une augmentation de son inverse, l'affinité K_M.

Il en résulte que pour bien préciser l'action d'un effecteur sur une phosphatase végétale des essais doivent être réalisés à différents pH d'activité de l'enzyme et avec plusieurs concentrations en effecteur. Il n'y a eu que peu de recherches conduites de façon aussi systématique.

(2) L'action d'un inhibiteur ou activateur ne peut être considérée comme vraiment caractéristique d'une enzyme que si elle est étudiée sur une préparation notablement purifiée de cette enzyme. Il nous faut reconnaître que dans les mémoires consacrés aux phosphatases végétales l'action des effecteurs a été le plus souvent étudiée sur des préparations d'une pureté assez moyenne. Avec les préparations d'un degré de purification moyen l'action de l'effecteur est susceptible d'être masquée ou accentuée par des substances présentes dans le milieu. Choisissons une fois encore comme exemple le cas de l'acide phytique, présent dans presque toutes les graines, et qui se rencontre dans les préparations phosphatasiques de graines moyennement purifiées.

L'acide phytique est en tout premier lieu un inhibiteur compétitif des phosphomonoestérases comme nous l'avons indiqué en p. 340. Mais

* $K_M = 1/Km$ où Km est la constante de dissociation moléculaire de la combinaison enzyme—substrat (constante de Michaelis).

il peut aussi exercer dans certains cas une action activatrice indirecte en bloquant des cations métalliques inhibiteurs (*163*).

Il n'en demeure pas moins que l'action des effecteurs chimiques sur les préparations de phosphatases végétales non purifiées présente un intérêt au point de vue métabolique. Ce ne sont pas des enzymes purifiées qui interviennent dans les processus cellulaires, les phosphatases agissent en présence de corps dont une partie subsiste dans les préparations impures.

b) Action des anions.

Orthophosphates. Les orthophosphates sont des inhibiteurs de toutes les phosphatases aussi bien alcalines qu'acides. Tous les auteurs ayant étudié l'action des orthophosphates sur les phosphomonoestérases végétales ont observé une action inhibitrice (une telle unanimité est exceptionnelle en enzymologie). Les orthophosphates exercent une inhibition compétitive, le phosphate tendant à se fixer sur l'enzyme aux lieu et place du substrat hydrolysable.

La combinaison active enzyme-substrat et la combinaison privée d'activité enzyme-phosphate sont dissociables. L'équilibre est réglé par la loi d'action de masse; de ce fait, pour une dose constante d'enzyme, l'inhibition sera d'autant plus élevée que le rapport phosphates/substrat non hydrolysé sera plus élevé. Il en résulte qu'au cours d'une hydrolyse phosphatasique, la proportion d'orthophosphates croissant au fur et à mesure que celle du substrat hydrolysé décroît, l'inhibition s'accentuera régulièrement au fur et à mesure que la réaction enzymatique progresse.

L'hydrolyse des glycérophosphates par les phosphatases de graines se ralentit très régulièrement et n'est pratiquement totale qu'après une longue durée de contact [Courtois (*44, 45*)]. Turner et Turner (*202*) ont envisagé une signification physiologique de cette action inhibitrice des phosphates, nous l'examinerons dans le dernier chapitre (p. 360).

Arséniates. Ils exercent une inhibition concurrente analogue à celle exercée par les phosphates (*44, 45*).

Fluorures. Ce sont des inhibiteurs énergiques des phosphatases acides. Toutes les phosphomonoestérases végétales sont inhibées par les fluorures. Les fluorures inhibent d'autres enzymes et dans quelques cas on a même pu élucider le mécanisme de leur action (*148, 205*).

Les phosphatases alcalines, elles aussi, sont inhibées par les ions F^-, mais cette action exige, pour être comparable, une concentration en fluorure beaucoup plus élevée que pour les phosphatases acides. De plus, le mécanisme d'action est différent car les phosphatases alcalines sont des enzymes à coenzyme métallique dissociable. Anagnostopoulos et Courtois (*7, 57*) ont étudié d'une façon détaillée l'inhibition fluorée des phosphatases végétales.

Dans le cas de la phosphatase de Moutarde-Blanche, l'inhibition n'augmente pas en proportion directe de la concentration en fluorure, ce qui serait le cas si le fluorure agissait comme réactif précipitant d'un ion métallique nécessaire à l'activité de la phosphatase.

Avec une quantité constante d'enzyme, réagissant dans des milieux plus ou moins dilués, l'inhibition est fonction de la concentration moléculaire en effecteur dans le milieu, mais non de la quantité de fluorure ajoutée. Il tend donc à se former une combinaison dissociable et non une précipitation d'un ion métallique. Cette combinaison phosphatase-fluorure paraît être fortement dissociée: (a) la quantité de fluorure combiné à des doses variables d'enzyme ne paraît pas diminuer de façon appréciable la concentration en fluorure dans le milieu; (b) l'enzyme totalement inhibée par addition de fluorure peut être réactivée soit par dialyse, soit par relargage au sulfate d'ammonium, soit par adsorption sur alumine puis élution. La phosphatase réactivée par dialyse conserve sa sensibilité initiale à l'inhibition fluorée; (c) dans les précipités acide phytique-protéine, en milieu fluoré on ne retrouve que des traces de fluorure, ce dernier ne paraît donc pas être fixé sur la protéine ou former un dérivé type fluorophosphate avec l'ester phosphorique (Barré, Courtois et Wormser, recherches inédites).

Enfin, il semble difficile d'envisager que le fluorure précipite ou bloque un cation métallique indispensable à l'activité de la phosphatase; les solutions saturées des fluorures de calcium, magnésium et zinc inhibent diverses phosphatases acides au même degré que des concentrations équivalentes de fluorure de sodium.

Molybdates. Les molybdates sans action sur les phosphatases alcalines I sont des inhibiteurs énergiques de toutes les phosphatases acides [Massart et Vermeyen (*138*)]. Cette inhibition est particulièrement marquée pour les phosphatases végétales; une concentration $8 \cdot 10^{-5}$ M en acide molybdique dans le milieu inhibe d'au moins 90% les phosphatases de Pomme de Terre, racine de Raifort, graines de Ricin et Soja [Courtois et Bossard (*32, 61*)]. Bossard (*32, 33*) a pu montrer que cette inhibition était relativement spécifique des phosphatases et ne s'observait pas avec un certain nombre d'autres enzymes végétales: glucidases diverses, lipase, peroxydase, laccase, uréase, déshydrogénases.

Le molybdate intervient en se fixant sur l'enzyme pour former une combinaison dépourvue d'activité enzymatique. La phosphatase est réactivable par dialyse, ou encore en ajoutant au milieu des substances capables de former également des complexes molybdiques et qui déplacent par action de masse l'acide molybdique de l'enzyme.

L'acide citrique réactivera ainsi l'enzyme. Si à un mélange glycérophosphate-molybdate-enzyme qui n'a pas libéré d'orthophosphate en 24 h à 37° et pH 5,2 l'on ajoute un tampon acide citrique-soude de pH 5,2, la phosphatase est réactivée et dans les 24 heures suivantes il y a libération d'acide phosphorique (*32, 56, 61*).

L'acide glycérophosphorique, qui forme des complexes molybdiques, réactivera aussi l'enzyme (*tableau* 7). L'action inhibitrice d'une dose constante de molybdate est ainsi d'autant plus accentuée que la dose de substrat est plus faible.

Tableau 7. Réactivation de la phosphatase de Moutarde inhibée. (Incubation à 37° et pH 5,2 de glycérophosphate $4 \cdot 10^{-3}$ M dans le milieu avec l'enzyme et l'effecteur $8 \cdot 10^{-5}$ M dans le milieu. Après 1 jour addition de glycéro-phosphate pour élever sa concentration à $12 \cdot 10^{-2}$ M dans le milieu.) [Courtois et Anagnostopoulos (*56*).]

	Effecteur:	o	ac. mo-lybdique	ac. phos-phomo-lybdique	ac. tung-stique	ac. phos-photung-stique	ac. méta-vana-dique
mg de P libérés (sous forme de PO_4H_3)	1 jour	1,96	0,0	0,0	0,0	1,74	1,42
	2 jours	20,0	20,0	20,0	16,1	19,6	20,0

Les complexes molybdiques de l'acide phosphorique sont bien connus, les orthophosphates libérés au cours de l'hydrolyse réactivent l'enzyme inhibé par le molybdate; ils exercent ainsi une action activatrice imprévue, qui s'oppose à leur propre action inhibitrice.

Courtois et Anagnostopoulos (*56*) ont généralisé ces résultats; d'autres acides minéraux à poids moléculaire élevé et également formateurs de complexes ont une action similaire: acides phosphomolybdique, tungstique, phosphotungstique et métavanadique.

Le mécanisme de l'inhibition est similaire à celui observé avec l'acide molybdique, en particulier réactivation par les acides citrique, ortho-phosphorique, β-glycérophosphorique (tableau 7).

L'action de ces anions ne semble pas se limiter aux phosphomono-estérases car Rothstein et Meier (*186*) ont signalé une inhibition de la pyrophosphatase de la levure et Naganna et coll. (*145*) ont trouvé que molybdates et tungstates inhibent la pyrophosphatase acide de Pomme de Terre sans toucher la pyrophosphatase alcaline associée. D'autre part, Spencer (*194*) a signalé que le molybdate exerce une inhibition compétitive sur l'hydrolyse de l'ATP par un extrait de Tomate.

Tous ces acides formateurs de complexes paraissent intervenir d'une façon analogue, ils se fixent sur les sites de l'enzyme où se fixe également l'ester monophosphorique, d'où une inhibition compétitive. Ces divers inhibiteurs sont des réactifs précipitants des protéines d'un emploi courant pour éliminer par insolubilisation les protéines dans les liquides biologiques.

Le fait que ces divers acides insolubilisent les protéines en se fixant sur leurs groupements basiques en milieu acide permet de supposer qu'ils interviennent de même avec les phosphatases. L'action inhibitrice, nulle sur les phosphatases alcalines, est d'autant plus accentuée sur les phosphatases acides que le pH est plus acide.

L'action inhibitrice pourrait être due au blocage de groupes basiques nécessaires à l'activité du centre actif. Il n'est pas encore possible de décider si ce blocage intervient dans le centre actif lui-même ou dans son voisinage; ce qui modifierait l'architecture de la protéine enzymatique et paralyserait le centre actif ainsi que le fait l'acide phytique comme nous l'avons indiqué dans le chapitre précédent. La seconde possibilité paraît être la plus probable. Des groupements basiques ne semblent pas constituer la partie réactionnelle du centre actif des phosphatases végétales. ANAGNOSTOPOULOS (4) a comparé l'action de divers réactifs des fonctions amine sur des phosphatases des différents types.

Les phosphatases alcalines animales sont inactivées par ces réactifs, la phosphatase II de la prostate est aussi inhibée mais de façon moins marquée. Les phosphatases II des graines de Blé et Moutarde-Blanche ne sont pas inactivées par une courte durée d'action du cétène. L'acide nitreux ne provoque qu'une faible inhibition; le formaldéhyde réduit modérément l'activité; l'isocyanate de phényle, qui réagit aussi avec les groupements guanidiques, exerce une plus forte inhibition.

La lente action inhibitrice du cétène et de l'acide nitreux sur les phosphatases acides apporte un argument non négligeable à l'appui de l'hypothèse émise par divers auteurs que les hydroxyles phénoliques sont indispensables aux activités phosphatasiques acide et alcaline.

Autres formateurs de complexes. Il n'apparaît pas que des cations bivalents soient indispensables à l'activité des phosphatases II car les oxalates n'ont pas d'action bien nette. L'acide éthylènediamine-tétracétique (versène) inactive totalement les phosphatases I où un cation bivalent est nécessaire à l'activité. Le versène active la phosphatase II de la prostate, activation analogue à celle de l'acide citrique [ANAGNOSTOPOULOS (5)]; le versène rentre en compétition avec le fluorure pour se fixer sur le même groupe de l'enzyme. A l'opposé les phosphatases acides des végétaux supérieurs ne sont que très faiblement activées par le versène et leur inhibition par le fluorure n'est pas influencée par la présence de versène (5, 6).

c) Action des cations.

L'action des sels de métaux lourds: Pb—Hg^{++} n'a rien de significatif; ils inhibent les phosphatases à de fortes concentrations. Le blocage des groupements SH par de faibles doses d'ion Hg^{++} ne paraît pas indiquer que ces groupements thiol soient d'une absolue nécessité pour les phosphatases acides.

Les sels alcalins exercent parfois une légère activation en modifiant la dispersion de la protéine enzymatique (*32, 62*). Les cations bivalents ont une action plus caractéristique.

Nous avons déjà indiqué à propos de la classification des phospho-monoestérases (voir p. 320) que les enzymes appartenant au type II n'ont pas besoin de Mg^{++} pour manifester leur activité hydrolytique, au contraire de ce qui se vérifie avec les phosphatases alcalines. Or nous devons préciser que même dans les phosphatases type II l'action du magnésium peut varier selon l'origine de l'enzyme. On peut ainsi trouver des phosphatases qui sont activées par les ions Mg^{++}, comme par exemple la phosphatase d'Avoine (*116*) ou celle d'Orange (*12*), à côté d'autres dont l'activité n'est pas modifiée (*187*). Il convient de remarquer que pour les enzymes activables par Mg^{++} cette activation est assez minime et demeure très inférieure à ce que l'on observe avec les phosphatases alcalines purifiées, dont certaines ont leur vitesse quintuplée à décuplée par les ions Mg^{++}.

A ce propos nous devons rappeler que la phosphatase type III trouvée par Courtois et Khorsand (*58, 68*) dans les feuilles de certaines plantes est nettement activée par les ions Mg^{++} et par les autres cations bivalents qui, agissant comme le Mg^{++}, sont classés parmi les activateurs réguliers des phosphatases alcalines. Cette identité d'activation nous laisse penser que la formation du complexe enzyme-substrat pourrait bien s'effectuer par l'intermédiaire des cations bivalents, tout comme cela a été déjà établi pour les phosphatases alcalines (*21, 23*).

Si l'action du Mg^{++} sur les phosphatases type II peut varier selon l'origine de l'enzyme, les autres cations bivalents se conduisent le plus souvent dans le sens de l'inhibition. Dans le *tableau 8* on trouve

Tableau 8. Action de quelques effecteurs sur différentes phosphatases. (La concentration en effecteur dans le milieu est de $2 \cdot 10^{-2}$ M pour tous sauf pour le F^- qui est de $1 \cdot 10^{-2}$ M. Le signe — correspond à une diminution d'activité de plus de 20% par rapport à l'activité sans effecteur. Le signe + correspond à une augmentation d'activité de plus de 20% par rapport à l'activité sans effecteur. Le signe o correspond à une variation de $\pm$ 20% par rapport à l'activité sans effecteur.) [Khorsand (*121*).]

Provenance de l'enzyme	Classification selon Roche et Courtois (*184*)	Effecteurs							
		F^-	$C_2O_4H_2$	Ca^{++}	Mg^{++}	Zn^{++}	Mn^{++}	Ni^{++}	Fe^{++}
Feuilles de Lierre ...	type III	—	—	—	+	+	+	+	—
Feuilles de Lierre ...	type II	—	o	—	+	—	—	o	—
Feuilles de Marronnier	type III	—	—		+	+	+		—
Feuilles de Marronnier	type II	—	o		+	—	—		—
Feuilles de Belladone .	type III	—	—	—	+	+	+	+	—
Feuilles de Belladone .	type II	—	o	—	+	—	—	o	—

Bibliographie: pp. 363—373.

un aperçu du différent comportement des deux types de phosphatases vis-à-vis de quelques effecteurs.

5. Réversibilité d'action.

L'action d'une phosphomonoestérase sur un substrat est représentée par l'équation générale :

$$R—O—PO_3H_2 + H_2O \rightleftarrows R—OH + PO_4H_3$$

Depuis le début de cette revue nous n'avons envisagé la réaction que dans le sens hydrolytique de gauche à droite ; en effet, en milieu aqueux dilué l'hydrolyse est pratiquement presque totale. Cependant la réaction est réversible comme pour les autres hydrolases. L'action biosynthétisante des phosphatases a surtout été étudiée avec le glycérol comme accepteur de phosphates.

Les phosphatases animales, réagissant sur une solution concentrée de glycérol et de phosphate disodique, estérifient très préférentiellement, sinon exclusivement, l'une des fonctions alcool primaire du glycérol et il se forme de l'α-glycérophosphate. L'action synthétisante des phosphatases végétales fut plus difficile à mettre en évidence. C'est en 1939 que Courtois (*46*) montra que la phosphatase d'Amande à pH 6,0 réagit sur une solution concentrée de glycérol et de phosphates mono- et disodiques pour synthétiser un ester. Ce dernier fut isolé et identifié à l'α-glycérophosphate. Cette estérification de la fonction alcool primaire est préférentielle. Un mélange fut réalisé ayant les rapports P minéral/P organique et P minéral/glycérol identiques à ceux atteints par biosynthèse avec au départ, glycérol et phosphates. Dans ce milieu le P organique était représenté par le β-glycérophosphate ; en ajoutant de la phosphatase d'Amande à ce mélange les rapports P minéral/P organique et P minéral/glycérol ne sont pas modifiés, mais le β-glycérophosphate est transformé en α-glycérophosphate. La forme β est hydrolysée et la forme α synthétisée pour retourner à l'équilibre (*46*).

Les préparations de phosphatase d'Amande ne réalisent que lentement la biosynthèse de l'α-glycérophosphate. Avec un mélange phosphates-éthanol la synth.se d'ester est très minime car l'enzyme s'inactive par séjour prolongé en milieu alcoolique concentré.

Avec les préparations non purifiées d'Amandes la synthèse est plus rapide qu'avec les préparations purifiées. Les préparations impures renferment des activateurs de la synthèse découverts par Nguyen van Thoai, Roche et Danzas (*155*). Ces activateurs dialysables, très labiles, sont détruits aussi bien par autolyse que par une courte durée de chauffage en milieu acide, ils sont insolubilisés par l'acétate de plomb.

Ces activateurs sont vraisemblablement des dérivés phosphorés riches en énergie. Meyerhof et Green (*141 a*) ont montré, avec les phosphatases animales, que des dérivés phosphorés riches en énergie accéléraient la synthèse par les phosphatases alcalines. Il s'agit de l'intervention de transphosphatases (phosphokinases) distinctes des phosphomonoestérases. Ces réactions secondaires amènent plus vite le milieu à son état d'équilibre final qui est le même soit que l'on utilise une phosphomonoestérase purifiée, soit que l'on ait recours à une phosphomonoestérase non purifiée qui est associée à des activateurs naturels, composés phosphorés riches en énergie.

L'action réversible des phosphomonoestérases est lente; elle exige des concentrations en composé hydroxylé accepteur et orthophosphate qui ne paraissent pas pouvoir être atteintes dans la cellule dans les conditions normales de fonctionnement. Il est douteux que la réversibilité présente une grande importance métabolique; c'est plutôt aux trans-phosphatases, qui agissent bien plus rapidement que l'on doit attribuer la synthèse des composés phosphorés. Mais en plus de ces trans-phosphatases une part de l'action peut revenir à l'activité phospho-transférasique des phosphomonoestérases.

6. Action transférante des phosphomonoestérases.

Nous avons indiqué au début que les transphosphatases étaient les enzymes transférant un reste orthophosphoryl d'un substrat donateur sur un substrat accepteur selon la réaction générale

$$R\text{—}PO_3H_2 + R'H \rightleftarrows R\text{—}H + R'\text{—}PO_3H_2$$

avec les phosphomonoestérases nous n'avons jusqu'ici envisagé que le cas où $R' = OH$. En réalité, lorsque dans le milieu se trouve un composé hydroxylé il fonctionne comme $R'H$ et se trouvera estérifié par l'acide phosphorique.

En 1948 Axelrod (13) signala que certaines phosphatases végétales, en particulier celle d'Orange, pouvaient catalyser le transfert du radical orthophosphoryl à partir des esters de phénols sur un certain nombre d'accepteurs à fonction alcool primaire.

Avec le p-nitrophénylphosphate donateur et le méthanol accepteur la réaction s'effectue à pH 6,0 selon l'équation (4):

$$NO_2\text{—}C_6H_4\text{—}O\text{—}PO_3H_2 + CH_2OH\text{—}H \rightarrow$$
$$\rightarrow NO_2\text{—}C_6H_4\text{—}OH + CH_3\text{—}O\text{—}PO_3H_2 \tag{4}$$

Axelrod (13) isola et identifia le méthylphosphate synthétisé. Il apporta ensuite la preuve que c'est le groupement orthophosphoryl du donateur qui se retrouve dans le méthylphosphate (14). En utilisant du p-nitrophénylphosphate dont le phosphore n'est pas marqué, et en opérant dans un milieu contenant de l'orthophosphate ^{32}P, le méthyl-phosphate qui apparaît ne renferme pas de ^{32}P.

Inversement avec comme donateur du p-nitrophénylphosphate marqué réagissant en présence d'orthophosphate non marqué, la radio-activité du méthylphosphate synthétisé est identique à celle du donateur.

La réaction de transfert s'accompagne toujours de la réaction d'hydrolyse (5). de l'ester de phénol, et aussi de l'hydrolyse ultérieure du méthylphosphate selon l'équation (6)

$$NO_2—C_6H_4—O—PO_3H_2 + HO—H \rightarrow NO_2—C_6H_4—OH + PO_4H_3 \qquad (5)$$

$$CH_3—O—PO_3H_2 + HO—H \rightarrow CH_2OH—H + PO_4H_3 \qquad (6)$$

Pour une durée de contact réduite la réaction (6) est en général restreinte; (4) et (5) sont suivis analytiquement en déterminant l'acide phosphorique libéré qui correspond à (5) et le composé phénolique libéré qui correspond à la somme de (4) et (5).

Si l'on exprime par Sc le pourcentage de substrat scindé, apprécié en déterminant le composé phénolique libéré, par a le pourcentage de phosphore du substrat libéré sous forme d'acide orthophosphorique, la différence $Sc — a$ correspond au pourcentage du phosphore transféré sur l'accepteur par rapport au phosphore initialement présent dans le donateur. Le pourcentage de transfert T est le pourcentage de phosphore transféré sur l'accepteur par rapport au substrat scindé

$$T = [Sc — a/Sc] \ 100.$$

Les valeurs de T varient selon les accepteurs (*tableau 9*). Avec les phosphatases d'Orange et de Pomme les activités des deux enzymes sont tout-à-fait voisines tant au point de vue de la spécificité qu'à celui de l'intensité du pouvoir transférant.

Tableau 9. Pourcentages de transfert à partir du phénylphosphate avec les préparations d'Orange et de Pomme. (Essais à pH 5,2 et 37°; Sc = pourcentage du substrat scindé; T = pourcentage du phosphore transféré sur l'accepteur par rapport au substrat scindé.) [SERVANT (*192*).]

Accepteur	Concentration moléculaire de l'accepteur dans le milieu	Phosphatase d'Orange		Phosphatase de Pomme	
		Sc	T	Sc	T
Méthanol	5,0	81	64	73	70
n-Propanol	2,5	28	50	16,5	31
Isopropanol.........	2,5	76,5	38	16	17
Éthylèneglycol	1,0	74,0	23	49	20
Glycérol...........	1,5	66,0	33	53	30
Butylèneglycol (2,3) .	1,0	87	10	20	10
Acide tartrique	0,5	77,5	4	19,5	12
Sorbitol	1,0	87	58	53	64
Saccharose.........	1,0	90	18	50,5	28
Raffinose ,..........	0,125	84	0	37	0

23*

Les alcools primaires: méthanol et *n*-propanol sont de meilleurs accepteurs que les alcools secondaires: isopropanol et butylèneglycol-2,3. L'acide tartrique est un médiocre accepteur. Les polyols acycliques sont de bons accepteurs ainsi que les hexoses; mais les saccharides sont de mauvais accepteurs, le saccharose est encore accepteur, alors que l'on n'observe aucun transfert sur raffinose et stachyose. Enfin il n'a pas été possible d'observer de transfert sur le mésoinositol. Avec les polyols acycliques le transfert s'effectue préférentiellement sur la fonction alcool primaire.

Courtois, Desjobert et Servant (65) utilisant la phosphatase de feuilles de Belladone ont isolé le corps formé par transfert à partir du *p*-nitrophénylphosphate sur le glycérol; ce corps fut identifié à l'α-glycérophosphate. Avec le phénylphosphate donateur et le sorbitol accepteur la même phosphatase de Belladone conduit à un sorbitolmonophosphate où l'une des deux fonctions alcool primaire du sorbitol est estérifiée.

L'activité phosphotransférasique varie selon l'origine des enzymes. Les enzymes des sucs d'Orange et Pomme, celle des feuilles de Belladone ont une action transférante marquée [Anagnostopoulos, Courtois et Servant (8)]; par contre ces auteurs n'ont décelé que de faibles transferts avec l'enzyme des graines de Moutarde-Blanche. La faible action transférante de l'enzyme de Moutarde par rapport aux trois autres phosphatases végétales paraît résulter de l'influence dans le même sens d'au moins deux facteurs: (a) écart peu accentué entre les hydrolyses phosphomonoestérasiques par la Moutarde des esters de phénols utilisés comme donateurs et des esters synthétisés à partir des alcools accepteurs; (b) plus grande sensibilité de la préparation de Moutarde à l'inhibition par de fortes concentrations en accepteur dans le milieu.

Par la suite, Anagnostopoulos et Lino (9) ont cherché à délimiter quelles étaient les phosphatases de graines douées d'une activité de „phosphotransférase". Ils ont constaté que les préparations possédant une forte activité de phosphotransférase sont, en général, des préparations

Tableau 10. Activité de phosphotransférase et de phytase de quelques préparations de graines [Anagnostopoulos et Lino (9)].

Origine de l'enzyme	Millimol de PO_4H_3 libérées par heure à partir du *p*-nitrophénylphosphate en l'absence d'alcool accepteur	Transfert % à partir du *p*-nitrophénylphosphate sur le méthanol 5 M	Millimol de PO_4H_3 libérées par heure à partir de l'acide phytique
Avoine	2,0	73,0	0,0
Brome des Prés	1,15	68,0	0,06
Tournesol	0,69	70,0	0,05
Blé	2,39	58,0	0,07
Orge	2,97	53,0	0,33
Maïs	2,08	19,5	0,0
Seigle	3,30	22,5	0,80
Luzerne	2,90	6,0	0,0

n'ayant qu'une action minime ou nulle sur l'acide phytique (*tableau 10*, p. 356).

Dans un cas (Seigle), où l'écart entre les activités de phosphomono-estérase et de phytase est relativement faible, l'activité de phospho-transférase est faible. Mais il existe également des préparations dépourvues de phytase qui n'ont qu'une faible activité de phosphotransférase: Maïs, Luzerne. Ces résultats suggéraient que les activités de phosphotransférase et de phytase appartenaient à deux enzymes distincts. Cela fut confirmé avec les enzymes du Seigle, qui furent soumis à un relargage fractionné par le sulfate d'ammonium. Le *tableau 11* montre que l'activité transférante accompagne la phosphomonoestérase banale tandis que les fractions contenant la phytase sont dépourvues d'activité transférante.

Tableau 11. Fractionnement des enzymes des graines de Seigle [Anagnostopoulos et Lino (9)].

Molarités en sulfate d'ammonium entre lesquelles la fraction est précipitée:	0,82 à 1,23	1,23 à 1,64	1,64 à 2,05	2,05 à 2,46	2,46 à 2,87
Rapport d'activité sur le *p*-nitro-phénylphosphate/activité sur l'acide phytique	11,1	16,0	4,8	3,3	2,75
Pourcentage de transfert, *p*-nitro-phénylphosphate donateur et méthanol accepteur	35,0	27,0	3,0	0	0

Dans le chapitre consacré à la phytase (p. 335) nous avions envisagé que la phosphomonoestérase banale avait une molécule plus importante et peut être aussi plus compacte que la molécule de la phytase.

La phosphomonoestérase banale possède indiscutablement un caractère hydrophobe plus marqué que celui de la phytase (ou phytophosphatase).

L'activité transférante exige-t-elle pour se manifester une certains taille de la molécule enzymatique? C'est un problème qui mériterait d'être étudié de façon plus approfondie. Il n'est pas invraisemblable que la phytase ayant une molécule moins lourde, la structure de cette dernière soit plus modifiée par l'accepteur dont la concentration est susceptible d'exercer une action dénaturante. En effet, Anagnosto-poulos et Lino (9) ont constaté que l'inhibition de l'hydrolyse du *p*-nitrophénylphosphate par le méthanol était plus accentuée pour les fractions contenant la phytase que pour celles contenant la phospho-monoestérase banale.

De plus, lorsque l'eau et l'alcool accepteur sont en compétition pour se fixer sur l'enzyme, l'on peut concevoir que l'équilibre résultant de cette compétition diffère selon les enzymes. Il pourrait être nettement en faveur de l'eau pour les phytases et plus favorable à l'alcool pour les phosphomonoestérases banales.

Les rapports entre les phosphatases douées d'action transférante et celles dépourvues de cette action sont susceptibles de varier au cours du développement de la plante. Anagnostopoulos et Lino (*10*) ont observé qu'au cours de la germination du Blé il y a augmentation considérable de l'activité de phosphotransférase dans les tissus en croissance: tigelle et radicules. L'activité de „phytase" augmente également mais elle est absente de ces tissus et demeure localisée dans le restant de la graine: cotylédon, albumen et téguments.

Aux côtés des phosphomonoestérases banales douées d'une activité de phosphotransférase non spécifique il faut joindre d'autres enzymes ayant une action transférante plus spécifique.

Brawerman et Chargaff (*35, 36*), avec des préparations végétales et animales ont obtenu le transfert du radical phosphoré du phényl-phosphate sur des nucléosides du ribose et du désoxyribose en réalisant ainsi la synthèse de nucléotides. Tunis et Chargaff (*200*) ont séparé une nucléoside-phosphotransférase de Carotte exempte d'activité hydrolytique; la séparation entre activités hydrolysante et transférante n'a été réalisée qu'à un niveau de purification relativement élevé tandis que dans les étapes antérieures le rapport des deux activités semblait constant. Cette synthèse de nucléotides paraît s'effectuer très spécifiquement et avec de faibles concentrations de nucléosides accepteurs.

Nous signalerons en dernier lieu les intéressantes observations de Takashima et Mitsui (*197*) sur le transfert photochimique de phosphate catalysé par les feuilles d'épinard avec formation d'un composé phosphoré très labile et de structure inconnue.

II. Les phosphodiestérases.

Les phosphodiestérases non spécifiques sont très peu répandues chez les Végétaux supérieurs et les recherches effectuées à ce sujet sont peu nombreuses, le substrat le plus fréquemment utilisé étant le diphényl-phosphate.

Une certaine activité phosphodiestérasique, toujours très faible, a été trouvée dans le son de Riz (*203, 212*), dans le Malt (*34*) et dans les Patates douces (*113*). Ricaud-Manouvrier (*179*) et Courtois (*51, 52*) n'ont pas décelé de diestérases active sur le diphénylphosphate dans les graines d'Amandier et le son de Blé.

Mais, à côté de ces données assez fragmentaires, il y a un ensemble d'études sur des phosphodiestérases spécifiques d'importance considérable: les nucléases. Un nombre de recherches furent effectuées sur les ribo- et désoxyribonucléases (RNAse et DNAse) (*12, 27, 34, 92, 98, 107, 120, 159, 166, 167, 176—178, 189, 191, 193*).

Les premières recherches se bornaient à déceler tout simplement l'activité sur les acides nucléiques d'extraits bruts ou qui, bien que partiellement purifiés, contenaient encore plusieurs autres phosphatases. Il faut attendre jusqu'en 1955 pour avoir une préparation végétale dépourvue d'activité sur le diphénylphosphate et sans propriété phosphomonoestérasique (*107*). Cependant cette préparation était encore capable d'attaquer les métaphosphates, l'ATP et le glucose-6-phosphate. Ensuite PIERPOINT (*167*) put obtenir, par chromatographie sur résine échangeuse de cations, une RNAse dépourvue de toute autre activité phosphatasique. Une autre préparation très pure de RNAse a été obtenue à partir des feuilles de Tabac (*92*); ici l'élimination de la phosphomonoestérase associée a été effectuée par une série successive d'adsorptions et élutions sur alumine C_γ.

III. Les pyrophosphatases.

Des activités pyrophosphatasiques plus ou moins importantes ont été décelées chez de nombreuses espèces végétales et pratiquement dans toutes les plantes où on les a cherchées (*12, 38, 39, 52, 76, 85, 86, 93, 96, 113, 114, 121, 127, 128, 136, 139, 145—147, 149, 164, 166, 180, 187, 199, 206, 208, 212*). La diffusion de ces enzymes paraît être aussi ubiquitaire que celle des phosphomonoestérases; on a retrouvé une activité pyrophosphatasique dans tous les tissus et organes végétaux: graines, tubercules, feuilles, fruits, etc.

Les pyrophosphatases accompagnent souvent les phosphomonoestérases; cette analogie de répartition apparaît encore plus marquée si on ajoute la difficulté de séparer les deux activités et la présence de différents types isodynames de pyrophosphatases, de façon analogue à ce qu'on a décrit pour les phosphomonoestérases. ROCHE et COURTOIS (*184*) ont d'ailleurs proposé de classer les pyrophosphatases en trois catégories correspondant dans leurs grandes lignes aux classes des phosphomonoestérases. On a ainsi les pyrophosphatases type I ou alcalines, à pH optimum vers 8,0; les pyrophosphatases type II à pH optimum vers 5,0 et celles du type III, à pH optimum vers 3,5—4,0.

La plupart des chercheurs se sont occupés des pyrophosphatases acides et se sont bornés à en déceler la présence dans les extraits enzymatiques bruts, sans essayer de les purifier. D'ailleurs, l'embarras le plus grand auquel on s'est heurté dans les recherches sur les pyrophosphatases c'est la difficulté, voire l'impossibilité, de séparer une enzyme à étroite spécificité de substrat, dépourvue de toute activité phosphomonoestérasique.

Les premières tentatives de purification sont, à notre connaissance celles de FLEURY et COURTOIS (*85, 86*). Ces chercheurs ont étudié d'une façon détaillée la pyrophosphatase des graines d'Amandier, mais ils n'ont pas réussi à obtenir une préparation dépourvue d'activité phosphomonoestérasique; par contre, l'inactivation en milieu alcalin détruisait complètement la pyrophosphatase en laissant encore une faible activité phosphomonoestérasique. Ensuite COURTOIS (*52*), dans une étude sur les enzymes de son de Blé, a décrit quelques faits permettant d'envisager l'existence d'une pyrophosphatase spécifique sans avoir pu la séparer de la mono-estérase associée.

Des essais de purification ont été effectués par Ito et Kondo (*113*) et puis par Ito, Kondo et Watanabe (*114*) avec des extraits de Patate douce. Ils arrivent à une purification de 500 fois, mais ni l'électrophorèse, ni l'adsorption sur colonne d'amidon ne leur donnent la possibilité de séparer les deux activités et ils concluent en faveur de l'identité des deux enzymes. Cependant, tout récemment, Lassota (*128*) paraît avoir réussi à séparer la pyrophosphatase de Haricots de la phosphomono-estérase associée (mais la préparation conserve encore une ATPase).

Les recherches rapportées jusqu'ici n'ont été l'objet uniquement que de la pyrophosphatase acide; or, Naganna et coll. (*145—147*), ont découvert dans plusieurs végétaux la présence d'une pyrophosphatase alcaline, très fragile, activée par le Mg^{++} et qui est aisément séparable de toute activité phosphomonoestérasique.

ATPases et apyrases. La capacité d'hydrolyser l'ATP a été retrouvée dans plusieurs plantes et par différents chercheurs, mais souvent il s'agissait d'essais effectués avec des préparations peu purifiées et par conséquent il est impossible de bien établir s'il s'agissait de „vraies" ATPases ou s'il y avait aussi des apyrases.

L'activité hydrolytique sur l'ATP a été décelée un peu partout dans les différents tissus et organes des plantes: dans les graines (*96*), tubercules (*113, 114, 206*), cotylédons de Petits Pois (*213*), feuilles (*140, 170, 208*), pollen (*149*). Forti en 1957 (*91*) a pu isoler des mitochondries de Petits Pois une ATPase dépourvue de toute activité sur l'ADP et sur les pyrophosphates inorganiques. Cette enzyme était activée par le Ca^{++}, le 2,4-dinitrophénol et l'acide déhydroascorbique. Les apyrases végétales ont été étudiées surtout dans les Pommes de Terre; des essais de purification, qu'ont effectués plusieurs chercheurs (*117, 118, 125, 126, 141, 156*) ont permis d'obtenir des préparations assez pures.

Kotelnikova, dans une série de recherches sur le rôle physiologique de l'apyrase des Pommes de Terre (*122—124*) a pu démontrer que l'activité apyrasique n'est pas influencée par des variations de culture, mais augmente en fonction directe de la maturation et de l'accroissement du taux d'amidon dans les tubercules.

IV. Rôle physiologique des phosphatases.

Le rôle physiologique des phosphatases végétales n'a pas encore été précisé d'une façon indiscutable. Il faut d'ailleurs reconnaître que le rôle des phosphatases animales demeure presque aussi imprécis. Il est reconnu que les phosphatases osseuses participent à la formation du sel de l'os, mais l'on n'a pas attribué de fonction physiologique dominante aux phosphatases des autres organes animaux. L'opinion la plus recevable est que les phosphatases animales remettent en circulation l'orthophosphate lié à des composés phosphorés synthétisés lors des processus métaboliques. C'est finalement la même interprétation qui semble la plus satisfaisante chez les Végétaux.

Bibliographie: pp. 363—373.

Variations d'activité. Il est indiscutable que chez les Végétaux l'on note une forte activité phosphatasique dans les tissus où les échanges métaboliques sont importants.

Nous avons signalé que dans les graines en voie de germination il y avait toujours un net accroissement de l'activité phosphatasique. Dans les feuilles de plantes annuelles (Haricot, Pomme de Terre, Radis) il existe de nettes variations d'activité [IGNATIEFF et WASTENEYS (*111, 112*)]: augmentation au cours du développement puis baisse régulière après que la plante a atteint sa maturité. Mais il existe cependant des exceptions, c'est le cas des arbres à feuilles annuelles. COURTOIS et TRAN VAN AI (*75, 199*) ont comparé les phosphatases de trois lots de feuilles de Marronnier d'Inde récoltés sur le même arbre dans un jardin parisien: (a) au milieu du Printemps à la fin de Mai, (b) au début Octobre, feuilles encore vertes, (c) en fin de végétation, feuilles complètement jaunies, légèrement adhérentes à l'arbre et prêtes à tomber, récoltées début Novembre au moment de la chute des feuilles. La proportion d'activité phosphatasique par rapport au poids de feuilles desséchées est sensiblement la même dans les feuilles en pleine végétation et dans les feuilles sur le point de tomber. Les courbes d'activité en fonction du pH des trois échantillons ne marquent pas de différences notables entre eux; les cations métalliques influencent de façon similaire les phosphatases des trois lots de feuilles.

La forte teneur en phosphatase des feuilles jaunies récoltées au moment de leur chute spontanée tend à indiquer que cette enzyme ne suit que faiblement ou même ne suit pas du tout l'évacuation automnale des substances protéiques, qui se manifeste dès que les feuilles jaunissent. En effet, au cours de la senescence des feuilles 50 à 70% de l'azote total est évacué vers les tiges et les racines, cette fraction azotée comprend la presque totalité des protéines. Il est possible d'envisager comme hypothèse que cette difficulté de migration de la phosphatase des feuilles de Marronnier est en relation avec sa grande insolubilité.

Action synthétisante possible. Nous avons signalé qu'il était des plus douteux que les concentrations en orthophosphate et dérivé hydroxylé accepteur dans la cellule permettent une biosynthèse notable par réversibilité d'action des phosphatases. Les multiples transphosphatases, dénommées le plus souvent phosphokinases, ont une efficacité infiniment supérieure pour synthétiser les constituants phosphorés. Une place doit cependant être accordée aux phosphatases intervenant en tant que phosphotransférases.

Les composés phosphorés riches en énergie apparaissent au cours de processus métaboliques où une première réaction dégage de l'énergie, tandis que dans une seconde réaction une fraction de cette énergie est captée dans la synthèse d'un composé phosphoré. Ces dérivés riches en énergie sont, soit dégradés directement, soit transformés en d'autres composés phosphorés moins riches en énergie.

Divers nucléotides, et spécialement l'acide adénosinetriphosphorique, sont de loin les principaux agents phosphorylants en tant que donateurs de phosphore. Une place modeste est susceptible d'être occupée par les phosphotransférases, contribuant à la biosynthèse de certains corps en transférant sur des accepteurs nouveaux le phosphore de donateurs synthétisés dans les processus dégageant de l'énergie.

Dans l'état actuel de nos connaissances il n'y a pas encore de réaction que l'on puisse attribuer avec certitude à l'intervention d'une phospho-monoestérase agissant en tant que phosphotransférase. Il sera d'ailleurs fort difficile de dissocier cette action de celle des diverses transphosphatases associées dans la cellule.

La principale forme de réserve phosphorée dans la cellule est l'acide phytique; il s'accumule dans la graine lors de la maturation et est utilisé lors de la germination. Cette utilisation est réalisée par des processus encore très mal connus, cependant l'acide phytique ne paraît pas intervenir comme donateur de phosphore pour l'action phosphotransférasique des monoestérases. Anagnostopoulos et Lino (9) ont constaté que de nombreuses préparations phosphatasiques, procédant à d'excellents transferts sur des alcools à partir du *p*-nitrophenylphosphate comme donateur, ne conduisaient à aucun transfert lorsque l'on substituait l'acide phytique à l'ester de phénol.

Action hydrolysante. C'est certainement le rôle physiologique essentiel des phosphatases végétales. Elles remettent en circulation les composés associés dans les esters phosphoriques. L'on a par exemple envisagé la participation de la phosphatase lors de la biosynthèse du saccharose; la phosphatase hydrolysant le saccharosephosphate en saccharose qui pourra migrer hors de la cellule et phosphate qui y reste pour être combiné dans un nouveau cycle métabolique.

Les esters phosphoriques sont des composés polaires qui tendent à rester attachés aux éléments basiques de la cellule; l'hydrolyse phosphatasique en sépare la fraction hydroxylée migrant bien plus aisément dans la plante.

Courtois et Tran Van Ai (75) ont émis l'hypothèse que la grande insolubilité de la plupart des phosphatases des organes chlorophylliens aurait des conséquences physiologiques. La phosphatase liée aux éléments figurés et très probablement à la membrane aurait pour rôle de libérer l'acide phosphorique des multiples dérivés phosphorés, étapes inter-médiaires de l'assimilation chlorophyllienne et des processus métaboliques qui en dérivent. Ainsi localisée la phosphatase ne perturberait pas les multiples enzymes provoquant ces transformations et qui sont situées dans d'autres territoires de la cellule.

La facile inhibition de la phosphatase par les orthophosphates qu'elle libère exercerait alors un contrôle à son action en limitant la quantité de composés qu'elle scinde. Cette forme d'inhibition représentant un frein puissant de la phosphatase. Cette inhibition par les phosphates paraît en effet avoir une signification métabolique pour D. H. et J. F. TURNER (202). Ces auteurs se basent sur le fait que les phosphates limitent la vitesse d'oxydation par les mitochondries des substrats de la respiration.

Lorsque la concentration en orthophosphates tombe à un faible taux qui ralentirait la glycolyse, il pourrait y avoir une tendance à ce que les glucosemonophosphates soient hydrolysés pour relever la concentration en orthophosphates. Inversement, si la concentration en orthophosphate s'accroît comme dans la phase anaérobie de la glycolyse, un effet retardateur proviendra de l'accumulation d'esters phosphoriques qui ne sont pas hydrolysés par la phosphatase inhibée. La phosphatase exercerait un contrôle pour stabiliser la concentration d'orthophosphates dans la cellule; cet effet serait selon les TURNER (202) analogue aux mécanismes de contrôle primitifs de Krebs (mécanismes qui ne sont pas contrôlés par les hormones ou le système nerveux central).

Enfin, un rôle de contrôle du pH de la cellule est susceptible d'inclure la participation des phosphatases. Les monoesters orthophosphoriques ont des pK'_1 et pK'_2 inférieurs à ceux de l'acide orthophosphorique.

L'hydrolyse des monoesters va donc alcaliniser le milieu, tandis que l'estérification de l'acide phosphorique l'acidifiera. Les phosphatases végétales n'agissant qu'en milieu acide, est-il permis de relier ces deux données? L'acidification provoquée par l'estérification favorisant l'activité de la phosphatase qui tendrait à régulariser le pH en hydrolysant ces esters.

Conclusions.

Nos connaissances sur les phosphatases végétales se sont accrues régulièrement depuis cinquante ans. Elles ont montré la vaste répartition de ces enzymes, une relative homogénéité de leurs caractères généraux. La structure, le rôle physiologique de ces enzymes demeurent encore mal connus. Le domaine des investigations possibles est encore fort vaste, il mérite de susciter l'intérêt des chercheurs.

Bibliographie.

1. ADLER, L.: Über die Phosphatasen im Malz. Biochem. Z. 70, 1 (1915).
2. — Gewinnung von Phytase aus Malz. Biochem. Z. 75, 319 (1916).
3. ALBAUM, H. G. and W. UMBREIT: Phosphorus Transformation during the Development of Oat Embryos. Amer. J. Bot. 30, 553 (1943).

4. ANAGNOSTOPOULOS, C.: Action comparée de quelques réactifs des fonctions amines sur diverses phosphatases acides et alcalines. Biochim. Biophys. Acta 4, 584 (1950).
5. — Action du versène sur l'activité phosphatasique. Bull. soc. chim. biol. (Paris) 35, 554 (1953).
6. — Mécanismes d'action et spécificité des effecteurs des phosphatases acides. Thèse, Paris, 1956.
7. ANAGNOSTOPOULOS, C. et J. COURTOIS: Recherches sur l'inhibition des phosphatases acides par les fluorures. II. Essais d'inhibition et de réactivation dans diverses conditions. Bull. soc. chim. biol. (Paris) 31, 1504 (1949).
8. ANAGNOSTOPOULOS, C., J. COURTOIS et J. SERVANT: Recherches sur l'action transférante de quelques phosphomonoestérases d'origine végétale. Bull. soc. chim. biol. (Paris) 36, 1581 (1954).
9. ANAGNOSTOPOULOS, C. et A. LINO: Étude comparative de l'activité de phospho-transférase et de phytase de certaines préparations de phosphatases de graines. Bull. soc. chim. biol. (Paris) 39, 781 (1957).
10. — — Activité de phosphotransférase non spécifique et de phytase dans les graines de blé et l'*E. coli*. Bull. soc. chim. biol. (Paris) 40, 1045 (1958).
11. ANDERSON, R. J.: The Hydrolysis of Phytin by the Enzyme Phytase Contained in Wheat Bran. J. Biol. Chem. 20, 475 (1915).
11a. ANDREU, M., E. FERNÁNDEZ ALVAREZ y M. LORA-TAMAYO: Investigaciones sobre fosfatasas. XVII. Purificación de la fosfatasa ácida de patata y caracterización de los grupos activos de la enzima. Anal. real soc. españ. fís. y quím. 56, 67 (1960).
12. AXELROD, B.: Citrus Fruit Phosphatase. J. Biol. Chem. 167, 57 (1947).
13. — A new Mode of Enzymic Phosphate Transfer. J. Biol. Chem. 172, 1 (1948).
14. — A Study of the Mechanism of "Phosphotransferase" Activity by Use of Radioactive Phosphorus. J. Biol. Chem. 176, 295 (1948).
15. AXELROD, B. and A. T. JAGENDORF: The Fate of Phosphatase, Invertase, and Peroxidase in Autolysing Leaves. Plant. Physiol. 26, 406 (1951).
16. BAGLIONI, T.: Amilasi e fosfatasi in radici di *Curcumis sativus* cresciute in presenza di alcuni fattori di crescita. Atti soc. ital. sci. vet. 10, 384 (1956).
17. BANERJEE, S. and N. NANDI: Effect of Germination on Phytin Content and Phytase Activity of Some Common Indian Pulses. Proc. Soc. exp. Biol. Med. 71, 588 (1949).
18. BARRÉ, R.: Recherches sur les protéines des graines d'amandiers et leurs combinaisons avec divers dérivés phosphorés. Thèse, Paris, 1953.
19. BARRÉ, R. et J. E. COURTOIS: Recherches sur les combinaisons des protéines avec les composés phosphorés. II. Étude des combinaisons de l'acide β-glycéro-phosphorique et de l'acide phytique. Bull. soc. chim. biol. (Paris) 35, 921 (1953).
20. BARRÉ, R., J. E. COURTOIS et G. WORMSER: Étude de l'action de quelques cations divalents sur les combinaisons acide phytique-protéine. In: Bio-chemistry of Nitrogen. Ann. Acad. Sci. Fennicae S A II. Série II, Chemica 60, 104 (1955).
21. — — Action de quelques cations divalents sur diverses préparations glycérophosphatasiques et phytasiques du son de Blé. Bull. soc. chim. biol. (Paris) 38, 387 (1956).
22. BARRÉ, R. et G. WORMSER: Les protéines des graines d'amandier. Ann. pharm. franç. 13, 736 (1955).
23. — — Recherches sur les phosphatases des feuilles de Belladone. Bull. soc. chim. biol. (Paris) 38, 377 (1956).

24. BELAVADY, B. and S. BANERJEE: Studies on the Effect of Germination on the Phosphorus Values of some Common Indian Pulses. Food Res. 18, 223 (1953).

25. BELFANTI, S.: Sopra alcune azioni della fosfatasi del riso. Giorn. chim. ind. appl. 13, 270 (1931).

26. BERNARDINI, L. e G. MORELLI: R. C. accad. naz. Lincei 21, 357 (1912) [d'après PEERS (160)].

27. BOLOMEY, R. A. and F. W. ALLEN: The Enzymic Hydrolysis of Ribonucleic Acid and its Relation to the Structure. J. Biol. Chem. 144, 113 (1942).

28. BOMAN, H. G. and L. E. WESTLUND: Chromatography of Horse-radish Enzymes: a Separation of two Acid Phosphatases. Arch. Biochem. Biophys. 70, 572 (1957).

29. BOOTH, R. G.: Cereal Phosphatases. I. The Assay of free Wheat Phospho-monoesterase and Characterization of free Phosphatases of Wheat. Biochemic. J. 38, 355 (1944).

30. BOROUGHS, H.: Studies on the Acid Phosphatases of Green Leaves. Arch. Biochem. Biophys. 49, 30 (1954).

31. — Separation of Phosphatase Activity from the Bulk Protein of Leaves. Arch. Biochem. Biophys. 53, 94 (1954).

32. BOSSARD, M.: Action des molybdates et de quelques autres effecteurs sur les phosphatases et divers enzymes. Thèse, Paris, 1946.

33. — Action des molybdates sur divers enzymes. Bull. soc. chim. biol. (Paris) 29, 218 (1947).

34. BRAWERMAN, G. and E. CHARGAFF: On a Desoxyribonuclease from Germinating Barley. J. Biol. Chem. 210, 445 (1954).

35. — — On the Synthesis of Nucleotides by Nucleoside Phosphotransferases. Biochim. Biophys. Acta 15, 549 (1954).

36. — — On the Distribution and Biological Significance of the Nucleoside Phosphotransferases. Biochim. Biophys. Acta 16, 524 (1955).

37. CALÈ, M. T.: Attività fosfatasica dei semi di Ricinus communis. Quaderni nutriz. 13, 324 (1953).

38. CARADANTE, G.: Fitasi e pirofosfatasi. Boll. soc. ital. biol. sperim. 19, 158 (1944).

39. CHANDRASEKHARA, M. R. and M. SWAMINATHAN: Enzymes of Ragi (Eleusine coracana), and Ragi Malt. III. Pyro and Glycerophosphatases. J. Sci. Industr. Res. (India) 13, 492 (1954).

40. COLLATZ, F. A. and C. H. BAILEY: The Activity of Phytase as Determined by the Specific Conductivity. Science (Washington) 51, 351 (1920).

41. COURTOIS, J.: Hydrolyse comparée des acides α- et β-glycérophosphoriques par diverses phosphatases végétales. I. Influence de l'origine du ferment et du pH. Bull. soc. chim. biol. (Paris), n⁰ spécial IVe Congrès Chim. Biol. 1933, 388.

42. — Influence de la réaction du milieu sur l'hydrolyse des acides α- et β-glycéro-phosphoriques par la taka-diastase. C. R. hebd. Séances Acad. Sci. 199, 95 (1934).

43. — Hydrolyse comparée des acides α- et β-glycérophosphoriques par diverses phosphatases végétales. V. Recherches sur la purification de la phosphatase d'Amande. Enzymologia 5, 273 (1938).

44. — Hydrolyse comparée des acides α- et β-glycérophosphoriques par diverses phosphatases végétales. VI. Étude de la phosphatase des graines de Moutarde-blanche. Bull. soc. chim. biol. (Paris) 20, 1359 (1938).

45. Courtois, J.: Hydrolyse comparée des acides α- et β-glycérophosphoriques par diverses phosphatases végétales. VII. Étude de la phosphatase d'Amande douce. Bull. soc. chim. biol. (Paris) **20**, 1376 (1938).

46. — Recherches sur l'action synthétisante de la phosphatase d'Amande. J. pharm. chim. **29**, 433 (1939).

47. — Action des phosphatases végétales sur les α- et β-glycérophosphates. Ann. fermentations **5**, 93 (1939).

48. — Recherches sur la phytase. I. Action comparée de diverses préparations fermentaires sur l'inositohexaphosphate et le glycérophosphate de sodium. Bull. soc. chim. biol. (Paris) **27**, 411 (1945).

49. — Recherches sur la phytase. III. Essais de séparation de l'activité glycérophosphatasique et de l'activité phytasique du son de blé. Biochim. Biophys. Acta **1**, 270 (1947).

50. — Recherches sur la phytase. V. Étude préliminaire de l'hydrolyse des inositohexaphosphates et glycérophosphates par les graines de Moutarde-blanche. Bull. soc. chim. biol. (Paris) **29**, 944 (1947).

51. — Recherches sur la phytase. VII. Action de préparations phosphatasiques purifiées de Moutarde-blanche sur divers esters phosphoriques. Bull. soc. chim. biol. (Paris) **30**, 37 (1948).

52. — Recherches sur la phytase. X. Action de préparations phosphatasiques purifiées de son de blé sur divers esters phosphoriques. Bull. soc. chim. biol. (Paris) **30**, 618 (1948).

53. — Les esters phosphoriques de l'inositol. Bull. soc. chim. biol. (Paris) **33**, 1074 (1951).

54. — Quelques récentes données sur le rôle du phosphore en Biochimie. Dans: M. Polonovski, Exposés annuels de biochimie médicale **14**, 233 (1952).

55. — Les phosphatases végétales. Farmaco (Pavia), ed. sci. **10**, 454 (1955).

56. Courtois, J. et C. Anagnostopoulos: Action de quelques acides minéraux formateurs de complexes sur les phosphatases. Enzymologia **13**, 183 (1949).

57. — — Recherches sur l'inhibition des phosphatases acides par les fluorures. I. Influence du fluorure de sodium sur la cinétique de la réaction. Bull. soc. chim. biol. (Paris) **31**, 1494 (1949).

58. Courtois, J., C. Anagnostopoulos et M. Khorsand: Recherches sur la phosphatase des feuilles de Belladone. Bull. soc. chim. biol. (Paris) **33**, 1813 (1951).

59. Courtois, J. et R. Barré: Influence de la réaction du milieu sur la précipitation de quelques esters phosphoriques avec une protéine des amandes. Bull. soc. chim. biol. (Paris) **31**, 740 (1949).

60. — — Recherches sur les combinaisons des protéines avec les composés phosphorés. I. Étude de l'amando-globuline, la légumine et l'insuline. Bull. soc. chim. biol. (Paris) **35**, 913 (1953).

61. Courtois, J. et M. Bossard: Action des inhibiteurs sur la phosphatase d'Amande. IV. Action des molybdates. Bull. soc. chim. biol. (Paris) **26**, 464 (1944).

62. — — Action des sulfocyanates sur les phosphatases. Bull. soc. chim. biol. (Paris) **27**, 406 (1945).

63. Courtois, J. et P. Denis: La phosphatase des feuilles d'Amandier. Enzymologia **5**, 288 (1938).

64. Courtois, J., A. Desjobert et P. Fleurent: Recherches sur la phytase. XVIII. Action de la phytase du blé sur quelques sels des acides inositophosphoriques. Bull. soc. chim. biol. (Paris) **34**, 691 (1952).

65. COURTOIS, J., A. DESJOBERT et J. SERVANT: Obtention d'acide α-glycéro-phosphorique par action transférante de la phosphatase des feuilles de Belladone. Bull. soc. chim. biol. (Paris) 39, suppl. II, 97 (1957).

66. COURTOIS, J. et G. JOSEPH: Recherches sur la phytase. VI. Action de diverses préparations phosphatasiques sur quelques esters phosphoriques de l'inositol. Bull. soc. chim. biol. (Paris) 29, 951 (1947).

67. — — Recherches sur la phytase. IX. Recherches sur la structure des acides inositoltriphosphorique et inositoltétraphosphorique obtenus par hydrolyse phytasique partielle de l'acide inositolhexaphosphorique. Bull. soc. chim. biol. (Paris) 30, 610 (1948).

68. COURTOIS, J. et M. KHORSAND: Sur quelques propriétés des phosphatases de feuilles. Biochim. Biophys. Acta 6, 175 (1950).

69. COURTOIS, J. et L. MANET: Recherches sur la phytase. XVII. Les phytases du Colibacille. Bull. soc. chim. biol. (Paris) 34, 265 (1952).

70. COURTOIS, J. et J. MANOUVRIER: Spécificité et champ d'action de la phosphatase d'Amande douce. II. Action de quelques composés chimiques sur l'hydrolyse de divers substrats. Enzymologia 6, 342 (1939).

71. COURTOIS, J. et CH. PÉREZ: Recherches sur la phytase. VIII. Teneur en inositophosphates et activité phytasique de diverses graines. Bull. soc. chim. biol. (Paris) 30, 195 (1948).

72. — — Recherches sur la phytase. XI. Essais en vue d'obtenir des matières premières dont l'activité initiale a été artificiellement accrue. Bull. soc. chim. biol. (Paris) 30, 631 (1948).

73. — — Recherches sur la phytase. XII. Essais de fractionnement des phosphatases du Radis. Bull. soc. chim. biol. (Paris) 31, 1220 (1949).

74. COURTOIS, J. et R. PERLÈS: Étude de la teneur en phosphore phytique de diverses graines de mils du Sénégal. Conséquences possibles sur l'absorption du calcium des aliments. Bull. Inst. Français Afrique Noire, sér. A 16, 379 (1954).

75. COURTOIS, J. et TRAN VAN AI: Sur quelques propriétés de la phosphatase des feuilles de Marronnier. Bull. soc. chim. biol. (Paris) 35, 721 (1953).

75a. COURTOIS, J. E. et H. VILLIERS-HUIBAN: Emploi de l'acide phytique pour déterminer l'activité de la pepsine du suc gastrique. Clin. Chim. Acta 2, 332 (1957).

76. COVELLO, M.: Il legame del fosforo nell'acido fitinico naturale. Boll. soc. ital. biol. sperim. 18, 158 (1943).

77. DALE, N. I.: Phosphatase and Hexosephosphatase in the Banana. Austral. J. Exp. Biol. Med. Sci. 14, 329 (1936).

78. DESJOBERT, A. et F. PETEK: Chromatographie des esters phosphoriques de l'inositol. Application à l'étude de la dégradation hydrolytique de l'inositol-hexaphosphate. Bull. soc. chim. biol. (Paris) 38, 871 (1956).

79. EHRENBERG, M.: Beziehungen zwischen Fermenttätigkeit und Blattbewegung bei Phaseolus multiflorus unter verschiedenen photoperiodischen Bedingungen. Planta 38, 244 (1950) [Chem. Abstr. 44, 10053 (1950)].

80. — Einfluß verschiedenen Licht-Dunkel-Wechsels auf die Rhythmik der Phosphataseaktivität in den Blättern von Kalanchoë blossfeldiana. Planta 43, 528 (1954).

81. — Über die Bestimmung von Fermenten in Gegenwart von Gerbstoffen. Biochem. Z. 325, 102 (1954).

82. EHRNST, L. E. and G. J. YAKISH: Phosphatase Activity of Malt. Amer. Brewer 77, 17 (1944).

83. ESSERY, R. E.: Phytase Activity and Phosphate Partition during Malting. J. Inst. Brewing 57, 125 (1951).

84. Euler, H. v. und B. v. Euler: Über die Spaltung organischer Phosphor-säureester. Z. physiol. Chem. (Hoppe-Seyler) **92**, 292 (1914).
85. Fleury, P. et J. Courtois: Recherches sur les pyrophosphatases végétales. I. Étude comparative des cinétiques d'hydrolyse des acides pyrophosphorique et β-glycérophosphorique dans diverses conditions. Enzymologia **1**, 377 (1937).
86. — — Recherches sur les pyrophosphatases végétales. II. Activités glycéro-phosphatasique et pyrophosphatasique de diverses préparations fermentaires d'Amande. Enzymologia **5**, 254 (1938).
87. — — Recherches sur la phytase. II. Cinétiques comparées de l'hydrolyse du glycérophosphate et de l'inositohexaphosphate par le son de blé. Biochim. Biophys. Acta **1**, 256 (1947).
88. Fleury, P., J. Courtois, C. Anagnostopoulos et A. Desjobert: Action de phosphatases de types distincts sur différents esters phosphoriques. Bull. soc. chim. biol. (Paris) **32**, 771 (1950).
89. Fleury, P. et Z. Sutu: Remarques sur l'hydrolyse comparée des acides glycérophosphoriques α- et β- par les agents chimiques et les ferments. Bull. soc. chim. France **39**, 1716 (1926).
90. Folley, S. J. and H. D. Kay: The Phosphatases. Erg. Enzymforsch. **5**, 159 (1936).
91. Forti, G.: Adenosinetriphosphatase Activity of Pea Mitochondria. Physiol. Plantarum **10**, 898 (1957).
92. Frisch-Niggemeyer, W. and K. K. Reddi: Studies on Ribonuclease in Tobacco Leaves. I. Purification and Properties. Biochim. Biophys. Acta **26**, 40 (1957).
93. Giri, K. V.: Über Pflanzenphosphatasen. I. Die Phosphatase gekeimter Sojabohnen *(Glycine hispida)*. Z. physiol. Chem. (Hoppe-Seyler) **245**, 185 (1937) [Chem. Abstr. **31**, 4348 (1937)].
94. — The Availability of Phosphorus from Indian Foodstuffs. Indian J. med. Res. **25**, 896 (1938).
95. Giri, K. V. und A. Sreenivasan: Die Änderung der amylolitischen Wirksamkeit beim Reiskorn während der Reifung und Keimung. Biochem. Z. **296**, 428 (1938).
96. Grillo, M. A.: Su una adenosintrifosfatasi dei cotiledoni di *Amygdalus communis*. Arch. sci. biol. (Bologna) **39**, 324 (1955).
97. Haeckel, A.: Beitrag zur Kenntnis der Pollenfermente. Planta **39**, 431 (1951).
98. Hara, M.: Deoxyribonuclease of Soybean Seedlings. Osaka Shiritsu Daigaku Igaku Zasshi **7**, 435 (1958) [Chem. Abstr. **53**, 480 (1959)].
99. Haskins, F. A.: Changes in the Activities of Several Enzymes during Germination and Seedling Development in Corn *(Zea mays)*. Plant Physiol. **30**, 74 (1955).
100. Hayashi, T. and Y. Takijima: Metabolism in the Roots of Crop Plants. I. Dephosphorization of Organic Phosphorus by Roots. J. Sci. Soil Manure Japan **21**, 185 (1951) [Chem. Abstr. **45**, 10321 (1951)].
101. Helferich, B. und G. v. Bruck: Zur Kenntnis der Kartoffelphosphatase. 4. Mitt. Z. physiol. Chem. (Hoppe-Seyler) **295**, 114 (1953).
102. Helferich, B., H. Stetter, W. Püschel, H. Söntgerath und G. v. Bruck: Zur Kenntnis der Kartoffelphosphatase. 3. Mitt. Z. physiol. Chem. (Hoppe-Seyler) **290**, 48 (1952).
103. Hill, R.: The Influence of Time, Temperature, pH, and Calcium Carbonate on the Activity of Phytase of Certain Cereals. J. Agr. Sci. **44**, 306 (1954).

104. HIRAI, T. and T. IMAIZUMI: Nature of Virus Infection in Plants. III. Phosphatase Activities of the Leaves during Infection. Virus (Osaka) **6**, 402 (1956) [Chem. Abstr. **51**, 11491 (1957)].

105. HOAGLAND, M. B.: Beryllium and Growth. III. The Effect of Beryllium on Plant Phosphatase. Arch. Biochem. Biophys. **35**, 259 (1952).

106. HOFMANN, ED.: Einfluß der Pflanzennährstoffe auf die Aktivität der Phosphatasen (Phosphomonoesterasen). Arch. Biochem. Biophys. **69**, 33 (1957).

107. HOLDEN, M. and N. W. PIRIE: Partial Purification of Leaf Ribonuclease. Biochemic. J. **60**, 39 (1955).

108. HORECKER, B. L., J. HURWITZ and A. WEISSBACH: The Enzymic Synthesis and Properties of Ribulose 1,5-Diphosphate. J. Biol. Chem. **218**, 785 (1956).

109. HORIUCHI, K.: Phytase and the Specificity of the Phosphoesterase. J. Biochemistry (Tokyo) **14**, 163 (1931) [Chem. Abstr. **26**, 745 (1932)].

110. IGNATIEFF, V.: A Note on the Relative Distribution of Phosphorus and Phosphatase Activity in the Floral Parts of *Nicotiana affinis, Petunia, Salpiglossis* and *Gladiolus*. Biochemic. J. **30**, 1815 (1936).

111. — Distribution of Phosphatase Activity and Analysis of Growth in Canada Wonder Bean. Biochemic. J. **31**, 1611 (1937).

112. IGNATIEFF, V. and H. WASTENEYS: Phosphatase Distribution in some Higher Plants. Biochemic. J. **30**, 1171 (1937).

113. ITO, E. and S. KONDO: Sweet-potato Phosphatase. Symposia Enzyme Chem. (Japan) **9**, 113 (1954) [Chem. Abstr. **49**, 13317 (1955)].

114. ITO, E., S. KONDO and S. WATANABE: Sweet-potato Phosphatase. I. Specificity. J. Biochemistry (Tokyo) **42**, 793 (1955).

115. JANECKE, H.: Studien über die Hafer-Phosphatase. Die Stärke **2**, 129 (1950) [Chem. Abstr. **44**, 7364 (1950)].

116. — Studien über die Haferphosphatase. 2. Mitt. Die Stärke **2**, 297 (1950).

117. KALCKAR, H. M.: Adenylpyrophosphatase and Myokinase. J. Biol. Chem. **153**, 355 (1944).

118. — Differential Spectrophotometry of Purine Compounds by Means of Specific Enzymes. III. Studies of the Enzymes of Purine Metabolism. J. Biol. Chem. **167**, 461 (1947).

119. KAY, H. D. and E. R. LEE: The Rate of Hydrolysis of α- and β-Glycerophosphate by Enzymes. J. Biol. Chem. **91**, 135 (1931).

120. KECK, K., W. FRISCH-NIGGEMEYER, D. ASCHER und O. HOFFMANN-OSTENHOF: Inhaltsstoffe des Knoblauchs und ihre Wirkungen. I. Über Substanzen, welche den Austritt des Chromatins aus Zellkernen im Wurzelmeristem von Allium-Arten bewirken. Monatsh. Chem. **82**, 755 (1951).

121. KHORSAND, M.: Recherches sur les phosphatases des feuilles. Thèse, Paris, 1951.

122. KOTELNIKOVA, A. V.: Variation de l'activité de l'apyrase des tubercules de pommes de terre. Comm. et Rapports 3ᵉ Congrès Inter. Biochimie, Bruxelles, 1955.

123. KOTELNIKOVA, A. V. and V. V. SOLOMATINA: Some Variety Peculiarities of Activity of Apyrase in Potato Tubers. Doklady Akad. Nauk (USSR) **95**, 325 (1954) [Chem. Abstr. **48**, 8335 (1954)].

124. — — Apyrase and Phosphatase Activities in Potato Tubers Grown under Different Conditions, at Different Stages of Growth Development, in Relation to the Process of Starch Formation. Biokhimiya (USSR) **19**, 144 (1954) [Chem. Abstr. **48**, 9486 (1954)].

125. Krishnan, P. S.: Studies on Apyrases. I. Purification of Potato Apyrase by Fractional Precipitation with Ammonium Sulfate. Arch. Biochem. Biophys. **20**, 261 (1949).

126. — Studies on Apyrases. II. Some Properties of Potato Apyrases. Arch. Biochem. Biophys. **20**, 272 (1949).

127. Kumar, K. S. V. Sampath, T. A. Sundararajan and P. S. Sarma: Dephosphorylation of Casein by Plant Phosphatases. Enzymologia **18**, 228 (1957).

128. Lassota, Z.: Pyrophosphatase und Adenosintriphosphatase der Bohnenkeimlinge. Acta Biochim. Polonica **6**, 3 (1959).

129. Lora-Tamayo, M.: La chimie de la phosphatase. Bull. soc. chim. biol. (Paris) **38**, 983 (1956).

130. Lüers, H. and L. Malsch: Phosphatases in Malt. Wochschr. Brau. **46**, 143 (1929) [Chem. Abstr. **23**, 4012 (1929)].

131. Lüers, H. and K. Silbereisen: Phytase of Malt. Wochschr. Brau. **44**, 263 (1927) [Chem. Abstr. **22**, 134 (1928)].

132. — — Phytase of Malt. Wochschr. Brau. **44**, 273 (1927) [Chem. Abstr. **22**, 2233 (1928)].

133. McCance, R. A. and E. M. Widdowson: Activity of the Phytase in Different Cereals and its Resistance to Dry Heat. Nature (London) **153**, 650 (1944).

134. McCollum, E. V. and E. B. Hart: The Occurrence of a Phytin-splitting Enzyme in Animal Tissues. J. Biol. Chem. **4**, 497 (1908).

135. MacDowall, F. D. H.: Absence of Acid Phosphatase from Chloroplast of Spinach and Iris. Plant Physiol. **28**, 317 (1953).

136. McGregor, S. M. and H. E. Street: The Carbohydrate Nutrition of Tomato Roots. IV. The Nature and Distribution of Acid Phosphatases. Ann. Botany **17**, 385 (1953).

137. Massart, L.: Action of Enzymes during the Course of Malting. Fermentation **1946**, 88 [Chem. Abstr. **42**, 2394 (1948)].

138. Massart, L. und K. Vermeyen: Über die Hemmungsmittel der Kartoffelphosphatase. Naturwiss. **30**, 170 (1942) [Chem. Abstr. **37**, 3799 (1943)].

139. Mayer, K. und M. Klinga-Mayer: Über das Stärkeverflüssigungsvermögen während der Keimung von Gerste. Z. physiol. Chem. (Hoppe-Seyler) **267**, 115 (1940) [Chem. Abstr. **35**, 3290 (1941)].

140. Mazelis, M.: Enzymatic Degradation of Adenosine Triphosphate to Adenine by Cabbage Leaf Preparations. Plant Physiol. **34**, 153 (1959).

141. Meyerhof, O.: The Origin of the Reaction of Harden and Young in Cell-free Alcoholic Fermentation. J. Biol. Chem. **157**, 105 (1945).

141a. Meyerhof, O. and H. Green: Synthetic Action of Phosphatase. II. Transphosphorylation by Alkaline Phosphatase in the Absence of Nucleotides. J. Biol. Chem. **183**, 377 (1950).

142. Morris, H. J., C. A. Weast and H. Lineweaver: Seasonal Variation in the Enzyme Content of Eleven Varieties of Carrots. Bot. Gaz. **107**, 362 (1946) [Chem. Abstr. **40**, 3799 (1946)].

143. Murthy, H. B. N., M. Swaminathan and V. Subrahmanyan: Starch-synthesizing Enzymes in Tapioca Roots. J. Sci. Industr. Res. (India) **13 B**, 223 (1954) [Chem. Abstr. **48**, 11573 (1954)].

144. Nagabhushanam, A., K. S. Srinivasan and M. Srinivasan: Oxidative Enzymes and Phosphatases in Agave Vera Cruz. J. Sci. Industr. Res. (India) **16 C**, 127 (1957).

145. Naganna, B., A. Raman, B. Venugopal and C. E. Sripathi: Potato Pyrophosphatases. Biochemic. J. **60**, 215 (1955).

146. NAGANNA, B. and C. E. SRIPATHI: Pyrophosphatases in Plants during Growth. Nature (London) **174**, 593 (1954).

147. NAGANNA, B., B. VENUGOPAL and C. E. SRIPATHI: Occurrence of Alcaline Pyrophosphatase in Vegetable Tissues. Biochemic. J. **60**, 224 (1955).

148. NAJJAR, V. A.: The Isolation and Properties of Phosphoglucomutase. J. Biol. Chem. **175**, 281 (1948).

149. NAKAMURA, G. R. and E. L. BECKER: Phosphatases of Ragweed Pollen. Arch. Biochem. Biophys. **33**, 78 (1951).

150. NĚMEC, A.: Über die Verbreitung der Glycerophosphatase in den Samenorganismen. Biochem. Z. **93**, 94 (1919).

151. — Zur Kenntnis der Glycerophosphatase der Pflanzensamen. Biochem. Z. **137**, 570 (1923).

152. — Glycerophosphatasewirkung von Pflanzensamen und Fermentsynthese. Biochem. Z. **202**, 229 (1928).

153. NEUBERG, C. und M. KOBEL: Studien an Tabak-enzymen. Biochem. Z. **229**, 455 (1930) [Chem. Abstr. **25**, 1545 (1931)].

154. NGUYEN VAN THOAI: Recherches sur les phosphatases végétales. IV. Caractères des préparations enzymatiques de soja et d'amande douce. Bull. soc. chim. biol. (Paris) **24**, ,,Travaux" 1369 (1942).

155. NGUYEN VAN THOAI, J. ROCHE et E. DANZAS: Sur les activateurs des systèmes enzymatiques. I. Sur l'action synthétisante des phosphomonoestérases. Bull. soc. chim. biol. (Paris) **26**, ,,Travaux" 1139 (1944).

156. NGUYEN VAN THOAI, J. ROCHE et TRAN-THI AN: Sur la purification et le mécanisme d'action de l'apyrase de pomme de terre. Bull. soc. chim. biol. (Paris) **36**, 529 (1954).

157. NIZHARADZE, A. N.: The Role of Phosphorus Compounds in Material Metabolism of Tea Leaf. Biokhimiya Chainogo Proizvodstva, Sbornik **5**, 67 (1946) [Chem. Abstr. **46**, 2630 (1952)].

158. OHMORI, Y.: Über die Phosphomonoesterase. Enzymologia **4**, 217 (1937).

159. PARKER, G.: A Study of the Breakdown of Ribonucleic Acid in Tobacco-leaf Extracts. Biochemic. J. **51**, 389 (1952).

160. PEERS, F. G.: The Phytase of Wheat. Biochemic. J. **53**, 102 (1953).

161. PÉREZ, CH.: Contribution à l'étude de la spécificité de la phytase. Thèse, Paris, 1949.

162. — Recherches sur la phytase. XIII. Essais de fractionnement des phosphatases du Seigle. Bull. soc. chim. biol. (Paris) **31**, 1229 (1949).

163. PERLÈS, R.: Recherches sur la phytase. XIX. Sur la possibilité de démasquer la phytase des graines par emploi de très faibles doses de substrat. Bull. soc. chim. biol. (Paris) **37**, 307 (1955).

164. PFANKUCH, E.: Über die Phosphatase der Kartoffel und Zuckerrübe. Z. physiol. Chem. (Hoppe-Seyler) **241**, 34 (1936).

165. PHILPOTT, M. W.: In: Report of the Rubber Research Institute of Malaya for the period September 1945 to December 1948. Chem. Div. p. 191 [Chem. Abstr. **45**, 9292 (1951)].

166. PIERPOINT, W. S.: The Phosphoesterase of Pea Plants *(Pisum sativum)*. Biochemic. J. **67**, 466 (1957).

167. — The Chromatography of Leaf Ribonuclease. Biochim. Biophys. Acta **21**, 136 (1956).

168. PILEGGI, V. J.: Distribution of Phytase in the Rat. Arch. Biochem. Biophys. **80**, 1 (1959).

169. PLUMEL, M.: Recherches sur les phosphatases acides animales. Thèse, Paris, 1948.

170. Poglazov, B. F.: Adenosinetriphosphatase Activity and Motor Reaction in Plants. Doklady Akad. Nauk (USSR) **109**, 597 (1956) [Chem. Abstr. **51**, 4505 (1957)].

171. Posternak, S. et T. Posternak: Sur la configuration de l'inosite inactive. Helv. Chim. Acta **12**, 1165 (1929).

172. Pratesi, P.: Sulla presenza della fosfatasi nelle foglie verdi. I. Ann. chim. appl. (Roma) **27**, 309 (1937).

173. — Sulla specificità della fosfatasi delle foglie verdi. II. Ann. chim. appl. (Roma) **27**, 321 (1937).

174. — Azione dell'aldeide formica sulla fosfatasi delle foglie verdi. III. Ann. chim. appl. (Roma) **27**, 382 (1937).

175. — Sull'azione delle fosfatasi nelle foglie verdi e sull'etere monofosforico della formaldeide. Enzymologia **4**, 242 (1937).

176. Ratner, E. I. and S. A. Samoilova: Extracellular Phosphatase and Invertase Activity of Roots. Fiziol. Rastenii **2**, 30 (1955) [Chem. Abstr. **51**, 14899 (1957)].

177. Reddi, K. K.: Studies on Tobacco Leaf Ribonuclease. II. Mechanism of Action. Biochim. Biophys. Acta **28**, 386 (1958).

178. — Action of Tobacco-leaf Ribonuclease on Polyuridylic and Cytidylic Acids. Biochim. Biophys. Acta **30**, 638 (1958).

179. Ricaud-Manouvrier, J.: Action de l'acide phytique sur les phosphatases. Bull. soc. chim. biol. (Paris) **35**, 889 (1953).

180. Roberts, D. W. A.: The Wheat Leaf Phosphatases. I. A Survey of the Inhibitors at pH 5.7. J. Biol. Chem. **219**, 711 (1956).

181. — The Wheat Leaf Phosphatases. III. A Survey of the Heat Stability of the Enzymes Active at pH 5.7. J. Biol. Chem. **226**, 751 (1957).

182. — The Wheat Leaf Phosphatases. IV. The Effect of Metal Ions on the Acid Phosphatase Activity of Dialyzed Juice. J. Biol. Chem. **230**, 213 (1958).

183. — Physiological and Biochemical Studies in Plant Metabolism. VIII. The Distribution of Invertase in the First Leaf of Khapli Wheat. Canad. J. Bot. **31**, 367 (1953).

184. Roche, J. et J. Courtois: Les phosphatases. Dans: M. Polonovski, Exposés annuels de biochimie médicale **4**, 219 (1944).

185. Rogers, H. T., R. W. Pearson and W. H. Pierre: The Source and Phosphatase Activity of Exoenzyme Systems of Corn and Tomato Roots. Soil Sci. **54**, 353 (1942).

186. Rothstein, A. and R. Meier: The Relationship of the Cell Surface to Metabolism. IV. The Role of Cell Surface Phosphatases of Yeast. J. Cellular Comp. Physiol. **34**, 97 (1949).

187. Sadasivan, V.: The Phosphatases in Coconut *(Cocos nucifera)*. Arch. Biochem. Biophys. **30**, 159 (1951).

188. Sarma, M. L. and K. V. Giri: Plant Phosphatases. IV. Occurrence of "Free" and "Bound" Phosphatases in Seeds. Ann. Biochem. exp. Med. **2**, 17 (1942).

189. Schlamowitz, M. and R. L. Garner: The Ribonucleinase of the Soy Bean. I. Isolation of the Enzyme. J. Biol. Chem. **163**, 487 (1946).

190. Schulerud, A.: The Phytin Contents in the Usual Flour Types and the Rate of Decomposition in Flour Suspensions. Acta Physiol. Scand. **8**, 259 (1944).

191. Semenenko, G. I.: Transformation of Nucleic Acids in Seeds. Fiziol. Rastenii **4**, 332 (1957) [Chem. Abstr. **52**, 1375 (1958)].

192. Servant, J.: Recherches sur l'action transférante des phosphatases végétales. Thèse, Paris, 1958.

193. SISAKYAN, N. M., N. A. VASILEVA and N. N. KOSHTOYANTS: Ribonuclease of Sprouting Wheat Grain. Doklady Akad. Nauk (USSR) **112**, 300 (1957) [Chem. Abstr. **51**, 9805 (1957)].

194. SPENCER, D.: The Effect of Molybdate on the Activity of Tomato Acid Phosphatases. Austral. J. Biol. Sci. **7**, 151 (1954) [Chem. Abstr. **48**, 12241 (1954)].

195. SUZUKI, U., K. YOSHIMURA and M. TAKAISHI: On the Occurrence of an Enzyme which Decomposes Anhydroxymethylenephosphoric Acid. Tokyo Chem. Soc. **27**, 1330 (1906) [Chem. Abstr. **1**, 1147 (1907)].

196. TAKASHIMA, S.: Effect of Light on the Phosphatase Activity of Green Leaves. Symposia Enzyme Chem. (Japan) 8, 87 (1953) [Chem. Abstr. **47**, 12437 (1953)].

197. TAKASHIMA, S. and A. MITSUI: Photochemical Phosphate Transfer in Green Leaves. J. Biochemistry (Tokyo) **41**, 443 (1954).

198. — — Effect of Light on Phosphatase. Symposia Enzyme Chem. (Japan) **9**, 31 (1954) [Chem. Abstr. **48**, 7082 (1954)].

199. TRAN VAN AI: Recherches sur la phosphatase de feuilles de Marronnier. Thèse, Paris, 1953.

200. TUNIS, M. and E. CHARGAFF: Separation of Nucleoside Phosphotransferase and Phosphatase Activities. Biochim. Biophys. Acta **21**, 204 (1956).

201. — — Nonparticipation of Inorganic Phosphate in the Enzymic Formation of Nucleotides by Nucleoside Phosphotransferases. Arch. Biochem. Biophys. **69**, 295 (1957).

202. TURNER, D. H. and J. F. TURNER: The Hydrolysis of Glucosemonophosphates by a Phosphatase Preparation from Pea Seeds. Biochemic. J. **74**, 486 (1960).

203. UZAWA, S.: Über die Kleienphosphoesterase. I. J. Biochemistry (Tokyo) **15**, 1 (1932).

204. — Über die Phosphomonoesterase und die Phosphodiesterase. J. Biochemistry (Tokyo) **15**, 19 (1932).

205. WARBURG, O. und W. CHRISTIAN: Isolierung und Kristallisation des Gärungsferments Enolase. Biochem. Z. **310**, 384 (1942).

206. WATANABE, S., E. ITO, I. TAKEUCHI and S. KONDO: Phosphatases of Sweet Potatoes. J. Chem. Soc. Japan **74**, 361 (1953) [Chem. Abstr. **47**, 10569 (1953)].

207. WATTS, J. H. and R. M. GRISWOLD: Enzyme and Ascorbic Acid Content of Fresh and Frozen Pineapple. Food Res. **18**, 162 (1953).

208. WILDMAN, S. G. and J. BONNER: The Proteins of Green Leaves. I. Isolation, Enzymatic Properties and Auxin Content of Spinach Cytoplasmic Proteins. Arch. Biochem. Biophys. **14**, 381 (1947).

209. WILSON, K. S. and V. M. CUTTER: The Distribution of Acid Phosphatases during the Development of the Fruit of *Cocos nucifera*. Amer. J. Bot. **39**, 57 (1952).

210. WRENSHALL, C. L. and W. J. DYER: Organic Phosphorus in Soils. II. The Nature of the Organic Phosphorus Compounds. A: Nucleic Acid Derivatives. B: Phytin. Soil Sci. **51**, 235 (1941).

211. YOSHIDA, A.: Phosphatases. J. Chem. Soc. Japan **71**, 115 (1950) [Chem. Abstr. **45**, 4750 (1951)].

212. — Phosphatases of Rice Bran. Symposia on Enzyme Chem. (Japan) **4**, 58 (1950) [Chem. Abstr. **45**, 7161 (1951)].

213. YOUNG, J. L. and J. E. VARNER: Enzyme Synthesis in the Cotyledons of Germinating Seeds. Arch. Biochem. Biophys. **84**, 71 (1959).

(Reçu le 30 août 1960.)

Namenverzeichnis. Index of Names. Index des Auteurs.

RUNDELL J. T. 297, 309, 312.
RUZICKA, L. 3, 4, 10, 30, 78, 111, 116, 122, 129, 130, 132, 133, 137–142, 145, 147, 162, 163, 173, 182, 237, 239, 248, 266, 267, 269, 271, 286, 288.
RYDON, H. N. 3, 4, 10, 28, 30.

SABO, E. F. 179, 232.
SACHSE, H. 168, 239.
SADASIVAN, V. 336, 337, 352, 359, 372.
SAMOILOVA, S. A. 333, 358, 372.
SAMPATH KUMAR, K. S. V. 359, 370.
SANCHEZ-HIDALGO, L. M. 204, 229.
SANDERS, H. 145, 164.
SANDERS, H. J. 146, 162.
SANDRIN, R. 52, 116.
SANDRIS, C. 222, 228, 234, 239.
SANNIÉ, C. 188, 239.
SARMA, M. L. 323, 330, 372.
SARMA, P. S. 359, 370.
SASAKI, Y. 59, 60, 119.
SATO, A. 42, 44, 49, 80–91, 115.
SATO, M. 279, 288.
SATO, T. 37, 44, 49, 81–83, 86, 88, 115.
SATO, Y. 39, 115.
SAUERS, R. R. 270, 286.
SAVARD, K. 188, 239.
SAWICKI, R. M. 128, 129, 139, 151, 152, 162, 163.
SCHADE, G. 13, 29, 78, 115.
SCHÄFER, E. 281, 290.
SCHALES, O. 259, 287.
SCHAPIRO, J. A. 294, 302, 314.
SCHECHTER, M. S. 137, 143, 144, 146, 147, 161, 163.
SCHENCK, G. O. 271, 288.
SCHINDLER, O. 198, 232.
SCHINZ, H. 268, 269, 286, 289, 290.
SCHLAMOWITZ, M. 358, 372.
SCHLITTLER, E. 264, 287.
SCHLUBACH, H. H. 291, 293–296, 300, 301, 303–314.
SCHMETTERLING, A. 252, 289.
SCHMID, R. W. 52, 116.
SCHMIDT, H. 148, 163.
SCHNEIDER, B. 78, 113.
SCHNEIDER, J. 52, 102, 112.
SCHNEIDER, W. 281, 285.
SCHNEIDER, W. G. 40, 109, 111, 116, 196, 197, 237.
SCHÖPF, CL. 250–253, 255, 256, 259, 264, 288, 289.
SCHREIBER, E. C. 144, 155.

SCHREIBER, J. 188, 239.
SCHUERCH, C. 211, 227.
SCHULERUD, A. 336, 372.
SCHULZE, E. 292, 294, 295, 314.
SCHULZE, J. 41, 80–83, 112, 117.
SCHUMACHER, E. 184, 206, 225.
SCOTONI, R., Jr. 48, 84, 86, 108, 202, 233.
SCOTT, A. I. 284, 286.
SCOTT, J. J. 275, 289.
SECOR. G. E. 303, 304, 307, 315.
SEDERHOLM, C. H. 170, 235.
SEIBT, S. 133, 163.
SEMENENKO, G. I. 358, 372.
SEMMLER, F. W. 266, 290.
SERVANT, J. 327, 328, 355, 356, 364, 367, 372.
SETO, S. 34, 36, 37, 39, 43, 44, 49, 50, 80–89, 92–95, 103, 115.
SHABICA, A. C. 197, 239.
SHAFIQ, M. 76, 109, 272, 285.
SHARPLESS, R. V. 148, 162.
SHAW, J. I. 203, 239.
SHEARER, H. M. M. 40, 116.
SHEEHAN, J. C. 252, 289.
SHINOHARA, T. 135, 160.
SHOOLERY, J. N. 196, 197, 239.
SHOPPEE, C. W. 175, 194, 207, 209, 223, 227, 231, 239.
SHRINER, R. I. 167, 239.
SICHER, J. 52, 117, 170, 187, 216, 240.
SIDDIQUI, S. 128, 163.
SIDHU, G. S. 149, 158.
SIEDEL, W. 275, 289.
SIGG, H. P. 179, 214, 236, 240.
SILBEREISEN, K. 336, 370.
SIM, G. A. 40, 71, 73, 76, 112, 116.
SIMON, W. 187, 240.
SIMONSEN J. L. 3, 30, 34, 52, 53, 55–58, 61, 78, 107, 108, 129, 163, 182, 240, 266, 289.
ŠIPOŠ, F. 216, 240.
SISAKYAN, N. M. 358, 373.
SKITA, A. 171, 240.
SLADE, P. 186, 236.
SLEEP, K. C. 132, 134–137, 154, 157, 159.
SLIFER, E. D. 304, 312.
SLOBODKIN, N. R. 35, 117.
SLOMP, G. 197, 240.
SLY, J. C. P. 209, 239.
SMITH, F. 304, 306, 307, 313.
SMITH, H. 223, 228.
SMITH, L. L. 197 240.

Sachverzeichnis. Index of Subjects. Index des Matières.

Manzsche Buchdruckerei, Wien IX.